Lecture Notes in Intelligent Transportation and Infrastructure

Series editor

Janusz Kacprzyk, Systems Research Institu  ᴬᶜᵃᵈᵉᵐʸ of Sciences, Warszawa, Poland

The series "Lecture Notes in Intelligent Transportation and Infrastructure" (LNITI) publishes new developments and advances in the various areas of intelligent transportation and infrastructure. The intent is to cover the theory, applications, and perspectives on the state-of-the-art and future developments relevant to topics such as intelligent transportation systems, smart mobility, urban logistics, smart grids, critical infrastructure, smart architecture, smart citizens, intelligent governance, smart architecture and construction design, as well as green and sustainable urban structures. The series contains monographs, conference proceedings, edited volumes, lecture notes and textbooks. Of particular value to both the contributors and the readership are the short publication timeframe and the world-wide distribution, which enable wide and rapid dissemination of high-quality research output.

More information about this series at http://www.springer.com/series/15991

Mohamed Ben Ahmed ·
Anouar Abdelhakim Boudhir ·
Ali Younes
Editors

Innovations in Smart Cities Applications Edition 2

The Proceedings of the Third International Conference on Smart City Applications

Volume 1

 Springer

Editors
Mohamed Ben Ahmed
Association Méditerranéenne des Sciences
et Technologies
Tangier, Morocco

Anouar Abdelhakim Boudhir
Association Méditerranéenne des Sciences
et Technologies
Tangier, Morocco

Ali Younes
Faculty of Sciences
Présidence UAE
Tétouan, Morocco

ISSN 2523-3440   ISSN 2523-3459   (electronic)
Lecture Notes in Intelligent Transportation and Infrastructure
ISBN 978-3-030-11195-3   ISBN 978-3-030-11196-0   (eBook)
https://doi.org/10.1007/978-3-030-11196-0

Library of Congress Control Number: 2018966834

This Springer imprint is published by the registered company Springer Nature Switzerland AG
The registered company address is: Gewerbestrasse 11, 6330 Cham, Switzerland

Preface

A significant research activity has occurred in the area of Smart City in recent years. This field attracted multiple disciplines (Academy, Economy, Industry, Healthcare, Government, Society, Water and Energy…) for the main goal focused on how to improve services and the well-being living of citizens in the cities.

This Book presents a second edition of the "Innovations in Smart Cities and Applications" Book published in 2018. It's a real continuation in advanced contributions and scientific works done on several axes and subarea of smart city applications and domain, especially on Computer Technologies, Energy management, Logistics, Governance, Policy, Healthcare, Economy and Environment, and more. Those editions, the past one and the future one's, are considered as an added value in the research for interested in developing the future smart cities. Thanks to participants and researchers who trusted to those series of conferences, this research area will grow and exist due to their high quality of presented papers.

This volume contains selected extended papers of The Third International Conference on Smart City Applications (SCA 2018) held on October 10–11, 2018 in Tetouan, Morocco. It regroups original research results, new ideas, and practical development experiences and includes papers from all areas of Smart City Applications. The scope of SCA 2018 comprises methods and practices that combine various emerging internetworking and data technologies to capture, integrate, analyze, mine, annotate, and visualize data in a meaningful and collaborative manner.

We thank all authors from across the globe for choosing SCA18 to submit their manuscripts. A sincere gratitude to all keynotes speakers for offering their valuable time and sharing their knowledge with the conference attendees. Specials thanks are addressed to all organizing committee members, to all program committee members, and to all chairs of sessions for their efforts and the time spent in order to success this event.

Many thanks to the Springer staff for their support and guidance. In particular, our special thanks to Dr. Thomas Ditzinger and Ms. Varsha Prabakaran for their kind support.

Mohamed Ben Ahmed
Anouar Abdelhakim Boudhir
Ali Younes

Committee

Conference General Chairs

Mohamed Ben Ahmed   FSTT, Abdelmalek Essaadi University, Morocco
Anouar Abdelhakim Boudhur   FSTT, Abdelmalek Essaadi University, Morocco

Conference Local Chair

Younes Ali   Faculty of Science Tetouan, Abdelmalek Essaadi
University, Morocco

Workshops Co-chairs

Abdellatif Medouri   ENSATE, Abdelmalek Essaadi University,
Morocco
Naoufal Raissouni   ENSATE, Abdelmalek Essaadi University,
Morocco
Samira Khoulji   ENSATE, Abdelmalek Essaadi University,
Morocco

Publications Co-chairs

Mohamed Ben Ahmed   FSTT, Abdelmalek Essaadi University, Morocco
Anouar Abdelhakim Boudhur   FSTT, Abdelmalek Essaadi University, Morocco

Bernadetta Kwintiana Ane   University of Stuttgart, Germany
Wassila Mtalaa   Luxembourg Institute of Science
                                      and Technology, Luxembourg

Tutorials Co-chairs

Othman Chakkor   ENSATE, Abdelmalek Essaadi University,
                                      Morocco
Mounir Arioua   ENSATE, Abdelmalek Essaadi University,
                                      Morocco

Panels Co-chairs

Yassine Tabaa   ENS, Abdelmalek Essaadi University, Morocco
Azza Lajjam   ENSATE, Abdelmalek Essaadi University,
                                      Morocco

Posters and Ph.D. Track Co-chairs

Mohammed l'Bachir   FS, Abdelmalek Essaadi University, Morocco
  El Kbiach
Loubna Bounab   ENSATE, Abdelmalek Essaadi University,
                                      Morocco

Web Chair

Mohamed Ben Ahmed   FSTT, Abdelmalek Essaadi University, Morocco

Publicity and Social Media Co-chairs

Mohamed Charyah   ENSATE, Abdelmalek Essaadi University,
                                      Morocco
Abderrahim El Mhouti   FSTH, Abdelmalek Essaadi University, Morocco

Ahmed Bendahman   ENS, Abdelmalek Essaadi University, Morocco
El Arbi Abdellaoui Alaoui   EIGSI, Casablanca, Morocco

Sponsorship and Exhibits Chair

Ahmed Ziani   FS, Abdelmalek Essaadi University, Morocco

Ph.D. Committee

Aziz Mahboub   FSTT, Abdelmalek Essaadi University, Morocco
Abdeltif El Ouahrani   FS, Abdelmalek Essaadi University, Morocco

Technical Program Committee

Program Committee Chair

Pr. El Amrani Chaker   FSTT, UAE, Morocco

Program Committee

Ahmad S. Almogren   King Saud University, Saudi Arabia
Abdel-Badeeh M. Salem   Ain Shams University, Egypt
Alabdulkarim Lamya   King Saud University, Saudi Arabia
Riccardo Accorsi   Bologna University, Italy
Jarallah Alghamdi   Prince Sultan University, Saudi Arabia
Ahmed Kadhim Hussein   Babylon University, Iraq
Mahasen Anabtawi   Al-Quds University, Palestine
Mounir Arioua   UAE, Morocco
Abdelali Astitou   UAE, Morocco
Zainab Assaghir   Lebanese University, Lebanon
Fatma Zohra   CERIST, Algeria
  Bessai-Mechmach
Benaouicha Said   UAE, Morocco
Sadok Ben Yahya   Faculty of Sciences of Tunis, Tunisia
Mohammed Boulmalf   UIR, Morocco
Ahmed Boutejdar   German Research Foundation, Bonn, Germany
Lala Saadia Chadli   University Sultan Moulay Slimane, Beni-Mellal,
                                     Morocco
Žarko Damir   Zagreb University, Croatia
Bernard Dousset   UPS, Toulouse, France

Keynotes

Gilles Betis
OrbiCité, France

Biography: Gilles Betis is the founder of OrbiCité, a consulting company dedicated to Smart Cities, Mobility, Innovation, and Entrepreneurship. He previously held various positions in high-tech international companies, transport services providers, and a European agency dedicated to digital technologies. In 2013, he co-founded the IEEE Smart Cities Initiative and has been chairing it until 2017 during its incubation phase. Since the end of the 80s, he has constantly been involved with prospective, innovation and design of complex systems, as well as entrepreneurship and value creation. Having an extensive industrial experience in Intelligent Transportation Systems and Smart Cities (mobility, data, security and resilience, and civic tech), he always linked up emerging behaviors and societal needs to innovative technological solutions, allowing smooth adoption by final users. Gilles Betis was graduated in 1987 from Ecole Supérieure d'Electricité in France.

MaaS—Mobility as a Service, a New Paradigm Implementation

Abstract: Focusing on the finalities rather than on the means to move people and goods, a new field of solutions is emerging now. Technological standards and business processes are ready, and new business models and new actors are emerging. Those models are not exclusive to developed countries but also have a great interest in developing countries. The lecture will go through:

1. issues and limitations
2. innovation dynamics
3. innovation in transportation technology
4. the transport experience
5. innovation in mobility, open data, and business models
6. political innovation and regulations.

Prof. Dr. İsmail Rakıp Karaşo
Karabuk University, Turkey

Biography: Dr. İsmail Rakıp Karaşo is a Professor of Computer Engineering Department and Head of 3D-GeoInformatics Research Group at Karabuk University, Turkey. He received his B.Sc. degree from Selcuk University, M.Sc. degree from Gebze Institute of Technology, and Ph.D. degree from GIS and Remote Sensing program of Yildiz Technical University, in 1997, 2001, and 2007, respectively, three of them from Geomatics Engineering Department. In 2002, he involved in a GIS project as a Graduate Student Intern at Forest Engineering Department, Oregon State University, USA. During the summer of 2014 and summer of 2010, he was a Visiting Researcher in 3D GIS Research Lab, Faculty of Geoinformation Science and Engineering, Universiti Teknologi Malaysia. Between 2000 and 2009, he was a Research Assistant at Geomatics Engineering Department of Gebze Institute of Technology. Since 2009, he has been in Karabuk University and taught undergraduate and graduate classes in Geoinformation and Computer Sciences. He has also carried out administrative duties such as Head of Computer Science Division of Department, Director of Safranbolu Vocational School of Karabuk University. Currently, he is the Dean of Safranbolu Fine Art and Design Faculty in the same university.

Artificial Intelligence Based Smart Evacuation System Design for the Complex Buildings of Smart Cities

Abstract: In this talk, 3D Network Analyses and Interactive Human Navigation System for indoor which consists of three components will be presented. The first component is used to extract the geometrical and 3D topological vector data automatically from architectural raster floor plans. The second component is used for network analysis and simulations. It generates and presents the optimum path in a 3D modeled building, and provides 3D visualization and simulation. And the third component is used to carry out the generation of the guiding expressions, and it also provides that information for the mobile devices such as PDA's, laptops, etc. via Internet.

In addition, an Intelligent Evacuation Model for Smart Buildings will be introduced in this presentation. The model dynamically takes into account environmental (smoke, fire, etc.) and human-induced (age, disability, etc.) factors and generates personalized evacuation route by performing network analysis interactively and in real time. Intelligent Control Techniques (Feed-Forward Artificial Neural Networks) have been used in the design of the model.

Prof. Dr. Daniyal Alghazzawi
King Abdulaziz University,
Jeddah, Saudi Arabia

Biography: Daniyal Alghazzawi is a Professor in the Computing Information Systems Department and the Head of the Information Security Group at King Abdulaziz University. He received his B.Sc. degree with honor in Computer Science from King Abdulaziz University in 1999. Then, he completed his M.Sc. and Ph.D. degrees in the field of Computer Science at the University of Kansas in the United States in 2007. He also received another M.Sc. degree in Teaching and Leadership from University of Kansas in 2004 which helped him to develop his teaching and leadership skills. This helped him to obtain a certificate in Management International Leadership (LMI). Since 2007, he served as the Head of Department for 5 years, and then he served as a Vice Dean of Development of the Deanship of Information Technology for 2 years. In 2017, he became an Honorary Lecturer at School of Computer Science and Electronic Engineering, at University of Essex in UK.

He organized number of domestic workshops and international conferences. He published more than 100 papers in international journals and conferences in the field of Smart e-Learning, Information Security, and Computational Intelligent. He is the Project Manager for two international collaborations in King Abdulaziz University, which are Smart Building with University of Essex and Multi-Agent Systems with University of Southampton. He served as a reviewer and an editor for international conferences, journals, workshops, and contests.

Smart Blockchain Model for IoT Security Challenges in Smart Cities

Abstract: Smart cities are the nearest future for most countries around the world, which involve growth in the Internet of Things devices. There are 34.8 billion IoT devices connected in 2018, and it is predicted to exceed 50.1 billion IoT devices by 2020. Based on Forrester's analysis, number of IoT Security Challenges are illustrated, such as IoT Encryption, IoT Authentication, IoT PKI, IoT Security Analytics, IoT Network Security, and IoT API Security. The world has already faced one of the biggest securities hijacked on more than 1 million IoT devices in October 2016 using Mirai Botnets. Therefore, a proposed model will be introduced to solve the most IoT Security Challenges using Blockchain and AI techniques.

Prof. Mohammed Bouhorma
Abdelmalek Essaadi
University, Tangier, Morocco

Biography: Mohammed Bouhorma is a Professor in the Department of Computer Science, Abdelmalek Essaadi University, where he has been since 1996. He received his Ph.D. degree in Communication Systems from the Polytechnic Institute of Toulouse, France in 1995. From 2006 to 2009, he served as Department Chair. He is responsible of the Master in Computer Sciences and Systems (Since 2006). He was a Visiting Professor at the Laboratoire d'Infomatique d'Avignon (France) and IEMN-Lille France. He was Head (and founder) of the Computer and Communication Systems Laboratory from 2013 to 2017.

He has been the supervisor of more than 20 Ph.D. students, and he has published more than 100 peer-reviewed publications. His research interests include computer security, wireless sensors network, cybersecurity, and serious games.

He has also served as a general chair, technical program chair, technical program committee member, organizing committee member, session chair, and reviewer for many international conferences and workshops. His research has been supported by several agencies.

The Role of Big Data and IoT in Smart City

Abstract: The expansion of big data and rapid advances in artificial intelligence (AI) and machine learning have played an important role in the feasibility of smart city initiatives. The evolution of Internet of Things (IoT) technologies and a combination of the IoT and big data is an unexplored research area that has brought new and interesting challenges for achieving the goal of future smart cities. Intelligent transportation systems based on vehicular ad-hoc network (VANET) communications will improve many services, expressively, related to transport, security, reliability, and management, including the assistance in the reduction of traffic congestion.

In this talk, we will present some recent research works in the smart city fields:

- Leveraging Smartphone Sensors to Detect Distracted Driving Activities
- IoT for ITS: A Dynamic Traffic Lights Control based on the Kerner Three Phase Traffic Theory
- ACO and PSO Algorithms for Developing a New Communication Model for VANET Applications in Smart Cities
- Smart Citizen Sensing: A Proposed Computational. System with Visual Sentiment Analysis and Big Data.

Contents

Big Data for Smart Cities

Smart Education

Security for Smart City Applications and Safe Systems

Smart Mobility

Smart Renewable Energy Management

Smart Cities

An Approach to the Garbage Collection's Simulation in the "Smart Clean City" Project

Olga Dolinina[1](✉) , Vitaly Pechenkin[1] , Nikolay Gubin[1] ,
and Vadim Kushnikov[2]

[1] Department of Information Systems and Technology, Yuri Gagarin State
Technical University of Saratov, SSTU, Saratov, Russia
odolinina09@gmail.com
[2] Institute of Precise Mechanics & Control, Russian Academy of Science,
Saratov, Russia

Abstract. Simulation modeling of complex dynamic systems is one of the most important elements for their analysis. In this paper there is considered a problem of modeling of the solid waste management process, which is a part of a wide range of "Smart City" concepts. Mathematical model for solving optimization problems and transport scheduling for the targeted garbage collection is suggested. The model is based on the apparatus of Petri nets with priorities. The priorities of Petri net transitions are considered as probabilistic characteristics of the live transitions in the network. Changing priorities allow to customize the Petri net behavior according to existing empirical data. Results allow to investigate the impact of the whole process quantitative characteristics on the solid waste disposal such as speed of filling of waste containers, the number of trucks used, the waiting time for the job assignment and some others.

Keywords: Smart city · Simulation modeling · Garbage collection
Source transition · Priority petri net

1 Introduction

The ecological situation of the modern city, especially, metropolis, depends on many factors that must be taken into account in order to choose a strategy for "Smart city" management system development. Smart city today is usually defined as this that use information and communication technologies, new urban infrastructure to improve service and in this way make life of common citizen more comfortable including improving of the ecological situation. Taking into consideration the growth of the world's population the priority is to find ways to make cities operate better for their population. One of the highest problems is the pollution the result of which is associated with problems in the health of city residents, degradation of urban infrastructure. The basic infrastructure smart city elements must include adequate water supply, assured electricity supply, efficient urban mobility and public transport, IT connectivity and digitalization, social services, sanitation, including garbage collection management. Also there are problems that are related to the social organization of the society, the level of understanding of the values and standards and attitudes towards the

© Springer Nature Switzerland AG 2019
M. Ben Ahmed et al. (Eds.): SCA 2018, LNITI, pp. 3–14, 2019.
https://doi.org/10.1007/978-3-030-11196-0_1

environmentally sound behavior. All of the above reasons lead to the fact that the solution of waste disposal problem requires an integrated approach.

The task of waste removal is a part of the overall problem of creating an environmentally friendly environment in the urban space, usually associated with the natural habitat, protecting the city's ecology from pollution [1]. "Smart Environment" can be considered as a part of the "Smart City" concept, the core of which is using the mobile communication technologies (Internet of Things). It is obvious that the targeted waste collection on time saves expenditures, fuel, reduces exhaust gas emissions.

When considering the essence of the "Smart City" concept it is important to identify a set of factors that are necessary for understanding the projects implemented within its framework. These factors can be divided into internal and external ones, which have an impact on the various stages of development, implementation and operation of solutions within the framework of the urban space intellectualization initiative. The project initiatives for the smart city should be focused on the creation of urban space infrastructure and organizational systems based on modern technologies that respond to emerging problems. Among the considered factors one must mention the following [2]:

- management and organization;
- technologies;
- policy;
- social communities;
- the economy;
- infrastructure;
- natural environment.

Typical task for the most of the projects implemented within the framework of the "Smart City" concept is the targeted managing of the removal process of solid industrial and domestic garbage. This task is directly related to the all characteristics of a typical project, given in the definition of "Smart City" [3]. As a rule, in the solution methods there are components responsible for the use of mobile technologies, information systems based on the intellectual intelligence approach using the knowledge bases. This approach involves information, communication (based on mobile communication) and Web 2.0 technologies, which make it possible to accelerate decision-making processes, apply innovative methods of city management, and improve the urban space environmental safety [4]. Similar tasks are relevant and at the present time methodological fundamentals and applied methods of their practical solution are being actively developed.

2 A Platform for Development "Smart Clean City" Applications

The general structure of the system "Smart Clean City" and optimization schedules algorithms for garbage collection trucks (GCT) are described in the article [5] where the system model is presented in the form of a dynamic graph, which weight functions depend on time and are determined by the current state of the area for garbage containers (AGC) and current road traffic. The structure of the platform for application development within the project "Smart clean city" is presented in Fig. 1.

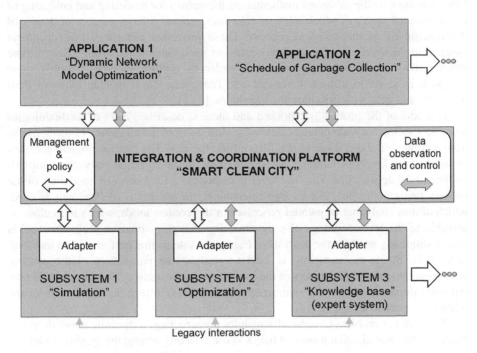

Fig. 1. Structure of "Smart Clean City" platform

The platform unites several subsystems, which allow to solve various problems of analysis and management of the system for collecting and exporting of the solid waste. These subsystems are combined by means of special adapters that allow you to synchronize data, manage various parts of the entire system. At an early stage of development such interaction was carried out directly between subsystems with the help of inherited interactions between them. Developed applications have a unified interface and allow to manage the data integrity model in various subsystems.

The subsystems are server applications that allow to model the system of collection and removal of waste (SUB_1), apply optimization algorithms for the schedule of trucks on several parameters (SUB_2), use in the decision-making system a knowledge base built on the experience of the enterprise organization of waste disposal (SUB_3), special expert knowledge of the traffic situations in the city. Developed applications are located both on the server part of the system, and in client applications using mobile platforms.

In the next section, there is described the mathematical apparatus of Petri nets used in the SUBSYSTEM_2, which is used for developing of the garbage collection system model and for analyzing of its effectiveness. Subsystem 2 "Simulation" of this structure is modeled by means of Priority Petri Net (PPN) framework.

3   Simulation of the Garbage Collection Process by Petri Nets

Petri nets are a well-established mathematical formalism for modeling and analyzing of the distributed systems, which allow to take into account a large number of details of the functioning of the analyzed process. These processes can range from technical systems to the systems of business processes, or social interactions [6–9]. Recent time many extensions have been proposed in order to capture specific, possibly quite complex, behavior in a more direct manner. These extensions include Inhibitor Petri nets [10] and nets with priorities for transitions [11, 12].

A model of the process is proposed and there is described a set of methodologies that allow to solve a wide range of issues for increasing of the efficiency of using trucks transporting household waste to landfills outside the city. The methodological complex includes three levels of building decision-making systems, based on various formal-mathematical approaches. The first level is responsible for stochastic modeling of the process being controlled. For its implementation, the apparatus of Petri nets is used, which allows analyzing of parallel processes in the contest mode, which is as close as possible to the real conditions of a dynamically changing situation [13]. This tool is also traditionally used for the analysis of transport systems that perform cargo handling tasks [14]. Stochastic modeling is used to analyze the parameters of the system, determine the optimal parameters of the loading and unloading cycle of trucks that take out solid domestic waste. The optimized model allows, in turn, to test the use of the other two approaches used.

To solve the problem of optimal schedules constructing, a dynamic network model based on the formalization proposed below is used. In this setting, the system model and method of solution are different from the existing ones [15] in that the network model of the transport system which used is dynamic, taking into account the actual state of the road network and traffic. In addition, in real time, information about the fullness of containers with waste is received and used in calculations. The third approach that allows to use the expert knowledge of specialists in expert systems for managing the logistics of waste disposal is the methodology for designing and applying knowledge bases, namely expert rule-based systems that allow you to evaluate the variants of automatically constructed schedules and choose among them the most realistic ones. Expert knowledge can include knowledge about typical traffic problems including the peak hours, time of the elimination of the road accidents in the various parts of the city and etc.

Garbage disposal is a complex technological process in which companies for the garbage collection are included. The effectiveness of this process is affected by a large number of factors that were described earlier and which make it difficult to obtain the optimal solution. To analyze the process of waste disposal, it is necessary to take into account random factors and complex dependencies between all subjects of the process as well. To implement this analysis, one can use the methodology of simulation.

Formally, a Petri net with priorities is defined as a 6-Tuple in the form:

$$PN = (P, T, F, m_0, w, \rho) \tag{1}$$

where

$P = \{p_1, p_2, \ldots, p_n\}$ is a finite set of places;
$T = \{t_1, t_2, \ldots, t_m\}$ is a finite set of transitions;
$F \subseteq (P \times T) \cup (T \times P)$ is a finite set of arcs;
$m_0 : P \rightarrow \{0, 1, 2, 3 \ldots\}$ is some initial marking.
$w : F \rightarrow \{1, 2, 3 \ldots\}$ is a weight function for arcs;
$\rho : T \rightarrow \{1, 2, 3 \ldots\}$ is a transition priority mapping;
$P \cap T = \varnothing$
$P \neq \varnothing$
$T \neq \varnothing$

Places p_i are represented by circles, transitions t_i by boxes, and the flow relation F by directed arcs. Places may carry tokens represented by filled circles or natural number near the place. An initial marking in a Petri net is a function m_0, mapping each place to some natural number (possibly zero). A current marking m is designated by putting $m(p)$ tokens into each place $p \in P$. For a transition $t \in T$ an arc (x, t) is called an *input arc*, and an arc (t, x) —an *output arc*.

A place $p \in P$ with an arc from itself to a transition $t \in T$ is an input place for t, and a place with an arc from a transition t to itself is an output place. A transition $t \in T$ is enabled in a marking m if

$$\forall p \in P \; m(p) \geq F(p, t) \tag{2}$$

A transition with no input places is referred to as a source transition and a transition with no output places is called a sink transition (see Fig. 2). A source transition is always enabled.

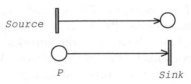

Fig. 2. Source and Sink transitions

An enabled transition t can or cannot fire. The firing of an enabled transition t removes $w(p, t)$ tokens from each input place p of t and adds $w(t, p)$ tokens to each output place p of t. An enabled transition t may fire yielding a new marking m' (denoted $m\, t \rightarrow m'$), such that

$$\forall p \in P \, m'(p) = m(p) - w(p, t) + w(t, p) \tag{3}$$

We say in this case that a marking m' is directly reachable from a marking m. A marking m is called dead if it enables no transition. An initial run in PN is a finite or infinite sequence of firings for some enabled transitions

$$m_0 \rightarrow m_1 \rightarrow m_2 \rightarrow \ldots \tag{4}$$

A marked Petri net with priorities is a Petri net together with a priority mapping ρ. The firing rule for a Petri net PN with priorities is defined as follows. Let m be a marking in PN, and $T' \subseteq T$ be the subset of enabled transitions in m (according to usual rules for Petri nets). Then probability of the firing for an enabled transition $t \in T'$ in m is equal to

$$\frac{\rho(t)}{\sum_{t_i \in T'} \rho(t_i)} \tag{5}$$

i.e. transitions with higher priorities have higher probability of firing over transitions with lower priorities.

Within the framework of the Petri net approach, it is proposed to build a model for the process of handling waste containers, which includes various states of vehicles and waste containers, technical specialists, drivers and the processes of their interaction. The tokens of the model are trucks for the removal of waste and a site for containers. The model fragment contains 16 places and 19 transitions as follows:

$$P = \{p_i | i = 1, .., 16\} \tag{6}$$

$$T = \{t_i | i = 1, .., 19\} \tag{7}$$

The transitions and places selected for modeling reflect the parallel processes taking place and take into account the various possible states depending on the factors influencing them. The obtained model is used to carry out a simulation experiment to analyze the characteristics of the process of removal of solid domestic waste. The set of places P and transitions T is presented in Tables 1 and 2 respectively.

We use the Grin software [grin-software.net] to implement our Priority Petri nets and to simulate their dynamic behavior. Grin provides a unifying graph theory framework with special extension for Petri net simulation. With this software tool one can design, animate, and simulate Petri nets models. It provides features that facilitate the implementation of our Petri nets in automatic simulation mode.

The software allows you to edit the network in interactive graphical mode, assign the attributes of places and transitions, run the simulation process with the specified quantitative restrictions (the number of transitions in one simulation cycle). It is important to note that for all transitions, priority is assigned, which is directly related to the probability of their triggering during the simulation procedure.

Figure 3 shows a fragment of the Petri net that is used to analyze the process of waste disposal. After the analysis, the network parameters are modified and a comparative analysis of the effectiveness of the organization of the process is carried out. The proposed model of the Petri net for the process of waste collection and removal allows a computational simulation experiment to be performed to determine an effective algorithm for managing the collection and removal of waste.

Table 1. Places of the Petri net for garbage collection modeling

Name	Interpretation
P_1	GCT in the parking (in the garage)
P_2	AGC is clear
P_3	GCT is waiting for assignment
P_4	AGC is filled
P_5	GCT is assigned to clean up the AGC
P_6	AGC is being processed ("is busy")
P_7	GCT is on the AGC
P_8	Confirmation of assignment GCT completion
P_9	Waiting for a message on the status of the GCT
P_{10}	GCT diagnostics
P_{11}	GCT is full
P_{12}	GCT is on the solid garbage dumps
P_{13}	AGC is empty
P_{14}	AGC driver requests for the next assignment
P_{15}	Completion of the working day schedule
P_{16}	Solid garbage dump

Table 2. Transitions of the Petri net for garbage collection modeling

Name	Interpretation
T_1	Driver receives permission for work
T_2	Driver does not receive a work permit permission for work
T_3	There is a partial filling of the AGC
T_4	There is a filling of the AGC (TSource)
T_5	Waiting for task assignment
T_6	GCT movement to the AGC
T_7	AGC maintenance
T_8	There is a complete AGC cleaning
T_9	There is a partial AGC cleaning
T_{10}	There is a GCT breakdown
T_{11}	There is a GCT filling
T_{12}	On-site GCT repair
T_{13}	Transportation of the GCT to the garage
T_{14}	GCT movement to the solid garbage dumps
T_{15}	GCT discharging
T_{16}	Finish of the working day for a GCT
T_{17}	GCT continues to work
T_{18}	GCT cannot continue working
T_{19}	The driver finishes working day

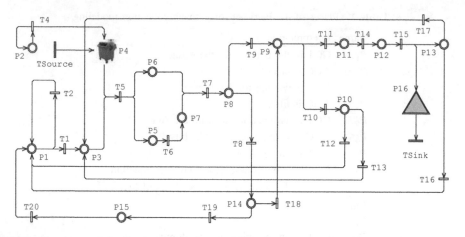

Fig. 3. Fragment of the Petri network for modeling the process of collection and removal of solid waste

The relevance of the Petri net model represented in Fig. 3 and described in the previous sections is demonstrated through several simulations made for different configurations of the system. They are defined according to the functions to be activated in the Petri net model of the system.

Since in the network shown in Fig. 3 there is a source transition (TSource), then it is alive at any moment, since PN has at least one enabled transition. The tokens corresponding to the trucks move through the network cyclically, returning either to the garage P1 (transition T16), or they queue up for a new job assignment (firing transition T17).

4   Results for the Process of Waste Collection and Removal Modeling

Let's consider the results of modeling the system. This article presents only some of the results obtained, which allow us to judge the effectiveness of the entire system. One of the important results of the presented analysis is an estimation of the number of trucks necessary for waste disposal at a given rate of appearance of filled containers necessary to reduce the number of untreated containers. It is clear that the increasing of the number of trucks makes it possible to minimize the number of such containers, but this, in turn, increases the waiting time for trucks when the next task is assigned.

Figure 4 shows the number of transitions firing when the network is started for simulation. The length of each column corresponds to the number of transitions triggered by the network for some common number (about 2000). The transition TSource, which forms the total number of filled containers, triggers about 7% of these events for the model.

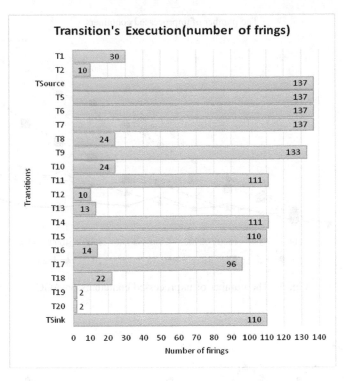

Fig. 4. Evolution of the number of firings in the garbage collection process

In Fig. 5, you can see the number of unprocessed containers after the determined number of firings of all model transitions, depending on the number of trucks participating in the removal of waste. The lower axis shows the number of transitions triggered.

When using a small number of trucks (5–10), this value remains significant and comparable to the total number of containers processed in the system. With an increase in the number of trucks in the system (15–25), the number of unprocessed filled containers is reduced to 10–20% (accumulated in P_4). But with the increase in the number of trucks, the waiting time also begins to increase for the next job assignment (accumulate in P_3).

In Fig. 6 it can be seen that the number of trucks waiting for the assignment of the task increases from 0 to 12–16 with the increasing of their numbers.

Figures 5 and 6 show the situation when using 15–25 trucks, the number of unprocessed containers is minimal, but the waiting time for the next job assignment is increased. The optimal solution in this case is a compromise between the reliability of the system (there is a certain number of trucks in reserve) and the speed of containers processing.

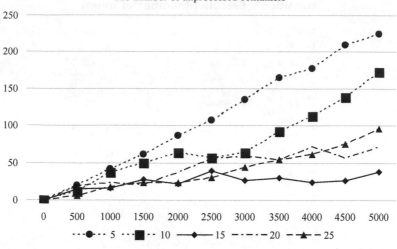

Fig. 5. The number of unprocessed containers on AGC

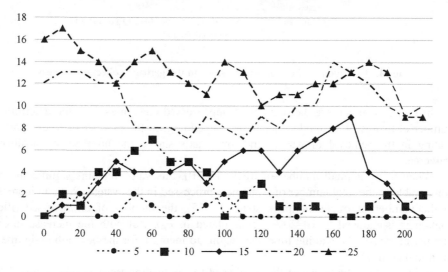

Fig. 6. Number of trucks waiting to be assigned jobs in the queue

5 Conclusion

To increase the effectiveness of waste collection, it is necessary to develop a dynamically managed targeted waste collecting system. The separate solutions, which are currently implemented in Russia, should be prepared for integration into the intellectual urban infrastructure.

The described system "Smart Clean City" can be considered as a targeted waste management system, it has been tested in Saratov (Russia), city with a population of approximately 1 million people. Pilot implementation of the system in the period from September 2015 to the present time shows 21% decrease in the processing time of containers compared to traditional manual planning, without taking into account the dynamic parameters of the system described above. Similar advantages are demonstrated by simulation modelling of the waste removal process based on the Petri nets and using several levels of optimization. The trucks company, which uses the system as a tool for preparing timetables for the cars, has 24 trucks in its composition.

Discussion aspect of the described approach is, for example, the lack of information on the degree of compressibility of cargo in containers and its weight. This does not allow to determine accurately the amount of waste that can be loaded into the truck and taken out of the site. If the load can be more tightly compressed, more containers can be placed in the truck, therefore it is difficult to estimate the exact number of containers that can be cleaned.

Comparative analysis has been carried out using real data and simulation results. The area of the city for which the model has been developed has about 250 containers at 56 sites. Two landfills for the removal of solid household waste are used. Each container has a capacity of 100 kg. The carrying capacity of the truck is estimated at 5000 kg (the actual capacity depends on the degree of compressibility of the waste).

The model of the system is universal and can be used to optimize the process of collecting and exporting solid domestic waste in the large metropolitan areas.

As for the discussion which requires further analysis, it is necessary to note the necessity to develop ways of taking into account the effect on the performance indicators of the processes of the subsystems listed in Fig. 1 in the proposed model. Of particular interest is the modeling of optimization based on the current traffic in the city, the inclusion of expert knowledge in the decision-making system, use of an automatic alarm system on the filling of containers. Such an analysis makes it possible to estimate the contribution of the optimization subsystems to the improvement of the whole model parameters.

References

1. Global Innovators: International Case Studies on Smart Cities.: Research paper number 135. https://www.gov.uk/government/publications/smart-cities-international-case-studies-global-innovators. Oct 2013
2. Chourabi, H., Nam, T. et al.: Understanding smart cities: an integrative framework. In: Proceedings of the 2012 45th Hawaii International Conference on System Sciences, IEEE Computer Society, pp. 2289–2296 (2012)
3. IEEE Smart City definition.: http://smartcities.ieee.org/about
4. Toppeta D.: The Smart City Vision: How Innovation and ICT Can Build Smart, "Livable", Sustainable Cities. The Innovation Knowledge Foundation (2010). https://inta-aivn.org/
5. Dolinina, O., Brovko, A., Pechenkin, V.: Method of the management of garbage collection in the "Smart Clean City" project. In: Communications in Computer and Information Science, Proceedings of 24th International Conference on Computer Networks, vol. 718, pp. 432–443 (2017)

6. Huang, Y, Chung, T.: Modeling and Analysis of Urban Traffic Lights Control Systems Using Timed CP-nets. J. Inf. Sci. Eng. **24**: 875–890 (2008). http://www.iis.sinica.edu.tw/page/jise/2008/200805_13.pdf

7. DiCesare, F., Kulp, P.T., Gile, M., List, G.: The application of Petri nets to the modeling, analysis and control of intelligent urban traffic networks. In: Valette R. (eds) Application and theory of Petri nets. Lecture notes in computer science, vol. 815 (1994). https://link.springer.com/chapter/10.1007%2F3-540-58152-9_2

8. Benarbia, T., Labadi, K., Moumen, D., Chayet, M.: Modeling and control of self-service public bicycle systems by using Petri nets. Int. J. Model. Ident. Control **17**, 173–194 (2012)

9. Malkov, M.V., Maligina, S.N.: Petri nets and modeling. In: Proceedings of the Kolsk Science Center RAS. vol 3, pp. 35–40 (2010). (in Russian)

10. Verbeek, H.M.W., Wynn, M.T., van der Aalst, W.M.P., Hofstede, A.H.M.: Reduction rules for reset/inhibitor nets. J. Comput. Sys. Sci. **76**(2), 125–143 (2010)

11. Best, E., Koutny, M.: Petri net semantics of priority systems. Theoret. Comput. Sci. **96**(1), 175–215 (1992)

12. Lomazova, I.A., Popova-Zeugmann, L.: Controlling Petri net behavior using priorities for transitions. Fundamenta informaticae **143**(1–2), 101–112 (2016)

13. Ryabtsev, V.G., Utkina, TYu.: Information technology for design of automated control of technological processes systems. Control, Commun. Secur. Syst. **1**, 207–239 (2016). (in Russian)

14. Naumov, V.S.: Petri nets in modeling the process of freight forwarding services. Road Trans. (Kharkov) **24**, 120–124 (2009). (in Russian)

15. Anagnostopoulos, T., Zaslavsky, A., Medvedev, A., Khoruzhnikov, S.: Top-k query based dynamic scheduling for IoT-enabled smart city waste collection. In: Proceedings of the 16th IEEE International Conference on Mobile Data Management (MDM 2015), Pittsburgh, US (2015)

Climate Change: An Environmental and Economic Challenge

Lamia Boukaya$^{(\boxtimes)}$ and Sahar Saoud

École Nationale de Commerce de Gestion, Agadir, Morocco
{lam.boukaya, sahar.saoud}@gmail.com

Abstract. This paper sheds light on the ecological transition phenomenon by suggesting a sustainable development model that demonstrates the different variables that impact the climate which in return suggests actions that ensures a renewal in our way of consuming, producing, working and living together. The paper then explains the factors that prevent the ecological transition in Morocco because the protection of the environment is conditioned by the implementation of preventive and corrective actions in order to remedy the harmful effects of overusing the existing natural resources. Then we explain the model that we will use to quantify and analyse the different weather variables. The first part is dedicated to explaining the components of economical model as well as the tools used in our research, the second part is dedicated to decorticating the model and its underlying mechanisms, and the last part explains the fuzzy logic behind it.

Keywords: Ecological transition · Fuzzy logic theory · The SES
Human behaviour · Natural resources · Sustainable development
Environment · Climate change · Economic challenge · Climatic and
non-climatic variable · Health · Biological diversity

1 Introduction

The ecological transition is a factor of innovation, competitiveness and economic development. It is the capacity to render resource development compatible with unpredicted climate changes. More precisely, the ecological transition is nowadays considered as a group of political, economical and social decisions in a given country. It is a concept that implicates an industrial, behavioural and intellectual revolution: it is a rebuilding of the society on new foundations:

- Improving the regulatory predictability of actors
- Complete the tools for mobilizing the public and private funding towards the ecological transition
- Reinforce the consideration of extra financial issues
- Reinforce the intellectual framework of financial actors' practices.

Current forecasts say that by the end of the 21st century, increases in atmospheric gas concentration will cause changes in all climate subsystems, but the question remains: Why worry about climate change?

© Springer Nature Switzerland AG 2019
M. Ben Ahmed et al. (Eds.): SCA 2018, LNITI, pp. 15–20, 2019.
https://doi.org/10.1007/978-3-030-11196-0_2

The answer is simple: warming temperatures or any change in climatic parameters can only result in environmental and socio-economic impacts. Sectors such as agriculture, forestry, ecosystems, infrastructure, fisheries, water management, tourism, economic activity, energy production and demand have all adjusted to historical climatic parameters. All these activities, already sensitive to natural climate variability, will necessarily be affected if climate statistics are not consistent with their history. Given this reality, while considering the uncertainties, several scenarios are possible:

- Pursue activities that cause massive greenhouse gas emissions and suffer the consequences (do nothing and respond after the fact);
- Decrease emissions and hope that the climate will not change despite the trends already initiated and the massive additions of greenhouse gases of the past 40 years (cut off the source of the problem and hope for it);
- Prepare and implement adaptation strategies in the face of a new climate reality (adapt to anticipated changes);
- Reduce emissions while implementing adaptation strategies in critical sectors.

The Moroccan environment favors the latter option for several reasons:

- This approach is by far the safest and most far-sighted
- Several events suggest that our socio-economic infrastructure is already poorly adapted to the natural variability of the climate. Better adaptation to climate is therefore beneficial, even without climate change
- It opens up opportunities for an economy of the future integrating sustainable development.

2 Main Components of the Model

From the 90 s the actors became fully aware of the fragility of the environment, many defendants of the subject of ecological transition claim that the responsibility for the protection of the environment is universal, the climate is considered then as being a complex system that encompasses not only multiple variables but also their effect and impact on nature. One can then raise a preoccupying issue which is: Why is ecological change so important than before?

On the other hand, human behaviour is not the only one responsible for the ecological damage, the integration of other moderating variables in the model is judged more or less logical so that the public authorities and the companies can understand it for better integration into their long-term ecological strategy, in other words companies, public authorities and organizations have become aware of their ecological responsibilities, they have confirmed that they are the first actors responsible for protecting the environment through the establishment of corrective and preventive actions.

2.1 Social Ecological System as a New Multidisciplinary Approach

Climate change and human behaviour are two inseparable and entangled variables, the explanation of this phenomenon depends on three types of factors according to several previous studies, which are: natural factors, social factors and political factors.

Other theoreticians consider the territoriality variable as an influencing factor, but the major problem is the inadequacy of scale contexts (American context is largely different from the Moroccan context). This difference remains in the ecological maturity of each country. It is for this reason that the design of a scale of measurement is important while taking into account: the awareness, the environmental culture, the education and the strategy of each region. The implementation of this multidisciplinary approach is imminent to build and forge an interaction between different disciplines in order to share the same vision.

2.2 Methodological Structure

To better understand the context of our article and to be aware of the underlying issues, our study will have a mathematical reference in the first place for the purpose of deciphering the state of the ecological transition in Morocco, modelling and studying the ecological transition Ecological requires the application of not only mathematics but also computer science (tool for data analysis and analysis of results).

At this point, we can deduce that applied mathematics is a vector of performance and efficiency, it allows to understand the ecological damage caused by human behaviour. It is better positioned compared to other sciences in terms of perspective. And for this reason we have considered it as the most reliable tool for measuring unquantifiable variables such as human behaviour.

We are convinced that it presents and provides a multitude of methods for analysing, modelling and even helping to make decisions to overcome future scenarios in terms of climate change, especially that human behaviour is a key element when it comes to climate change. Moreover, the analysis of this model will allow us to understand its complexity and the diversity of the disturbing elements of the climate.

3 The Social Ecological System Model

3.1 Climatic and Non-climatic Variable

According to multiple studies and works dealing with the subject of ecological transition, the problem is always oriented in a mono-direction or double-direction fashion, in other words, the ecological transition is often explained by one or two variables, which seemed illogical for we can not explain such a phenomenon without taking into account all the variables complicit in climate change. The socio-ecological model is a corrective action that will take into consideration any possible variable without forgetting the principle of territoriality.

4 The SES: The Social Ecological System

It is a dynamic system that transforms inputs that are part of the GNEH list into outputs over a specified period of time. As a first proposition, the model below is the typical design that we have proposed to schematize the existing links between the different variables and elements that we mentioned earlier (Fig. 1).

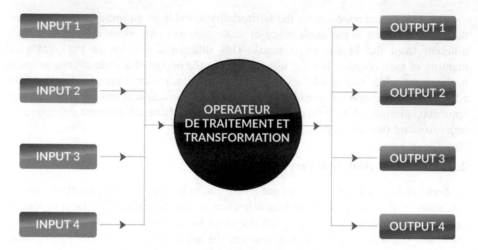

Fig. 1. The SES model

The measurement of human behaviour includes items relating to the three components of the attitude namely: cognitive, affective and conative. The first represents the body of subjective knowledge about environmental problems and the measures taken to improve them. The second reflects the emotional responses to ecological problems while the third component is explained by the added value of improving the environment. According to several studies that affirm that there is a positive link between the protection of the environment and the ecological behaviour of consumers, even if the intensity for some is more or less low, while for others, the relationship is almost strong. This divergence can be explained by two factors:

- Environmental protection is multidimensional, encompassing both cognitive, affective and conative attitudes, which explains the complexity of explaining the ecological behaviour of consumers.
- the measures taken into account this different from one study to another, they generally depend on the context of the study.

5   Biodiversity and the Ecosystem

Internationally, climate change will contribute to the disruption of the earth's natural ecosystems, potentially leading to the extinction of indigenous wildlife. In Morocco, it is mainly the impact of climate change on habitat and ecosystems that will dictate the changes of biodiversity. Several studies show impressively how climate change can seriously disrupt Moroccan habitat, health, behavior and economic growth. However, the provision of climate information should be able to reduce costs but only if it is available in a timely manner. From an economic perspective, it is quite reasonable to invest in adaptation in order to limit or eliminate the anticipated negative impacts while exploiting the new opportunities that will arise. The risks of climate change and natural disasters and recurrent climatic disturbances are, moreover, only contributing to the weakening of ecosystems and the degradation of natural resources, with serious impacts on economic and social development. The relationship between biodiversity and the functioning of ecosystems is one of the most current issues of ecology today. In addition to the intellectual challenges of understanding and simulating the dynamics of ecosystems, since the end of the 1980s, crucial environmental issues have emerged. Indeed, there is a positive link between biodiversity and ecosystem health that has been expressed in terms of biological insurance: the more species there are, the more likely it is that one or more of them adapted to the new environment induced by an exceptional or extreme event. It is for this reason that Morocco has engaged a participatory process in a long-term strategy that will enable it to meet its international obligations in this area. Morocco has initiated several programs for the rational management of water, the fight against desertification, the protection of biological diversity and the initiation of an energy policy promoting energy efficiency and the use of renewable energies. insufficient efforts to support the ecological transition. It will be necessary to focus on institutions and programs of monitoring, prevention and disaster management, mitigation and adaptation to climate change, development of natural and cultural resources as well as targeted communication strategies and activities. sensitization and mobilization of the population, taking into account the different needs and perspectives of men and women in these areas.

6 The Ecological Transition

The ecological transition consists of a set of principles and practices resulting from experiments on local self-government in a context of dependence on oil and other sources of energy. It is the transition from the current mode of production and consumption to a more ecological mode. The ecological transition is not just a simple layer of green paint on our current society, but is a change of economic and social model that will brutally transform our ways of consuming, producing, working and living together. These changes in our ways of consuming, working, producing or cohabiting are intended to serve sustainable development. The ecological transition as a whole is composed of several interdependent components, such as: the food transition, favorable to a more organic and peasant agriculture.

7 Conclusion

In this paper we proposed a model for describing the process of ecological transition in Morocco. It is therefore a scientific base and a reference for any new research in the field. It will also allow to model this transition via mathematical methods that have already given satisfaction in other domains. It is an essential decision-making tool for all stakeholders with regard to climate change and sustainable development. At the strategic level, an integrated modelling will favour originality and creativity from a climatic and ecological perspective; it will thus help in the implementation of thoughtful actions for the Moroccan territory. That's why we decided to put in place a general model that guarantees an ecological transition to a stable state. The mistake is to believe that the ecological transition is a simple industrial issue that is transparent to the people. In reality, it is above all a question of society and of change of modes of consumption that is strongly linked to consumer behavior. We need to put tools in place to promote consumer communication and awareness nationally and internationally. Some actions are simple, others less because they will affect elements of comfort. For example, the transport sector is the biggest contributor to climate change: It could be decided to limit the fuel consumption of new vehicles by law so that avoided greenhouse gas emissions are sufficient to meet our climate goals. but it would affect the performance of vehicles (speed, payload…). This is why it is essential to involve the population in the energy transition, whose impact on their lifestyle they will only accept if they understand its harmful effects. Climate change is probably the biggest threat to our survival in the coming decades. It comes from what makes the comfort of our societies. Climate change will not be easy and will require a significant effort on our part. Without this, the ecological transition will continue to be a failure, and we will all pay for it.

References

1. Massé, E, Edem, R.: Livre Blanc sur le Financement de la Transition Écologique, Novembre, Pages 5–38
2. Laouina, A.: Gestion Durable Des Ressources Naturelles et de la Biodiversité AU MAROC, Mars 2006. Pages 4–118
3. Saoud, S., Mahani, Z., Daadaoui, L.: Climate Change-Human Behavior: Towards Integrated Modeling publié dans le International Journal of Science and Engineering Investigations, 2017. Pages 183–186
4. Picard, F., Tanguy, C.: Innovation et transition techno-écologique ISTE Edition: UK, Juin 2017. Pages 19–111
5. Yves, L.: Initiation à Méthodologie de recherche en SHS, Janvier 2015. Pages 3–80
6. Redman, C.L., Grove, J.M., Kuby, L.H.: Integrating social science into the Long-Term Ecological Research (LTER) network: Social dimensions of ecological change and ecological dimensions of social change. Ecosystems 7, 161–171 (2004)
7. Lemos, M.C., Morehouse, B.J.: The co-production of science and policy in integrated climate assessments. Glob. Environ. Change 15(2005), 57–68 (2005)
8. Karsky, M., Adamo, M.: Application de la Dynamique des Systèmes et de la Logique Floue à la Modélisation d'un Problème de Postes en Raffinerie, Actes du Congrès de l'AFCET. Edition Hommes and Techniques 2, 479–491 (1977)
9. Fankhauser, S., Smith, J.B., Tol, R.S.J.: Weathering climate change: some simple rules (1999)
10. Walker, B., Holling, C.S., Carpenter, S.R., Kinzig., A.: Resilience, adaptability and transformability in social—Ecological Systems. Ecol. Soc. 9(2), 5 (2004)
11. Ostrom, E., Janssen, M.A., Anderies, J.M.: Going beyond panaceas. Proc. Natl. Acad. Sci. 104(39), 15176–15178 (2007)
12. Liu, J., Dietz, T., Carpenter, S.R., Alberti, M., Folke, C., Moran, E., Pell, A.N., Deadman, P., Kratz, T., Lubchenco, J., Ostrom, E., Ouyang, Z., Provencher, W., Redman, C.L., Schneider, S.H., Taylor, W.W.: Complexity of coupled human and natural systems. Science 317, 5844 (2007)

Compressive Strength of Concrete Based on Recycled Aggregates

Khaoula Naouaoui[✉], Azzeddine Bouyahyaoui, and Toufik Cherradi

Mohammadia School of Engineers, Mohamed V University of Rabat, Rabat,
Morocco
{naouaoui.khaoula, bouyaz2, tcherradi}@gmail.com

Abstract. Recycled aggregate concrete is considered the next generation in the field of construction: it respects the environment, solves the problem of debris management and is economically profitable. In order to better adapt its use, technical studies, experimental studies and simulations are carried out in all research centers around the world in order to define its field of application. Our study falls within this framework. It is concerned with the study of the mechanical characteristics of recycled aggregate concrete essentially the compression test for various percentages of replacement. The purpose of this study is to confirm the results of studies by other researchers and to find techniques that will maximize the replacement of natural aggregates with recycled aggregates. The concrete chosen for these tests is an old building in the region of Rabat, Morocco which has been built more than 40 years and demolished in the year of 2017. The tests carried out showed a decrease in the compressive resistance noted when the replacement rates exceed 50% rate. The first improvement methods were put in place and being tested: the partial replacement of cement with pozzolan (20% rate) known by his improving of the compressive strength for ordinary concrete, the partial replacement of the large proportion [12.5–31.5] only in recycled concrete and work with natural gravels. Other improvements will be proposed as the studies progress.

Keywords: Recycled aggregates (RA) · Compressive strength
Recycled aggregates concrete (RCA) · Laboratory tests · Replacement
percentage

1 Introduction

The use of recycled aggregates is not new but rather goes back to the 1990s of the previous decade.

The most common examples are in the field of roads or superstructures. This is due to a poor knowledge of the results of the use of recycled aggregates concrete in terms of mechanical, physical and chemical characteristics.

Recycled aggregates are based on crushing old concrete (usually with unknown characteristics). Thus it is natural aggregates surrounded by cement paste that why we can't predict the result of its interaction with the components of the new concrete, mainly water and cement.

© Springer Nature Switzerland AG 2019
M. Ben Ahmed et al. (Eds.): SCA 2018, LNITI, pp. 21–31, 2019.
https://doi.org/10.1007/978-3-030-11196-0_3

In the domain of the recycled aggregates concrete, many studies have been done to determinate the effect of using RA instead of NA in the concrete's properties.

The results of these studies are aleatory, he majority of these have proven, based on experimental studies, that the compressive strength is lower for concrete made with RCA but some ones declare the contrary([1–3]).

An another study [4] even went from experimental studies to analytical studies and performed statistical analysis based on the collected data from literature and reported that it is possible to develop a model to predict the strength decrease in concrete containing RCA for different replacement level.

2 Test Procedure and Results

2.1 Materials Used

Two types of aggregates are used:

- Natural aggregates with the following dimensions: [5–12.5] and [12.5–31.5]
- Crushed aggregates obtained during the demolition of a building in RABAT, MOROCCO built in the 80 s.

   After demolition, the concrete was sorted and cleaned and crushed in a jaw crusher.

   The proportions chosen are similar to those of the natural aggregates namely [5–12.5] and [12.5–31.5].

   The sand used is a natural with a fineness of 2.66 and a sand equivalent of 95%.

   The cement used is CPJ 45 cement, produced by Asment Temara (Portuguese group Cimpor) and available in Rabat, Morocco;

   The water used is tap water.

2.2 Aggregates Properties

Before testing on recycled aggregate concrete, a characterization of recycled aggregates is required. In this sense, a study of the physical characteristics: density, water content, water absorption capacity, granulometric curve … was made in the laboratory.

Particle size analysis. The granulometric curve of the two aggregates is as follows:

Our study is based on a replacement of natural gravels by recycled gravel so a comparison between the two must be made.

The first criterion of comparison is the granular squelette, this is due to its importance in the filling of the voids and the compactness of the concrete.

The following tables summarize the results of the passage of our various samples in the columns of the test of particle size analysis (Table 1).

In order to compare the granular shape of the various aggregates, the design of the granular curves is essential (Fig. 1).

The interpretation of the granular curves show that:

- The shape of GR1 and GN1 is almost identical, at the 10 mm opening, for example, the percentage of the cumulative sieve of GN1 is almost 50% and the GR1 is 60%.

Table 1. Result of particle size analysis

Sieve size	% passing			
mm	GN1: natural gravel size [5–12.5]	GN2: natural gravel size [12.5–31.5]	GR1: recycled gravel size [5–12.5]	GR2: recycled gravel size [12.5–31.5]
0.08			0.00	0.00
5	0.00		2.78	0.60
6.3	6.23		16.15	1.23
8	17.67			
10	50.89		67.90	2.54
12.5	100.00	3.77	97.33	6.64
16		5.76		35.10
20		21.81		70.35
25		70.87		31.41
31.5		100.00		99.13

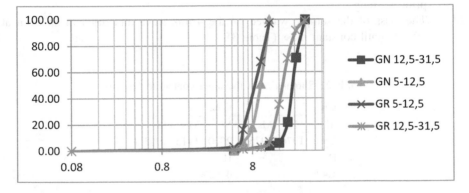

Fig. 1. Granometric curve for natural and recycled aggregates used in experimental study

- The distribution of the various sizes of GR2 and GN2 aggregates is different; This distribution is favorable for GR2 ca allows to have less vacuum and pore in our concrete and to compensate for the defects of the recycled aggregates.

Absolute Density for the recycled aggregates. It is the quotient of the dry mass of the specimen by the volume of the solid matter of it (excluding pores).

There are several ways to determine these densities according to the desired accuracy and the nature of the aggregate; in our case we use the method of the graduated test tube.

The test specimen used has a diameter of 1.6 cm so the $V = \Pi * (1.6)2 * h$ [cm^3].

The test was carried out on three samples (Table 2):

Table 2. Summarize of the absolute density test results

	Specimen 1	Specimen 2	Specimen 3
Weight of the sample (g)	300	300	300
Difference in height after and before addition of gravel (cm)	14.65	15.59	15.04
Volume of the gravel (cm³)	117.85	125.34	120.97
Absolute Density (g/cm³)	2.54	2.4	2.48
Calculated average absolute density (g/cm³)	2.48, We consider 2.5		
Chosen absolute density (kg/m³)	2500		

Water content (w %). The water content is equal to the ratio of the mass of water contained in the sample by its dry mass. It is determined according to the standard "NF P 18-554". $\eta = 100 * (M - Ms')/Ms$.

With:

M   The mass of the sample

M's   The mass of the sample dried in an oven at 105 °C to constant mass without prior washing.

Ms   The mass of the sample washed on the 4 mm sieve and dried in an oven at 105 °C until constant mass (Table 3).

Table 3. Summarize of the water content test results

GR1			GR2		
M	M's	Ms	M	M's	Ms
2000	1976	1975.2	2000	1915	1963.8
H			H		
1.49%			4.32%		

Water absorption rate (ab%). The water absorption by definition is the quotient of the mass of a sample immersed in water for 24 h at 20 °C and at atmospheric pressure, by its dry mass. It is determined according to standard standards "NF P 18-554, 18-555, EN 1097-3, EN 1097-6".

$$Ab = 100 * (Ma - Ms)/Ms$$

With:

Ms   The mass of the sample washed on the 4 mm sieve and dried in an oven at 105 °C until constant mass.

Ma   The mass of the sample immersed in water for 24 h at 20 °C at atmospheric pressure and sponged thoroughly with an absorbent cloth (Table 4).

Table 4. Summarize of the water absorption test results

GR1		GR2	
Ms	Ma	Ms	Ma
1975.32	2194.8	1963.8	2148.4
Ab			
11.11%		9.4%	

2.3 Concrete Formulation

The concrete chosen to study is an ordinary concrete with a fc28 compressive strength at 28 days) of 25 MPA, and an average slump of 7 cm.

The chosen quantities using the dreux gorisse method is thus (for $1 m^3$ of concrete) (Table 5):

Table 5. Summarize of the concrete formulation

Component		Quantity for 1 m^3
Cement		350 kg
Water		193L
Sand		473 kg
Aggregates	Gravillon 5–12.5	402 kg
	Gravier 12.5–31.5	512 kg

In order to compare the replacement effect of natural aggregates with recycled aggregates, several replacement ratios were studied: 0–20–50–75–100%.

2.4 Result of the Tests

The results of the compressive strength test for standardized 16 * 32 test specimens which have been removed at 24 h and left in wet cure until they are fouled are as follows (in average value) (Fig. 2).

3 Methods of Improving Test Results

In introduction, the modification of some parameters (w/ c, cleaning condition, initial moisture,) has shown their effect in increasing the compressive strength of the RCA, but this improvement remains limited. Thus the researchers have worked out more sophisticated methods to improve the performance of recycled aggregates.

Some of these methods are detailed below:

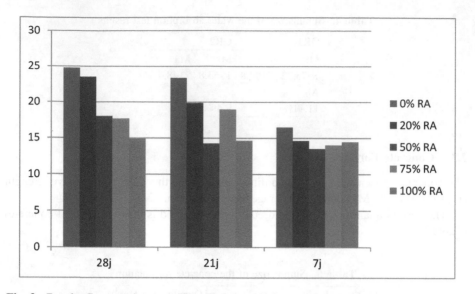

Fig. 2. Results Compressive strength at 7j, 21jet, 28j for recycled aggregates proportions of 0, 20 50, 75 and 100% (MPa)

3.1   Improvement of Mechanical Properties of Recycled Aggregate Concrete Basing on a New Combination Method Between Recycled Aggregate and Natural Aggregate [5]

This study is based on the fact that aggregates occupies more than 70% of the volume of concrete and plays an important role in deciding properties of concrete.

Coarse aggregate forms the matrix of particles as a main frame bearing the load in concrete structure.

Fine aggregates and cement paste files the volume of void among coarse aggregate particles.

The current aggregate combination method in RAC based on the replacement percentage of the entire of coarse aggregate mixture including all particle sizes in coarse aggregate particles or fine aggregate combined with all particle sizes of NA. Nonetheless, in this new method, the authors replace only large size of RA particle by NA in coarse aggregates.

The objectives of this study were to investigate the influence of new combination method on the mechanical properties of RAC and to compare with a conventional combination method between RA and NA with different replacement proportions.

The result of compressive strength test in 28 days for four replacement proportions of RA which are 30, 50, and 100% respectively are showed in the figure below (Fig. 3):

We notice, in this graph, the improvement in compression for the proportions 30, 50 and 70% by respectively 15.2, 22.7 and 11.4%.

Thus this solution is conceivable to improve the quality of the mixture.

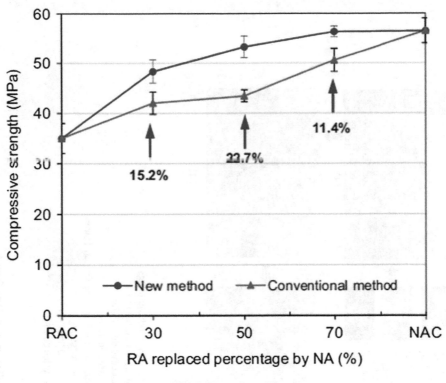

Fig. 3. Data from [5]

3.2 Performance of Recycled Aggregate Concrete Based on a New Concrete Recycling Technology [6]

The study is based on recycling the concrete in accordance with the European C2CA process who developed a new concrete recycling technology purely mechanically and in situ.

The technology consists of a combination of smart demolition, gentle grinding of the crushed concrete in an autogenous mill, and a novel dry classification technology called ADR to remove the fines (Fig. 4).

The feasibility of this recycling process was examined in demonstration projects involving in total 20,000 ton of End of Life (EOL) concrete from two office towers in Groningen, The Netherlands.

After recycling, an experimental study was carried out on mechanical and durability properties of produced Recycled Aggregate Concrete (RAC) compared to those of the Natural Aggregate Concrete (NAC) to understand the importance of RA substitution, w/c ratio and type of cement to the properties of RAC.

Two series of concrete, representing typical concrete types in the Dutch market were produced:

– C1, corresponding to C25/30 S3 D16 (produced by Holcim), and
– C2, corresponding to C45/55 S3 D16 (produced by Heidelberg Cement).

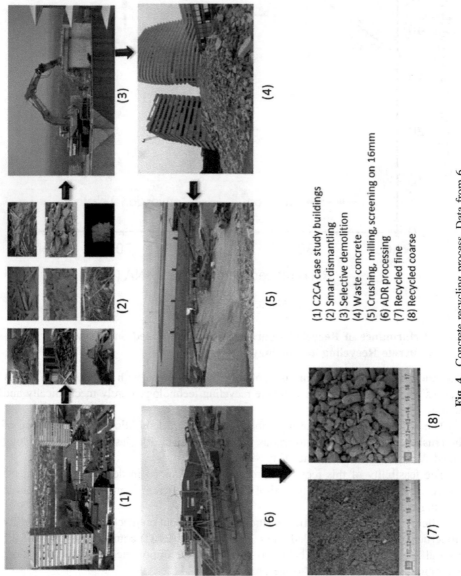

Fig. 4. Concrete recycling process. Data from 6

(1) C2CA case study buildings
(2) Smart dismantling
(3) Selective demolition
(4) Waste concrete
(5) Crushing, milling, screening on 16mm
(6) ADR processing
(7) Recycled fine
(8) Recycled coarse

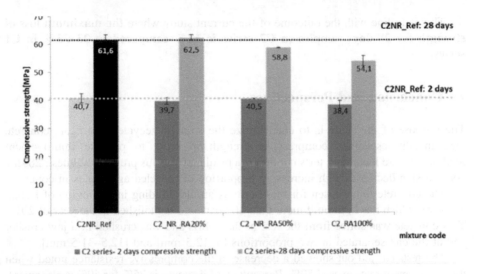

Fig. 5. Compressive strength results in 2 days and 28/days for the C2 serie

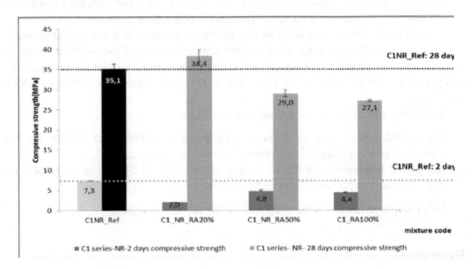

Fig. 6. Compressive strength results in 2 days and 28/days for the C1 serie

The measured compressive strength of concrete corresponds well to the targeted concrete class.

The results show a slight loss in compressive strength with increasing RA replacement in C2 series. However, compressive strength loss in C1 series is higher. The influence of recycled aggregate could be more negative for concretes with high w/c. It is reported that the loss of compressive strength due to the incorporation of RA is less for mixes with higher targeted compressive strength (45 and 65 Mpa) compared to mixes with lower targeted compressive strength (20 Mpa) (Figs. 5 and 6).

This is in line with the outcome of the current study where **the maximum loss of strength for 28 days samples in C2 series is 12% compared to 23–33% in C1 series.**

4 Conclusion and Perspectives

The purpose of the study is to characterize the strength recycled aggregate concrete mechanically especially compressive strength in order to propose improvement methods in case laboratory tests confirm the results of various previous studies.: 28-day compression decreases with increasing proportion of recycled aggregates in concrete.

The concrete thus chosen for these tests is an old building in the region of Rabat, Morocco which has been built more than 40 years and demolished in the year of 2017. This concrete was sorted from the rest of the debris and was crushed in a jaw crusher and at the end separated in two proportions [5–12.5 mm] and [12.5–31.5 mm].

The tests carried out showed a decrease in the compressive resistance noted when the replacement rates exceed 50%. Percentage of decrease is 25% for 50% replacement, 30% for 75% and 40% for 100% replacement rate.

Knowing its environmental value, its economic value and the growing need for its use, improving the quality of recycled aggregates concrete is essential.

Several studies have been done in this direction, some of it was detailed above.

The prospects for this work are to experiment with some of the methods found in the literature, to propose new ones and to combine them in order to increase the compressive strength for concrete based on recycled aggregates proportions of 50% and more.

In the first place, the ameliorations tested will be:

- The partial replacement of cement with pozzolan (20% rate) known by his improving of the compressive strength for ordinary concrete,
- The partial replacement of the large proportion [12.5–31.5] only in recycled concrete and work with natural [5–12.5] gravels.

Other improvements will be proposed as the studies progress.

References

1. Poon, C.S., Shui, Z.H., Lam, L., Fok, H., Kou, S.C.: Influence of moisture states of natural and recycled aggregates on the slump and compressive strength of concrete. Cem. Concr. Res. **34**(1), 31–36 (2004)
2. Gesoglu, M., Güneyisi, E., Öznur, H., Taha, I., Taner, M.: Failure characteristics of self-compacting concretes made with recycled aggregates. Constr. Build. Mater. **98**, 334–344 (2015)
3. Kurad, R., Silvestre, J.D., de Brito, J., Ahmed, H.: Effect of incorporation of high volume of recycled concrete aggregates and fly ash on the strength and global warming potential of concrete. J. Clean. Prod. **166**, 485–502 (2017)

4. Silva, R.V., De Brito, J., Dhir, R.: Properties and composition of recycled aggregates from construction and demolition waste suitable for concrete production. Constr. Build. Mater (2014)
5. Bui, N.K., Satomi, Tomoaki, Takahashi, H.: Improvement of mechanical properties of recycled aggregate concrete basing on a new combination method between recycled aggregate and natural aggregate. Constr. Build. Mater. **148**(2017), 376–385 (2017)
6. Lotfi, S., Eggimann, M., Wagner, E., Mróz, R., Deja, J.: Performance of recycled aggregate concrete based on a new concrete recycling technology. Constr. Build. Mater. **95**(2015), 243–256 (2015)

Constraints Facing the Implementation of a Smart Water Management in Moroccan Rural Area

Abdesselam Ammari$^{(\boxtimes)}$, Mohammed Ammari, and Laila Ben Allal

Research Team: Materials, Environment and Sustainable Development (MEDD),
Faculty of Sciences and Techniques of Tangier Morocco, Tangiers, Morocco
ammari.abdesselam@yahoo.fr, {m.ammari,l.benallal}
@fsst.ac.ma

Abstract. The water sector suffers of a complex interaction between water resources and the socio-economic and environmental systems. The range of stakeholders is huge, public and private, from global to local companies, supported by national, regional and again local authorities. This different nature in stakeholders and the several schemes for water governance are the principal reasons for current market fragmentation in water management solutions [1]. In This paper, we describe the different mutilations that a drinking water supply project, in a Moroccan rural area, undergoes before studies and design until its operation. This study highlights several constraints that each project is subject. Those constraints could be technical, regulatory, financial, economic and social. If we go beyond these constraints we can develop a smart model. Finally, recommendations are provided on how water managers can upgrade drinking water supply and contribute by anticipating the challenging circumstances inherent to transform an ancient water network to a SMART one.

Keywords: Smart water management · Operation · Constraints

1 Introduction

Water is a vital resource for life, and its management is a key issue nowadays [1].

Water distribution systems (WDSs) are one of the major infrastructure assets of the society [2]. A drinking water system is divided into two subsystems: Production and Distribution.

The production is made up of the part from the water resource (borehole, well, dam, rivers) comprising the production structure (treatment plant, pumping station) and going via a pipeline and supply tanks and up to distribution tanks.

As for Distribution, it is composed of all the pipes and network equipment used for protection or good management.

However, good management of a drinking water system requires good design (network equipment by different operating devices) and a good knowledge of all its components:

© Springer Nature Switzerland AG 2019
M. Ben Ahmed et al. (Eds.): SCA 2018, LNITI, pp. 32–43, 2019.
https://doi.org/10.1007/978-3-030-11196-0_4

Tanks: capacity, population served, altitude, power supply,

Pipeline or distribution network: Nature of the pipe material, linear, density of the population served, leakage history and failure, equipment available and its condition (sectoring and disconnecting valves, pressure reducers, relief valves, hydraulic meters,...).

As a result, knowledge of the region and the environment is essential.

Water management is defined as the activity of planning, developing, distributing and managing the optimum use of water resources. Water and sanitation are at the very core of sustainable development, critical to the survival of people and the planet. Goal 6 not only addresses the issues relating to drinking water, sanitation and hygiene, but also the quality and sustainability of water resources worldwide [3].

Challenges and problems in drinking water supply are generally more complex and daunting in poorer countries which are at the same time typically less well equipped (both financially and with technical and policy expertise) to deal with the complexity [4]. For that SMAT WATER MANAGEMENT is necessity. In deed, An efficient water management system requires thousands of constraint devices (sensors and/or actuators) to be deployed across the water distribution network to enable near real time monitoring and control of the water grid components [5].

In reality, the rural Moroccan does not have a suitable platform for a smart water management.

Through this study, we will try to highlight the different constraints that can block the development of a smart model.

This study provides the different mutilations that a drinking water supply project, in a Moroccan rural area, undergoes before studies until its operation. A drinking water supply project, for a territorial collectivity, before its commissioning, goes through several stages: Study, works and operation. Indeed, each project is subject to several technical, regulatory, financial, economic and social constraints.

In fact, a study of drinking water provides, for a distant horizon, the most adequate and comprehensive way drinking water supply of the territorial collectivity object of the project. However, for the realization of the works, several modifications will be made. In fact, for the future management of the drinking water service, management and co-financing agreements will be established and subsequently determine the part of the study that can be financed and subsequently carried out.

In addition, for the designation of the companies in charge of the realization and installation, calls for tenders are launched followed by the phase of judgment and establishment of the corresponding contracts. This phase generally lasts more than six months.

The realization of the works (installation) then knows constraints: opposition of the population, unconfirmed water resources, bad weather, business failure....etc. These constraints increase the turnaround time which generally exceeds one year in the best cases.

On the other hand, the operation, which is considered as the daily task of the managers, is found with a system of supply of drinking water deprived of several protectives equipment or poorly dimensioned.

The overall goal is to elucidate the multiple barriers that a drinking water project should win to be realized. The first part talks about the study and its different components. The second part concerns constraints and recommendations. The third part describes operation and its daily struggle. In last part, a related works description.

2 Studies and Design

Drinking water studies for a local community (rural area) are divided into four main parts:

- Pre-project summary **PPS**,
- Topographic works **TW**,
- Detailed preliminary project **DPP**,
- Business consultation files **BCF**.

Each step requires a particular importance in the establishment of the study of drinking water supply. Indeed, the Pre-Project Summary is the first step, followed by detailed preliminary project and in the end the business consultation files.

A study in general lasts more than one year (preparation of the special prescription file, judgment, realization the study of the PPS-DPP-BCF and launch call for tender for installation).

2.1 Pre-project Summary

This stage is the first phase of a drinking water study. In this phase the consulting engineer (design office) collects all the necessary information related to the field, topology, monograph, population, economic mode, housing. Pre-project summary includes:

Generalities on the study area: The consulting engineer is required to carry out in-depth surveys of the project site and to approach populations, services concerned (administrations) and local authorities, to bring out: Activities, social and economic level of inhabitants, Number of households and the population of douars (villages) and their structures, housing typology, existing basic infrastructures (access, electricity, roads, sanitation, school... etc.).

Description of current drinking water supply systems DWS: The consulting engineer must approach the provincial services (Water Service, Basin Agency...) to collect all the information concerning the existing DWS systems (Plans, statistics, water resources sheets, data sheets, management method, potential problems, water rights...) and also should carry out all the necessary investigations for the inventory and the recognition of the water resources as well as the existing installations of production and distribution of drinking water in the zone of study.

Study of water requirements: The consultant engineer must define the water needs for a horizon of thirty years and make comparison between the population on that date and the available water resources. If that resource isn't sufficient, the consultant engineer should propose another alternative for water supply.

Diagram of water supply: The consultant engineer describes all possible variants to adequately supply the study area. So, he must highlights the scenarios as well as the paths taken by the water pipe, the location of protective devices such as pressure reducers, discharging valve, sectoring valves, as well as tanks and their capabilities.

Economic evaluation of the project: This phase is considered the most important. Indeed, the economic and financial evaluation is considered as the first element of negotiation of the budget with the donors. So specific agreements are concluding depending of the kind of the supply adopted (individual connection or public fountain). Then the project is divided into many parts production (civil engineering and pipe), distribution (pipe) and network inside the study area.

2.2 Topographic Works

The topographic works are divided into two parts. The first concerns the topography of the tracks and paths for future passages and plans of structures to be built. While the second concerns the parcel studies defining the owners of the land to initiate the procedure of expropriation or temporary occupation as the case may be.

2.3 Before Detailed Project

The purpose of the detailed design is to prepare the justifications and detailed descriptions of installation from the previous file and to give a detailed estimate. The detailed project file consists of three parts:

A Description of the context of the center (main village of the local authority) and douars (secondary village). In general, main village beneficiate of the intervention of the national office of Electricity and drinking water before the other villages since it has a compact population and contains the main part of the project.

B Justification and description of installation by component: the consultant engineer studies in detail all the technical and economic aspects for a good sizing of pipes, tank, pumps,...

C Establishment of the bill and Quantitative and Estimated Quote.

During this mission, the consultant engineer will carry out a detailed study of the main installations according to the variant adopted at the end of the first part—Before summary project—which takes into account the drinking water supply of the douars (villages). A detailed economic and financial evaluation of the project will be carried out according to the possible scenarios.

2.4 Business Consultation File

As part of this mission, the Consultant Engineer will be responsible for the study and the preparation of tender documents and/or competitions for the realization of the installation resulting from the mission of the detailed preliminary project. this file contains the financial, administrative and technical terms to be respected by company that desire to realize the project.

3 Range of Constraints on Drinking Water Projects

3.1 Water Resources

Confirmation of the quality and quantity of water intended for human consumption is an essential phase. Indeed, the duration between a pumping test and the operation of a water resource can last several years, which can affect the available information. Consequently, realizing a project on an uncertain resource opens the door for a temporary intervention that then lasts in time pending the confirmation of an alternative and safe resource.

In that situation moving towards surface resources, especially Dams would be a safe alternative to low-flow drilling.

3.2 Water Rights

Water sources in Morocco are affected by the right of water that allows farmers in the region to benefit from the operation of water emerging sources. This right is used for the blockade of several drinking water projects by the concern of these farmers to suffer water shortages that can be generated by the use of water and their impacts on their crops.

3.3 Opposition of Residents

Like any infrastructure project, drinking water projects are exposed to different forms of opposition. Indeed, the water resources are a target frequented by the opposition of residents who request either to benefit from the project before "their" water is intended for another population, or a categorical refusal for the use of water of the source in question. Also, the owner of the land containing the resource (especially the drilling) opposes the operation of his land before being compensated (the procedure of expropriation is long) or refuses the amount of expropriation proposed by the commission of expropriation (commission regrouping several bodies of state).

3.4 Proposed Mode of Service

All the studies are established on an individual connection service mode. However, its realization is subject to financing conditions. In fact, according to the agreements signed with the beneficiary territorial municipality, the national office of electricity and drinking water programs its intervention and defines the project to be realized. Thus, for lack of awareness of the beneficiary population the drinking water project is facing fierce opposition of adhesion that can manifest itself at any phase of the project (Studies, works, operation).

3.5 Economic Evaluation

The realization of a drinking water project is conditioned by eligibility criteria limiting any financing proposal and as a result of implementation. In addition, to bend the project, changes made to make this project eligible. In fact, the hydro-mechanical equipment is eliminated, especially the shut-off and sectoring valves, taps, pressure reducers, suction cups, drain valves, etc.

That equipment is essential and serves to better lighten the drinking water network and allow better operation. That equipment is removed from the project in order to supply drinking water to the target population because of limited budget.

3.6 Expropriation for Reasons of Public Utility

In most cases, drinking water studies provide for the passage along existing roads and tracks to avoid private land. Nevertheless, the typology of the rural world requires the passage through these lands whose property belongs to individuals. In its second mission, parcel work, technical files and the expropriation procedure are considerably delayed for their implementation or for lack of legally defined owners.

3.7 Funders

To make a project eligible it is necessary to satisfy the conditions of the funders. Thus, a restructuring or division of a large project is generally applicable. Indeed, this way of doing is a mutilation of the project.

3.8 Judgment Phase and Award of the Contract

This phase, which is between the study and the realization, generally lasts more than six months, which constitutes an additional delay (Fig. 1);

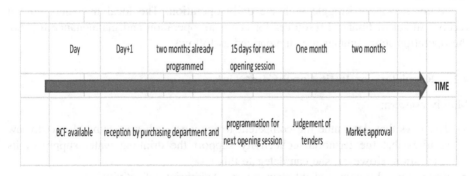

Fig. 1. Details of the time lag between a business consultation file and the development of the corresponding market

For example if Business consultation files **BCF** is available in that day and has been received by purchasing department at the second day, and if it's supposed that there is no mistake, it needs at least five months and half in the best case to have approved market. Unfortunately, administrative procedure is too long and very complicated.

3.9   Installation

A drinking water project includes reservoirs, pumping stations and network pipes. The installation of such a project suffers from a number of obstacles. The opposition generated by the population is the major obstacle blocking the completion of a drinking water supply project. Indeed, this opposition manifests itself in the form of non-adherence to the initial conception, which aims, for example, supplying population by fountain and when individual connection is strongly solicited. Also, opposition to passage through private land before compensation or sometimes categorical refusal of passage is a big problem.

4   Operation

The intervention of the national office of electricity and drinking water succeeds following an agreement signed with the territorial commune wishing the transfer of the management of this service of water. In fact, the new manager is confronted with two types of situations: an existing system or the absence of a water distribution system.

In the first case, in the majority of the cases the ancient managers do not have information concerning the transferred system: plans of network, State of operability of works (valves, pressure reducers,…) nature of driving, operating history. This situation makes the operation more and more difficult and forces the operator to constitute himself his state of health of his system which is are often degraded or not properly functioning.

In the second case where the intervention is new, the water distribution system is victim of all the sieving carried out before arriving at the actual situation which it's mutilated.

Many managers struggle with the daily operation. The local managers do not receive sufficient financial resources for adequate operation and maintenance [6]. So, the operation shortcomings are summarized as follows:

4.1   Regulatory or Institutional Constraints

Mainly concern:

- The laws that determine the competent to manage such a system: Moroccan law requires that the territorial collectivity support the drinking water supply of its population. However, she can delegate this task.
- Regulatory obligations of intervention in a territorial collectivity.
- Slow administrative procedures (appearance in the newspapers, launch of the call for tenders, judgment, establishment of the contracts, follow-up of the contracts, Expropriation,…).

4.2   Technical Constraints

Mainly concern:

- The network: no shut-off or sectoring valves, emptying, suction cups, pressure reducer, network saturation, etc.
- The lifting station: equipment failure: anti-ram ball, suction cup, pressure switch, non-return valve.

4.3   Financial or Economic Constraints

There is a considerable delay in budget programming of a project. For example, if you have a project ready to go, budget planning is done the following year. If this project is retained, it is necessary to add an additional deadline for launching the calls for tenders, judging and carrying out the works, which reach, in the best cases, an additional year. So, an overall delay of more than two years. This programming approach does not differentiate between project categories and their urgency. Thus, the operator is obliged to face a cumulative failure.

In addition, the funding criteria required by funders do not take into account the context of the region and generally do not adapt with the market.

4.4   Social Constraints

Mainly concern:

- Non-adherence of the population to the project: the rural population has lost confidence in the local authority and its deputies. The population requires the start of drinking water works before paying for its participation and sometimes categorically refuses the type of service that can be by fountain and asks for its replacement by individual connections.
- Population accustomed to free service during the management by the Territorial Commune: before the intervention of the National office of electricity and drinking water in some region, the population used to consume drinking water without having to pay in counterpart; this situation causes an abstinence or reluctance of the adhesion of the population to the project.
- Lack of awareness of the population.

4.5   Acts of Sabotage

The Moroccan rural area has an open and scattered character, which makes it difficult to control. Indeed, water works are victim of any type of sabotage, namely:

Water theft:

Some Illicit water connections are detected at the network. This situation distorts most of the interpretations made by the operator, and involves additional work such as the search for non-apparent leaks, analyses of all data of production and distribution. it is the wrong way forward for operator. Also, when the illicit connection is detected, recourse to justice is mandatory, which a long way.

Pipe sabotage:

In order to water their cattle, some shepherds resort to breaking the drinking water pipe during the night. also, during the summer, fires produced in the fields, following high temperatures or other reasons, the residents break the pipe in order to extinguish the fire.

Valve opening:

In period of water scarcity, a system of distribution is operated in order to supply drinking water to the whole population and also to manage length of time of water service interruption. However, maneuvering of the sectorisation valves are operated by the population. However, the operator is affected by the claims, whether they come from the population or by the local authorities.

Deterioration of water works:

A lot of manholes are sabotaged. Indeed, covers of most manholes are stolen. In addiction devices in side are sabotage which causes a huge loss of water and oblige the change of defective device.

Bad weather or force majeur:

Bad weather (particularly rain or snow storms or a rise in temperature) or any case of force majeure causes disruption of water supplies. For example, in some area, the landslide causes a lot of damage to the water network (pipe and water works).

4.6   Lack of Benchmarking

The water systems need a huge and reliable database that can provide a comparison between similar systems [7]. One can compare state of health of the network, resources, tank and any component of the network water supply. Unfortunately, in Morocco the operation manager is still worrying, of the main goal, that is just giving water to people.

4.7   Comparison Between a SMART Management Model and Moroccan Rural Management Model

See Table 1.

5   Recommendations

Given the current situation, to establish a smart model one needs to:

- Upgrade of drinking water supply to networks.
- Sectorization of the network with implementation of a system of metering by sector (at least use of mechanical meter).
- Installation of the electromagnetic flow meter to the entire production system and also tanks as well (In case of non-electrification, provide a solar energy flow meter).
- Setting up a set of devices that informs about the state of the network (like manometer).
- Updating network plans and digitizing them.
- Setting up a system for archiving all events and not just use the human memory.

Table 1. Comparison between SMART and rural model

Designation	Smart Model	Rural Model
Information	In real time	Claim by a third-party or routine visit
Operation	Smart	Archaic
Leakage loss	Minimal	Considerable
Network	Known and mastered	Known roughly
Flow of water consumed	Known and mastered by sector	global
Production	Known in real time	global
Distribution	Known and mastered by sector	global
Pressure	Known in all the network	Measurable in situ
History of data	Recorded on the data base	Staff memory
Decision	Based on Scientific data	archaic
Benchmarking	Possible	Possible
Treatment	Machine to machine	Manual

6   Related Works

For the united Nation, in its report "Progress towards the Sustainable Development Goals", access to drinking water is a right of every citizen and is a prerequisite for all fundamental rights. In this perspective the whole world is mobilized, through governments and organizations, and this for decades, to implement strategies of watch for the rational management of the water resources and to make benefit population of clean water. Effective water and sanitation management also depends on the participation of stakeholders. According to a 2013–2014 Global Analysis and Assessment of Sanitation and Drinking-Water survey, 83 per cent of the 94 countries surveyed reported that procedures for stakeholder participation were clearly defined in law or policy. In the Sustainable Development Goals, the focus is being refined to also include the participation of local communities, which will be captured in the next cycle of Global Analysis and Assessment of Sanitation and Drinking-Water monitoring.

In this sense, the Moroccan legislator is no exception. Indeed, a panoply of laws governing the management of the drinking water service has been put in place (Law 10–15, 36–15, national water plan PNE,…).

On the other hand, the national context is subject to major constraints, notably the difficulty of mobilizing and protecting water resources and financing problems.

According to the Moroccan communal charter, the communal council decides on the management of communal public services, notably the sector of supply and distribution of drinking water. In this sector, the national office of electricity and drinking water is presented as the first operator.

N. Ntuli et al. in their article "A Simple Security Architecture for Smart Water Management System" explain that water scarcity and water stress issues have become clear threat to the global population. This makes water management a critical aspect to ensure sustainable water. An efficient water management system requires thousands of constraint devices (sensors and/or actuators) to be deployed across the water distribution network to enable near real time monitoring and control of the water grid components.

Also, Mudumbe M. et al. Confirmed, in their article "Smart Water Meter System for User-Centric Consumption Measurement", that the traditional management, manual one, is furthermore inconvenient and time consuming, and it wastes resources. This method is also unable to manage the sustainable water resources effectively since it requires efficient, accurate and reliable monitoring techniques that enable the utilities sector and consumers to know the level of water consumption in real-time. Real-time is essential and constitute a key component of the water management system. A smart water-monitoring system will make users mindful of their water consumption and help them to reduce their water usage. At the same time, users will be alerted to abnormal water usage to reduce water loss.

For R. S Rautu et al. in their paper "Use of Benchmarking For the Improvement of the Operation of the Drinking Water Supply Systems" confirm that water systems need a wide and reliable database that can provide an image of efficiency of a system by comparing its performance level with another similar system. That Benchmarking increases organizational performance, Creates a knowledge package and encourage performance culture.

7   Conclusions

In conclusion, each drinking water supply system presents management challenges that are cumulative failures from the initial study through the installation to the operation. Also, a lack of benchmarking makes sure that each department does not receive feedback for a future improvement. Indeed, smart water management can be the solution for a future planning and a reel tool of decision and of course a history data bank that could be shared with all stakeholders. Smart water management Technology is seen as a promising solution for resolving recent critical global water problems [8].

References

1. Robles, T., Alcarria, R., Martin, D., Navarro, M., Calero, R., Iglesias, S., Lopez, M.: An IoT based reference architecture for smart water management processes. J. Wirel. Mob. Netw. Ubiquit. Comput. Dependable Appl. **6**(1):4–23 (2015)
2. Mala-Jetmarova, H., Sultanova, N., Savic, D.: Lost in optimisation of water distribution systems? A literature review of system design. http://www.mdpi.com/journal/water (2018)
3. United Nations. Report of the Secretary-General: Progress towards the Sustainable Development Goals, E/2016/75. https://sustainabledevelopment.un.org/sdg6
4. Bartramp, J., Howard, G.: Drinking-water standards for the developing world. In: Handbook of Water and Wastewater Microbiology (2003)
5. Mudumbe, M., Abu-Mahfouz, AM.: Smart water meter system for user-centric consumption measurement. In: Proceedings of the IEEE International Conference on Industrial Informatics, p. 993–998. Cambridge, UK (2015)

6. Cuppens, A., Smets, I., Wyseure, G.: Identifying sustainable rehabilitation strategies for urban wastewater systems: a retrospective and interdisciplinary approach. Case study of coronel oviedo. J. Environ. Manage. **114**, 423–432 (2013)
7. Rautu, R.S., Racoviteanu, G., Dinet, E.: Use of benchmarking for the improvement of the operation of the drinking water supply systems. In: Urban Subsurface Planning and Management Week, SUB-URBAN 2017, Procedia Engineering, vol. 209, pp. 180–187. Bucharest, Romania (2017)
8. Lee, S.W., Sarp, S., Jeon, D.J., Kim, J.H.: Smart water grid: the future water management platform. Desalin. Water Treat. **55**, 339–346 (2015)

Environment Monitoring System for Smart Cities Using Ontology

Nisha Pahal[(✉)] [ID], Deepti Goel [ID], and Santanu Chaudhury [ID]

Indian Institute of Technology Delhi, New Delhi, India
{nisha23june, deeptigoyal2003, schaudhury}@gmail.com

Abstract. This research focuses on an ontology-based context-aware framework for providing services such as smart surveillance and intelligent traffic monitoring, which employ IoT technologies to ensure better quality of life in a smart city. An IoT network combines the working of Closed-circuit television (CCTV) cameras and various sensors to perform real-time computation for identifying threats, traffic conditions and other such situations with the help of valuable context information. This information is perceptual in nature and needs to be converted into higher-level abstractions that can further be used for reasoning to recognize situations. Semantic abstractions for perceptual inputs are possible with the use of a multimedia ontology which helps to define concepts, properties and structure of a possible environment. We have used Multimedia Web Ontology Language (MOWL) for semantic interpretation and handling inherent uncertainties in multimedia observations linked with the system. MOWL also allows for a dynamic modeling of real-time situations by employing Dynamic Bayesian networks (DBN), which suits the requirements of an intelligent IoT system. In this paper, we show the application of this framework in a smart surveillance system for traffic monitoring. Surveillance is enhanced by not only helping to analyze past events, but by predicting anomalous situations for which preventive actions can be taken. In our proposed approach, continuous video stream of data captured by CCTV cameras can be processed on-the-fly to give real-time alerts to security agencies. These alerts can be disseminated via e-mail, text messaging, on-screen and alarms not only to pedestrians and drivers in the locality but also the nearest police station and hospital in order to prevent and decrease the loss incurred by any event.

Keywords: Multimedia ontology · Internet of things (IoT) · Dynamic bayesian network (DBN)

1 Introduction

A smart city involves a city where multiple information and communication technology and IoT solutions are aggregated in a secured manner to govern assets of a city to facilitate smarter delivery of services like traffic, surveillance, governance, policing etc. The basic idea behind development of smart cities is to ensure excellent infrastructure, better quality of life to its citizens, clean and livable environment and application of smart solutions. To secure these smart cities, it is imperative that a city be equipped with smart monitoring capabilities and safety systems like video surveillance.

© Springer Nature Switzerland AG 2019
M. Ben Ahmed et al. (Eds.): SCA 2018, LNITI, pp. 44–56, 2019.
https://doi.org/10.1007/978-3-030-11196-0_5

Surveillance systems have analog to digital converters and high levels of computation within the video sensors. This advancement has greatly benefited the way video data is processed today. Existing video surveillance systems were introduced to assist the security agencies in monitoring crimes and other such events, and for collecting evidence when such events happen. They could also help prevent crimes if constant monitoring of the video data was done. Manual monitoring of these video streams is possible till the data is confined to a site or building and CCTV output comes from a few devices. But when surveillance scales up to the levels of a locale or a city, and to hundreds and thousands of surveillance devices, it gets impossible to do manual processing of the huge amount of data. Typically in existing surveillance systems, the streams are analyzed for forensic evidence after the occurrence of an incident. Even then, a large volume of video footage needs to be processed to arrive at the exact temporal location of the incident. This causes wastage of time and effort and very often evidence can be missed entirely. Another obvious disadvantage of using such systems is that they *are not* designed to prevent a crime or any other harmful event from happening. In our world today, intelligent devices with considerable sensing and networking capabilities have become more and more pervasive. These devices are used to continuously monitor and interpret the environment and thus form the basis for an IoT network. To enable the **smart** working, a huge web of IoT is employed. A smart traffic surveillance system would be one that exploits IoT technologies, by using sensors and cameras which are not only interconnected, but also connected to the people, vehicles, machines which are part of the secured area. This working is routed wirelessly through a cloud. In context to smart surveillance for intelligent traffic monitoring it would be possible to monitor people's action, movement of vehicles, road accident, and congestion to determine whether an anomalous situation such as hit-and-run is arising based on the current movement patterns of vehicles. A strong communication is required to be established between the three subsystems of a traffic system namely, vehicle, driver and infrastructure in order to cooperate in real-time traffic management in a smart city environment. The infrastructure comprises of:

- Communication of sensors deployed at diverse locations
- Real-time data aggregation and processing
- Taking necessary action based on the assessed situation.

Sensing and data analytics sits in the heart of a smart technology as depicted in Fig. 1. It specifies the various IoT applications for smart cities to monitor, manage and control the environment remotely. The gap between the sensed event and probable event is bridged by semantic reasoning. Thus, in IoT applications, data in different formats from sensor observations needs to be stored in a common format for semantic integration and dynamic processing. To seamlessly integrate these heterogeneous entities (devices, humans etc.), as well as the data that is generated, an abstraction mechanism like an ontology needs to be employed for semantic interpretations. As the observations are basically in various media formats, a multimedia ontology which can encode media manifestations and perceptual modeling of concepts would be ideal. The MOWL encoded ontology models the real world dynamics based on a reasoning framework for tracking situations as the system context changes over time. This reasoning engine takes as input, not just the current context, but all the past information about earlier situations as well. MOWL reasoning for situation tracking is based on DBN model. A DBN can incorporate prior knowledge and has excellent modeling ability to infer the dynamic dependencies between entities involved in a visual interpretation. The proof of concept with a smart traffic surveillance application provide valuable recommendations to monitoring agencies based on changing situations and altering system context. For instance, a vehicle traveling at high speed and breaking a traffic signal, can be automatically detected from a scene, increasing the probability of a potential road accident. An alert for preventive action like identifying and stopping such a vehicle before the accident happens can thus be generated by the smart traffic surveillance system. The trigger can not only be sent to pedestrians and drivers in the locality but also the nearest police station and hospital. Such a system guarantees better quality of life, security and improved commerce, and thus manifests a smart city objective.

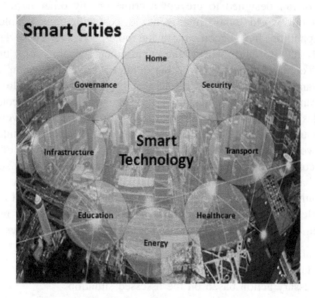

Fig. 1. Components of a smart city

The significant advantage of intelligent traffic surveillance in a smart city environment includes:

- Preventing a road accident by triggering timely alerts in case a speeding car which has broken a traffic signal is detected.
- Preventing and catching a crime in case a pedestrian walking on the streets with an object say gun in his hand and showing suspicious behavior is identified.
- Improving the efficiency of traffic flows since real-time information is available for traffic control and management system.
- Saving a life on road by regulating the traffic smartly in case the surveillance system detects an ambulance on its way to the hospital or a person in need of one.

Application of IoT in smart city environment such as smart traffic surveillance systems such as one presented in our work has undergone a tremendous change in recent years. Zanella et al. in [1] presented a comprehensive survey of the enabling technologies, protocols, and architecture for an urban IoT. They also discussed the technical solutions and best-practice guidelines adopted in the Padova Smart City project as a relevant example of application of the IoT paradigm to smart cities. The authors in [2] presented a smart city framework for processing large-scale IoT data streams by enriching data streams with semantic annotations, enabling adaptive processing, aggregation and federation of data. They also provided samples of the use-case scenarios that are being developed in the CityPulse project. Gaur et al. [3] proposed a Multi-Level Smart City architecture based on semantic web technologies and Dempster-Shafer uncertainty theory. The authors have used a reasoning approach for extracting knowledge and combining information from different smart city domains like transport, home, infrastructure etc. Ontology based frameworks in IoT applications have been used in [4]. Fensel and Rogger [5] presented a semantic approach to enhance security at a port. Their architecture makes use of an ontology-based approach to reduce noise in sensor data, cope with data heterogeneity, pattern detection and data fusion to provide real-time decision support in the future. The authors in [6] integrated different types of knowledge in an ontology for detecting the objects and events in a video scene. They discussed a case-study wherein the events have been detected in the Underground video-surveillance domain using ontology. The authors in [7] developed an intelligent video surveillance system to enable remote monitoring of real time scenarios. Their system intelligently analyzed single person activity and abnormal behavior to enhance the security system in home. Their system maintains the security situation at home and this reduces the incidence of burglary cases and enhances social stability. Zhang et al. [8] used DBN based approach to derive context, suited for applications where the decision is taken from dynamically available information. The work in [9], proposed BeAware!'s ontology in traffic domain, which introduced the concept of spatio-temporal primitive relations among observed real-world objects, increasing situation awareness. The authors in [10] developed an ontology based model for development of an integrated architecture of road traffic but the papers lacked to predict situation causing traffic. The overall challenge in these systems is a lack of reasoning with uncertainties which cannot be avoided when handling multimedia observations.

With these bottlenecks in mind, we have proposed a smart traffic surveillance system, employing IoT services, with an ontology-based framework to interpret data, and offer context-based alerts and recommendations. Different from the aforementioned techniques, our proposed ontology-based context-aware IoT framework for smart traffic monitoring automatically identifies the context information from a video footage, and based on that assesses or even predicts the situation. The dynamic reasoning capability of the framework identifies situations of interest in smart city environments. It can assist in understanding and prediction of suspicious activity or traffic situations in real-time. The proposed system can continuously monitor people, vehicles and other objects in an environment by gathering sensor observations; process them to obtain a high-level representation; and finally recommend an action to be executed by actuators available to it.

2 Ontology for Smart Environment Monitoring

Conventional ontology languages, such as OWL, use natural language constructs to represent entities and their relations, making it convenient to use them for processing textual information. An attempt to use the conceptual model to interpret multimedia data gets severely impaired by the semantic gap that exists between the perceptible media properties and high-level abstractions which describe these entities. The key to semantic processing of media data lies in harmonizing the seemingly isolated conceptual and perceptual worlds. The observation of perceptible media properties forms the basis of concept recognition in the real-world as well as in multimedia recordings. Thus, we need ways and means to reason with perceptual models of real world concepts, which are not efficiently provided by existing ontology languages like Web Ontology Language (OWL). Different manifestations of a concept in media documents may have significant variations. Ontology representations and their reasoning schemes need to cope with these uncertain observations, but description logic and crisp logic reasoning of OWL are unable to do so. We have used the Multimedia Web Ontology Language for ontology representation and reasoning in our framework of smart surveillance system for traffic monitoring to automatically detect unannounced activities such as vehicle over-speeding, collision and so forth. Different sensor and other IoT devices like motion sensors, cameras, microphone, photo sensors produce outputs such as videos, audio, text in different media formats. Thus MOWL, in many ways, provides the basic infrastructure for encoding knowledge about an IoT domain with multimedia inputs from multi-modal sensor devices which can have partial and uncertain observations. MOWL exploits DBN model to encode a system that is dynamically changing and evolving over time. This model allows the user to observe and update the system, and even predict its future behavior. In this work, we outline a DBN based extension to existing capabilities of MOWL that allows for modeling a dynamic world with changing situations. This DBN capability of MOWL has been discussed in [11]. As the domain is the IoT, context data is available through sensor and other device inputs.

2.1 MOWL Encoded Ontology in Smart Traffic Surveillance Environment

The events captured in a smart surveillance environment need to be digitized in audio-visual formats. MOWL representation of a multimedia ontology provides the constructs and ability to integrate, unify, and relate diverse digital assets which belong to smart surveillance environment, through perceptual abstractions. The key to semantic processing of media data lies in harmonizing the seemingly isolated conceptual and perceptual worlds. A multimedia ontology is required to represent the knowledge corresponding to the smart traffic domain which also includes IoT products and services. A different kind of knowledge needs to be integrated into this knowledge graph for detecting various objects, people and events in a video surveillance scene. Prior information related to every possible video event is required to formalize the knowledge relevant for video analysis.

The multimedia ontology facilitates evaluation and understanding of a scene by representing the specific events leading to a particular situation for detection of expected media patterns. That is, a video shot might belong to different contexts - identification of which plays a crucial role in detection and classification of events. The context of an event gets defined with respect to an instance which comprises of temporal, geographical and domain related aspects. The multimedia ontology snippet represented using MOWL constructs, shown in Fig. 2 has 3 parts: *user ontology*, *event ontology* and *device ontology*. User profiles in the user ontology encode specific user information such as age, gender etc. that the system needs to be aware of. The event ontology represents a set of circumstances that happen at a given location and time. The device ontology incorporates information regarding various sensors deployed at diverse locations. For instance, an event named *RoadAccident* can take place at location *Location* which involves many people. The relation between these nodes is shown by a *TakesPlaceAt* relation. A concept can be recognized on the basis of accumulated evidential value as a result of detection of a number of expected media patterns with a closed-world assumption. The implicit knowledge can be derived through automated inference by making use of DBN as explained in the next section.

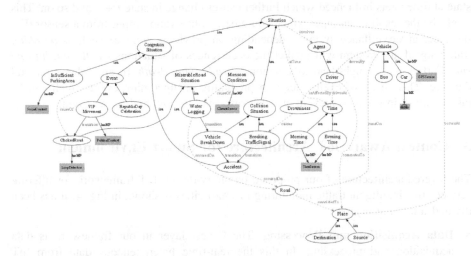

Fig. 2. An Ontology snippet for smart environment

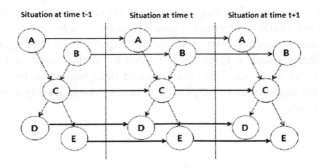

Fig. 3. Dynamic bayesian network

2.2 Dynamic Bayesian Network

The dynamic network consists of a sequence of ontology sub-graphs (OMs), each representing the system at a particular point or interval in time (time-slice). The time slices are interconnected by transition links between different situations in the system. As a situation is predicted, an appropriate action is recommended and carried out by the system. The change in context due to the action may be a trigger for another change in situation. MOWL ontology provides the universe of discourse for the changing situation as well as possible transitions and their conditional probabilities to the DBN. An example dynamic situation is *Stampede*, which is a critical crowd situation. It is modeled by its superconcept *AbnormalCrowdCondition*, related concepts like *Congregation* and *BottleNeckSituation*, and through observable media patterns *AbnormalCrowdMovement*, *FallenPeople*, *FastMovingCrowd*, *ScreamingNoise*, and so on. The situation can be predicted on the observation of one or more media patterns which increase the probability of Stampede.

A context-aware system can make use of constructs provided by MOWL ontology to track situations. Figure 3 shows different time-slices as snapshots of the system as it evolves over time. That is, based on the changes in the situation state at time $t - 1$, the state at time t gets influenced which further causes change in state $t + 1$ and so on. This might be the result of an input which could be an observation input from a sensor. For instance, in surveillance systems, depending on previous evidence such as *breaking* traffic signal, the system will predict the occurrences of situations - *collision, hit and run*. Further these might lead to situations - *traffic jam, death*, or any other collateral damage. This string of events helps in predicting an upcoming event which can be taken care of in order to prevent and decrease the loss incurred by an event.

3 Context-Aware IoT Architecture for Smart Environment

The layered architecture of our proposed ontology-driven IoT framework for offering services like intelligent traffic monitoring in smart cities is shown in Fig. 4. It has been divided into three layers:

- **Data Acquisition and Processing**: The lowest layer in our framework is data acquisition and processing. In this the real-time heterogeneous data from IoT devices which are deployed at various locations is collected using heterogenous sensors such as camera, WSN, GPS and other hardware components. This data can be in different formats like csv, database schemas, text messages, images, videos etc. Next, the valuable information from raw data is collected by performing data analytics and processing. Here, the feature extraction module applies computer vision techniques such as GMM and SPP-Net to monitor real-time traffic flow and other vehicular activities. This data is passed onto next layer for semantic interpretation and exploitation by employing semantic web technologies as discussed in the next phase.

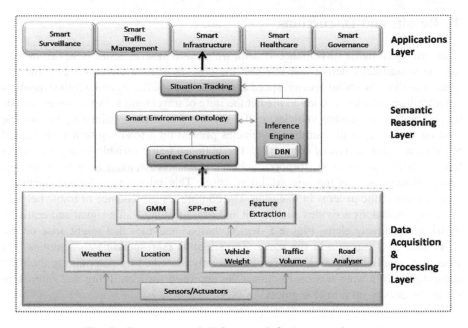

Fig. 4. Context-aware IoT framework for smart environment

- **Semantic Reasoning Layer**: To semantically exploit the collected information, it is required that the information be presented in a standard semantic web format such as RDF. It is considered as one of the most common data representation formats to exchange information over the web and it also helps in describing the meta-data about the resources. The data obtained from the previous layer is converted into a common format such as RDF which builds up the knowledge in the smart city domain aiming to minimize the ambiguity of the derived contexts. The obtained high-level context information is further exploited to track situations intelligently and dynamically to offer context-aware services and trigger appropriate actions. To achieve this, a multimedia ontology encoded in MOWL which provides reasoning with DBN has been utilized for timely exploitation of domain-specific concepts.
- **Application Layer**: In this layer, there are domain-specific applications which interface with the users, to seamlessly provide smart services to enhance the performance of IoT systems for improving the quality of life of citizens n a smart city environment. These applications can range from smart home to smart surveillance for intelligent traffic monitoring to smart governance to disaster management. In this paper, we have focused on application of this framework on an intelligent traffic monitoring system in smart city environment. The inferred knowledge from the previous layer can be utilized for predicting situations and making appropriate recommendations in many ways such as generating automated alerts, warnings etc. The alerts can be disseminated via e-mail, text messages, and alarms. Continuous monitoring and analysis of sensor information in a smart city domain can create a more productive and safer environment for individuals.

4   Results and Discussions

In an intelligent traffic surveillance system, we need to monitor traffic behavior in real-time to semantically derive and interpret the context of road networks. For this, we need to analyze the traffic density, speed analysis, road conditions and weather report as basic traffic parameters needed to predict the state of travel routes. Detection of various traffic parameters (moving vehicles, traffic volume..) facilitates estimating the vehicle load on road. Larger the number of vehicles present on a road segment with limited movement higher the risk of congestion. The obtained traffic variables along with other traffic related data (weather, location, vehicle speed..) as obtained from heterogenous sensor streams acts as the context information. This information further facilitates prediction of traffic patterns in real time. The continuous surveillance of traffic helps in detecting anomalous activities such as hit and run, breaking traffic signal and collision by triggering timely alerts. Figure 5 depicts various activities that might arise on the occurrence of a situation in the scene. The situation corresponding to activities in a video are usually ambiguous, but can be correctly identified by determining the right context. The following section discusses the approach for identifying and estimating the traffic density on a road segment.

fighting, running, breaking traffic signal, meeting	Hit and Run
playing, holding gun, firing, fighting, running	Gun Fire
crowd, scuffle, snatch-ing, attack, hitting	Stampede

Fig. 5.   Different activities corresponding to a situation

4.1   Estimating the Traffic Density

For measuring the traffic volume on a road segment, a simple approach of counting the stopped vehicles has been used. The measured count is then compared to a threshold to analyze the road area covered which results in an estimation of traffic load [12].

- GMM: Gaussian Mixture Model aims at detection of objects from a video parsed into frames i.e. objects in motion can be detected. Thus, to track the moving vehicle count over a sequence of frames segmentation of foreground and background is done, generating a blob of foreground images. By per-forming blob analysis, count for moving vehicles can be detected as shown in Fig. 6.

Fig. 6. Estimating count for moving vehicles

- SPP-net: Spatial Pyramid pooling makes use of deep convolutional neural networks to identify the objects. For this a feature map of the extracted image is done and marked with strong responses of objects with a probability as shown in Fig. 7 which shows count of total vehicles detected.

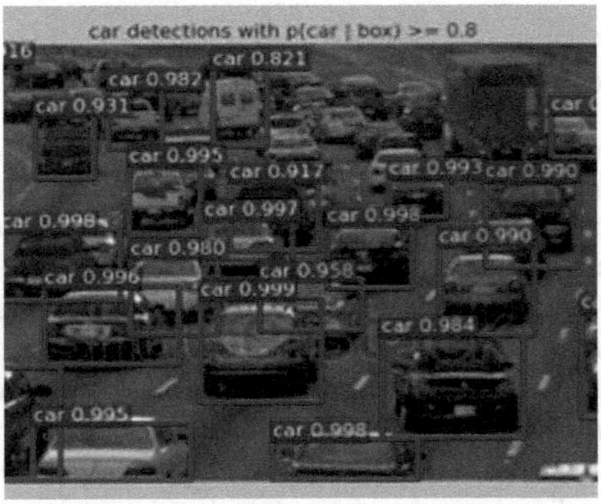

Fig. 7. SPP-net result estimating count of total vehicles

Further, the traffic load estimation is done as follows:

$$\mathfrak{C}_s = \mathfrak{C}_t - \mathfrak{C}_m \tag{1}$$

where, $\mathfrak{C}_s$ indicates the count of stopped vehicles, $\mathfrak{C}_t$ is the total number of vehicles and $\mathfrak{C}_m$ is the count of moving vehicles. Figure 8 depicts the scatter plot of actual and predicted vehicle load based on the number of stopped vehicles for the sequence of 254 videos. Here, the overlap marks the correctness of the result and the slight deviations mark the presence of error. It has been observed that the efficiency of the proposed framework for identifying the load is promising as compared to the traffic dataset provided by Statistical Visual Computing Lab, University of California (http://www. wsdot.wa.gov).

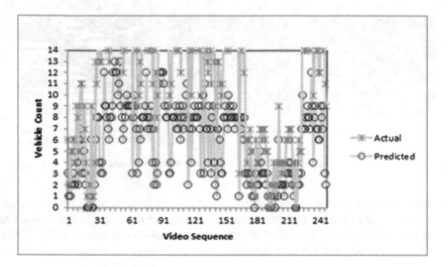

Fig. 8. Comparison of actual and predicted traffic load

4.2   Situation Tracking

The traffic situation is altered by activities such as accidents, congestion, vehicle breakdown or changes in weather. To illustrate context aware traffic monitoring services, the traffic measuring parameters, and other related attributes are analyzed to interpret semantic behavior. There are certain events which also results in congestion thus, to exploit such scenarios, MOWL based DBN model is build to effectively predict the current situation reasoning with contexts. Figure 9 shows the DBN model. In this, **case 1** depicts *congestion* based on *miserableroadsituation* while **case 2** shows *congestion* based on *collisionsituation*. We cannot predict a situation by considering just the current context instead the probabilities required for defining the DBN are required. That is, it is difficult to predict the next *congestion* situation only on the basis of current context, but if the context of previous situation is propagated through DBN, then based on belief propagation the next situation can be easily predicted. Let us consider an example where we have not considered the transition link and DBN for situation tracking. Based on context inputs the posterior probability computed for different congestion situations are P (*InsufficientParkingArea*) = 0.76 and P (*Accident*) = 0.31. Corresponding to only given two inputs, the system therefore assumes situation as *InsufficientParkingArea*, based on which suggestions for improving parking space will be given which is incorrect. Addition of the transition links in the MOWL encoded ontology provides prior conditional probabilities for transition between dynamic situations. Let the prior probability of P (*Accident | BreakingTrafficSignal*) be 0.83, P (*InsufficientParkingArea | BreakingTrafficSignal*) be 0.22, and P (*BreakingTrafficSignal*) be 0.93. Using these prior probabilities the belief is propagated in the DBN which will result in the updated posterior beliefs as: P (*Accident*) = 0.88 and P (*InsufficientParkingSpace*) = 0.29. This infers that the current inferred situation will be *Accident* based on the previous depicted situation *BreakingTrafficSignal* based on which accurate recommendations can be given. This approach gave promising results of prediction of future events with estimating load efficiency of 83.8% and DBN prediction accuracy of 83.3%.

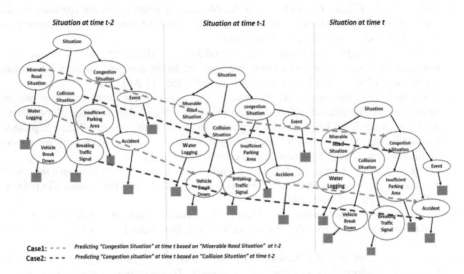

Case1: - - -   Predicting "Congestion Situation" at time t based on "Miserable Road Situation" at t-2
Case2: - - -   Predicting "Congestion situation" at time t based on "Collision Situation" at time t-2

Fig. 9. Situation tracking in smart environment using DBN

5 Conclusions

We have presented a multimedia ontology-based smart IoT framework that provides intelligent and dynamic recommendations by continuously monitoring real-world scenarios, obtaining constant stream of inputs from IoT devices in the network, and reasoning to interpret and predict situations. Uncertainty aspect in a smart city service has been handled using the DBN capability offered by the ontology representation language MOWL. The dynamic reasoning capabilities of DBN helps in computing the probability of an expected event, say a road accident based on the current movement patterns and previous activities of the observed vehicle. Using the proposed approach, it would be possible to monitor people's actions, movement of vehicles and determine if a critical or violent situation is arising. The efficacy of the proposed framework has been illustrated in a smart surveillance application. It can be applied to other smart city initiatives which engage IoT services and require reasoning for dynamic situation prediction. As smart cities develop, unforeseen technology challenges are presented. In this scenario, frameworks like the one proposed here, which employ intelligent networks and provide for context-awareness, will be the enablers of future that will survive the vestiges of time.

References

1. Zanella, A., Bui, N., Castellani, A., Vangelista, L., Zorzi, M.: Internet of things for smart cities. IEEE Internet Things J. **1**(1), 22–32 (2014)
2. Tönjes, R., Barnaghi, P., Ali, M., Mileo, A., Hauswirth, M., Ganz, F., Ganea, S., Kjærgaard, B., Kuemper, D., Nechifor, S., Puiu, D.: Real time iot stream processing and large-scale data analytics for smart city applications. In: Inposter session, European Conference on Networks and Communications 2014
3. Gaur, A., Scotney, B., Parr, G., McClean, S.: Smart city architecture and its applications based on IoT. Procedia Comput. Sci. **1**(52), 1089–1094 (2015)
4. Barbero, C., Dal Zovo, P., Gobbi, B.: A flexible context aware reasoning approach for iot applications. In: Mobile Data Management (MDM), 12th IEEE International Conference on 2011. IEEE, vol. 1, pp. 266–275, 6 Jun 2011
5. Fensel, A., Rogger, M., Gustavi, T., Horndahl, A., Martenson, C.: Semantic data management: sensor-based port security use case. In: Intelligence and Security Informatics Conference (EISIC), 2013 European. IEEE, pp. 155–158, 12 Aug 2013
6. SanMiguel, J.C., Martinez, J.M., Garcia, Á.: An ontology for event detection and its application in surveillance video. In: Advanced Video and Signal Based Surveillance. AVSS'09. Sixth IEEE International Conference on 2009. IEEE, pp. 220–225. 2 Sep 2009
7. Sivarathinabala, M., Abirami, S.: An intelligent video surveillance Framework for remote Monitoring. Int. J. Eng. Sci. Innovative Technol. **2**(2) (2013)
8. Zhang, Y., Ji, Q., Looney, C.G.: Active information fusion for decision making under uncertainty, Information Fusion. In: Proceedings of the Fifth International Conference. IEEE, vol. 1, pp. 643–650 (2002)
9. Baumgartner, N., Gottesheim, W., Mitsch, S., Retschitzegger, W., Schwinge, W.: Be Aware!-situation awareness, the ontology-driven way. Data Knowl. Eng. **69**(11), 1181–1193 (2010)
10. Samper, J.J., Tom´as, V.R., Martinez, J.J., Van den Berg, L.: An ontological infrastructure for traveller information systems, In: Intelligent Transportation Systems Conference. ITSC'06. IEEE, IEEE, 2006, pp. 1197–1202 (2006)
11. Mallik, A., Tripathi, A., Kumar, R., Chaudhury, S., Sinha, K.: Ontology based context aware situation tracking. In: Internet of Things (WF-IoT), 2015 IEEE 2nd World Forum on. IEEE, pp. 687–692, 14 Dec 2015
12. Goel, D., Pahal, N., Jain, P., Chaudhury, S.: An ontology driven context aware framework for smart traffic monitoring. In: IEEE Region 10 Symposium (TENSYMP), 2017. IEEE, pp. 1–5, July 2017

Exploring Causes of Wastes in the Moroccan Construction Industry

Mohamed Saad Bajjou$^{(\boxtimes)}$ and Anas Chafi

Department of Industrial Engineering, Faculty of Sciences and Techniques,
Sidi Mohamed Ben Abdellah University, Fez, Morocco
mohamedsaad.bajjou@usmba.ac.ma

Abstract. Waste generation has negative impact on the environment, productivity, budget, and project completion schedule. Nevertheless, no study has been
carried out in Morocco aiming at assessing the common causes of waste generation. In the current study, 25 causes of wastes were extracted from previous
studies and, then, validated through semi-structured interviews with Moroccan
construction experts. Subsequently, a questionnaire method was used to survey
330 Moroccan construction practitioners. The frequency method was employed
to measure the level of occurrence of each item. Statistical Analysis of the
collected data indicated that the five most recurrent causes of wastes are: 1- late
payment, 2- lack of training for employees, 3- lack of waste management
strategy, 4- inefficient planning and scheduling, and 5- lack of collective planning. These findings enable academics and engineering managers to benefit
from increased awareness regarding the most common causes of wastes and,
hence, develop appropriate strategies for improved performance through waste
reduction.

Keywords: Causes of wastes · Construction industry · Survey
Statistical analysis · Morocco

1 Introduction

Construction is an indispensable industry contributing to socio-economic growth [1]. It
provides the required infrastructure and facilities for economic activities such as
business, public services, and utilities. In addition, it also creates employment opportunities and strengthens the national economy by supporting foreign and local
investment [2]. However, the construction industry is confronted with the issue of the
generation of construction waste. As reported by [3], in the united kingdoms, more than
50% of the waste in a typical landfill is filled up by construction wastes. Almost 9% of
all materials purchased had ended up in a landfill [4]. As indicated by [5], the estimated
total generation of construction and demolition (C&D) waste was 1.6 million ton/year.

The concept of waste not only includes physical waste, which is due to the loss of
materials, but also nonphysical waste related to others factors such as time, cost, labor,
and equipment [6–9]. Besides that, waste generation can reduce the productivity,
effectiveness, efficiency, value of construction process [10, 11]. Therefore, Construction waste issues have become one of the predominant global challenges that need to be

© Springer Nature Switzerland AG 2019
M. Ben Ahmed et al. (Eds.): SCA 2018, LNITI, pp. 57–64, 2019.
https://doi.org/10.1007/978-3-030-11196-0_6

seriously addressed by researchers and construction practitioners. However, there has been little research to assess and analyze the root causes of waste generation, especially in developing countries.

As a result, this research attempts to fill this knowledge gaps by providing, for the first time, the scientific construction community with new insights from Morocco related to the most common causes of wastes through conducting a survey study with a large public of Moroccan construction professionals.

2 Literature Review

It is important to prevent and control the root causes contributing to construction wastes generation. Prior to its control, it is crucial to determine the factors generating waste in the construction field. In the literature, the concept of construction waste was defined in different ways. [12] stated that a construction waste is:

> A material which needed to be transported elsewhere from the construction site or used on the site itself other than the intended purpose of the project due to damage, excess or non-use or which cannot be used due to non-compliance with the specifications, or which is a by-product of the construction process.

However, under Lean Construction philosophy, [13] reported that:

> Construction wastes include both the incidence of material losses and the execution of unnecessary work, which generates additional costs but does not add value to the product.

Subsequently, this assertion was supported by [14], which defined construction waste as:

> Any inefficiency that results in the use of equipment, materials, labor, or capital in larger quantities than those considered as necessary in the production of a building.

This implies that besides physical waste (material loss), overrun cost and time, unnecessary works, and quality defects can also be considered as waste. It can be produced as a result of inefficient activities such as overproduction, waiting time, and over-processing, but adds no value to the client.

[15] mentioned fifteen potential causes of waste that were related to resource use, administration, and the information system. Besides, [15] also listed three major causes of wastes classified into three categories: conversion, flow activities, and management. Therefore, in accordance with [7], the causes of wastes can be assigned to five categories: (1) Information/communication-related causes; (2) Management/administration/finance-related causes; (3) Execution/performance-related causes; (4) Material/Equipment-related causes; and (5) People-related causes, as shown in Table 1.

3 Research Methodology

This study opted for a questionnaire survey for the aims to assess the critical causes of wastes according to Moroccan construction professionals' perceptions. The first step was designed to extract the major causes of wastes that have been identified in previous

Table 1. Potential causes of construction wastes generation

N°	Label	Waste causes	Group
1	A1	Lack of communication between the client and the main contractor	Information/communication
2	A2	Late or incorrect information	
3	B1	Lack of waste management strategy	Management/administration/finance
4	B2	Lack of coordination among project stakeholders	
5	B3	Conflicts between project stakeholders	
6	B4	Lack of involvement of the client/supplier in the construction process	
7	B5	Lack of commitment from top management	
8	B6	Unrealistic duration of project	
9	B7	Inefficient planning and scheduling (Errors or changes in planning)	
10	B8	Lack of collective planning	
11	B9	Insufficient financial resources to accomplish the work	
12	B10	Late payment	
13	C1	Lack of quality control on construction site	Execution/performance
14	C2	Inadequate storage of materials/equipment	
15	C3	Delay in delivery of materials/equipment	
17	C4	Improper handling of materials/equipment	
18	C5	Errors and inconsistencies in design documents	
19	C6	Obsolete design software	
	D1	Insufficient quality of materials/equipment	Material/equipment
20	D2	Changes in material types and specifications during the construction period	
21	D3	Change in prices of basic materials	
22	E1	Unskilled labors	People
23	E2	Lack of employee experience	
24	E3	Low labor productivity	
25	E4	Lack of training for employees	

investigations carried out in different countries. Subsequently, the relevance and the reliability of the designed questionnaire has been tested through semi-structured interviews with Moroccan experts have been performed. The interviewees included two academics, four on-site managers, and four project managers with high experience in the construction field (more than 10 years working experience). Hence, the interviews reveal that the extracted items were found to be adequate to the Moroccan context. As a result, the first list of wastes' causes was validated.

The questionnaire encompasses two major parts. The first part is designed to explore the respondents' profiles following two main clusters; clustering of individuals (qualification, experiences) and clustering of organizations (field of specialization, organization type, organization size and field of activities). The second part aimed to assess the level of occurrence for each cause based on a five-point Likert scale; starting from 1 for "never" to 5 for "always" and the middle position (3) was Sometimes.

A random sampling was adopted in order to target Moroccan construction practitioners from different location in Morocco. The survey started in March 2017 and several distribution techniques of the questionnaire were used such as email, LinkedIn, phone, and direct contact. The final questionnaire was sent to 440 Moroccan construction professionals and 330 valid and completed questionnaire forms were gathered representing a response rate of 75%.

The data obtained were analyzed by using the statistical package for social sciences (SPSS V 25.0 for Windows). In order to assess the level of occurrence for each item, the Mean Item Score (MIS) was adopted by using the following equation:

$$MIS = \frac{\sum_{i=1}^{5} a_i.x_i}{\sum_{i=1}^{5} x_i}$$

where: i indicates the response category index, such as: $1 =$ never, $2 =$ rarely, $3 =$ sometimes, $4 =$ frequently, $5 =$ always. In the numerator, a_i reveals the numerical value affected to the i-th response, ranging between 1 and 5, and x_i indicates the frequency of occurrence of i-th response in the total responses given to i.

Cronbach's alpha was employed with the aim to test the internal consistency of the 25 items adopted in the current study. Its value varies between 0 and 1. According to [16], a higher value indicates a high degree of consistency between the analyzed items. An alpha value ranging from 0.70 to 0.95 indicates a high internal consistency and it is considered acceptable, as mentioned in previous studies [17–19].

4   Data Analysis

4.1   Respondents' Profiles

This study has reached diversified respondents' characteristics from different locations in Morocco. The survey targeted a mixture of respondents at the aims to cover a large population and provide a balanced view of the research topic. Construction organizations were categorized into small companies (less than 50 employees), medium companies (50–200 employees), and large companies (more than 200 employees). The sample includes 53.0% participants from private sector and 47.0% from public sector; 44.5% are involving in the building field and 55.5% of respondents are working in civil engineering field; 43.0% of respondents are from consultant firms and 57.0% are from contractors; 48.8% of the respondents were from small-sized organizations, 24.0% from medium-sized organizations, and 27.2% were from large-sized organizations. It worth noting that most of the surveyed participants have a great experience in the construction field (54.6% have Master's degree, 11.5% have Ph.D. degree). Moreover, the surveyed respondents have a great working experience (49.1% have more than 10 years of working experience).

4.2 Ranking of Causes of Wastes

The respondents were asked to evaluate the level of occurrence for each cause of wastes among the 25 items that have been extracted from previous studies and then validated via semi-structured interviews with Moroccan construction experts. The MIS

Table 2. MIS analysis results of causes of waste within the whole survey sample (Cronbach's Alpha = 0.950)

Label	N	Min	Max	MIS	S.D.	Rank	Group
A1	330	1.00	5.00	3.024	1.156	18	Information/communication
A2	330	1.00	5.00	3.036	1.083	16	Information/communication
B1	330	1.00	5.00	3.533	1.193	3	Management/administration/finance
B2	330	1.00	5.00	3.161	1.061	8	Management/administration/finance
B3	330	1.00	5.00	3.221	1.090	7	Management/administration/finance
B4	330	1.00	5.00	3.091	1.194	11	Management/administration/finance
B5	330	1.00	5.00	3.009	1.131	19	Management/administration/finance
B6	330	1.00	5.00	3.155	1.163	9	Management/administration/finance
B7	330	1.00	5.00	3.318	1.097	4	Management/administration/finance
B8	330	1.00	5.00	3.312	1.209	5	Management/administration/finance
B9	330	1.00	5.00	3.070	1.174	15	Management/administration/finance
B10	330	1.00	5.00	3.652	1.147	1	Management/administration/finance
C1	330	1.00	5.00	3.076	1.232	13	Execution/performance
C2	330	1.00	5.00	2.779	1.073	25	Execution/performance
C3	330	1.00	5.00	3.079	0.958	12	Execution/performance
C4	330	1.00	5.00	2.912	0.981	24	Execution/performance
C5	330	1.00	5.00	2.967	1.076	21	Execution/performance
C6	330	1.00	5.00	2.979	1.251	20	Execution/performance
D1	330	1.00	5.00	2.942	0.967	22	Material/equipment
D2	330	1.00	5.00	2.915	1.025	23	Material/equipment
D3	330	1.00	5.00	3.024	1.066	17	Material/equipment
E1	330	1.00	5.00	3.245	1.126	6	People
E2	330	1.00	5.00	3.076	1.062	14	People
E3	330	1.00	5.00	3.103	1.047	10	People
E4	330	1.00	5.00	3.636	1.162	2	People

of each cause of waste is calculated in line with the scores given by the surveyed respondents. The descriptive statistic for the 25 items is presented in Table 2. For this study, the Cronbach's alpha coefficient is 0.950, which indicates a high internal consistency between the 25 wastes' causes.

Due to the high standard deviation (S.D = 1.109), the scores given by the respondents were found to be fluctuating quite large. Wastes' causes with MIS higher than the overall mean value (3.133) were considered as critical items. Within the total sample, the whole respondents believed that the five most significant causes of wastes are as follows: the first item was B10 "Late payment" (MIS = 3.652), which is thus considered as an extremely recurrent cause of wastes; followed by E4 " Lack of training for employees" (MIS = 3.636); the third is B1 "Lack of waste management strategy" (MIS = 3.533); The fourth- and fifth-ranked factors are B7 "Inefficient planning and scheduling (Errors or changes in planning)" (MIS = 3.318) and "Lack of collective planning" (MIS = 3.312), respectively.

As part of its objectives, this study assesses and ranks the typical waste categories of Lean Construction (LC) philosophy. Figure 1 shows the MIS of each waste category according to the scores given by the whole sample.

The items with MIS higher than the overall mean score (3.095) were identified as critical wastes' categories. Figure 1 indicates that the whole respondents perceived that the critical wastes' causes' categories are as follows: 1- People (MIS = 3.265), 2- Management/administration/finance (MIS = 3.252).

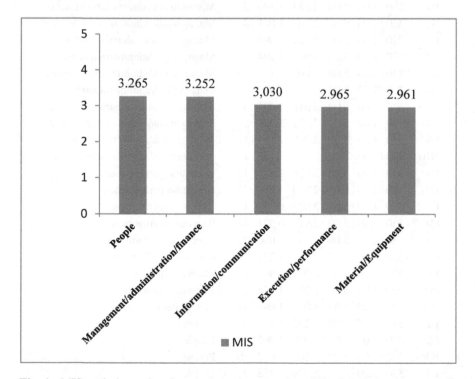

Fig. 1. MIS analysis results of wastes' causes categories within the whole survey sample

5 Discussions and Conclusions

This work aims to assess the perceptions of Moroccan construction professionals involving in both public and private sectors regarding the level of frequency of the most common causes of wastes. The first step of the current study was designed to identify a list of 25 causes of wastes determined on the basis of deep literature review and, then, confirmed through semi-structured interviews with Moroccan construction experts. Subsequently, a total of 440 questionnaire forms were administered to Moroccan practitioners and a total of 330 completed questionnaires were gathered, representing a response rate of 75%. The Mean Item Score (MIS) was used to measure the level of occurrence of the 25 causes of waste. Within the total sample, the most recurrent cause of waste is "late payment" (MIS = 3.652); followed by lack of training for employees (MIS = 3.636); the third was "lack of waste management strategy" (MIS = 3.533); the fourth- and fifth-ranked items s were "inefficient planning and scheduling (Errors or changes in planning)" (MIS = 3.318) and "lack of collective planning" (MIS = 3.312), respectively.

These findings indicate that among the five most frequent causes of wastes, four items are associated to management/administration/finance's category, which implies that more attention should be directed to improving the current management system as well as administrative and financial issues. Besides, "People category" (MIS = 3.264) was found to be the most frequent causes' category which entails an urgent need to improve both managerial and technical skills of the construction staff. Training programs are recommended to be ensured to workers from different level for the aims to spread the required techniques for waste minimization such as collaborative planning, root causes analysis (RCA), and total quality management.

The findings could provide managers with a more in-depth understanding of the critical causes contributing to waste generation, and thus impact negatively the effectiveness of the construction work process. By addressing the root causes of waste in current and future projects, construction managers can successfully manage waste in their projects. Therefore, a construction project can be carried out more effectively, especially if all employees know how to fully inspect, monitor, and prevent the waste in each activity on-site. Since then, the benefit of construction projects can be increased by reducing or eliminating the sources of waste.

Further research should be performed to develop appropriate strategies aiming at eliminating the root causes of wastes in the Moroccan construction. In addition, there is an increasing need to introduce Lean Construction practices into the Moroccan construction context in order to test its practical impacts on waste reduction.

References

1. Bajjou, M.S., Chafi, A., Ennadi, A., El Hammoumi, M.: The practical relationships between Lean construction tools and sustainable development: a literature review. J. Eng. Sci. Technol. Rev. 10(4), 170–177 (2017)
2. Bajjou, M.S., Chafi, A., En-Nadi, A.: The potential effectiveness of Lean construction tools in promoting safety on construction sites. Int. J. Eng. Res. Afr. 33, 179–193 (2017)

3. Ferguson, J.: Managing and Minimizing Construction Waste: A Practical Guide. Thomas Telford, London (1995)
4. Bossink, B.A.G., Brouwers, H.J.H.: Construction waste: quantification and source evaluation. J. Constr. Eng. Manage. **122**(1), 55–60 (1996)
5. Kartam, N., Al-Mutairi, N., Al-Ghusain, I., Al-Humoud, J.: Environmental management of construction and demolition waste in Kuwait. Waste Manage. **24**(10), 1049–1059 (2004)
6. Khanh, H.D., Kim, S.Y.: Evaluating impact of waste factors on project performance cost in Vietnam. KSCE J. Civil Eng. **18**(7), 1923–1933 (2014)
7. Khanh, H.D., Kim, S.Y.: Identifying causes for waste factors in high-rise building projects: a survey in Vietnam. KSCE J. Civil Eng. **18**(4), 865–874 (2014)
8. Khanh, H.D., Kim, S.Y.: Development of waste occurrence level indicator in Vietnam construction industry. Eng. Constr. Architectural Manage. **22**(6), 715–731 (2015)
9. Bajjou, M.S., Chafi, A., En-Nadi, A.: A comparative study between lean construction and the traditional production system. Int. J. Eng. Res. Afr. **29**, 118–132 (2017)
10. Bajjou, M.S., Chafi, A.: Barriers of lean construction implementation in the Moroccan construction industry. In: AIP Conference Proceedings vol. 1952(1), p. 020056. AIP Publishing, Apr 2018
11. Bajjou, M.S., Chafi, A.: Towards implementing lean construction in the Moroccan construction industry: survey study. In: 4th International Conference on Optimization and Applications (ICOA). IEEE, p. 1–5. Apr 2018
12. Skoyles, E.F.: Material wastage: A misuse of resource. Batiment Int. Build. Res. Pract. **4**(4), 232–243 (1976)
13. Koskela, L.: Application of the new production philosophy to construction, Center for Integrated Facility Engineering. pp. 1–81 (1992)
14. Formoso, C. T., Isatto, E.L., Hirota, E.H.: Method for waste control in the building industry. In: Proceedings IGLC, vol. 7, p. 325 July 1999
15. Alarcon, L.F.: Tools for the identification and reduction of waste in construction projects. Lean Constr. **5**, 365–377 (1997)
16. Cronbach, L.J.: Coefficient alpha and the internal structure of tests. Psychometrika **16**(3), 297–334 (1951)
17. Bajjou, M.S., Chafi, A.: Empirical study of schedule delay in Moroccan construction projects. Int. J. Constr. Manage. https://doi.org/10.1080/15623599.2018.1484859
18. Enshassi, A., Mohamed, S., Abushaban, S.: Factors affecting the performance of construction projects in the Gaza strip. J. Civil Eng. Manage. **15**(3), 269–280 (2009)
19. Bajjou, M.S., Chafi, A.: Lean construction implementation in the Moroccan construction industry: awareness, benefits and barriers. J. Eng. Des. Technol. **16**(4), 533–556 (2018). https://doi.org/10.1108/JEDT-02-2018-0031

Investigation into the Critical Sources of Wastes Influencing the Performance of Construction Projects in Morocco

Mohamed Saad Bajjou[✉] and Anas Chafi

Department of Industrial Engineering, Faculty of Sciences and Techniques, Sidi Mohamed Ben Abdellah University, Fez, Morocco
mohamedsaad.bajjou@usmba.ac.ma

Abstract. Construction projects worldwide are significantly affected by waste occurrences which negatively influence their performance. Nevertheless, limited studies have been performed aiming at waste assessment, especially in developing countries. In the current study, 24 sources of wastes were identified through a review of previous works, consolidated by interviews with Moroccan construction experts. Subsequently, a structured questionnaire was used to survey 330 Moroccan construction professionals to explore the critical sources of wastes in Morocco. "Activity start delays" was found to be the most significant source of waste according to the perceptions of the surveyed respondents. Furthermore, the statistical analysis revealed that the top three ranked waste categories were: 1—unused employee creativity, 2—delay and waiting, and 3—defects/errors/corrections. These findings are expected to help practitioners and managers of construction projects to focus their resources and interest on the main sources of wastes and to establish, therefore, more effective strategies for waste reduction.

Keywords: Critical sources of wastes · Survey · Construction industry Morocco

1 Introduction

The construction industry is a strategic sector that significantly contributes to the economic growth of both developed and developing countries [1, 2]. In Morocco, the field of construction is considered amongst the most promising and the most dynamic of the Moroccan economy as it reinforces other economic sectors through ensuring the required infrastructure and by decreasing the unemployment rate [3]. It contributed by 6.3% of the total value added and employs almost a million people (9.3% of the active population) [4].

As a common worldwide, waste generation is a key challenge being faced by most construction projects, especially in developing countries. Hence, several studies have been carried out in different countries like Vietnam [5–7], Palestine [8], Malaysia [9], and Egypt [10]. Similarly, the Moroccan construction industry is also suffering from many critical issues such as low productivity, cost and time overrun, and poor quality [11, 12]. However, to date, no study has been carried out to assess the main sources of

© Springer Nature Switzerland AG 2019
M. Ben Ahmed et al. (Eds.): SCA 2018, LNITI, pp. 65–73, 2019.
https://doi.org/10.1007/978-3-030-11196-0_7

waste that are frequently occurring in the Moroccan construction industry. The current paper is expected to fill this knowledge gap by conducting a structured survey among public and private construction companies for objective to explore the perception of Moroccan construction professionals regarding the most recurrent sources of waste in Morocco.

2 Literature Review

In the past, construction professionals perceived waste as being directly associated with the loss of materials on construction sites [5, 7]. Evidence of wastage of materials has been reported in several previous studies. Reference [13] reported that 9% of the total amount of purchased materials eventually ends up as construction waste, and 1–10% of each purchased construction material is landfilled as solid waste in the Dutch construction industry. Reference [14] indicated that approximately 21–30% of cost overruns in construction projects were incurred as a result of waste materials.

Nowadays, under Lean philosophy, there is an increasing and emerging requirement to adopt a broad view of waste including not only the waste due to materials loss but also the waste associated with other inputs such as time, equipment and labor [15].

According to [16], waste is defined as any inefficiency that results in the use of materials, workforces, equipment, and money in a higher amount than those considered as needed in the construction of a building. Furthermore, in the United States, Ref. [16] estimated that quality defects accounted over 12% of construction project costs. Dupin (2014) indicated that 68% of the total time spent on construction site is expended by non-value added activities. According to [17], 90–95% of the on-site work is made up of two kinds of activities: non-value adding but required (NVAR) and pure non-value adding (NVA).

A deep and rigorous literature review resulted in identifying 22 sources of waste classified according to the typical 7 waste categories of lean philosophy (excessive motions, defects, over-processing, excessive conveyance, over-production, excessive inventory, and delay), as shown in Fig. 1.

3 Research Methodology

This study opted for a quantitative approach based on a questionnaire survey in order to provide a holistic overview of the critical sources of wastes in the Moroccan construction industry. The first stage aimed at identifying the main sources of wastes that were reported in previous studies carried out in a similar context. Subsequently, semi-structured interviews with Moroccan experts have been performed to investigate the relevance of the designed questionnaire and to evaluate the adequacy of the extracted items to the Moroccan context. The interviewees consisted of four on-site managers, two academics, and four project managers and with more than 10 years working experience. Hence, the first list of wastes' sources was validated. In addition, it was recommended to add 'unused employee creativity' and 'accidents during construction'

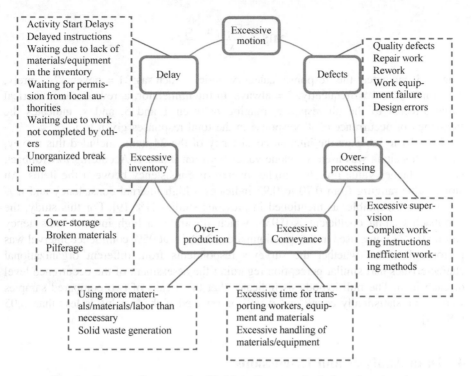

Fig. 1. Sources of wastes classified into the seven typical waste categories

to the final questionnaire version as they perceived that these items are also frequently prevalent in the Moroccan construction industry and should be evaluated.

The questionnaire included two main parts. The first part aimed at collecting general information about the surveyed participant according to two groups; clustering of organizations (field of specialization, organization type, organization size and field of activities) and clustering of individuals (qualification, experiences). The second part is designed to assess the level of waste occurrence for each item using a five-point Likert scale; starting from 1 for "never" to 5 for "always" and the middle position (3) was Sometimes.

To target Moroccan construction practitioners a random sampling method was employed and several techniques were used such as email, direct contact, fax, and LinkedIn. The survey has been started in March 2017. The final questionnaire was administrated to 440 Moroccan professional and 330 valid responses were received, which represents a response rate of 75%.

The data obtained is computed by using the statistical package for social sciences (SPSS V 25.0 for Windows). The Mean Item Score (MIS) was used to reveal the perceptions of the Moroccan construction professionals about identified items. The MIS was computed using the following equation:

$$MIS = \frac{\sum_{i=1}^{5} a_i \cdot x_i}{\sum_{i=1}^{5} x_i}$$

where: i indicates the response category index, such as: 1 = never, 2 = rarely, 3 = sometimes, 4 = frequently, 5 = always. In the numerator, a_i reveals the numerical value affected to the ith response, ranging between 1 and 5, and x_i indicates the frequency of occurrence of ith response in the total responses given to i.

In order to measure the internal consistency of the 24 items included this survey, Cronbach's alpha was used. Its value varies between 0 and 1. According to Cronbach (1951) a higher value reveals a higher degree of consistency between the items. An alpha value ranging from 0.70 to 0.95 indicates a high internal consistency and it is considered acceptable, as mentioned in previous studies [18, 19]. For this study, the Cronbach's alpha coefficient is 0.935, which illustrates a high internal consistency among the 24 items used in the questionnaire. ANOVA at 95% confidence interval was performed to test whether the survey's respondents from different organizational characteristics had similar perception regarding the assessment of the occurrence level of each item. The null hypothesis indicates that the means of two compared samples were to be statistically equal, and it will be rejected if the p-value is less than 0.05 [18, 19].

4   Data Analysis and Discussions

4.1   Respondents' Profiles

Figure 2 shows the profile of the surveyed respondents. It worth noting that most of the surveyed participants have a great experience in the construction field (62.0% have more than 5 years' experience). In addition, they are highly educated (66.1% have a master degree and above) which increase the reliability of the results.

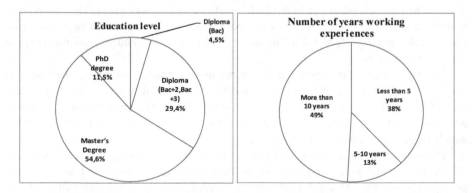

Fig. 2. Distribution of the sample in percentage (clustering of individuals)

4.2   Ranking of Sources of Wastes

The respondents were asked to assess the level of waste occurrence for each item among the 24 sources of waste that have been identified based on literature review and then validated through semi-structured interviews with Moroccan construction experts. The MIS of each source of waste is computed based on the responses of all the surveyed participants. The ranking of the 24 items is presented in Table 1. The overall

Table 1. MIS analysis results of sources of waste within the whole survey sample (Cronbach's Alpha = 0.935)

Sources of wastes	MIS	S.D.	Rank	Anova (p-value)			
				Group1	Group 2	Group 3	Group 4
Quality defects	3.233	0.997	8	0.982	0.514	0.343	0.311
Repair work	3.233	0.969	7	0.613	0.112	0.939	0.895
Rework	3.415	1.078	2	0.945	0.914	0.930	0.585
Work equipment failure	3.100	1.025	9	0.949	0.462	0.221	0.872
Design errors	2.746	1.044	18	0.442	0.346	0.411	0.788
Activity start delays	3.455	1.094	1	0.397	0.513	0.863	0.956
Delayed instructions	3.242	0.965	6	0.598	0.097	0.686	0.696
Waiting due to lack of materials/equipment in the inventory	2.885	1.037	12	0.996	0.302	0.134	0.295
Waiting for permission from local authorities	3.355	1.192	4	0.617	0.145	0.117	0.577
Waiting due to work not completed by others	3.255	1.023	5	0.992	0.926	0.370	0.363
Unorganized break time	2.779	1.199	15	0.177	0.436	0.051	0.093
Over-storage	2.497	1.114	23	0.929	0.798	0.293	0.841
Broken materials	2.770	1.061	17	0.787	0.363	0.693	0.811
Pilferage	2.533	1.149	21	0.640	0.944	0.263	0.462
Excessive supervision	2.506	1.152	22	0.292	0.181	0.022[a]	0.018[a]
Complex working instructions	2.573	1.115	20	0.183	0.741	0.376	0.035[a]
Inefficient working methods	3.052	1.162	10	0.631	0.798	0.393	0.571
Excessive time for transporting workers, equipment and materials	2.773	1.119	16	0.134	0.672	0.424	0.028[a]
Excessive handling of materials/equipment	2.812	1.098	14	0.330	0.500	0.320	0.679
Using more materials/materials/labor than necessary	2.488	1.141	24	0.379	0.825	0.414	0.604
Solid waste generation	2.861	1.174	13	0.332	0.865	0.484	0.880
Unnecessary labor move	2.988	1.151	11	0.321	0.902	0.121	0.345
Unused employee creativity	3.373	1.294	3	0.133	0.349	0.151	0.814
Accidents during construction	2.621	1.091	19	0.482	0.936	0.009[a]	0.299

[a]ANOVA results indicate there is a significant difference between mean values (p-value less than 0.05)

sample was divided into four groups according to the organizational characteristics; group 1 according to the field of specialization (building, road and bridge projects), group 2 according to the type of organization (contractor, consultant), group 3 according to the size of organization (small, medium, large), and group 4 according to the major clients (public, private). Table 1 also displays the ANOVA analysis of perceptions of these 24 items for the four groups.

Items with MIS higher than the overall mean value (2.939) were identified as critical sources of waste. Table 1 shows that 11 out of the initial 24 CWFs have mean scores greater than 2.939. Within the total sample, the whole respondents perceived that the five most significant sources of wastes are as follows: the first item was "Activity Start delays" (MIS = 3.455), which is thus considered as an extremely frequent source of waste; followed by "Rework" (MIS = 3.415); the third is "Unused employee creativity" (MIS = 3.373); The fourth- and fifth-ranked factors are "waiting for permission from local authorities" (MIS = 3.355) and "Waiting due to work not completed by others" (MIS = 3.255), respectively.

As shown in Table 1 all p-values were found to be higher than 0.05 for group 1 and 2, suggesting that the field of specialization and the type of organization did not affect respondents' perceptions of the level of occurrence of the 24 items listed in this study. However, for group 3, the respondents have different perceptions for two items (excessive supervision and Accidents during construction) since the p-value is lower than 0.05 (0.022 and 0.009, respectively). Similarly, for group 4, the respondents have different perceptions for three items (excessive supervision, complex working instructions, and excessive time for transporting) since the p-value is lower than 0.05 (0.018, 0.035, and 0.028, respectively).

As part of its objectives, this study assesses and ranks the typical waste categories of lean philosophy. Figure 3 shows the MIS of each waste category according to the scores given by the whole sample.

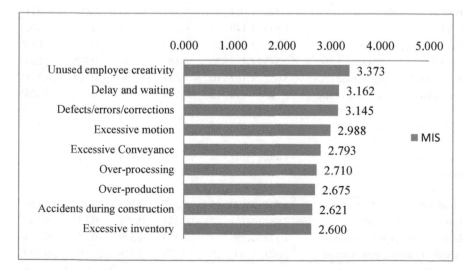

Fig. 3. MIS analysis results of wastes' categories within the whole survey sample

The items with MIS higher than the overall mean score (2.896) were identified as critical waste categories of waste. Figure 3 indicates that the whole respondents perceived that the critical waste categories are as follows: 1—Unused employee creativity (MIS = 3.373), 2—Delay and waiting (MIS = 3162), and defects/errors/corrections (MIS = 3145).

5 Discussions

This work was designed to investigate the points of view of public and private construction firms regarding the most common sources of wastes that are frequently prevalent in the Moroccan construction industry. The current study identified the key sources of wastes through a structured questionnaire survey. The survey is based on items extracted from previous works that have been carried out in other countries, together with additional items identified through semi-structured interviews with Moroccan experts and which were specific to the Moroccan context. The Mean Item Score (MIS) was employed to assess the level of occurrence of the identified items (24 items). Within the total sample, "Activity start delays" was found to be the most frequent source of waste (MIS = 3.455). In addition, the whole respondents believed that the most recurrent wastes' categories are ranked as follows: 1—unused employee creativity (MIS = 3.373), 2—Delay and waiting (MIS = 3162), and defects/errors/ corrections (MIS = 3145). These findings are in accordance with the Lean thinking as they reveal that waste is not only associated with the waste due to materials loss but also included other sources of wastes such as unused labor creativity, waste of time, and rework. The statistical test of ANOVA indicated that organizations size and Field of activities can impact the perceptions of the surveyed respondents as it has noticed that they had different points of views for several items such as excessive supervision and Accidents during construction for group 3, and excessive supervision, complex working instructions, and excessive time for transporting for group 4.

6 Conclusions

This paper develops a better understanding of sources of wastes in construction projects in Morocco which represent a significant contribution in identifying the key issues that must be addressed by Moroccan construction professionals for the aim to increase the project efficiency. Although this study is specific to the Moroccan industry, its findings could help construction managers from other developing countries, especially those facing similar sources of wastes in their construction projects. There are no straightforward solutions to reduce waste in Moroccan construction projects. However, there are systematic steps that can be taken: For the specific items identified in the current study, the following recommendations are suggested:

- There is an increasing need to improve both technical and managerial skills of the construction staff. Training programs should be assured to workers from different level in order to spread the required skills and techniques for waste minimization such as cost and time control, scheduling, and risk analysis.
- It has become crucial for Moroccan industry to adopt more innovative management systems. For instance, Lean Construction can provide a number of benefits to the construction industry, especially at the level of quality improvement, workplace accidents reduction, and deadline compliance.
- Contractors are recommended to adopt Total Quality Management (TQM) in order to continuously monitor the quality of construction activities and, therefore, that may significantly minimize quality defects, rework and mistakes.
- Construction managers are recommended to adopt new planning techniques. Last Planner System (LPS) may lead to more collaborative planning as it involves all the project stockholders (client, architect, the main contractor, subcontractors, and suppliers). Therefore, it is expected to help to reduce waiting times, synchronization problems and communication difficulties among the various project collaborators.
- The use of the 5S process as well as visual management may lead to a well-organized construction site which promotes the safety and the productivity of the workforce and reduce damaged materials and equipment.

References

1. Shehu, Z., Endut, I.R., Akintoye, A., Holt, G.D.: Cost overrun in the Malaysian construction industry projects: a deeper insight. Int. J. Project Manage. 32(8), 1471–1480 (2014)
2. Bajjou, M.S., Chafi, A.: Lean construction implementation in the Moroccan construction industry: awareness, benefits and barriers. J. Eng. Des. Technol. 16(4), 533–556 (2018). https://doi.org/10.1108/JEDT-02-2018-0031
3. Bajjou, M.S., Chafi, A., En-Nadi, A.: The potential effectiveness of lean construction tools in promoting safety on construction sites. Int. J. Eng. Res. Africa 33, 179–193 (2017)
4. Tableau de bord sectoriel: Ministry of Economy and Finance, 88p (2015)
5. Khanh, H.D., Kim, S.Y.: Development of waste occurrence level indicator in Vietnam construction industry. Eng. Constr. Architect. Manage. 22(6), 715–731 (2015)
6. Khanh, H.D., Kim, S.Y.: Identifying causes for waste factors in high-rise building projects: a survey in Vietnam. KSCE J. Civil Eng. 18(4), 865–874 (2014)
7. Khanh, H.D., Kim, S.Y.: Evaluating impact of waste factors on project performance cost in Vietnam. KSCE J. Civil Eng. 18(7), 1923–1933 (2014)
8. El-Namrouty, K.A., Abushaaban, M.S.: Seven wastes elimination targeted by lean manufacturing case study "gaza strip manufacturing firms". Int. J. Econ. Finance Manage. Sci. 1(2), 68–80 (2013)
9. Nikakhtar, A., Hosseini, A.A., Wong, K.Y., Zavichi, A.: Application of lean construction principles to reduce construction process waste using computer simulation: a case study. Int. J. Serv. Oper. Manage. 20(4), 461–480 (2015)
10. Garas, G.L., Anis, A.R., El Gammal, A.: Materials waste in the Egyptian construction industry. In: Proceedings IGLC-9, Singapore (2001)

11. Bajjou, M.S., Chafi, A.: Towards implementing lean construction in the Moroccan construction industry: Survey study. In: 4th International Conference on Optimization and Applications (ICOA), pp. 1–5. IEEE, April 2018
12. Bajjou, M.S., Chafi, A.: Barriers of lean construction implementation in the Moroccan construction industry. In: AIP Conference Proceedings, vol. 1952, No. 1, p. 020056. AIP Publishing (2018)
13. Bossink, B.A.G., Brouwers, H.J.H.: Construction waste: quantification and source evaluation. J. Constr. Eng. Manage. **122**(1), 55–60 (1996)
14. Oko John, A., Emmanuel Itodo, D.: Professionals' views of material wastage on construction sites and cost overruns. Organ. Technol. Manage. Constr. Int. J. **5**(1), 747–757 (2013)
15. Bajjou, M.S., Chafi, A., En-Nadi, A.: A comparative study between lean construction and the traditional production system. Int. J. Eng. Res. Africa **29**, 118–132 (2017)
16. Koskela, L. Application of the new production philosophy to construction (Vol. 72). Stanford, CA: Stanford university, 1992
17. Bajjou, M.S., Chafi, A., Ennadi, A., El Hammoumi, M.: The practical relationships between lean construction tools and sustainable development: a literature review. J. Eng. Science & Technol. Rev. **10**(4), 170–177 (2017)
18. Jin, R., Hancock, C., Tang, L., Chen, C., Wanatowski, D., Yang, L.: Empirical study of BIM implementation-based perceptions among chinese practitioners. J. Manage. Eng. **33**(5), 04017025 (2017)
19. Bajjou, M.S., Chafi, A.: Empirical study of schedule delay in Moroccan construction projects. Int. J. Constr. Manage. https://doi.org/10.1080/15623599.2018.1484859

Marketing and Smart City: A New Model of Urban Development for Cities in Morocco

Asmaa Abyre[1]([⊠]), Kaoutar Al Haderi[2], and Mohamed El Kandili[3]

[1] Faculty of Law, Economic and Social Sciences, Kenitra, Morocco
Abyre.asmaa@gmail.com
[2] Faculty of Law, Economic and Social Sciences, Fez, Morocco
Kaoutar.alhaderi@gmail.com
[3] Fez Business School, Private University of Fez, Fez, Morocco
m.elkandili@gmail.com

Abstract. A smart city is an urban space that uses information and communication technologies (ICT) and other means to improve the quality of life, the efficiency of urban operations and services, and competitiveness, while ensuring that the needs of present and future generations are met in economic, social and environmental terms. The city of Casablanca occupies a prominent place at the national level, hence the state's desire to integrate it into the global competition market between territories. This requires the need to integrate an appropriate territorial marketing approach.

Keywords: Smart city · Urban development · Marketing · Moroccan cities Casablanca

1 Introduction

Morocco has known implementation of a new structuring project, launched by His Majesty King Mohammed VI, during his speech on advanced regionalization, on January 3, 2010. In this project, the territory is part of a logic of national and international competitiveness. National competitiveness is the result of competition between local authorities at the national level. International competitiveness concerns certain national territories that have assets that allow them to compete in a global competition with certain internationally renowned territories (Belkadi 2015).

The city of Casablanca occupies a prominent place at the national level, hence the state's desire to integrate it into the global competition market between territories. This requires the need to integrate an appropriate territorial marketing approach. After the launch of the new territorial brand of the city "We Casablanca" the challenge of digital continues with the launch of a major turning point: The smart city "e-Madina", which works to make the city of Casablanca more attractive, more effective and more competitive, for companies, citizens and visitors through Public, Private, and Citizen and through the use of technologies.

This article aims to take stock of the experience the smart city "e-Madina". Thus, at first, we will present the review of the literature of the smart city by focusing on all

© Springer Nature Switzerland AG 2019
M. Ben Ahmed et al. (Eds.): SCA 2018, LNITI, pp. 74–83, 2019.
https://doi.org/10.1007/978-3-030-11196-0_8

related concepts. And then in a second part, we will present the methodology and results of an exploratory survey on the Casablanca experience.

2   Smart City: A Successful Tool for Territorial E-Marketing

An intelligent city is a city using information and communication technologies (ICT) to "improve" the quality of urban services. One potential way of cushioning the costs of implementation and management is to use this equipment together for local communication and advertising communication: marketing has a role to play in the design of this new model of urban development.

2.1   Digital Marketing

E-marketing, digital marketing or digital marketing are just designations for a typology of marketing that includes all the marketing practices used on digital media and channels (Ovazza 2015). In general, this concept is associated with the Internet: it is called webmarketing or cybermarketing (Gayet and Marie 2016). With the current technological advance, its field of intervention has however expanded. Today, other tools and media are also used, such as mobile applications and video games (Scheid and Vaillant 2014). With each new trend, marketing specialists are forced to review their working methods. This makes its various fields, techniques and tools on which digital marketing is based.

Contrary to what the author Christian Dussart claims: "Traditional marketing and digital marketing are so different in nature that their necessary integration is a real headache (Marketing Decisions 2012, p. 83), personally and as confirmed (Lombardi 2015) "digital marketing is no exception to the rules of classical marketing", I find that digital marketing is an extension of traditional marketing, imposed by the circumstances in which the company operates but also the territory of 'today. These entities led to act against the vagaries of new technologies, which represents by its power and also by its popularization a double-edged sword.

2.2   Territorial Digital Marketing

Nowadays, fierce and rigid competition between territories leaves no room for the weak to survive, while the search for the grail for these territories consists in following the contemporary digital revolution. As a result, the follow-up of digital marketing trends is essential for the development of territories (Al Haderi 2016).

E-marketing has become an essential means of communication for the territories. Today more than ever, they need to heal their e-reputation and establish permanent interaction with their targets if they want to have a place in this new digital age (Flores 2012). Soon, almost all players and customers in a territory will only use digital channels to communicate and make their choices. What makes a non-digitized territory, an outdated entity.

Digital has two undeniable advantages for marketing: the reduced cost compared to other media and the exceptional targeting capacity of digital media (Decressac 2016).

The success of territorial marketing lies largely in the ability of the different actors of the territory to work together to build differentiated marketing strategies according to the target audiences. Indeed, the synergy of the speeches is necessary to create a coherent and attractive image.

2.3   The Marketing Aspect of the Smart City

The literature emphasizes that the concept of the smart city comes from the territorial marketing or said urban, that is to say that a city self-proclaims "smart city", without there being behind this speech of real tangible elements (Douay and Henriot 2016). D or the example of the Chinese case, which according to Douay and Henriot (2016) "this' smart 'staging' makes it possible to hide the limits of the territorial project and to reinforce its social acceptability by masking controversies about the beneficiaries of the project who will be able to benefit from the valorization of the district. Thus, even if the use of technologies is rather limited or at least not more than in other projects, the digital and intelligent theme is central in the communication of the project.

The smart city is in itself a marketing project established by local actors to promote their city to different targets. Regardless of the targeted target (Investors, residents or tourists) (Desdemoustier and Crutzen 2015), the territorial actors have no other issues than that of taking into account this channel that will allow them to control their reputation and their reputation online and also to maintain permanent and engaged contact with current and potential customers.

In the academic literature, the scientific research works deal with the concept of "Smart City" according to various interpretations where the word "Smart" is replaced by other qualifying adjectives namely: smart city, digital city and smart city. There is also a new development approach that integrates new information and communication technologies in order to optimize the management of the city in an efficient way. Indeed, there is not one model of smart city nor a unique definition of it. On the other hand, cities are considered as smart cities and take different forms and aspects related to architecture, construction, town planning, urban aesthetics, etc... through the integration of Information Technologies and of Communication (ICT). Public actors are now trying to integrate these ICTs into the different phases of urban planning projects. From the design of planning, planning to implementation, technology integration is possible and can take many forms. The example of "Smart Buildings" or "Smart Building" is very relevant in this case. Also, the environmental approach in urban management has become an undeniable asset in valuing the concept of "Smart City". Indeed, the integration of environmental management allows the actors to direct the development of the soil towards a sustainable and friendly development to the local environment. The ISO 14000 standard and the LEED for Neighborhood Development (LEED-ND) certification are important references for the success of this concept.

In another register, the social approach in designing a "Smart City" is also important. It complements the two aspects mentioned above, namely technology and the environment. In today's cities intelligence is mainly based on the role of large technology companies and public authorities for the integration of new technologies in the city and, as a result, the solutions to the challenges of the city are proposed mostly top-down and remain focused solely on efficiency, innovation and transparency.

It lacks the more human aspects that are those of individuals, communities and small businesses such as their aspirations, anxieties and skills. Indeed, the concept of Smart City comes to answer the urban problems lived in ordinary cities. However, many problems related to the development of urban agglomerations are generally solved through creative means, human capital, cooperation of stakeholders and innovation or technology. Indeed, the ownership and social acceptance by individuals is crucial for the development of integrated technology in the city.

In summary of this introduction, we note that the concept of "Smart City" relies heavily on the following three pillars:

- **Technology**: The smart city is based on technologies to ensure its development and become attractive to local residents and visitors.
- **Sustainability**: The smart city is intimately linked to respect for its environment.
- **Humanity**: The social and human aspect in the design of the smart city is important for local ownership, the response to expressed needs and the ability to innovate continuously.

The concept of the smart city makes us discuss its relationship with the tourism sector. Otherwise, it will be necessary to prove the importance of ICTs in the promotion of tourism and, consequently, tourist destinations. We can talk as well, of smart destination or "Smart Destination".

To tell the truth, it will be necessary to understand the role that the technology can play in the configuration of the destination and its image. Can it be considered as a "smart" destination? How to qualify and characterize it? What assets can it have to distinguish itself?

These are all legitimate questions relating to our subject. We will endeavor to explain the concept of "Smart Destination", its attributes and characteristics, and also to identify, as far as possible, the actors who contribute to the formation of this concept and the relationships that may exist between them.

At first glance, it is important to explain that the introduction of the concept of "Smartness" in tourist destinations requires a dynamic interconnection between stakeholders through a technological platform on which information on tourism activities could be exchanged. fast and instant.

The platform is an important support in the management of huge information flows exchanged by users (tourists). They find a way to exchange information and instantly share the experiences they have experienced in places or spaces in the destination (Zhang et al. 2012 cited in Wang et al. 2013).

The goal would be to improve the tourism experience and improve the efficiency of resource management to maximize destination choice, competitiveness and consumer satisfaction while demonstrating sustainability over an extended period.

To achieve this goal, it must be noted that the implementation of smart tourism relies on three essential forms of ICT namely: Cloud Computing, Internet of Things (IoT) and end-user of the System (Wang and Xiang 2012).

2.4 Moroccan City and Smart City: A New Model of Urban Development in Question

Digital touches all sectors. No sector of activity is spared by the digital revolution that modifies and changes uses. The same is true for transportation, the environment and citizenship and its way of life (Letribot 2015). The digital city or smart city has become the new model of urban development. In our framework of study in the cities of Morocco, we take the city of Casablanca as a model and basic example.

3 Methodology

The positioning of the Casablanca metropolis like a Smart City, coped with competition on the African level. Many initiatives are to be carried out to make it possible the town of keep its length in advance (Mokhliss 2017). It is around this idea that we built our empirical study, which consist with a panoramic straightening out of the various sectors carrier of the town of Casablanca an also with the presentation of the projects dedicated to the digitalization of these sectors, and this within the framework of the initiative: "Smart city", "e-Madina". Wich works to return to the town of Casablanca gravitational, more effective and more competitive, for the companies, the citizens and the visitors within the framework of partnerships Public, Deprived, Citizen and through the use of technologies.

By taking into account the characteristic of our fields application, and seen that the project that we go studies is still in phase of launching, our research will be described as exploratory because, according to Git (2008), it aims at looking further into knowledge on the problems tackled by providing levels of familiarity with the broached subject. It is also descriptive, it understands a set of tchniques of investigation of primary data, to characterize the various implied aspects (Dencker 2007). The theoretical at framework was based on the research of the literature, is based on the contributions of many authors on the covered subjects (Gonsalves 2003). Besides the above mentioned sources, it is advisable to mention that we have to choose a directing maintenance, semi with the only entity who take over of surface a this subject and which is the "e-Madina Cluster Smart Cities".

4 Why the Town of Casablanca?

The town of Casablanca represents the Moroccan city of contrast par excellence, indeed it requires a greatest caution and much creativity, in the marketing field and in particular in the field of urban development. Actually, the city plays the economic part of engine of morocco. it occupies the first place as regards attraction of the investments in the country with 31% from the production facilities and 38% of employment in industry. This place of foreground at the economic level hides sever at disparities on the human plan and social (Belkadi 2015).

4.1 Technology and the Issue of Urban Mobility in Casablanca

The In the areas of transport, digital plays and will play a fundamental future role, with the multiplication and improvement of available mobility data, and the gains made on information processing tools. The insertion of technology on the urban mobility of citizens is a crucial process towards the development of the city of Casablanca towards an intelligent city. At first a structural organization of the city by technology is indispensable. Only, this one requires important infrastructures to put in place. The start-up solution is conceivable. The advent of Big Data, a phenomenon that tends to favor the emergence of startups in the transport sector, coming to compete with the big known companies or with an indisputable notoriety using important means in term of infrastructure and financing important. Young computer scientists are primarily there to understand the different jobs of their employees in order to develop solutions tailored to their demands. Thus, the implementation of applications in the field of transport whose purpose is to optimize the duration of travel. These encourage intermodal action by the user in the smart city since what interests him is to be able to move from point A to point B, and preferably in an efficient way, thereby introducing the notion of "urban mobility" (Fig. 1).

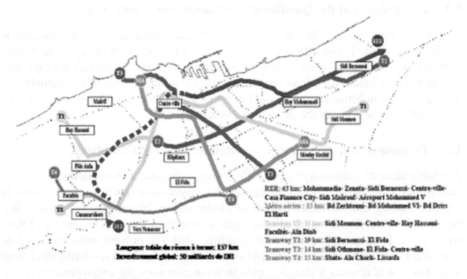

Fig. 1. Global transport network to 2020. *Source* Casa transport

The autonomous vehicle system is necessary to reinforce existing public transport lines, particularly with respect to the first and last kilometers between the end of the use of public transport and the actual destination of the citizen. Especially in the urban environment, which concentrates most of the mobility problems and their consequences, the digital mobility tools, also known as Intelligent Transport Systems (ITS), intelligence to be understood within the meaning of the word "information"—are particularly effective in meeting the needs of local communities, residents and enabling digital businesses to grow strongly. "Mobility is no longer thought of in terms of displacement from A to B, but in terms of travel, and the journey must be a pleasant, friendly and connected experience" (Orfeuil 2018).

4.2 Technology and the Issue of Urban Mobility in Casablanca

Taking into consideration the city's environment through technology is a key element towards smart urban development. This environment refers to the use of new technologies to protect and conserve the city's environment. For example, the installation of sensors measuring air quality, as is already the case in some capitals would regulate vehicle traffic, encourage citizens to use renewable energy.

Urban design has often evolved through the introduction of new technologies. The agrarian control and the canalization of the water allowed the construction of the first cities, the building of the roads, their putting in commercial networks, the industrialization, the development of the modern functionalist city (rails, infrastructure, zoning, etc.). With sustainable development, technical ambition, especially environmental and digital, must instead preserve the planet, people and civilizations. Technology changes status. It is no longer for the progress of man, but for the perpetuity of the world.

4.3 Technology and the Question of the Lifestyle of Casablanca

The adaptability of the lifestyle of citizens of Casablanca to new technologies is paramount. Called "Smart Lifestyle", it's a way of life using technology. This pattern of behavior will depend on the quality of housing conditions, the existence of cultural facilities, educational institutions and the quality of social cohesion. For example, for quality housing in Morocco, trustees of buildings and houses should, within five years, be operational and truly in the service of residents.

4.4 Technology and the Question of Administration of Casablanca

The integration of technology in prefectural and communal administrations is important. Smart administration refers to transparent governance that involves citizens in decision-making and provides a number of public and social services. We suggest that the administration install universal road signs to help the road user, regardless of their origin or destination, to live the route as a trip. This will prevent some motorists to discover, as for the first time, some signs almost reserved for local residents. Intelligent administration is also, as it already exists in some countries, the empowerment of the traffic actors in Morocco.

4.5 Technology and the Question of Companies Located in Casablanca

Businesses based in Casablanca are an important part of the transition process from a normal city to an intelligent city. A smart economy refers to the "smart factory" that is to say the industrially smart city, where companies are developing in information and communication technologies (ICT) and where industries are integrating ICT into their production processes. Indeed, Singapore is an example of a model city, and where the smart economy is already presence, his government invests heavily in ICT: cameras are ubiquitous for the safety of citizens and public buildings are all equipped with Real-time warning sensors in the event of an earthquake, for example. This allows decision-makers to react as quickly as possible.

5 E-Marketing: Tool Associated with the Urban Development of Casablanca Towards the Digital City?

The promotion and development of an intelligent city is driven by local players and major operators, such as telecom operators, IT companies, information systems integrators, electricity network operators, energy distributors, energy companies and public works, real estate developers, transport companies, telecommunications equipment suppliers. The technology market, which underpins smart city projects, has a strong market share.

Territorial marketing using technology contributes to the development of cities and to encourage them to take part in the competition of cities on an international scale. Thus, the marketing and communication provisions must be involved for each training area of an intelligent person.

A smart city must appeal to economic actors for the creation of value and contribute to the process of development of the city through technologies. "The smart city is the one that knows how to both attract and retain companies employing highly qualified labor" (Bouinot 2018).

6 Conclusions

Territorial marketing using technology contributes to the development of cities and to encourage them to take part in the competition of cities on an international scale. The concept of "smart city" aims, among other things, to meet the different needs of a city in the quest for a better quality of life for citizens and businesses. Marketing is a tool for developing and disseminating smart city culture. Moroccan cities must now devote efforts to a process of transition from a city without technology to an intelligent and digital city.

Morocco has launched a major development project of its economic capital "Casablanca" to promote it at the level of major international cities, businesses, citizens and visitors. This initiative is part of public-private partnerships.

However, what about the digitization of the city of Casablanca? The question of territorial e marketing of this city is essential. Territorial officials need to create a brand "Casablanca" based on an identity and on a strong image.

The objective of this work revolves around the realization of an exploratory study of the state of the existing smart city of Casablanca. The study of the digital projects established for the benefit of the various sectors of the city has allowed us to guide the design of a territorial offer apt to build a good project of urban development of the city.

We have also concluded that there is a strong willingness of local actors to support this initiative of digitalization of the city of Casablanca, and that by concretizing several ideas. Among these ideas is the concept of the intelligent village (or e-douar), the virtual museum, the transformation of the Al Amal complex into Living lab, etc.

References

Bakıcı, T., Almirall, E., Wareham, J.: A smart city initiative: the case of Barcelona. J. Knowl. Econ. **4**(2), 135–148 (2013)

Belkadi, E.: Marketing Territorial de Casablanca: Etude de l'image de Marque. Int. J. Innov. Appl. Stud. (2015)

Boes, K., Buhalis, D., Inversini, A.: Conceptualising Smart tourism destination dimensions. In: Tussyadiah, L., Inversini, A. (eds.) Proceedings of the International Conference on Information and Communication Technologies in Tourism, pp. 391–404. Switzerland: Springer International Publishing (2015)

Buhalis, D., Amaranggana, A.: Smart tourism destinations. In: Xiang, Z., Tussyadiah, I. (eds.) Information and communication technologies in tourism, pp. 553–564. Heidelberg: Springer (2014)

Dameri, R.P.: Searching for smart city definition: a comprehensive proposal. Int. J. Comput. Technol. **11**(5), 2544–2551 (2013)

Decressac, P.: Le digital au service du marketing territorial. Vision. Market. (2016)

Dencker, A.M.: Méthodes et techniques de recherche en tourisme. 9. ed Future (2007)

Desdemoustier, J., Crutzen, N.: Smart Cities en Belgique: Analyse qualitative de 11 projets Étude scientifique. Etude, HEC-Ulg (2015)

Douay, N., Henriot, C.: La Chine à l'heure des villes intelligentes. l'information géographique **80** (3), 89–102 (2016)

Elbanna, S.: Strategic decision making: process perspectives. Int. J. Manage. Rev. **8**(1), 1–20 (2006)

Flores, L.: Mesurer l'efficacité du marketing digital. Dunod (2012)

Gayet, C., Marie, X.: Web marketing et communication digitale "outils pour communiquer efficacement auprès de ses cibles". vuibert, p. 32 (2016)

Giffinger, R., Gudrun, H.: Smart cities ranking: an effective instrument for the positioning of cities? Architect. City Environ. (ACE) **4**(12), 7–25 (2010)

Gil, A.C.: Métodos e técnicas de pesquisa social, 6 ed. Atlas, São Paulo (2008)

Höjer, M., Wangel, J.: Smart sustainable cities: Definition and challenges. In: Hilty, L.M., Aebischer, B. (eds.) ICT Innovations (2015)

Institut technologies de l'information et sociétés (ITIS) (2012) Villes intelligentes: un bref survol. https://www.itis.ulaval.ca/cms/site/itis/page81388.html

Khomsi, M.R.: The smart city ecosystem as an innovation model: lessons from Montreal. Technol. Innov. Manage. Rev. **6**(11), 26–31 (2016). http://timreview.ca/article/1032

Koo, C., Gretzel, U., Hunter, W.C., Chung, N.: The role of IT in tourism. Asia Pacific J. Inform. Syst. **25**(1), 99–104 (2015)

Lamsfus, C., Wang, D., Alzua-Sorzabal, A., Xiang, Z.: Going mobile defining context for on-the-go travelers. J. Travel Res. (2014)

Letribot, M.: La révolution du digital dans le secteur de l'assurance. Publication Euro group (2015)

Mokhliss, B.: La ville de Casablanca veut être parmi les métropoles les plus intelligentes du continent. Africa's Smart City (2017)

Ovazza, Y.: Comment construire une stratégie digitale? (2015). www.butter-cake.com

Scheid, F., Vaillant, R., De Montegu, G.: Le marketing digital: Développer sa stratégie à l'ère numérique, Groupe Eyrolles, p. 6 (2014)

Sigala, M., Christou, E., Gretzel, U. (eds.).: Social Media in Travel, Tourism and Hospitality: Theory, Practice and Cases. Ashgate Publishing, Ltd. (2012)

Numerical Simulation of the Coal and Straw Co-firing in Swirling Stabilized Burner

Nadia Rassai[(✉)] and Noureddine Boutammachte

Energy Department, ENSAM – Meknès (Ecole Nationale Supérieure d'Arts et Métiers de Meknès), Meknes, Morocco
rassai.nadia@gmail.com

Abstract. The aim of this paper is to valorize straw waste by co-firing coal and straw in a swirl-stabilized reactor in order to produce a clean energy. To fulfill this purpose, a comprehensive computational fluid dynamics (CFD) modeling of co-firing coal and straw was presented, in which the pulverized straw particles and coal particles are independently injected into the burner through two concentric injection tubes. The numerical approach is based on Reynolds averaged Navier–Stokes (RANS) approach using the realizable k–ε turbulence model for turbulence, the non-premixed combustion model for gas phase combustion, the Lagrangian approach (DPM) for the discrete second and the DO for radiation. The finding results show a good agreement with numerical and experimental data.

Keywords: Coal · Straw · Pulverized · Co-firing · Swirling

1 Introduction

With the increasing energy demand and Morocco's dependency on fossil energy, Morocco makes tremendous efforts to reduce this dependency on imported fossil energy and it has set a goal of having 20% renewable energy by 2020 like solar, wind turbine and biomass. Several projects are conducted on solar and wind but the valorization of biomass is still modest even if biomass fuels have an important heating value. Nowadays, using biomass as a fuel or co-firing biomass with coal is a feasible alternative to produce clean energy. However, the choice of appropriate parameters for the combustion process is rather challenging due to the heterogeneous composition of the biomass. Hence, the idea of this project which consists in the numerical modeling of co-firing of coal and straw in swirling stabilized burner in order to select the suitable parameters for combustion and predict temperature, emissions, and ash deposition.

Computational Fluid Dynamics (CFD) has a crucial role for this project; it may help to understand deeply the combustion process and it can give a good approximate overview of gas and particle combustion mechanism as well as the parameters affecting temperature, turbulence, velocity and emission. Indeed, In literature there are just only a few numerical simulations of co-firing pulverized coal and biomass, Bonefacic et al. [1] presented a paper in which they summarized numerical models of co-firing pulverized coal and biomass in a vertical cylindrical laboratory furnace (20 Kw) and Elorf

M. Ben Ahmed et al. (Eds.): SCA 2018, LNITI, pp. 84–93, 2019.
https://doi.org/10.1007/978-3-030-11196-0_9

et al. [2] investigated the effect of the swirl motion on flame dynamic of pulverized olive cake in a vertical furnace.

Also Tabet et al. [3] published a review of numerical models used for co-firing biomass with pulverized coal under air and oxy-fuel conditions. In addition, Yin et al. [4] presented an article contains a state-of-the-art of the different sub- models used to model an oxy-fuel combustion of pulverized fuels (PF),they concluded that the standard k–ε is robust [5, 6], the realizable k–ε model is suitable for industry-scale oxy-firing and the SST k–w and RNG k–ε model is more appropriate for lab-scale oxy-firing but the inconvenient of SST k–w model that it is computationally expensive (e.g., in [7–9]).

The purpose of this paper is to model the co-firing of pulverized coal and straw in 150 kW swirling stabilized burner by the mixture fraction approach and the Discrete Particle Method (DPM), the finding results are compared with numerical and experimental results obtained by Yin et al. [4].

2 Configuration

2.1 Geometry

Figure 1 shows the 2D scheme of the combustion chamber, the inlet (a), and the outlet (b). The high of the furnace h = 240 cm, the diameter d = 75 cm.

Fig. 1. Geometry of the furnace

2.2 Operating Conditions

Table 1 presents the proximate analysis and particle Rosin Rammeler parameters of fuels.

Table 2 shows mass fuel rate of coal, straw and air.

3 CFD Modeling

3.1 Gas-Solid Fluid Dynamics Model

CFD modeling of biomass combustion processes describes fluid flow, heat and mass transfer and chemical reaction process.

Following is the basic equations for mass, momentum, energy, and species [10].

Continuity equation

Table 1. Characteristics of coal and straw

	Coal	Straw
Proximate analysis		
Moisture [wt%, as received]	2.1	7.7
Fixed carbon [wt%, dry]	51.5	15.6
Volatiles [wt%, dry]	40.6	79.5
Ash [wt%, dry]	7.89	4.91
HHV [kJ/kg, dry]	30,731	18,493
Particle rosin Rammler parameters		
Min diameter [µm]	25	50
Max diameter [µm]	200	1000
Mean diameter [µm]	110.4	451
Spread parameter	4.40	2.31

Table 2. Mass flow rate of fuels and air

Center fuel (straw) [kg/s]	Annular fuel (coal) [kg/s]	Center air [kg/s]	Annular air [kg/s]	Secondary air [kg/s]
0.004194	0.002083	0.0025	0.0033	0.0444

$$\frac{\partial \rho}{\partial t} + \nabla(\rho \bar{u}) = S_P \tag{1}$$

Momentum equation

$$\frac{\partial(\rho \bar{u})}{\partial t} + \nabla(\rho \bar{u}\bar{u}) = -\nabla p + \nabla(\mu \nabla \bar{u}) + S_N \tag{2}$$

Energy equation

$$\frac{\partial(\rho H)}{\partial t} + \nabla(\rho \bar{u}H) = \nabla(\lambda \nabla T) + S_H \tag{3}$$

Species transport equation

$$\frac{\partial(\rho Y_f)}{\partial t} + \nabla(\rho \bar{u}Y_f) = \nabla(D\nabla(\rho Y_f)) + S_Y + R_f \tag{4}$$

3.2 Combustion

The global biomass combustion reaction takes the following form, where the first reactant compound is a biomass fuel [11]:

$$\mathbf{C_{X1}H_{X2}O_{X3}N_{X4}S_{X5} + n \cdot (3.76N_2 + O_2) \rightarrow a\ CO_2 + b\ H_2O + C\ O_2 + dN_2}$$
$$\mathbf{+ e\ CO + f\ NO_x + g\ SO_x}$$

Because simulation of combustion process is expensive and needs high computational resources, often only considers the most basic of species and a limited number of reactions.

These species are: N_2, O_2, H_2O, CO_2, CO and CH_XO_Y.

These reactions are:

$$CH_XO_Y + a\ O_2 \rightarrow CO + b\ H_2O$$
$$CO + 0.5O_2 \rightarrow CO_2$$

In CFD many models of combustion are developed; following is the available models (Table 3).

Table 3. CFD model of combustion process

Fast chemistry models	Finite rate (slow) chemistry models
Eddy dissipation	Laminar finite rate
Premixed model	Eddy dissipation concept
Equilibrium model	Composition PDF
Steady laminar flamelet	
Flamelet generated manifold	
Partially premixed	
Non-premixed model	

3.3 Numerical Method

The pressure-based Navier-Stokes solution algorithm is adopted for computing a solution, The SIMPLE algorithm is applied to the pressure–velocity coupling, the flow is considered steady, incompressible, and turbulent. Realizable k–ε and model was chosen for turbulence, the DO model for the radiation, the Lagrangian approach for the discrete second phase and the diffusion limited char combustion model (Diffusion-limited, DLCCM) for the char combustion. There are many models of gas phase combustion process: Eddy dissipation, Premixed model, Equilibrium model, Steady laminar Flamelet, Flamelet generated manifold, Partially premixed,Non-premixed model,Laminar Finite rate, Eddy dissipation concept, Composition PDF.

In this case the non-premixed (mixture fraction) model is used because the fuel and the oxidizer enter the reaction zone in different streams, in this approach the mixture fraction f, is the local mass fraction of burnt and unburnt fuel stream elements (C, H, etc.) in all the species (CO_2, H_2O, O_2, NO, etc.), atomic elements are conserved in chemical reactions. In addition, the Probably Density Function (PDF) is a conserved scalar quantity; hence its governing transport equation does not have a source term.

Under this hypothesis, the combustion is simplified to a mixing problem, and the difficulties related to the closing non-linear mean reaction rates are avoided [12].

Equilibrium system for coal and straw co-firing calculated by PDF consists of 17 species (C, C(s), H, H_2, O, O_2, N, N_2, CO, CO_2, OH, H_2O, CH_4, C_2H_2, NO, HCN, H_2O).

3.4 DPM Approach

In literature, there are three methods used for the numerical modeling of the multiphase flow: Euler-Lagrange, Euler-Euler approach and Discrete Element Method CFD [13].

The Euler-Lagrangian approach is used here to model the multiphase flow; the primary phase is treated as a continuum by solving the time-averaged Navier Stokes equations and the discrete phase is solved in a Lagrangian frame of reference by tracking a large number of particles, bubbles, or droplets through the calculated flow field. The injection diameter follows a Rosin–Rammler distribution.

The trajectory of a discrete phase particle is predicted by integrating the force balance on the particle, which is written in a Lagrangian reference frame [14].

$$\left(\frac{d\vec{u}_p}{dt}\right) = F_D(\vec{u} + \vec{u}_p) + \vec{g}(\rho_p - \rho)/\rho_p \tag{5}$$

where $F_D(\vec{u} + \vec{u}_p)$ is the drag force per unit particle mass.

F_D is expressed by:

$$F_D = \frac{18\mu}{\rho_p D_P^2} \frac{C_D Re_p}{24} \tag{6}$$

And

$$R_{e_p} = \frac{\rho D_p |u_p - u|}{\mu} \tag{7}$$

CD is the drag coefficient and Rep is the relative Reynolds number.

3.5 Devolatization

The simplified approaches define devolatilization rate with single- or two-step Arrhenius reaction schemes.

The reaction kinetic rate is:

$$k = A^{(-Ea/RT)} \tag{8}$$

The devolatilization rate is [15]:

$$\frac{-dm_p}{dt} = k[m_p - (1 - f_{v,0})m_{p,0}] \tag{9}$$

Here, A and E are numerical constants of reacting substances.

$f_{v,0}$ is the initial volatile fraction and $m_{p,0}$ is the initial particle mass (kg).

4   Results and Discussion

The validation of the numerical model was based on the comparison with the numerical and experimental data from Chenguin Yin et al. The experimental results obtained from FT-IR and a Horiba gas analyzer.

Table 2 resumes the five cases presented by M. Chenguin Yin and the case studied in this paper (non-premixed) (Table 4).

Table 4. Models used in the six cases

Case Label	Volatile combustion	Particle conversion	Turbulence model	Radiation model
Traditional	Eddy dissipation model	Default DPM law	Realizable k–ε	DO
SKE-default	Eddy dissipation concept	Default DPM law	Standard k–ε	DO
SKE-custom	Eddy dissipation concept	Custom DPM law	Standard k–ε	DO
RKE-default	Eddy dissipation concept	Default DPM law	Realizable k–ε	DO
RKE- custom	Eddy dissipation concept	Custom DPM law	Realizable k–ε	DO
Non-premixed	Non premixed model	Default DPM law	Realizable k–ε	DO

In this part, the results given by non-premixed model will be compared firstly to CFD predicted values (Traditional, RKE default) and secondly to experimental results (Horiba, FT-IR) on line y = 0.075 m (i.e. r = 30 cm or r = 45 cm).

Figure 2 shows the CO_2, H_2O, O_2 and CO emissions for the non-premixed combustion model and the models proposed by M. Chenguin Yin on the line y = 0.075 m (i.e. r = 30 cm or r = 45 cm).

At the inlet ($0.1 \leq x \leq 0.5$) the emissions given by the non-premixed combustion model are lower than those given by the traditional and RKE default models except for CO which has molar fractions slightly higher than the M. Chenguin results (Traditional case), in the core of the furnace and at the outlet the three models (Traditional, RKE default and non premixed) give almost the same molar fractions except for H_2O which the shape of H_2O curve of non premixed model in the core region is different from others.

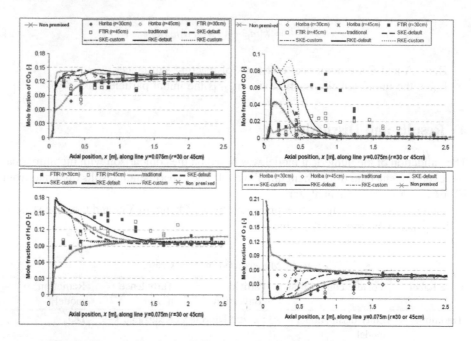

Fig. 2. Non premixed versus Chenguin Yin results on the line y = 0.075 m

The same figure presents the experimental results given by the two gas analyzers, it is seems clearly that the non premixed model provides results close to the experimental values especially in the core region and at the outlet of the burner and more accurately to those obtained by Horiba analyzer. The H_2O predicted emissions are different from FTIR results in core region and slightly different at the outlet of the burner, Horiba analyzer results are absent for H_2O emissions.

Figure 3 presents the profile of temperature in the furnace, there is a modest difference between the maximum value given by M. Chenguin Yin (1900K) and those obtained by the non-premixed model (1970K) but the profile is the same, at the inlet of the combustion chamber the gas is cold and it starts to warm up more until it reaches 1970K, the gas is hotter inside the chamber than on the near wall zone.

Fig. 3. Temperature contours

After the validation of the volatile combustion model, in this section we will present the results of this study; contours of swirl velocity (Fig. 4), contours of mean mixture fraction of straw (Fig. 5) and contours of secondary mean mixture fraction of coal (Fig. 6) the mean mixture fraction, denoted by f is equal to 0 when the fuel is absent.

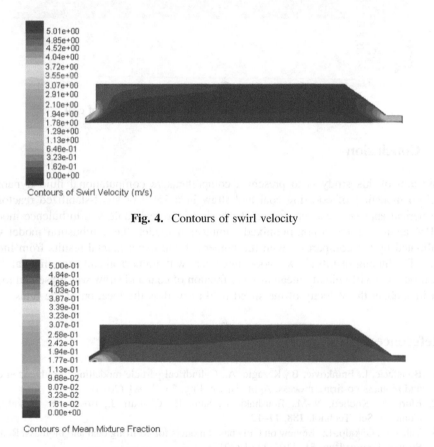

Contours of Swirl Velocity (m/s)

Fig. 4. Contours of swirl velocity

Contours of Mean Mixture Fraction

Fig. 5. Contours of mean mixture fraction of straw

At the Inlet of the furnace the swirl velocity reaches his maximum 5 m/s and starts decreasing in the core region of the furnace, approaching to the outlet the swirl velocity starts increasing due to the diminution of the section, in addition straw fuel is concentrated at the inlet of the furnace while the coal fuel is well distributed along the furnace, this observation can be explained by the difference of density of the two fuels.

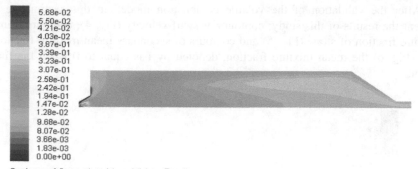

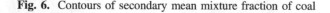

Contours of Secondary Mean Mixture Fraction

Fig. 6. Contours of secondary mean mixture fraction of coal

5   Conclusion

The aim of this study is to present a comprehensive computational fluid dynamics (CFD) modeling of co-firing coal and straw in a 150 kW swirl-stabilized reactor, a numerical approach was chosen on the basis of the realizable k–ε turbulence model, DPM approach, and a non-premixed combustion model, the combustion model was validated by the comparison with the numerical and experimental results from literature. The finding results show a good agreement with numerical and experimental data. Contours of swirl velocity mean mixture fraction of coal and straw was presented to get an idea about the velocity of the speed field as well as the location of the fuels.

References

1. Bonefacic, I., Frankovic, B., Kazagic, A.: Cylindrical particle modelling in pulverized coal and biomass co-firing process. Appl. Therm. Eng. **78**, 74–81 (2015)
2. Elorf, A., Koched, N.-M., Boushaki, T., Sarh, B., Chaoufi, J., Bostyn, S., Gökalp, I.: Combust. Sci. Technol. **188**, 11–12
3. Tabet, F., Gökalp, I.: Review on CFD based models for co-firing coal and biomass. Renew. Sustain. Energy Rev. **51**, 1101–1114 (2015)
4. Yin, C., Kaer, S.K., Rosendahl, L., Hvid, S.L.: Modeling of pulverized coal and biomass co-firing in a 150 kW swirling stabilized burner and experimental validation. In: Proceeding of the International Conference on Power Engineering-09 (ICOPE-09), Kobe, Japan, 16–20 Nov 2009
5. Porteiro, J., Collazo, J., Patiño, D., Granada, E., Moran Gonzalez, J.C., Míguez, J.L.: Numerical modeling of a biomass pellet domestic boiler. Energy Fuels **23**, 1067–1075 (2009)
6. Zhang, X., Chen, Q., Bradford, R., Sharifi, V., Swithenbank, J.: Experimental investigation and mathematical modelling of wood combustion in a moving grate boiler. Fuel Process. Technol. **91**, 1494–1499 (2010)
7. Álvarez, L., Yin, C., Riaza, J., Pevida, C., Pis, J.J., Rubiera, F.: Biomass co-firing under oxy-fuel conditions: a computational fluid dynamics modeling study and experimental validation. Fuel Process. Technol. **120**(22), 33 (2014)

8. Chen, L., Ghoniem, A.F.: Simulation of oxy-coal combustion in a 100 kWth test facility using RANS and LES: a validation study. Energy Fuels **26**, 4783–4798 (2012)
9. Muto, M., Watanabe, H., Kurose, R., Komori, S., Balusamy, S., Hochgreb, S.: Largeeddy simulation of pulverized coal jet flame—effect of oxygen concentration on NO_x formation. Fuel **142**, 152–163 (2015)
10. Shi, A., Pang, Y., Xu, G., Li, C.: Numerical simulation of biomass gasification in a fluidized bed. In: International Conference on Applied Science and Engineering Innovation (ASEI 2015), pp. 1585–1591 (2015)
11. ANSYS, Ansys Fluent theory guide, Release 14.5, ANSYS
12. User's Guide, Ansys 14.5. Modeling Non-premixed Combustion
13. Singh, R.I., Brink, A., Hupa, M.: CFD modeling to study fluidized bed combustion and gasification. Appl. Therm. Eng. **52**, 585–614 (2013)
14. Korytnyi, E., Saveliev, R., Perelman, M., Chudnovsky, B., Bar-Ziv, E.: Computational fluid dynamic simulations of coal-fired utility boilers: an engineering tool. Fuel **88**, 9–18 (2009)
15. Badzioch, S., Hawksley, P.G.W.: Kinetics of thermal decomposition of pulverized coal particles. Ind. Eng. Chem. Process Des. Dev. **9**, 521–530 (1970)

Outdoor Air Purification Based on Photocatalysis and Artificial Intelligence Techniques

Meryeme Boumahdi$^{(\boxtimes)}$ and Chaker El Amrani

Laboratoire Informatique Systèmes et Télécommunication (LIST), Université Abdelmalek Essaadi – FST de Tanger, Tanger, Morocco
Boumahdi.meryeme@gmail.com, ch.elamrani@fstt.ac.ma

Abstract. Multiphysics simulation had progressed significantly in the recent years so that predictions of flow around and inside complex geometries are now possible. In this work multiphysics is used to test the efficiency of a new, innovative solution for outdoor air purification using photocatalysis technology. Photocatalysis has received considerable attention in recent years with a huge potential in air purification applications. The work focusses on the semi-active use of photocatalytic surfaces in streets as an innovative method for removing anthropogenic pollutants (especially volatile organic compounds or VOCs) from urban air. This study combines different scientific fields: Physics, Chemistry and Computer engineering. The objective of this work is to design an air purified solution to be implemented in four streets in Tangier. To achieve this goal several techniques were evaluated using COMSOL and Matlab. Artificial intelligence methods allowed finding out optimized purification scenarios for different streets aspect ratio and therefore to design best purification strategies for a cleaner city, taking into account the decrease of the pollution concentration at lower energy cost. The method is based on lamellae, coated with photocatalyst (TiO_2), arranged horizontally at the walls of street canyons and lightened with UV light. A constant initial VOC background concentration was considered in the simulations, with a continuous source of VOCs to simulate the street traffic.

Keywords: Pollution · Photocatalysis · COSMOL · ANN

1 Introduction

Outdoor air pollution is increasing dramatically in Morocco due to rapid urbanization, burning of fuel and release of chemicals [1, 2]. The pollutants have harmful impacts on human health and natural ecosystems [3, 4]. Manufacturing industries release large amount of carbon monoxide, hydrocarbons, organic compounds, and chemicals into the air thereby depleting the quality of air [5, 6]. The development of photocatalysis has been the focus of considerable attention in recent years with photocatalysis being used in a variety of products across a broad range of research areas, including especially environmental and energy- related fields. Of the many different photo-catalysts, titanium dioxide(TiO_2) has been the most widely studied and used in many applications

© Springer Nature Switzerland AG 2019
M. Ben Ahmed et al. (Eds.): SCA 2018, LNITI, pp. 94–103, 2019.
https://doi.org/10.1007/978-3-030-11196-0_10

because of its strong oxidizing abilities for the decomposition of organic pollutants, chemical stability, long durability, nontoxicity, low cost, and transparency to visible light [7]. In PCO, a photosensitive semiconductor (mostly titanium dioxide, TiO_2) is activated by UV light, forming protons in the valance band and electrons in the conduction band:

$$TiO_2 + hv \rightarrow h^+ + e^- \tag{1}$$

Photocatalysis (PCO) technology exposes a catalyst, mostly titanium dioxide (TiO_2), to UV light to produce reactive hydroxyl radicals (OH^*) and superoxide anions (O_2^{*-}) that are able to mineralize harmful VOCs into H_2O and CO_2 [7] Designing a photocatalytic (PCO) reactor for outdoor air purification presents a challenge toward a large number of parameter designs, street characteristics, the traffic and aspect ratio that change in every building. As a first novelty in this work, we present an innovative solution for outdoor air purification based on coated lamellae placed horizontally at the walls of street canyons, using forced convection with a face to force polluted air over the system, and heating the coated surface to create a natural convection in purpose of reducing the energy cost. Full-scale experimental investigations are costly and some variables such as heat transfer coefficients are impossible to measure directly. Computational fluid dynamics (CFD) can be used to obtain pollutant concentration, adsorption and desorption, air velocity and the maximum adsorption capacity of the photocatalyst.

2 Methodology

2.1 Geometry

The geometry of the simulation was based on the real size of every street. The proposed method was based on lamellae coated with photocatalyst (TiO_2), with a dimension of 8 cm * 48 cm * 1 m, lightened with UV light and arranged horizontally at the wall of the streets. The internal walls are important in the system to keep the air flow generated by the fan, and forced the purified air to go back to the street. Several rotation degree were used for the lamellae, and 45° gives the best results, the distance between the lamellae is 1 m. Three different fans were tests, with different pressure and flow rate. 0.1, 0.5 and 1 m^3/s with a pressure of 100 Pa. A concentration of 10^{-4} mol/m^3 was considered as a VOCs background and for open boundaries also. The second approach was to use natural convection instant of the fans as a generator of the air flow in order to reduce the energy cost. The lamellae were heated with a temperature of 325.15 K (Fig. 1).

2.2 Multiphysics Modelling

Air Flow Modeling. COMSOL Multiphysics 5.3 software was used to develop the models. Using Computational fluid dynamics (CFD), heat transfer and transport of diluted species modules. 2D models were used instant of 3D models to save the

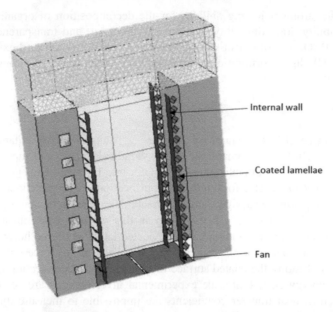

Fig. 1. Photocatalytic lamellae in a street canyon

calculation time. The mesh of the model affect strongly the results. The average of the mesh is 0.87 and it's used by the user. Finer mesh was required in the lamellae and the fans. A total of 53,680 triangles were needed to guarantee a good mesh for the model. A laminar air flow model of incompressible fluid was used. A steady-state for the air flow velocities in the street was generated using a stationary solver for the first approach using a fan. The second approach the lamellae were heated thus create natural convection. This problem was solved using the weakly compressible flow that neglects the influence of pressure waves, which are seldom important in natural convection. This option also eliminates the description of pressure waves, which requires a dense mesh and small time steps to resolve, thus also a relatively long computation time. The transport of acetaldehyde was then modelled by coupling the time dependent advection and diffusion equation to the velocity field vector u:

$$\frac{\partial C_{Acal,bulk}}{\partial t} = \nabla \cdot \left(D\nabla C_{Acal,bulk}\right) - \mathbf{u} \cdot \nabla C_{Acal,bulk} \qquad (2)$$

With $C_{Acal,bulk}$ the bulk concentration of acetaldehyde [mol m^{-3}] and D the mass diffusion coefficient of acetaldehyde in air [m^2 s^{-1}]. The latter value can be found in literature and was set to 10^{-4}m^2 s^{-1}. This equation accounts for the evolution of the acetaldehyde concentration in the flow due to diffusion (first term on the right-hand side) and advection (second right-hand side term) of acetaldehyde molecules.

Adsorption/desorption of acetaldehyde

Same equation used from other article [8]. We will present some equations.

An important precursory step in photocatalysis is the adsorption of pollutants on the TiO_2 surface. The fractional surface coverage θ_{Acal} is determined as the ratio of the surface concentration of adsorbed molecules [mol/m^{-2}] $C_{Acal,ads}$ and Γs, as defined the maximum surface coverage. Also θ_{Acal} can be determined by the bulk concentration $C_{Acal,bulk}$ and the Langmuir equilbrium constant K, with K is the ratio of the adsorption and desorption rate constants k_{ads}/k_{des}.

$$\theta_{Acal} = \frac{KC_{Acal,bulk}}{1 + KC_{Acal,bulk}} = \frac{C_{Acal,ads}}{\Gamma_s} \tag{3}$$

The use of independent value for both k_{ads} and k_{des} is required to simulate the evolution of the pollutant concentration in the system. Adsorption and desorption reactions drift away from equilibrium resulting in transient behavior whenever the pollutant concentration changes during operation.

Acetaldehyde adsorption was modelled as a species flux N_{ads} across the coated surface of the lamellae [mol m^{-2} s^{-1}] from the bulk to the surface; desorption, on the other hand, was a species flux N_{des} across the same boundaries but in opposite direction (Fig. 2).

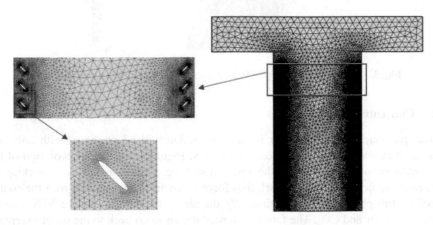

Fig. 2. Representation of the mesh used in the CFD simulations

$$-n.(-D\nabla C_{acal,bulk} + u.C_{Acal,bulk}) = -N_{ads} + N_{des} \tag{4}$$

3   Modelling Results

3.1   Air Flow Modelling

Three different flow rates were tested in this work. A typical steady-state air flow velocity profile is plotted (Fig. 3). The highest velocity is located on the fans in the two sides (resp 0.4, 2.3 and 3.92 m/s). In the boundaries of the lamellae the velocity varies between 0.4 and 1.6 m/s. The street recorded the lowest velocity in the model (less and 0.4 m/s). A typical laminar velocity profiles are observed in the three flow rates.

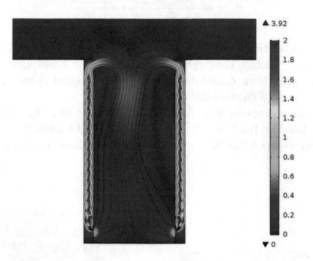

Fig. 3. Modelled velocity profile in the street in the forced convection case

3.2   Concentration

In this paragraph only results for forced convection will be presented with only one case, a street canyon with an aspect ratio of 1.8. Figure 4 present the evolution of the concentration of VOCs during the simulation time. Once the fan start working, it creates an air flow inside the model, thus force more air to get in touch with the coated lamellae, this process purify continuously the air by breaking down the VOCs molecules into water and CO_2. The fans also forced the air to go back to the street to replace the polluted air with a purified one, following the same streamlines of the air flow.

The concentration average decrease differently with each case (38% of purification with flow rate of 0.1 m^3/s). Using a fan with low flow rate, a low percentage of purification can be achieved, while high flow rate can gives high percentage. If a powerful fan used in this case the percentage of purification could get up to 65%. The concentration average get lower and lower until a stationary state is established. The concentration decreases with time, in a different ways. Depending on the used flow rate. Couple of aspect ratio were tested in this work. Figure 6 shows the purification percentage for different aspect ratio with a constant concentration of VOCs. According

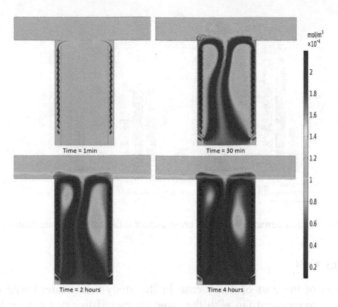

Fig. 4. Acetaldehyde concentration evolution during time

to this results, using a powerful fan gives better results. More the flow rate is High more the percentage is high also. The highest purification percentage was 65% using fan with 1 m³/s (Fig. 5).

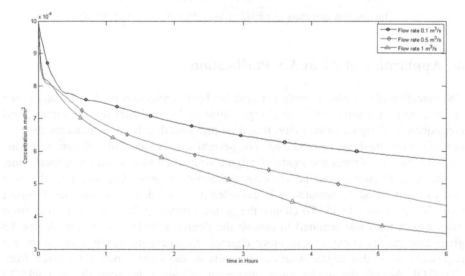

Fig. 5. Average concentration of VOCs with forced convection using different flow rate

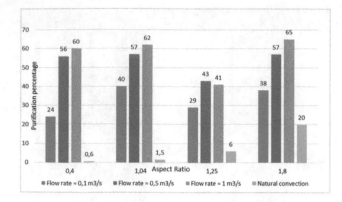

Fig. 6. Purification percentage for different aspect ratio with four purification scenarios

3.3   Energy Cost

A comparison of the cost of every case in the study was made, based on the consumption of a commercial fan with the same characteristics that can be found in the market, taking into account the energy cost of every fan. That Table 1, present a comparison between the three different flow rates with the purification percentage.

Table 1. Purification percentage and the energy cost for each method

Method	0.1 m^3/s	0.5 m^3/s	1 m^3/s
Purification percentage	38%|250 W	57%|370 W	65%|780 W

4   Application of AI in Air Purification

The objective of this part of study is to predict best purification method for any given street aspect ratio using polynomial regressions and neural network algorithms, and consequently, design a cleaner city. Based on previous data from simulations, the best purification method can be predicted. The several simulations with different scenarios could be base for a predicted model. Design a simulation model and run it takes time, so the idea of using machine learning techniques to win time and find the best purification method. Concentration is modelled as an nth degree polynomial in aspect ratio. N was from 1 to 15. To choose the optimal degree polynomial other data from other aspect ratio was required to choose the degree with less error. The degree 13 gives best results comparing to other degrees. Compare the result given from the regression using this degree with other results of the same aspect ratio given from COMSOL Model, the model gives prediction efficiency between (97 and 99.9%) (Fig. 7).

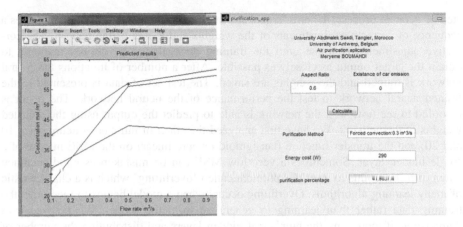

Fig. 7. Prediction of the best purification method and percentage

The Neural Network Fitting Tool GUI nntool available in MATLAB (R2017a) is used to carry out the analysis on the purification data using Artificial Feed-Forward Neural Network with back-propagation principles. The research begin with a simple feed-forward back propagation model and have explored the importance of the ANN at its unit level that is the artificial neurons. These artificial neurons are the building blocks of all such ANNs and to understand their potential has been our main study.

4.1 Dataset Details

Data used to train the models were collected from simulation result in COMSOL Multiphysics. The final data set consists of (x, y) coordinates and the concentration of VOCs taken each minute for six hours. Over 21 million samples were used to train the network. 70% samples called the training data are randomly selected by nntool for training the neural network. 15% samples called the validation data measure the generalization of the network by feeding it with data it has not seen before. The remaining 15% samples called the test data give an independent measure of the performance of the neural network in terms of MSE (Mean squared error). It is the square of the difference of the predicted value and the target, hence always positive.

4.2 Neural Network Model

A typical feed forward with back propagation network should have at least three layers, an input layer, a hidden layer, and an output layer. Appropriate selection of number of hidden layers and the number of neurons in each of them needs experimentation. We train the ANN using the Levenberg-Marquardt algorithm, a standard training algorithm from the literature. The algorithm is terminated according to the early stopping procedure. The validation set used in the early stopping procedure is chosen in a somewhat unusual manner. Finally, the training function produce forecast results on the basis of MSE (Mean square error) minimization criteria. The output signal is compared with the desired response or target output and consequently an error signal is produced. In each step of iterative process, the error signal activates a control mechanism which applies a sequence of corrective adjustments of the weights and biases of the neuron. The corrective adjustments continue until the training data attains the desired mapping to obtain the target output as closely as possible. After a number of iterations the neural network is trained and the weights are saved. The test set of data is presented to the trained neural network to test the performance of the neural network. The result is recorded to see how well the network is able to predict the output using the adjusted weights of the network. We have first analyzed the effect of number of neurons (5, 10 and 20) and the transfer function (tansigmoid or pure-linear) on the MSE in case of a single hidden layer. Sometimes a very low MSE can be mistaken as good accuracy when in fact it points to a serious problem called 'overfitting' which is a characteristic of many learning algorithms. Overfitting occurs when a model begins to memorize the training data rather than learning to generalize from the model. We then seek to demonstrate if increasing the number of hidden layers and distributing the number of neurons in them can contribute in any way to overcome this problem without compromising on performance.

Nine training algorithms were used to compare the results based on the training time, number of iterations and Mean Squared Error.

Used algorithms

trainLM: Levenberg-Marquardt
trainBFQ: BFGSQuasi-Newton
trainRP: Resilient Back propagation
trainSCQ: Scaled Conjugate Gradient
trainCQB: Conjugate Gradient with Powell/Beale Restarts
trainCQF: Fletcher-Powell Conjugate Gradient
trainCQP: Polak-Ribiére Conjugate Gradient
trainOSS: One Step Secant
trainQDX: Variable Learning Rate Backpropagation

Figure 8 shows a comparison between nine algorithms used to train the network. Levenberg-Marquardt gives the less mse comparing to other algorithms but it takes 38 min to train the network using a computer with 4 cores and 12 Gb in RAM.

Fig. 8. Training algorithms and related performance

5   Results and Discussion

Different training algorithms can be used for ANN. The graphs below shows a comparisons between nine algorithms, taking into account mean squared error and training time. Levenberg-Marquardt training algorithm gives lowest MSE, it takes 48 min. So a neural network with this algorithm will be used to predict new street aspect ratio. We noticed that the use of neural network algorithms allows reducing extremely the simulation time. For instance simulation of the concentration took 48 min COMSOL, while it take just 20 s using Levenberg-Marquardt ANN algorithm with mean squared error of 3.1027×10^{-8}. It represent 144 times acceleration ratio. Therefore, the use of ANN to simulate emission concentration would be an original and good option to design an air purification strategy for different streets aspect ratio, and to propose consequently an optimal solution for cleaner city.

6   Conclusion

The goal of this work was to simulate an outdoor air purification system, using different strategies and techniques. Photocatalysis was used in order to reduce the concentration of VOCs in the air. The convection with two types was used to generate the air flow in the system. Several physics and chemistry equations were combined in this system. Modeling can be beneficial to simplify and reduce calculation time, and sometimes it can be the only solution in some cases. Also it's cheaper comparing it with an experiences with real size. In the end of the simulation tasks it was advantageous to use Machine leaning and artificial intelligence to get more results. Artificial intelligence methods allowed finding out optimized purification scenarios for different streets aspect ratio and therefore to design best purification strategies for a green city. The cost of the system was taking into account and it was an important factor to choose between scenarios.

References

1. Künzli, N., Tager, I.B.: Air pollution: from lung to heart. Swiss. Med. Wkly. **135**(47-48), 697–702 (2005)
2. Cohen, A.J., et al.: The global burden of disease due to outdoor air pollution. J. Toxicol. Environ. Health Part A **68**(13–14), 1301–1307 (2005)
3. Kampa, Marilena, Castanas, Elias: Human health effects of air pollution. Environ. Pollut. **151**(2), 362–367 (2008)
4. Nakata, Kazuya, Fujishima, Akira: TiO_2 photocatalysis: design and applications. J. Photochem. Photobiol. C **13**(3), 169–189 (2012)
5. Sherwood, T.K., Pigford, R.L., Wilke, C.R.: Mass Transfer. McGraw-Hill, New York (1975)
6. Sauer, M.L., Ollis, David F.: Photocatalyzed oxidation of ethanol and acetaldehyde in humidified air. J. Catal. **158**(2), 570–582 (1996)
7. Vorontsov, Alexandre V., Dubovitskaya, Vera P.: Selectivity of photocatalytic oxidation of gaseous ethanol over pure and modified TiO_2. J. Catal. **221**(1), 102–109 (2004)
8. van Walsem, J., et al.: CFD investigation of a multi-tube photocatalytic reactor in non-steady-state conditions. Chem. Eng. J. **304**, 808–816 (2016)

Perceptions and Attitudes of the Rural Population of Morocco Towards EcoSan Latrines UDDTs

A. Taouraout[1,2(✉)], A. Chahlaoui[2], D. Belghyti[1], M. Najy[1], I. Taha[2], and A. Kharroubi[3]

[1] Laboratory of Agrophysiology, Biotechnology, Environment and Qualities, Faculty of Sciences, University IbnTofail of Kenitra, BP: 133, 14000 Kenitra, Morocco
aziz73.pam@gmail.com

[2] Laboratory of Natural Resources Management and Development Team, Health and Environment, Faculty of Science, Moulay Ismail University, Meknes, Morocco

[3] RU: Applied Hydrosciences Research Unit, Higher Institute of Water Sciences and Techniques, University of Gabès, Campus universitaire, 6072 Gabès, Tunisia

Abstract. In Morocco, most of the rural population is not connected to sewerage systems and many households and public buildings in rural areas lack any kind of sanitation facilities. This lack of sanitation, constitute a real burden on the health of the population and environment. Traditional methods of defecation such as the simple pit latrines and the open air are still widely used in rural areas of Morocco. In response to resolve this situation, the first EcoSan latrines (UDDTs) were introduced in a rural village called 'Ait Daoud Ou Moussa' in Morocco in December 2009. In this study, a questionnaire survey, a group discussion and an interview with stakeholders in the field of sanitation were taken to assess the socio-cultural acceptance of the EcoSan latrines in this village. Survey results showed that these eco-toilets have been generally accepted by almost all users and non-users. However, the use of excreta after treatment in their agriculture has not been accepted (psychic effect).At the end of this study, we propose some recommendations to encourage the village's population to reuse the treated excreta in their own fields.

Keywords: Acceptance · EcoSan latrines · Rural areas · Morocco

1 Introduction

According to the United Nations [1], 4.9 billion people globally used an improved sanitation facility but 2.4 billion did not. In addition, almost 1 billion people practice open defecation, of whom 90% live in rural areas [2]. Consequently, over than 2000 million tons of human waste go untreated in the environment each year [3]. Indeed, many developing countries do not have the means to generalize costly centralized sewage systems and rely on individual disposal systems (e.g. Pit latrines) and release the greywater in the nature [4]. These practices still to present a major risk to public health, wellbeing and the environment; exposing many people in developing countries to infection and disease [1, 5–7]. According to Jewitt [8] food, drinking and cooking water can be contaminated via human hands and flies. Moreover, Knowledge and practice of critical hygiene behaviors, such as hand washing after toilet use, are also widely lacking especially in the rural areas. Consequently, the ingestion of faecal pathogens from contaminated food and water resources as well as faecal-oral transmission are a leading cause of disease and preventable death, especially in children under five years [9]. Indeed, an estimated 1.5 million preventable deaths each year occur due to a lack of access to clean water and pathogens associated with human excreta [10, 11].

In Morocco for example, most of the rural population is not connected to sewerage systems and many households and public buildings in rural areas lack any kind of sanitation facilities. According to HCP [12], Morocco's rural environment continues to lag behind in the sanitation sector, since the share of households with access to a sewage system is only 2.8% in 2014. The WHO/UNICEF [13] indicates that, the gap between urban and rural sanitation is wide in Morocco. Indeed, in the urban part, 89% of the population had access to sanitation in 2015 and 2% of this population did not use toilets; however, only 38% of the rural population had access to improved sanitation facilities and 19% of this population still went for open defecation in 2015. This insufficient treatment of human excreta is threatening the quality of groundwater and watercourses already scarce water resources, resulting in risks for human health and the environment [12, 13].

The response to these challenges is so-called "ecological sanitation", which is based on the idea of preventing pollution rather than attempting to control it ex post factum; further sanitizing the urine and feces and finally using safe products to improve the agricultural purposes [14].

1.1 Ecological Sanitation (EcoSan)

EcoSan is characterized by a 'sanitize and recycle' process based on preventing pollution, sanitizing urine and feces and recovering nutrients for food production [14, 15]. EcoSan takes an integrated water and natural resource management approach to sanitation. Furthermore, the EcoSan latrine is characterized by the separation of excreta (feces, urine) before their treatment. First, the entire feces must be held for several days at a temperature of more than 50 °C to ensure the destruction of pathogens [16]. The use of ash, wood chips or sawdust reduce unpleasant odors and facilitate drying feces but mainly mitigate the potential exposure to bacteria and other pathogens. So, ach has been found to be more effective in urine-diversion dehydration toilets than sawdust for the destruction of *Escherichia coli* and *Enterococcus* spp.; because it has a high pH and low moisture [17]. This is a relatively cheap option to implement requiring limited tools and resources. In contrast to feces, urine is generally safer to use in agriculture without further treatment but does require storage for up to six months before use [14].

Urine-Diversion Dehydration Toilets (UDDT):

The urine diverting dry toilet (UDDT) is designed to separate Urine and Faeces to allow the Faeces to dehydrate and/or recover the Urine for beneficial use. It operates without water. Figure 1 gives an overview of this system.

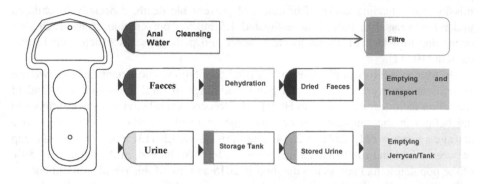

Fig. 1. System overview of a Double Vault Urine-Diverting Dry Toilet (UDDT) [18] (adapted)

This system is a sustainable sanitation option for households in rural areas where reliable water supply and wastewater management is missing. It can be used anywhere, but is especially appropriate for rocky areas where digging is difficult. A dry, hot climate can also considerably contribute to the rapid dehydration of the Faeces. The success of this system depends on the efficient separation of Urine and Faeces, as well as the use of a suitable cover material [18].

Despite all the advantages of UDDT, acceptance of EcoSan systems among users and farmers is a challenge in some contexts due to socio-cultural barriers relating to reuse of excreta [19–21]. Indeed, the successful introduction of such a new technology requires a change in behavior and must be accompanied with awareness raising, training and motivated local partners. The main drivers for people to become a beneficiary for a UDDT were the dissatisfaction with the pit latrine, especially limited comfort and hygiene (cold, windy, smell and flies) [22].

1.2   Psychological Factors in Focus

The acceptance of a new system of sanitation by users is a key to the success of this system and a user/consumer acceptability survey is a must before introducing a new technology. According to Avvannavar and Mani [23], the development of EcoSan programs can be hindered by a number of factors and these include public fear of human excrement or "faecophobia". Therefore, educational strategies (e.g. participatory learning, sharing information and skill training) are needed to raise public awareness, change attitudes and increase acceptance of EcoSan systems and behaviors [20, 24]. Winblad and Simpson-Hebert [14] classified the population into three groups: faecophobic, faecophilic, and the third class is between these two extremes. Indeed, faecophobic cultures are common among people with a tradition of Hinduism and in Africa south of the Sahara. In contrast, China's people used human excreta as fertilizer for several thousand years (faecophilic culture) [25].

Consumer/farmer attitudes and perceptions towards UDDTs in various countries have been investigated by conducting user surveys. Indeed, a survey of Swiss farmers revealed that urine based fertilizer was accepted by 57 and 42% were willing to purchase urine fertilized products [26]. The survey conducted in various focus groups in Switzerland showed that more than 60% expressed the willingness to purchase urine-fertilized products [27]. Similar results were also found in a review of surveys about urine diversion in 7 European countries: 80% liked the idea of urine separation (UDT), 85% regarded urine as a good fertilizer, and 70% would buy food/products grown using urine-based fertilizer [28]. Similar findings have been reported in South India due to improved soil quality and potential cost savings [29]. In contrast, A study conducted in 4 villages (Muslim communities) in the North West Frontier Province in Pakistan revealed that people consider urine less offensive than fecal matter and farmers, no matter how poor they are, usually avoid direct handling and use of excreta in crops but opt to channelize sewage in farms [30]. Another study conducted in peri-urban farming community of Efutu (Ghana) revealed, in general, negative attitudes towards the handling and use of human waste [31]. In these areas, tradition has no use of human excrement in agriculture. Traditionally, feces were dropped on the ground, and the smell of other people's feces was perceived as a warning signal (faecophobic culture).

To the best of our knowledge, no psycho–sociological study on farmer attitudes to the issue has been performed in Morocco. But, in general, the population of Morocco belongs to faecophobic culture because the subject of the defecation is rarely treated in the Moroccan society (taboo).The data published by HCP [12] were very explicit: only 38% of the rural population had access to improved sanitation facilities and 19% of this population still went for open defecation in 2015.

In order to resolve the sanitation problems in rural areas, Morocco has drawn up a national rural sanitation plan (PNAR), whose main objective is to improve sanitation in rural communities. Most of the techniques recommended by this plan are part of ecological sanitation.

1.3   The First Eco-toilets in Morocco

The first EcoSan project promoted in Morocco was installed in rural areas on a pilot scale in 2009. Indeed, the first EcoSan latrines installed were the Urine-Diversion Dehydration Toilet (UDDT) at Ait Daoud Ou Moussa near Ifrane city (latitude: 33 W340 N, longitude: 4550 W and altitude: 1665 m) (see Fig. 2). This village is chosen to be the pilot site of the EcoSan project because, first, it lacked sanitation infrastructure that a Moroccan village could face; Second, it is part of the "Ifrane Natural Park" area to be protected [32].

The project teams developed different strategies and tools to raise awareness and conduct training to support implementation of the EcoSan toilets. These approaches all emphasized the benefits of EcoSan systems for agricultural production as well as the potential health risks associated with open defecation to varying degrees. Furthermore, the purchase of the raw material to build the EcoSan toilets was covered by the project,

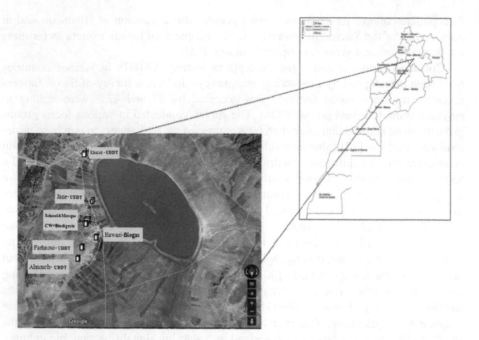

Fig. 2. Position of Ait Daoud Ou Moussa village in Morocco with the principal EcoSan structures

including the necessary equipment. However, the labor was the responsibility of the beneficiary household [32].

EcoSan latrine installed in this village is composed of two pits functioning alternately; EcoSan latrine installed in this village is composed of two pits functioning alternately, a wash-hand basin as well as an urinal and shower (see Fig. 3). The privacy of the users is ensured by a superstructure built using bricks and equipped with stairs serving as access to the cabin [33]. The fresh excrement will be covered with a mixture of ash as a better additive at destruction of *Escherichia coli* and *Enterococcus* spp. [17].

The adaptability and feasibility of diverting toilets as an alternative to conventional sanitation systems seems to be well established. This is evident through the huge number of installations of diverting toilets across the world [34]. In this process, the relatively harmless urine is not contaminated by the feces and can be used directly on crops as fertilizer. Containment and sanitization then form essential building blocks in the barrier against the spread of disease [14]. After numerous years of research and field experience [35] the conclusion reached is that dry methods diverting urine and faces kill pathogens more effectively than other commonly used methods.

1.4   Novelty of This Study

Seven years ago since the installation of the ecological sanitation (EcoSan) project in the village of Ait Daoud Ou Moussa, it is necessary to evaluate the degree of appreciation the new concept of sanitation among the population.

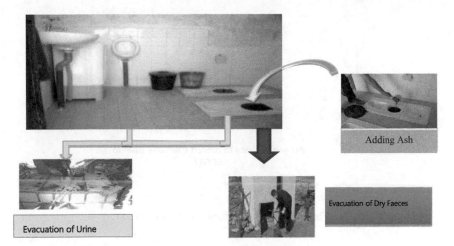

Fig. 3. UDDT installed at Ait Daoud Ou Moussa village

Photo 1. Lost wells: environment and groundwater pollution

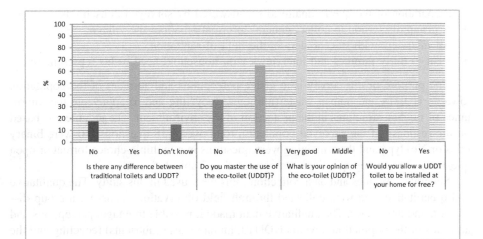

Fig. 4. Evaluation of the functional status of EcoSan latrines

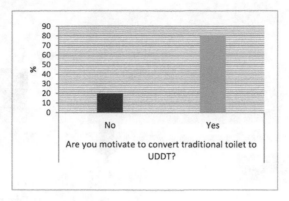

Fig. 5. Convert traditional toilet to UDDT

The main objectives of this study are:

- To assess the socio-cultural acceptance of the EcoSan latrines in this village;
- To propose some solutions to improve the development of this new EcoSan approach in this village and in other rural areas of Morocco.

2 Materiel and Methods

In this study, several materials were used; a summary map of recognition of the place of study, a digital camera for shooting, data sheets of the EcoSan latrines installed in this village, satellite images....

The main steps achieved could be summarized as follows:

- Preparation of the questionnaire to be answered by surveyed households;
- A group discussion to collect general information about the village;
- An interview with stakeholders in the field of sanitation;
- Visit to thirty families and interviews with household head or his representative.

The purpose of this survey is to gauge perceptions and attitudes of the population towards UDDTs and human waste (nutrients) recycling and to appreciate the importance of EcoSan structures in the protection of the environment. The judgment is based on the response of the surveyed population to above of 18 questions (10 were binary (Yes or No) type and the remaining were questions in a multiple-choice format or open question). The survey questions are presented in Table 1.

Several techniques and data collection tools were used in this study. The qualitative and quantitative data were collected through field observation, a survey, a group discussion and interviews. The qualitative data made it possible to gauge: perceptions and attitudes of the population towards UDDTs, human waste (nutrients) recycling and the importance of EcoSan structures in the protection of the environment and the promotion of sustainable development at this village. As for the quantitative data, they were collected at the level of the families and institutions benefiting of this project.

Table 1. The survey questions

	Survey questions	Answer choices
1	Are you beneficed or not of the EcoSan project?	2 (no, yes)
2	Have you a traditional toilet?	2 (no, yes)
3	Do you often use toilet?	2 (no, yes)
4	Do you know what happens and where the toilet wastes go when you flush it down the drain?	2 (no, yes)
5	Have you heard of (UDDT) toilet systems before this project?	2 (no, yes)
6	Is there any difference between traditional toilets and UDDT?	3 (no, yes, don't know)
7	Do you master the use of the eco-toilet (UDDT)?	2 (no, yes)
8	What is your opinion of the eco-toilet (UDDT)?	3 (very good, middle, bad)
9	What is the frequency of maintenance of UDDT per week?	–
10	what are the difficulties you conferred during the first use of UDDT?	–
11	Have you the means to install UDDT at your house?	2 (no, yes)
12	Would you allow a UDDT toilet to be installed at your home for free?	2 (no, yes)
13	Are you motivate to convert traditional toilet to UDDT?	2 (no, yes)
14	Are you aware of the nutrient value of human waste?	3 (no, yes, don't know)
15	What is your information source about the advantage of UDDT?	2 (awareness days, discussions between the villagers)
16	Do you feel the impact of the EcoSan project in your village?	3 (positive, negative, don't know)
17	Sanitized (treated) human waste can be used as fertilizer?	3 (no, yes, don't know)
18	Would you use sanitized human waste as a fertilizer?	2 (no, yes)

About 15% of the village population has been surveyed. The choice of these interviewees was based on the mass and the quality of information that they were likely to provide on this subject of study. Statistical software Excel 2010 was used for all statistical analyses.

3   Results

3.1   Sanitation System at Ait Daoud Ou Moussa

The sanitation system at Ait Daoud Ou Moussa is an autonomous system. Indeed, 76% of the population have a toilet connected to a lost wells and the gray water (shower, dishes, laundry) are released into nature without any treatment (Photo 1). While the remaining 24% do not have the financial possibility to dig the lost wells and defecate in the wild or in stables of cattle to hide the privacy. These results are consistent with those published by HCP [12].

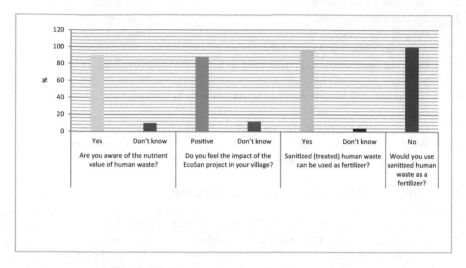

Fig. 6. The importance of sanitized human waste

This village is fed by a network of fountain that is fed with a well managed by a village's association. The average endowment of the population at the fountain is 12.5 L/inhabitant. Unfortunately, this well is exposed to significant sources of pollution such as agricultural pollution and domestic pollution (lack of adequate sanitation) [4]. Therefore, these practices generate an important source of waterborne diseases, especially for children and women taking contact with water, as well as groundwater and surface water pollution [1, 4, 36].

3.2   Acceptability and Adoption of the UDDT at Ait Daoud Ou Moussa

The survey results (see Fig. 4) show that about 65% of the household heads surveyed, master the good functioning of the UDDT of which 70% are informed in the sensitization sessions at the youth center in 2009 [32], while the remaining 30% are informed in the discussions between the villagers. In addition, 85% of the surveyed population gave the agreement for the adoption of the UDDT.

Households with UDDT are impressed by the cleanliness of these toilets, their operation making it easy to manage and handle products (urine, feces ...), reduced water consumption by using ashes instead of hunting water. It has been found an ease of acceptance to step backwards for anal cleansing because even in the wild, people retreat after defecation to move away from their excrement.

It should be noted that the populations of Ait Daoud Ou Moussa give an important place to the maintenance of these toilets. Indeed, it is clearly noticeable since the frequency of maintenance is especially high at household EcoSan latrines (2–3 times per week). This situation shows that people adhere to the logic of improving their living environment by regularly participating in the maintenance of the structures. All eco-toilets visited are relatively well maintained. Furthermore, most respondents say that UDDT is good for rural Morocco.

The using frequency of UDDTs by household members is dependent on sex and time. Indeed, only members available at home (women, girls ...) attend this toilet regularly. On the other hand, men and boys only use them during the night and cold days. While during the times when they are outside the house, they do their needs in nature. This shows that these villagers are used to defecating in nature.

The Households surveyed with no toilet: more than 90% showed a real interest in taking ownership of a UDDT. On the other hand, 80% of those who have lost wells want to convert to UDDT for the following reasons (see Fig. 5):

- Reduction of water consumption;
- Good waste management (urine, feces ...). Even people, who do not want to reuse the waste of these toilets, want to convert to the separation system given the ease of maintenance and emptying of pits, which is difficult in the case of the lost wells;
- Compatibility of the system with rocky soils;
- Reduction the flies and gastrointestinal diseases such as diarrhea, stomach ache, etc.

The results also show that beneficiaries and non-beneficiaries of UDDT hoped to increase the number of EcoSan toilets in their areas in the future. They were even interested in participating in future EcoSan projects in their area to help replicate that kind of toilets by providing labor. However, the adoption of UDDT toilet may be hampered by the low income of the village households. For this purpose, a subsidy must be provided for the construction of this type of toilet.

The survey shows that most respondents are convinced of the beneficial effect of re-using sanitized excreta in soil fertilization. This is mainly due to the good results obtained by demonstration tests which were installed at their village. However, no one of this population was able to retake the same test in his own field (see Fig. 6).

4 Discussions

The main results for this study show a real change in the habits of hygiene and sanitation practices after the introduction of EcoSan latrines in this village. According to the survey, beneficiaries are generally satisfied with the services provided by these structures. In all, it is clear that the EcoSan latrines meet the expectations of the beneficiaries because about 95% positively value this eco-sanitation system and believe that this new technology helps protect the water table and have a healthy living environment and clean. However, some socio-cultural and socio-economic challenges remain to be overcome for the development of this type of sanitation in rural Morocco.

4.1   The Main Challenges of UDDT in Rural Areas of Morocco

Before the installation of the EcoSan project almost everyone in this village expressed their fear due to social and religious beliefs which remain as a major challenge for adoption of the technology in other village of Morocco. Fortunately, the project teams developed different strategies and tools to raise awareness and conduct training to support implementation of the EcoSan toilets at Ait Daoud Ou Moussa. A second challenge is financial: about 97% of respondents consider UDDT are expensive than other toilets (most of the Households surveyed are waiting for the second installment of the project to benefit from these structures). A thirth one is the reuse of excreta after treatment: despite the fact that more than 95% the village's population is convinced of the beneficial effect of the reuse of sanitized excreta in soil fertilization, however, no one of this population was able to retake the same test in his own field (see Fig. 6).

The same result is reported by Drangert and Nawab [30] that raised that in 4 villages (Muslim communities) in the North West Frontier Province in Pakistan revealed that people consider urine less offensive than fecal matter and farmers, no matter how poor they are, usually avoid direct handling and use of excreta in crops. Another study conducted in peri-urban farming community of Efutu (Ghana) revealed, in general, negative attitudes towards the handling and use of human waste [37]. Indeed, according to Winblad and Simpson-Hebert [14] faecophobic cultures are common among people with a tradition of Hinduism and in Africa south of the Sahara. In these areas, tradition has no use of human excrement in agriculture. At this point, then, there is a purely psychological problem that must be dealt by the actions of explanation and sensitization of the population.

4.2   Some Approaches to Solution

In order to overcome the main challenges mentioned above, it is important to:

- strengthen the public awareness system on the importance and effective implementation of the measures relating to the use of EcoSan structures;
- generalize EcoSan toilets in this village to minimize criticism and reduce the psychic effect in terms of the use of excreta valued in agriculture;
- favor institutions such as: schools, health centers, public places, etc. in the realization of these EcoSan toilets;
- develop environmentally-friendly, on-demand, affordable and sustainable sanitation systems in the local context, in collaboration with beneficiary populations and other development actors;
- create an institutional environment conducive to innovative approaches to sanitation, by integrating in the legislative framework, the national policy, the development plan of all the rural communes of Morocco the duplicate "sanitize and recycle";
- support and promote the participation of the State and the local private sector in the provision of sanitation services, particularly with regard to ecological systems;
- develop the financing options for households: alternative financing approaches, such as micro-finance could be developed [38]. Public funds can also be used to pay for loans (for subsidized rates), spreading the cost of a latrine over a few years, which can be a catalyst for increased coverage [39].

5 Conclusion

The main results for this study showed that these eco-toilets have been generally accepted by almost all users and non-users. However, the reuse of excreta after treatment in agriculture has not accepted because this population is afraid of reputability of their production (psychic effect). Indeed, the choice solution of sanitation problem is sometimes more socio-cultural than technical or economic. The actions of explanation and sensitization of the population, especially in areas using individual and autonomous systems, must be improved through all conventional channels (media, rally, relay persons, etc.), but also, on every occasion, by the different associative actors, the politicians, the representatives of the administrations, etc. On the other hand, the various administrative and collective buildings (schools, dispensaries, mosques, "hammams", etc.) will have to be equipped with adequate systems, thus setting an example for the population and allowing their acceptance to be accepted.

At the end, the success of a sanitation project in the rural areas requires, on the one hand, the effective participation of the population concerned in the different phases of the planning process, and the taking into account of its opinions on the choice of technologies will sometimes be fundamental.

Acknowledgements. We would like to thank the survey participants for providing critical information and all who contributed directly or indirectly to this study.

References

1. United Nations: Agenda items 5, 6 and 18 (a) High-level segment High-level political forum on sustainable development, convened under the auspices of the Economic and Social Council Economic and environmental questions: sustainable development Progress towards the Sustainable Development Goals Report of the Secretary-General (2016)
2. UNICEF and WHO: Progress on sanitation and drinking water: 2015 update and MDG assessment. UNICEF and World Health Organization (2015)
3. UN-HABITAT: State of the World's Cities—Harmonious Cities! 2008/9. UN-HABITAT (2009)
4. Taouraout, A., Chahlaoui, A., Belghyti, D., Mansouri, B., Kharroubi, A.: The evaluation of the impact of the first EcoSan project installed in Morocco. Adv. Res. J. Multi-Disc. Discov. I 26.0(I). ISSN NO.: 2456-1045
5. Watson, R.T., Zakri, A.H.: Living beyond our means. Natural Assets and Human Well-being. Statement from the Board. Millennium Ecosystem Assessment (2008)
6. Sutton, M.A., Howard, C.M., Willem Erisman, J., Billen, G., Bleeker, A., Grennfelt, P., van Grinsven, H., Grizzetti, B. (eds.) European Nitrogen Assessment: Sources, Effects and Policy Perspectives. University of Cambridge Press, Cambridge (2011)
7. Galloway, J.N., Dentener, F.J., Capone, D.G., Boyer, E.W., Howarth, R.W., Seitzinger, S.P., Asner, G.P., Cleveland, C.C., Green, P.A., Holland, E.A., et al.: Nitrogen cycles: past, present, and future. Biogeochemistry **70**, 153–226 (2004)
8. Jewitt, S.: Poo gurus? Researching the threats and opportunities presented by human waste. Appl. Geogr. **31**, 761–769 (2011). This paper provides a good overview of the importance of human waste (as both a threat and an opportunity) in different spatial, historical and cultural contexts and highlights potential areas of interest for applied geographical research in the future

9. UNICEF and WHO: Progress on drinking water and sanitation. 2012 update. WHO/UNICEF Joint Monitoring Programme for Water Supply and Sanitation (2012). http://www.wssinfo.org/fileadmin/user_upload/resources/JMP-report-2012-en.pdf
10. UNICEF/WHO: World Health Organization and United Nations Children's Fund Joint Monitoring Programme for Water Supply and Sanitation (JMP): Progress on Drinking Water and Sanitation: Special Focus on Sanitation (2008)
11. Farmer, A.M.: Reducing phosphate discharges: the role of the 1991 EC Urban Wastewater Treatment Directive. Water Sci. Technol. **44**, 41–48 (2001)
12. High Commissioner of Planning: The Morocco between millennium goals for the development and objectives of sustainable development achievements and challenges. National Report, Aug 2015
13. WHO/UNICEF: Drinking Water, Sanitation and Hygiene Update and SDG Baselines (2017)
14. Winblad, U., Simpson-Hebert, M.: Ecological Sanitation—Revised and Enlarged Edition. Stockholm Environment Institute (2004)
15. Werner, C., Panesar, A., Rüd, S.B., Olt, C.U.: Ecological sanitation: principles, technologies and project examples for sustainable wastewater and excreta management. Desalination **248**, 392–401 (2009)
16. WHO: Guidelines for the Safe Use of Wastewater, Excreta and Greywater Use in Agriculture and Aquaculture. World Health Organisation (2006)
17. Niwagaba, C., Kulabako, R.N., Mugala, P., Jönsson, H.: Comparing microbial die-off in separately collected faeces with ash and sawdust additives. Waste Manage. **29**, 2214–2219 (2009)
18. Tilley, E., Ulrich, L., Lüthi, C., Reymond, Ph., Zurbrügg, C.: Compendium of sanitation systems and technologies, 2nd revised edn. Swiss Federal Institute of Aquatic Science and Technology (Eawag), Dübendorf, Switzerland (2014)
19. Andersson, E.: Turning waste into value: using human urine to enrich soils for sustainable food production in Uganda. J. Clean. Prod. **96**, 290298 (2015). https://doi.org/10.1016/j.jclepro.2014.01.070
20. Dellstrom Rosenquist, L.E.: A psychosocial analysis of the human-sanitation nexus. J. Environ. Psychol. **25**, 335–346 (2005)
21. Nawab, B., Nyborg, I.L.P., Esser, K.B., Jenssen, P.D.: Cultural preferences in designing ecological sanitation systems in North West Frontier Province, Pakistan. J. Environ. Psychol. **26**, 236–246. http://dx.doi.org/10.1016/j.jenvp.2006.07.005
22. Wendland, C., Deegener, S., Jorritsma, F.: Experiences with urine diverting dry toilets (UDDTs) for households, schools and kindergarten in Eastern Europe, the Caucasus and Central Asia (EECCA). Issue 6 (2011)
23. Avvannavar, S.M., Mani, M.: A conceptual model of people's approach to sanitation. Sci. Total Environ. **390**, 1–12 (2008)
24. Kjellén, M., Pensulo, C., Nordqvist, P., Fogde, M.: Global review of sanitation system trends and interactions with menstrual management practices. Stockholm Environment Institute (2011)
25. Winblad, U., Kilama, W.: Sanitation Without Water. Macmillan, London
26. Lienert, J., Haller, M., Berner, A., Stauffacher, M., Larsen, T.A.: How farmers in Switzerland perceive fertilizers from recycled anthropogenic nutrients (urine). Water Sci. Technol. **2003**(48), 47–56 (1985)
27. Pahl-Wostl, C., Schonborn, A., Willi, N., Muncke, J., Larsen, T.A.: Investigating consumer attitudes towards the new technology of urine separation. Water Sci. Technol. **48**, 57–65 (2003)
28. Lienert, J., Larsen, T.A.: High acceptance of urine source separation in seven European countries: a review. Environ. Sci. Technol. **44**, 556–566 (2010)

29. Simha, P., Lalander, C., Vinnerås, B., Ganesapillai, M.: Farmer attitudes and perceptions to the re-use of fertiliser products from resource-oriented sanitation systems—the case of Vellore, South India. Sci. Total Environ. **581**, 885–896 (2017). https://doi.org/10.1016/j. scitotenv.2017.01.044

30. Drangert, J.-O., Nawab, B.: A cultural-spatial analysis of excreting, recirculation of human excreta and health—the case of North West Frontier Province, Pakistan. Health Place **17**, 57–66 (2011)

31. Mariwah, S., Drangert, J.-O.: Community perceptions of human excreta as fertilizer in peri-urban agriculture in Ghana. Waste Manage. Res. **29**(8), 815–822 (2011). https://doi.org/10. 1177/0734242x10390073

32. Abarghaz, Y.: Assainissement écologique rurale - Projet pilote du douar de DayetIfrah. Master en génie et gestion de l'eau et environnement, p. 88. Faculté des sciences Rabat, Maroc (2009)

33. Gonidanga, S. B.: Contribution to the implementation of the ecological sanitation (Ecosan) in the African context: Study of the hygienisation process of the urine for a healthy use in agriculture. Thesis, EcolePolytechniqueFédérale de Lausanne, Switzerland, pp. 15–20 (2004)

34. Simha, P., Ganesapillai, M.: Ecological Sanitation and nutrient recovery from human urine: How far have we come? A review. Sustain. Environ. (2016). www.journals.elsevier.com/ sustainableenvironmentresearch/

35. Schönning, C., Stenström, T.-A.: Guidelines on the Safe Use of Urine and Faeces in Ecological Sanitation Systems. Book January 2004with16 Reads. ISBN: 9188714934

36. Esrey, S., Gough, J., Rapaport, D., Sawyer, R., Simpson-Hébert, M., Vargas, J., Winblad, U.: Ecological Sanitation. SIDA, Stockholm (1998)

37. Abeyuriya, K., Mitchell, C., White, S.: Can corporate social responsibility resolve the sanitation question in developing Asian countries? Ecol. Econ. **62**, 174–183 (2007)

38. Uddin, S.M.N., Tempel, A., Adamowski, J.F., Lapegue, J., Li, Z., Mang, H.-P.: Exploring alternative sources of funding for deploying sustainable sanitation technologies and services in Mongolia. Int. J. Water Resour. Dev. **32**, 881–894 (2016). https://doi.org/10.1080/ 07900627.2015.1121137

39. Trémolet, S., Kolsky, P., Perez, E.: Financing on-site sanitation for the poor. A six country comparative review and analysis (2010)

Smart Cities and Entrepreneurship: A New Challenge for Universities

Domingos Santos[✉]

Polytechnic Institute of Castelo Branco; CICS.NOVA—Social Sciences
Interdisciplinary Centre, Av. Pedro Álvares Cabral, 12, 6000-084 Castelo
Branco, Portugal
domingos.santos@ipcb.pt

Abstract. The world population is increasingly urbanized. There is a need to
analyze current smart urban challenges and opportunities, and to study how to
incorporate smart systems into an agenda of urban competitiveness. The smart
city approach is a recent theme, aiming at improving the quality of life of
citizens in urban territories. Born like a bottom-up movement, it is now
becoming critical in urban planning and development in cities all over the world.
The smart city success depends on how cities achieve higher innovation profiles
and competitiveness levels—on being entrepreneurial smart cities. Universities
are one of the most significant cities' agents on building up active strategies to
reinforce an entrepreneurial urban dynamics. Grounded on a vast and deep
literature review on both scientific papers and practitioner or institutional
reports, the chapter tries to address the role the entrepreneurial university might
bring to the formulation and implementation of smart cities strategies, pointing
out to some instruments that can be used to enhance urban competitiveness and
sustainability.

Keywords: Smart city · Entrepreneurial university · Entrepreneurship
Innovation · Urban policy

1 Introduction

During the last three decades, the urbanization has expanded all over the world. People
moves from rural and more peripheral areas to cities to find better chances for living,
working, studying and developing their entrepreneurial projects. Cities have always
been economic and cultural poles, but recently they have been augmenting their
political prominence and their capability to contribute to the wealth of nations. City
attractiveness is a competitive factor differentiating the better urban areas and driving
localization of RD&I infrastructures, firms, cultural entities, and so on [46, 54]. The
smart city approach emerged as an urban strategy that using high technology, and
especially ICT, for supporting city attractiveness, as well as upgrading their cohesion
levels, their competitiveness profiles and their sustainability. Competitiveness on a
global scale is demanding new answers from the territories. The challenges cities and
regions are facing require they must be capable of developing additional key competences in the use of their assets and strengths. Smart cities are a new approach to the

© Springer Nature Switzerland AG 2019
M. Ben Ahmed et al. (Eds.): SCA 2018, LNITI, pp. 118–131, 2019.
https://doi.org/10.1007/978-3-030-11196-0_12

problem of urban competitiveness and differentiation in the global arena, assuming that the "one size fits all" archetype cannot be the solution to the problem [53, 57, 62, 70].

Up to now, smart cities have been implemented mainly applying a spontaneous, bottom-up process; municipalities, enterprises, not-for-profit organizations and the citizens themselves pursue the smartness of their city suggesting or directly implementing smart projects, initiatives, solutions. Although involved on an ad hoc basis, universities, nonetheless, have been giving important contributions as S&T providers as well as partners a diverse range of projects.

To accomplish this, higher education institutions are key partners as a drive to innovation, knowledge and social capital building and valuation. The links between the entrepreneurial university paradigm and the implementation of smart specialization strategies are thus discussed as a means to deepen the understanding of this institutional, entrepreneurial and territorial dynamics that is evolving towards a new competitive paradigm based on knowledge, innovation and differentiation.

The aim of this chapter is three-fold. On the one hand, the first objective of the chapter is to contextualize and clarify the concept of smart city that has been gaining a growing importance in terms of urban policy. On the other hand, the chapter will also address, both theoretically and operationally, the challenges higher education institutions are facing in the context of smart city strategies—this will be done through the theoretical and methodological lens of the entrepreneurial university paradigm. The third objective is to formulate a set of recommendations about the role of entrepreneurial universities might bring to a better design and implementation of smart city strategies.

2   Smart Cities: The Entrepreneurial Challenge

The tag Smart City has appeared in recent years as both a policy and business notion, and is gradually growing also into a full-fledged research topic. Initial characterizations of this policy construct mainly focused on ICTs as the main support around which a city should build its smart upgrading trajectory. In fact, the Smart City concept has been used to identify a large spectrum of heterogeneous solutions and city programs, involving different types of technologies and aiming to reach a very large set of different and not well-defined goals [27, 76].

Yet, it seems crucial to ensure that discussions over smart urban planning and development do not become entrenched in an equivocal idea that smart cities are necessarily positive advances in the urban domain [1, 3, 51]. It is important for key concepts such as *smart* not to be simply void signifiers, and to question and discuss what is smart in a way which allows urban politics to prosper, and not to be removed from the need to rethink urban competitiveness.

Our core argument resides on the apparent assumption that digital technology is being seen as an end on itself, assuming that the ICT investments will automatically and optimistically turn the cities into more innovative and competitive world arenas. Some authors already criticize the deficit of social and governance matters on the construction of the smart city model. It is, nonetheless, uncommon to question the importance that smart city success also needs to address innovation and entrepreneurial interventions [43, 44]. Till now, on this ambit, the smart city basically configures a classic top-down approach similar to the linear innovation model, which prioritizes S&T research as the basis of innovation, and plays down the role of political, academic and economic players in the innovation process [2, 5, 7, 13, 25, 36, 58, 61].

We, consequently, characterize a smart city as one with a comprehensive commitment to innovation and entrepreneurship in technology, as well as in management and policy making. There is a gap in the orthodox literature of the smart city. The entrepreneurial smart city approach must have the central task of stimulating the competitiveness of the productive system in a context of globalization of economic relations and the acquisition of competitive advantages resulting from the ability to innovate [72]. In fact, it should be understood as a means to reinforce urban competitiveness. This way, the Smart City concept it is not an end point, but rather a process. Thus, business-led urban development has to be emphasized as a smart city key characteristic. A gradual transformation in the urban governance from a managerial to an entrepreneurial focus has to be accomplished. After all, there is a critical need for businesses in a smart city: public investments are often too marginal to be effective for a cost-intensive smart and competitive urban growth. Companies representing private capital markets are needed to supply the city with enough amount of money. These companies comprise small- and medium-sized enterprises (SMEs) as well as large corporations. To attract them, city governments have to supply advantageous conditions for businesses [4, 12, 18, 26, 39, 41, 50]. The necessity for constant private capital should make the smart city attractive for new businesses and what they mean for a smart economy. This is based on the idea of a smart city offering an innovative spirit, which is particularly important for entrepreneurship [11, 15, 30, 55, 76], making a smart city an entrepreneurial *locus* which delivers new economic opportunities. Smart cities must increasingly function as seedbeds for creativeness, innovation and entrepreneurship [20, 29]. Knowledge spillovers are likely to arise as the availability of above the average skilled labour force is high. To implement a Smart City initiative does not only mean to reach technological success, but to utilize technology to create public and economic value.

As the economy evolves into new models, more and more knowledge-intensive, that may erode traditional manufacturing jobs and those requiring routine cognitive skills, the establishment of new sources of employment and growth is paramount to maintaining competitiveness, reducing poverty, and increasing shared prosperity [19]. That is the reason why universities are critical in guaranteeing that the city competitiveness is pursued by their active role as *glocal* stakeholders, assuring simultaneously its embeddedness on the urban set and its pivotal role on the networking at an international scale.

3   Entrepreneurial Universities: An Emergent Paradigm

Nowadays, with the increasing global competitiveness of economies, universities are called to assist more directly in the new challenges that cities face, to become an important actor and partner [36, 37]. In fact, higher education is facing unprecedented challenges in the definition of its purpose, role, organisation and scope in society and the economy. The information and communication technology revolution, the emergence of the knowledge economy, the turbulence of the economy and consequent funding conditions have all thrown new light and new demands on higher education systems all over the world.

Universities are also challenged to be entrepreneurial. Fostering the entrepreneurial potential of a university is a major challenge which requires an approach that identifies knowledge as an asset which can be created, developed, transmitted and valued. Etzkowitz [27] adopted the designation of entrepreneurial university to precisely describe the many changes that reflect a more active role of universities in the direct and indirect promotion of knowledge transfer and scientific and technological know-how that is associated with academic research but also in the processes of collective learning, in a proactive attitude towards change and an anticipative and prospective vision. More concretely, an entrepreneurial university is considerate such as an organization that adopts an entrepreneurial management style, with members (faculty, students, and staff members) who act entrepreneurially, and that interacts with its outside milieu (community/region) in an entrepreneurial modus [21, 39], assuming innovative profiles throughout its research, knowledge exchange, teaching and learning, governance and external relations.

Traditional studies of entrepreneurial university tend to take a narrow view of industry-university relations, concentrating on the commercialization of research outcomes and on mechanisms of technology transfer, such as science parks and incubators, liaison offices, or intellectual property [38, 42, 75]. It is important, though, to stretch the role of the universities in urban innovation beyond just being a source of indigenous R&D city capability [73].

The development of an entrepreneurial higher education institution can be achieved in three ways that are not alternatives but rather complementary and interlocking. First, promoting entrepreneurship in universities involves creating a sense and a culture of risk, search and discovery. The promotion of this institutional cultural change necessarily comprises, for instance, innovation in teaching and learning but also in other organizational dimensions, like support services and reward systems.

Universities can thus create a virtuous circle in which the requirement of teachers may be oriented towards more creative, innovative and appealing environments and teaching and learning processes [20, 34, 48, 52]. It is also essential that an entrepreneurial culture may be stimulated by way of training programs on detection of market opportunities, implementing business solutions or the promotion of new start-ups [17, 22]. Entrepreneurship education is now so necessary to a student from management or economics, as for a student of humanities, social sciences or engineering. The awareness, the creation of training packages, modules and courses should be clearly oriented to provide frameworks and tools that facilitate the creation of conditions for self-employment and intrapreneurship.

Finally, a third dimension requires higher education institutions, as organizations, to be entrepreneurial, not only in their training programs, but also in the way they operate. They should, for example, be able to commercialize some of their assets, which are based on the production of codified, scientific and technological knowledge. Without jeopardizing its primary mission, universities need to rethink their offer and their markets [2, 7, 25, 45, 69]. It is fundamental that universities avoid being constrained by their own organisational and functional structures, making it more difficult to carry out the types of entrepreneurial activities which support their strategic objectives.

These three converging paths to develop an entrepreneurial university approach are, all of them, very important, as summarized in Fig. 1.

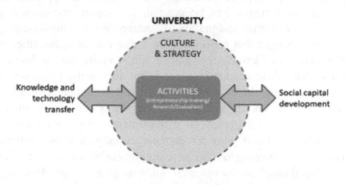

Fig. 1. Entrepreneurial university paradigm [64]

The growing social pressure on universities to extend their traditional missions and adopt a more proactive involvement in the economic development of their hosting territory led them to the definition of a third mission—a challenge that the concept of entrepreneurial university tries to incorporate. The defy being that universities, in the context of a thorough redesign of the pillars that have founded the society and the knowledge economy, may give a complete answer and be up to the challenges that are permanently raised. It is, in fact, also a matter of survival: the need for adaptation to the rapidly changing world cannot stay apart of their missions, as sometimes they were rightly accused—the so-called ivory tower syndrome [28, 29, 35].

4   Strengthening Smart City Entrepreneurship: University Policies and Instruments

As contended, entrepreneurial universities can be of great significance to the implementation and success of smart cities strategies. Because of their crucial role in shaping urban economy through knowledge advancement, universities will have an increasing role to play in the making and development of these competitive local innovation ecosystems and in assuring sustainable trajectories of their territories in the future.

The construction of an university with an entrepreneurial profile clearly seems to constitute an organizational and incremental innovation process—the relentless pursuit of an entrepreneurial culture that is incessantly attuned to the societal and economic challenges, ensuring the passage of a *non-schumpeterian* neo-classical paradigm to a *schumpeterian* model of entrepreneurship, whose first strategic aim lies on the innovation production, in its multiple modalities [26, 33, 56]. If higher education institutions do not become agents of innovation, i.e. entrepreneurial institutions, they certainly will become more part of the problem (the *status quo*) than part of the solution in terms of smart city development [6, 9, 30, 31, 63].

The definition of a smart city strategy requires an increasing role and an active participation and involvement from universities (and other key players), and, thus, we argue that the entrepreneurial university paradigm can be useful in this process. Table 1 tries to capture the perceived influence of the entrepreneurial university paradigm and the smart city capabilities impact. Besides, universities can play a key role in defining a smart city strategy by contributing to a rigorous assessment of the region's knowledge assets, capabilities and competencies, including those embedded in the university's own departments as well as local businesses [10, 23, 30, 32, 68].

Table 1. The entrepreneurial university and its impacts on smart cities

Entrepreneurial university functions	Smart city impacts
High level labour force provision	Retaining potential labor pools; assuring the normal flux of specialized labor to meet the dynamics of the business demand
Talent attraction	Attracting talents (faculties, R&D labs) who are entrepreneurially minded; promotion of the international networking
Business collaboration	Attracting cutting-edge industrial R&D cooperation with universities; FDI investment attraction
Spin-offs/transfer	Enabling or facilitating high-tech venture creation by the university; promotion of intellectual property rights
Entrepreneurs	Creating the favourable milieu for emerging talented graduates (professors and researchers) with entrepreneurial mindsets; launching new technology-based firms; increasing the city potential for disruptive economic trajectories
Venture ecosystem	Attracting risk and venture capital/angel investors & other venture professionals; network promotion; reinforcing city capability to invest in innovation
Bridging initiatives	Creating and professionalising TTOs; implementing S&T parks; collaborating in clusters
Governance	Inducing new formats of a more collaborative and participative smart city policy making; reinforcing the strategic role of entrepreneurship and innovation as flagships of smart cities
Territorial marketing	Improving smart cities image; reinforcing the attractiveness of smart cities

Certainly that alongside the more traditional established universities are newer universities, mainly universities of applied sciences, created with a clear territorial mission and solid links to local businesses. These often have a deficit of research capacity to produce new knowledge to underpin a high technology strategy and new technology-based firms creation but nevertheless are able to play a key role in supporting the diffusion of general purpose technologies especially concerning traditional industries, an approach which may be the smartest way ahead for some cities in need of productive restructuring [8, 17, 22, 47, 59, 60].

Although in this chapter the focus is centered on the role of universities, one cannot ignore the role of other stakeholders in the success of smart cities. In fact, this is not just a process internal to the university but one where a pluralistic approach is requires in providing access to internal and external opportunities and expertise. More than an entrepreneurial university alone, an entire entrepreneurial ecosystem must be the basis upon which innovation networks continually renew territorial assets, helping therefore (re)affirming city competitiveness and sustainability.

University leaders are working more and more closely with a wider spectrum of community stakeholders—city government, companies, venture capitalists, entrepreneurs, and workers—to improve access to university based assets and to implement urban innovation and economic strategies. Universities use a variety of collaborative models, including infrastructural answers, such as research parks, university corridors, startup accelerators, shared laboratory space, incubators, and innovation and manufacturing clusters. These venues try to bring together infrastructure and intellectual capital to address innovation and business challenges and to develop urban economic dynamics [9, 23]. Therefore, the role of innovation policies and, especially, the tools used to promote companies and institutions' ability to innovate do not solely depend on the entrepreneurs, as also communities, and especially cities, have an effect on innovation processes.

If we agree that the intervention by the city authorities should give priority to the implementation and strengthening of a relational culture, then policies have to comply with the existing overall network architecture and its specific territorial assets, rather than focus more on punctual and atomized actions. It should aim, thus, to reinforce the mechanisms for horizontal coordination and partnership, as well as interface management, avoiding political intervention supported in sectorial logics or fragmented actions [10, 29, 44]. Networking, design of value-added dialogue platforms and the opening up of new interfaces between innovation support infrastructures and industry, such supply aspects should thus be fostered, particularly between private and public spheres.

Many knowledge intensive structures may surround universities such as incubators and science parks—Smart Cities, on their global strategies, should have mechanisms in place to capitalise on this knowledge stock [34, 41]. These efforts, if effectively articulated, provide a cost-effective and productive means for conducting research, developing technology, and spurring new markets and businesses.

The rise of smart cities brings many new opportunities for higher education, and the different stakeholders they serve. Universities must work in close collaboration with city leaders and businesses to develop and test new smart city technologies, turn ideas into prototypes and unlock and analyze urban data for the good of the civic and economic communities. This is not just a process internal to the university but one where a pluralistic approach is indispensable in providing access to internal and external opportunities and expertise [3, 39]. The result will be a city that is more attractive to work and live in and is driven by a strong knowledge economy. It may be assumed that regional partnership organisations can effectively articulate territorial visions and are hence the best arenas for cooperative relations.

There is, nevertheless, the need to observe that partnership governance might hinder resolution of the conflicts that emerge reconciling the perspectives of different parties. Allied to this is that there may be a lack of vision amongst the partners that can produce a lowest common denominator strategic vision rather than aggressively targeting available resources to build up creative and differentiated urban environments [23, 31, 51]. The potential danger of institutional misalignment or drift should oblige to adopt a progressive and pedagogical learning framework—this really seems the context for an experimental policy design, implementation and evaluation.

As contended, entrepreneurial universities can be of great significance to the implementation and success of smart cities strategies. They can be protagonists on building an environment that is more accepting of failed businesses and encourages entrepreneurs to take meaningful risks to scale their start-ups. Cities that create effective positions for promoting and supporting innovation invest them with real problem-solving capabilities as well as the resources they need to effect change. This sends a clear message to the entrepreneurial community that the city government is serious about boosting innovation-driven growth [22, 34, 48].

There is a need, to promote city innovators and recognize entrepreneurs as legitimate experts in the area where they work. It is absolutely vital to support them and help upscale innovative and successful urbanl actions. To do all this, it is also urgent to reconsider Smart Cities strategies towards the urban world. How to have a positive approach, that starts from, but is not restricted to local ideas, which focuses on local communities's strengths and explores the particular opportunities open to them—rather than dwelling on their weaknesses and problems? How to move from problem-based programs towards stregthening the solutions to be found in city communities, building upon existing stregths and initiatives and supporting innovation? How can institutional resistance, experienced as a brake to policy changes, be overcome? Urban reality is not what it was, is constantly changing, that is for sure, however, some dimensions that possess a strucutral dimension do have to be adequately addressed, with innovative tools and strategies—higher education institutions are, in this context, a *sine qua non* condition for the success of such strategies [17, 37, 41, 74].

Because of their crucial role in shaping society through knowledge advancement, entrepreneurial universities will have an increasing role to play in the making and development of these competitive urban ecosystems and in assuring sustainable growth trajectories. Furthermore, the theoretical approach behind the interplay between the smart city and the entrepreneurial university is very promising—it needs, however, additional attention for it still seems to need a more intense and coherent theoretical background that may allow to provide more effective instruments for public policy intervention [14, 40, 49, 67]. On this ambit, the research agenda should also encompass a wider perspective, in order to diminish the gap that still exists between the theoretical approach of entrepreneurial universities and the way they may accomplish a growing role as drivers of a more innovative, competitive and sustainable city development trajectory.

Special attention should be paid to the design of the intervention policy, trying to avoid the classical functional top-down and supply-side approach, the classical repertoire of some innovation policies; on this ambit, innovation-led smart city policies must basically address the questions of enhancing the territorial capabilities to foster interaction among urban actors, of engaging them in processes of collective learning conducive to entrepreneurship and innovation.

5 Conclusion

Universities are, nowadays, increasingly scrutinized by society—their missions and their roles are being quizzed. There is a growing social force for universities to reexamine and redesign their traditional approaches to the challenges global knowledge economy faces. Policy-makers must acknowledge the interdependencies between universities and the cities, aiming at the alignment of universities' and urban authorities' strategies. While urban development organizations can contain their activities within specific boundaries, universities are porous and permeable institutions, with highly mobile staff and student populations.

Universities operate in national and international arenas for research, staff and student recruitment, knowledge production and dissemination, business creation. This should be recognized as an anchor and a key asset for the smart city, in fact, one of the main innovation drivers of city competitiveness. There is an increasing recognition that universities need to be more flexible and more adaptable to contemporary social, economic and environmental challenges.

Urban competitiveness policy in the knowledge economy has consequently become focused on attracting and developing knowledge-intensive services and technology-based industries, talented people, and the amenities that help attracting and retaining qualified human resources and firms. Increasingly this includes recognition of universities' roles, as key knowledge institutions, attractors of talent and contributors to civic amenity. City authorities are therefore increasingly looking to harness higher education institutions to support their economic development ambitions both as global players and urban providers of knowledge and skills.

However, trying to restrict the activities of universities to the needs of the immediate place would be to the advantage of neither [65, 66, 74]. It should be accepted that some of the activities and outcomes will inevitably locate elsewhere and accordingly spill over urban geographic borders. This can also work to the benefit of cities as universities can act as magnets to capture knowledge, skills, investments and networks into the local area from other territories.

The entrepreneurial university constitutes a paradigm transformation that is suggested to more adequately address these challenges, where culture, teaching and valuation of existing research and knowledge have a central role to play in terms of local and regional development [53, 71]. Regarding smart city strategic planning, the challenge is how to mobilize the strengths of the university effectively to support the city while building sufficient flexibilities to urban strategies to admit a certain amount of leakage in activities beyond its geographic boundaries. Only a few cities have comprehensive strategies to support innovation and entrepreneurship. While it's possible to develop sound policies on an ad hoc basis, sustaining them in a characteristically complex, politically charged city setting usually involves something more: an established strategy that articulates how activities across city government are organized and coordinated, and who is accountable for that collective effort [13].

Thus, it seems very important encouraging and empowering city authorities to develop the vision and leadership to provide solutions to their own problems. However, there must be some caution. While it is certainly instructive to examine and learn from successful Smart Cities, policy makers should be wary about treating them as exemplars that can be easily replicated or emulated in their own urban settings. Policies rarely travel well: successful strategies in one city do not transplant easily into other territories. In fact, given that many of the sources of city competitive advantage are locally rooted and embedded, policies necessarily have to respond to, and take account of, urban idiosyncrasies. It is unlikely that there is one size fits all recipe for promoting city entrepreneurship, innovation and competitiveness.

In a nutshell, smart universities increase the chance of having smarter cities—but the relationship is naturally expected to be reciprocal. The entrepreneurial smart city concept is therefore a challenge for the universities that needs to be constantly and adequately addressed.

References

1. Ahvenniemi, H., Huovila, A., Pinto-Seppä, I., Airaksinen, M.: What are the differences between sustainable and smart cities? Cities **60**(A), 234–245 (2017)
2. Albino, V., Berardi, U., Dangelico, R.M.: Smart cities: definitions, dimensions, performance, and initiatives. J. Urban Technol. **22**(1), 1–19 (2015)
3. Allwinkle, S., Cruickshank, P.: Creating smarter cities: an overview. J. Urban Technol. **18** (2), 1–16 (2011)
4. Angelidou, M.: Smart cities: a conjuncture of four forces. Cities **47**, 95–106 (2015)
5. Anholt, S.: Competitive Identity: The New Brand Management of Nations, Cities and Regions. Palgrave, New York (2007)
6. Anthony, A.: Entrepreneurial universities and regional development: policy origins, progress, and the future, with a focus on Poland. Polish Polit. Sci. Rev. **2**(1), 70–83 (2014)
7. Arancegui, M., Querejeta, M.J., Montero, E.: Las estrategias de especialización inteligente: una estrategia territorial para las regiones. Cuadernos de Gestión **12**, 27–49 (2012)
8. Asheim, B., Grillitsch, M., Trippl, M.: Specialization as an Innovation-driven Strategy for Economic Diversification: Examples from Scandinavian Regions. Lund University, Lund (2016)
9. Athey, G., Nathan, M., Webber, C., Mahroum, S.: Innovation and the city. Innov. Manage. Policy Practice **10**(2–3), 156–169 (2008)
10. Audretsch, D.: From the entrepreneurial university to the university for the entrepreneurial society. J. Technol. Transfer **39**(3), 313–321 (2012)
11. Borja, J.: Counterpoint: intelligent cities and innovative cities. Universitat Oberta de Catalunya Papers E-J. Knowl. Soc. **5** (2007). Available at http://www.uoc.edu/uocpapers/5/dt/eng/mitchell.pdf
12. Boulton, A., Brunn, S.D., Devriendt, L.: Cyberinfrastructures and "smart" world cities: Physical, human, and soft infrastructures. In: Taylor, P., Derudder, B., Hoyler, M., Witlox, F. (eds.) International Handbook of Globalization and World Cities. Edward Elgar, Cheltenham (2011)
13. Bradford, N.: Place matters and multi-level governance: perspectives on a new urban policy paradigm. Policy Opt. **25**(2), 39–45 (2004)

14. Camagni, R., Capello, R., Lenzi, C.A.: Territorial taxonomy of innovative regions and the European regional policy reform: smart innovation policies. Scienze Regionalli **1**(37), 69–105 (2014)
15. Caragliu, A., Del Bo, C., Nijkamp.: P. Smart cities in Europe. In: Proceedings of the 3rd Central European Conference in Regional Science. Košice, Slovak Republic, 7–9 Oct 2009
16. Cho, M.-H.: Technological catch-up and the role of universities: South Korea's innovation-based growth explained through the Corporate Helix model. Triple Helix J. Univ. Ind. Govern. Innov. Entrepren. **1**(2) (2014)
17. Couchman, P. K., McLoughlin, I., Charles, D. R.: Lost in translation? Building science and innovation city strategies in Australia and the UK. Innovation. Manage. Policy Practice **10** (2–3), 211–223 (2008)
18. Cresswell, A.M., Pardo, T.A., Canestraro, D.S., Dawes, S.: Why assess information sharing capability? Center for Technology in Government, Albany, NY (2005). Available at http://www.ctg.albany.edu/publications/guides/why_assess/why_assess.pdf
19. Cromer, C.: Understanding Web 2.0's influences on public e-services: a protection motivation perspective. Innov. Manage. Policy Practice **12**(2), 192–205 (2010)
20. Dawes, S.S., Cresswell, A.M., Pardo, T.A.: From "need to know" to "need to share": Tangled problems, information boundaries, and the building of public sector knowledge networks. Public Adm. Rev. **69**(3), 392–402 (2009)
21. Dawes, S.S., Pardo, T.A., Simon, S., Cresswell, A.M., LaVigne, M.F., Andersen, D.F., Bloniarz, P.A.: Making Smart IT Choices: Understanding Value and Risk in Government IT Investments, 2nd edn. Center for Technology in Government, Albany (2004)
22. Dirks, S., Gurdgiev, C., Keeling, M.: Smarter Cities for Smarter Growth: How Cities Can Optimize Their Systems for the Talent-Based Economy. IBM Global Business Services, Somers, NY (2010). Available at ftp://public.dhe.ibm.com/common/ssi/ecm/en/gbe03348usen/GBE03348USEN.PDF
23. Dirks, S., Keeling, M. A.: Vision of Smarter Cities: How Cities Can Lead the Way into a Prosperous and Sustainable Future. IBM Global Business Services, Somers, NY (2009). Available at ftp://public.dhe.ibm.com/common/ssi/ecm/en/gbe03227usen/GBE03227USEN.PDF
24. Dirks, S., Keeling, M., Dencik, J.: How Smart is Your City?: Helping Cities Measure Progress. IBM Global Business Services, Somers, NY (2009). Available at ftp://public.dhe.ibm.com/common/ssi/ecm/en/gbe03248usen/GBE03248USEN.PDF
25. Dobbs, R., Smit, S., Remes, J., Manyika, J., Roxburgh, C., Restrepo, A.: Urban World: Mapping the Economic Power of Cities. McKinsey Global Institute (2011)
26. Drucker, P.: Innovation and Entrepreneurship. Harper Business, New York (1993)
27. Etzkowitz, H.: Entrepreneurial scientists and entrepreneurial universities. Am. Acad. Sci. Minerva **21**, 198–233 (1983)
28. Etzkowitz, H.: The evolution of the entrepreneurial university. Int. J. Technol. Global. **1**(1), 64–77 (2004)
29. Etzkowitz, H., Webster, A., Gebhardt, C., Terra, B.: The future of the university and the university of the future: evolution of ivory tower to entrepreneurial paradigm. Res. Policy **29**, 313–330 (2000)
30. European Commission: Connecting Universities to Regional Growth. European Commission, Brussels (2011)
31. European Commission: European innovation partnership on smart cities and communities (EIP-SCC). Online: https://eu-smartcities.eu/ (2016)
32. Gibb, A., Haskins, G., Robertson, I.: Leading the Entrepreneurial University. Meeting the Entrepreneurial Needs of Higher Education Institutions. University of Oxford, Oxford (2009)

33. Giffinger, R., Fertner, C., Kramar, H., Kalasek, R., Pichler, N., Meijers.: ESmart Cities: Ranking of European Medium-Sized Cities. Centre of Regional Science, Vienna University of Technology, Vienna, Austria (2007)
34. Gjerding, A., Wilderom, C., Cameron, S., Scheunert, K.-J.: L'université entrepreneuriale: vingt pratiques distintives. Politiques et Gestion de l'Enseignement Supérieur **3**(18), 95–124 (2006)
35. Goddard, J., Robertson, D., Vallance, P.: Universities, technology and innovation centres and regional development: the case of the North East of England. Camb. J. Econ. **36**, 609–628 (2012)
36. Goddard, J., Robertson, D., Vallance, P.: The civic university: connecting the global and the local. In: Capello, R., Olechnicka, A., Gorzelak, G. (eds.) Universities, Cities and Regions: Loci for Knowledge and Innovation Creation. Routledge, London (2012)
37. Goddard, J., Kempton, L., Vallance, P.: Universities and smart specialisation: challenges and opportunities for innovation strategies of European regions. Ekonomiaz **83**(2), 82–101 (2013)
38. Grimaldi, R., Kenney, M, Siegel, D., Wright, M.: 30 years after Bayh–Dole: Reassessing academic entrepreneurship. Res. Policy **40**, 1045–1057 (2011)
39. Hadjuk, S.: The concept of a smart city in urban management. Bus. Manage. Educ. **14**(1), 34–49 (2016)
40. Hall, R.E.: The vision of a smart city. In: Proceedings of the 2nd International Life Extension Technology Workshop, Paris, France, 28 Sept 2000
41. Harrison, C., Eckman, B., Hamilton, R., Hartswick, P., Kalagnanam, J., Paraszczak, J., Williams, P.: Foundations for smarter cities. J. Res. Develop. **54**(4) (2010)
42. Hollands, R.G.: Will the real smart city please stand up? City **12**(3), 303–320 (2008)
43. Hospers, G.-J.: Governance in innovative cities and the importance of branding. Innov. Manag. Policy Practice **10**(2–3), 224–234 (2008)
44. Isaksen, A., Tödtling, F., Trippl, M.: Innovation policies for regional structural change: combining actor-based and system-based strategies. Institute for Multilevel Governance and Development, Vienna University of Economics and Business, Vienna (2016)
45. Johnson, B.: Cities, systems of innovation and economic development. Innov. Manage. Policy Practice **10**(2–3), 146–155 (2008)
46. Kourtit, K., Nijkamp, P.: In praise of megacities in a global world. Reg. Sci. Policy Practice **5**(2), 167–182 (2013)
47. Kraus, S., Richter, C., Papagiannidis, S., Durst, S.: Innovating and exploiting entrepreneurial opportunities in smart cities: evidence from Germany. Creat. Innov. Manage. **24**(4), 601–616 (2015)
48. Lee, J.H., Hancock, M.G., Hu, M.-C.: Towards an effective framework for building smart cities: Lessons from Seoul and San Francisco. Technol. Forecast. Soc. Chang. **89**(1), 80–99 (2014)
49. Leydesdorff, L., Deakin, M.: The triple-helix model of smart cities: a neo-evolutionary perspective. J. Urban Technol. **18**(2), 53–63 (2011)
50. Lombardi, P., Giordano, S., Caragliu, A., Del Bo, C., Deakin, M., Nijkamp, P., Kourtit, K., Farouh, H.: An Advanced Triple-Helix Network Model for Smart Cities Performance". In: Erkoskun, O.Y. (ed.) Green and Ecological Technologies for Urban Planning: Creating Smart Cities, pp. 59–73. IGI Global, Hershey (2012)
51. Luke, B., Verreynne, M., Kearins, K.: Innovative and entrepreneurial activity in the public sector: the changing face of public sector institutions. Innov. Manage. Policy Practice **12**(2), 138–153 (2010)
52. Marceau, J.: Introduction: innovation in the city and innovative cities. Innov. Manage. Policy Practice **10**(2–3), 136–145 (2008)

53. Markkula, M., Kune, H.: Making smart regions smarter: smart specialization and the role of universities in regional innovation ecosystems. Technol. Innov. Manage. Rev. **5**(10), 7–15 (2015)
54. Marsal-Llacuna, M. L., Colomer-Llinàs, J., Meléndez-Frigola, J.: Lessons in urban monitoring taken from sustainable and livable cities to better address the Smart Cities initiative. Technol. Forecast. Social Change **90**(B), 611–622 (2015)
55. Martin, R., Simmie, J.: Path dependence and local innovation systems in city-regions. Innov. Manage. Policy Practice **10**(2–3), 183–196 (2008)
56. Neirotti, P., De Marco, A., Cagliano, A.C., Mangano, G., Scorrano, F:. Current trends in smart city initiatives: some stylized facts. Cities **38**, 25–36 (2014)
57. Pardo, T.A., Burke, G.B.: Government Worth Having: A Briefing on Interoperability for Government Leaders. Center for Technology in Government, Research Foundation of State University of New York, Albany, NY (2008)
58. Pardo, T.A., Nam, T., Burke, G.B.: Egovernment interoperability: interaction of policy, management, and technology dimensions. Social Sci. Comput. Rev. (2016)
59. Paroutis, S., Bennett, M., Heracleous, L.: A strategic view on smart city technology: the case of IBM Smarter Cities during a recession. Technol. Forecast. Soc. Chang. **89**(1), 262–272 (2014)
60. Romano, A., Passiante, G., Delvecchio, P., Secundo, G.: The innovation ecosystem as booster for the innovative entrepreneurship in the smart specialisation strategy. Int. J. Knowl. Based Develop. **5**(3), 271–288 (2014)
61. Santos, D.: Política de inovação: filiação histórica e relação com as políticas de desenvolvimento territorial. Revista Portuguesa de Estudos Regionais **3**, 25–40 (2003)
62. Santos, D.: Teorias de inovação de base territorial. In: Costa, J.S., Nijkamp, P. (eds.) Compêndio de Economia Regional - Teoria, Temáticas e Políticas, pp. 319–352. Cascais, Principia (2009)
63. Santos, D., Caseiro, N.: Empreendedorismo em Instituições de Ensino Superior: um estudo de caso. In: Proceedings of 14° Workshop da APDR—Empreendedorismo e Desenvolvimento Regional. APDR, Setúbal (2012)
64. Santos, D., Caseiro, N.: The challenges of smart specialization strategies and the role of entrepreneurial universities: a new competitive paradigm. In: Farinha, L., Ferreira, J., Lawton Smith, H., Bagchi-Sen, S. (eds.), Handbook of Research on Global Competitive Advantage Through Innovation and Entrepreneurship, pp. 978–987. IGI Global, Hershey (2015)
65. Santos, D., Simões, M.J.: Regional innovation systems in Portugal: a critical analysis. Investigaciones Regionales **28**, 149–168 (2014)
66. Santos, D., Simões, M.J.: Challenging orthodoxies: territorial inonovation policies in peripheral regions. In: Proceeding of the 24th Workshop APDR—Entrepreneurship and Performance in a Regional Context, pp. 38–50. APDR, Lisboa (2016)
67. Simões, M.J., Santos, D.: Cidades e Regiões Digitais - uma oportunidade perdida? In: Baudin, G., Vaz, D. (eds.) Transação Territorial - novas relações cidade-campo, pp. 87–104. Húmus, V. N. Famalicão (2014)
68. Slavtchev, V., Laspita, S., Patzelt, H.: Effects of entrepreneurship education at universities. Jena Econ. Res. Papers **25** (2012)
69. Stratigea, A., Papadopoulou, C.-A., Panagiotopoulou, M.: Tools and technologies for planning the development of smart cities. J. Urban Technol. **22**(2), 43–62 (2015)
70. Tödtling, F., Trippl, M.: One size fits all? Towards a differentiated regional innovation policy approach. Res. Policy **34**, 1203–1219 (2005)

71. Toppeta, D.: The Smart City vision: how innovation and ICT can build smart, "Livable", sustainable cities: The Innovation Knowledge Foundation (2010). Available at http://www.thinkinnovation.org/file/research/23/en/Toppeta_Report_005_2010.pdf
72. Tranos, E., Gertner, D.: Smart networked cities? Innov. Eur. J. Social Sci. Res. (2012). Available at SSRN: https://ssrn.com/abstract=2032620
73. Urbano, D., Guerreo, M.: Entrepreneurial universities: economic impacts of academic entrepreneurship in a European region. Econ. Develop. Q. **27**(1), 40–55 (2013)
74. Uyarra, E.: Conceptualizing the regional roles of universities, implications and contradictions. Eur. Plan. Stud. **18**, 1227–1246 (2010)
75. Washburn, D., Sindhu, U., Balaouras, S., Dines, R.A., Hayes, N.M., Nelson, L.E.: Helping CIOs understand "Smart City" initiatives: defining the Smart City, its drivers, and the role of the CIO. Forrester Research Inc, Cambridge (2010)
76. Zygiaris, S.: Smart city reference model: assisting planners to conceptualize the building of smart city innovation ecosystems. J. Knowl. Econ. **4**(2), 217–231 (2013)

Smart Companies: Digital Transformation as the New Engine for Reaching Sustainability

Wail El Hilali$^{(\boxtimes)}$ ⓘ and Abdellah El Manouar

ENSIAS, Mohammed V University, Rabat, Morocco
wailelhilali@gmail.com, a.elmanouar@um5s.net.ma

Abstract. The concept of sustainability has increasingly gained traction in both business and academic worlds. Finding a balance between economy, society and environment requires implementing changes at the strategic level, particularly at the business model level. Moreover, and to survive to this digital revolution led by cutting-edge startups, companies have started to transform digitally their businesses by enhancing their digital capabilities. Since the two concepts need the implementation of radical changes regarding how the value is created and captured, analyzing how to incorporate sustainability while digitally transforming the business could be a promising road to explore. This paper is an attempt to discuss how to seize the opportunity of digital transformations in order to reach sustainability. It starts by defining the concepts of digital transformation and sustainability. After, it answers, using real world examples, how companies, during a digital transformation, could increase profitability and their social footprint while reducing their negative environmental externalities. The paper discusses after, a survey, in the Moroccan context, of the relationship between digital transformation and sustainability among 15 companies (from 40 companies contacted) from different sectors. It also explains the Moroccan actual context using a game theory approach.

Keywords: Digital transformation · Sustainability · Digital strategies
Big data · Shared value · Game theory · Prisoners' dilemma

1 Introduction

As the world is undergoing radical change, the way we do business has to change as well. The lack of resources, the rise of customers' awareness, the fierce competition between rivals, and the risk of market disruption, all of these push and drive companies to change how they think and how they interact with customers, partners, rivals and suppliers. The Digital transformation concept comes as all fresh hope for companies to survive in this uncertain and volatile world. It involves transforming key business operations to exploit and integrate digital technologies. It also affects products and processes, as well as organizational structures and management concepts [1]. Furthermore, companies are still looking to find the ultimate way to reach sustainability, a sort of an equilibrium between three dimensions, which are economy, society and environment. Finding a way to remain competitive and doing more with less while increasing the social footprint and reducing the environmental impacts is the main

© Springer Nature Switzerland AG 2019
M. Ben Ahmed et al. (Eds.): SCA 2018, LNITI, pp. 132–143, 2019.
https://doi.org/10.1007/978-3-030-11196-0_13

challenge that companies are facing nowadays. Multiple research studies were published, discussing reaching sustainability from different viewpoints. This paper is an attempt to enrich the published literature by discussing how to integrate sustainability at the strategic level of companies during a digital transformation. The discussion will be illustrated with examples from the real world and will include an empirical study from the Moroccan industrial context. A modelling of the Moroccan situation using the game theory will be used to clarify more the empirical data gathered from the field.

2 Digital Transformation: The Road to Enhanced Performance

There is no consensus in the literature on what digital transformation means [2], reference [3] defines this concept as an environment where everything for an organization is connected, creating "digital imperatives" for companies to proceed to a transformation impacting their business' functional areas, including operations, customer experience, how value is created and captured. Reference [4] sees it as "the application of digital capabilities to processes, products, and assets to improve efficiency, enhance customer value, manage risk, and uncover new monetization opportunities". So digital transformation is not only about digitalizing existing processes, it about using digital capabilities (ex: real-time data, cloud native applications...) to create and capture new kind of value. Reference [5] enhances this definition by seeing it as "a sustainable, company-level transformation via revised or newly created business operations and business models achieved through value-added digitization initiatives, ultimately resulting in improved profitability".

The most important result of a conducted digital transformation is the impacts on companies' business models. The change could concern either some of the business model's elements or the entire business model. The change could be marginal or radical [6].

3 The Sustainability Mindset: Another Way to View Business

Sustainability, sustainable development, sustainable use and green, all these are feel-good and attractive words that are used a lot nowadays in situations relating business to society and/or environment. The first appearance of the concept of sustainability can be traced back to 1987, in the famous report entitled 'Our common future' (known also as Brundtland report), sponsored by several countries for the count of the United Nations [7]. The report defined sustainable development as "meeting the needs of the present without compromising the ability of future generations to meet their own needs". Another well-known definition of sustainability links it to three pillars, which are economy, society and environment. If any one pillar is weak then the system as a whole is considered unsustainable [8]. Sustaining a business is equivalent to find some kind of equilibrium between the three dimensions. It is about answering the three questions as shown in Fig. 1:

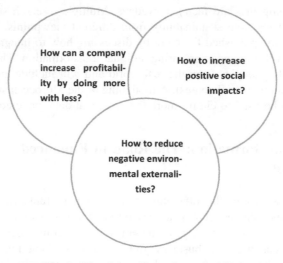

Fig. 1. The three pillars of sustainability

4   Digital Transformation: Creating and Capturing a New Kind of Value

Many companies see digital transformation as a way out from the existing competitive crowded markets. It is a way to move from aggressive red ocean environments to new blue ocean markets devoid of competition. Digital transformation helps companies to create new productivity frontiers, striking out into new markets full of opportunities to seize [9].

Digital transformation gives companies a way to create new kind of value through new business models. Netflix for example has avoided rigidity and succeeded in transforming digitally their business. In fact, and thanks to their new digital subscription model, they overturned Blockbuster's bricks-and-mortar approach and avoided a digital distribution that threatened the entire industry at that time [10]. Amazon, the online bookstore, has seen it a way to diversify its portfolio. In fact, by launching Amazon web service business unit, the giant retailer did not only make use of excess computing capacity, but, it had created a new business that dominates about one third of the worldwide Cloud computing market share [11]. Apple also has succeeded to transform itself from a manufacturing company to a digital business that revolutionized many industries, especially the music industry with ITunes and its subscription business model.

Customer relationships are one of the areas that companies deal with during a digital transformation. In the digital era, customers are no longer seen as a market to segment but as a dynamic network instead [12]. They also contribute to the product content creation. In some examples, this network could be the key asset, if not the main competitive advantage, that the company could possess. Waze is a good example of a peer-to-peer application that collects data from it users to provide real time traffic information [13]. The huge number of the application's users explains why Google paid $1.3 billion to acquire the company [14]. Customer experience is also redefined during the digital transformation. Reference [15] affirms that switching costs in the digital era are becoming very low, which means that the users could switch to a competitor in case of any negative customer experience that can go "viral".

Digital transformation is also the opportunity to move into adjacent markets. Take for example the Telecom operators who started to invest in e-banking businesses, exploiting their customer data, their well-established brand image as a strategic asset and their technological superiority while competing with traditional banking companies. Orange, the French telecom, has launched a 100% mobile-based banking named "Orange bank", which is considered as a key element in orange diversification strategy [16]. To improve their customers' digital experience, Orange bank has recently employed a virtual advisor power by IBM Watson, based on artificial intelligence technology [17]. With this digital mindset, the digital bank has succeeded to attract more than 100,000 customers in its first quarter [18].

Moreover, Big data deluge is changing the game rules by bringing new perceptions, helping companies to better reframe problems, understand customers, improve operational efficiency and discover new opportunities. Companies are more and more interested in this field, taking advantage of the storage capacity offered nowadays, which is no longer an obstacle. Reference [19] was not exaggerating when the author affirmed that "Data are to this century what oil was to the last one: a driver of growth and change".

The cloud computing is also a key driver in the digital era, companies are now focusing more on their business core rather than becoming distracted by IT management and consideration. Cloud solutions offer scalability, flexibility and an opportunity to reduce costs such as capital expenditure for servers and related hardware, maintenance and replacement costs, security, as well as IT support costs.

Competition also has taken another dimension in the digital era. In fact, the rivalry of companies in some fields does not impede their collaboration in other fields. Take for example the collaboration between VMware and Google, which reflects a new air of co-opetition among major cloud computing service suppliers. Every business opportunity is a one to seize in the digital age, as the ultimate goal is to sustain growth, which is more important than sustaining a competitive advantage [20].

A digital transformation of a company will give it also some sort of flexibility to pivot in case of any hurdle. Failure for a digital company is an opportunity to learn and to improve the business, as testing ideas is easy, cheap, fast and could be conducted constantly in the digital era [12].

5   Digital Transformation: Increasing Impacts on Society

Criticisms of business and its practices have become commonplace in recent decades. With the rise of the stakeholders' awareness, companies are trying nowadays to fill their social responsibilities in order to increase their customers' loyalties and to improve their brand images.

In the digital era, we believe that the first step towards a true reconciliation between business and society is to discover and understand the customers' social needs. With the abundance of data, there is no better way to do that than digging in the unstructured data coming from social media and connected devices. There are many IT solutions and tools that should be implemented to store, transform, model data in order to extract valuable information on customers' needs.

The relationship between society and business is no longer put under the umbrella of philanthropy. A relatively new concept was introduced by the well-known strategist Michael Porter [21] called "shared value" redefines the role of business in the society. The concept argues that companies should prioritize solving social and societal problems while still capturing value and doing business. Cisco for example has adopted this mind-set and invested in education through "Cisco Networking Academy" [22], a MOOC (massive open online course) solution that helps students to design, build, troubleshoot, and secure computer networks, increasing their access to career and economic opportunities [23]. By doing that, the company is building loyal and potential future customers of Cisco products, a win–win situation where business has the opportunity to increase its social footprint.

In the digital era, the creation of shared value beneficiates from the possibilities that technology offers. Reference [24] argues that intangible services in response of social and societal needs are key building blocks of the shared value innovation concept in the digital age.

Sharing economies are also a way to increase social footprints of business. Uber and Airbnb are examples of digital companies that offer to their customers a possibility to share extra capacity for money. Even these companies have disrupted some traditional and established industries [9], they are contributing in lowering unemployment and improving living conditions of their adopters.

Sharing economy is not an exclusive concept of Internet based companies. Businesses in classical industries could adopt this perspective. Telecom operators for example could seize the opportunity of digital transformation in order to create a new on-line sharing service. In fact, Telcos could allow their customers, who have accumulated and unused airtime credits or data buckets to be shared for money. Operators could also charge fees on any sharing transaction done capturing by that a new kind of value.

The opportunities that digital transformation offers to companies are endless. It is up to businesses to exploit this potential in order to resolve social and societal needs of communities [25]. A digital transformation will drive companies for sure to change its value proposition [12]. In our point of view, it is the opportunity for businesses to include a social dimension in their value proposition, letting down the greedy way of doing business and announcing a new approach of creating value.

6   Digital Transformation: Lowering Environmental Negative Externalities

The sensitivity of customers concerning sustainability issues continues to increase. According to a Nielsen survey [26], 66% of customers would pay more for a product or service if the company was committed to reduce its carbon footprints, which is higher than the result of 50% published in the same survey two years before.

Digital transformation is an opportunity for companies to renew their commitments and pledges in favour of environment. As a first step, companies could reduce the number of brick and mortar stores in favour of an online presence. Free mobile, the French telecom operator, is an example of a company that reduced to its minimum, the number of physical stores, lowering by that its carbon footprints [27]. It initially sold its sim cards exclusively online thereby avoiding the cost of running physical stores and their impacts on environment, before introducing self-service kiosks that helped to broaden distribution while keeping costs and environmental impacts low.

Furthermore, and with the rise of big data and Internet of things, companies are having now the possibility to understand and measure their impacts on environment [28]. Assessing ecological risks is also a contribution of big data for a more sustainable world. Many tools have emerged that monitor and assess risks using various parameters from the field [29].

Optimizing the use of resources is also an added value of big data [30]. Resources' scarcity is a global security threat and a concern of all the stakeholders. With information extracted from big data, companies nowadays have all the possibilities to follow a smarter approach regarding resources' allocation.

Regulation also could beneficiate from the use of big data. Better environmental regulation could be ensured if data collected from sensors were used while deciding on government policies related to environment.

7 Digital Transformation and Sustainability, Case of Moroccan Industries

To understand and assess the relationship between the digital transformation and sustainability concepts, a survey was conducted among 40 leading and big-size companies from different industries in the Moroccan context. We received 15 answers from managers inside leading companies in real estate and construction, telecommunication, wholesale and distribution, banking, consulting, pensions fund and logistics.

A Digital Transformation Initiative has been taken up or is under consideration in 87% of companies questioned in the survey, which shows how important is this trend inside Moroccan industries. The vast majority of companies link their digital transformations to customer relationship management (CRM), the use of cloud computing, the use of the mobile as a channel to reach customers, and the use of big data. Below a chart that summarizes the found results (Fig. 2):

The benefits of conducting a digital transformation are numerous, the majority of companies saw it a way to improve the operational efficiencies, enhance the customer experience, to save costs, and to contribute to Sustainability efforts of the organization. Figure 3 represents the most important benefits of conducting a digital information from Moroccan companies point of views.

Furthermore, 10 from the 15 interrogated companies admitted to have a corporate social responsibility program (CSR). Managers listed improving operational efficiencies, health and safety, transparency, brand differentiation and building competitive advantage as main drivers of their adhesion to sustainability mind-set.

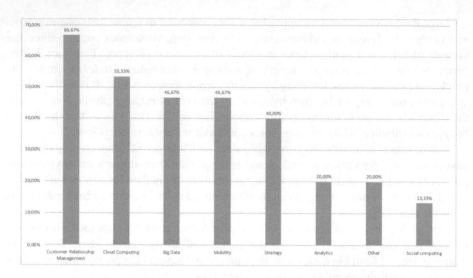

Fig. 2. Areas of digital transformation inside Moroccan industries

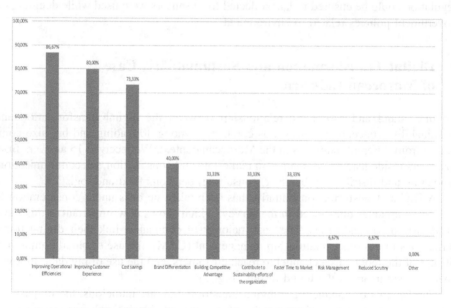

Fig. 3. Benefits of conducting a digital information

In a question that asked managers to indicate any relation that is visible between their companies' digital transformation efforts and corporate sustainability initiatives at the organization level, the majority answered that there is a link between sustainability and strategy (as a component of their digital transformation) from one hand, and sustainability and CRM from the other hand.

Reference [31] argues in a MIT article that strategy is what drives digital transformation instead of technology. The ultimate goal of a digital transformation is to be able to craft a digital strategy where IT becomes a business enabler and a creator of value [32]. It is very logical to incorporate sustainability on the strategic level during a digital transformation.

Moreover, linking sustainability to CRM while digitally transforming a company was expected. The two concepts are customer-centric ones. With customers seeking to play a significant role in the change towards more sustainable lifestyles, companies need to respond to that by cultivating markets that balance profitability with environmental and social responsibility [33]. If CRM is at the heart of change during a digital transformation, it is the opportunity to review the companies' priorities regarding sustainability.

8   Digital Transformation and Sustainability: A Game Theory Model

To understand better the relationship between sustainability and digital transformation, we decided to have recourse to a game theory modeling approach. The selected game is called "the prisoner's dilemma". The ultimate goal is to explain why companies could be reluctant to incorporate the three sustainability pillars while crafting digital business strategies during digital transformations. Three major players are identified, which are companies, customers and regulators (that represent the government).

The prisoner's dilemma game is "a paradox in decision analysis in which two individuals acting in their own self-interests do not result in the optimal outcome" [34]. The purpose of this game is to show that cooperation and the involvement of the three stakeholders are key parts of the puzzle in order to find a win-win situation [35]. Let's consider the following variables:

P0: Price of ordinary products
P1: Price of sustainable products, resulted from a digital transformation
Assuming P0 < P1
α: Sensitivity of customers to social value created by companies
SOC: Social value, the positive impact that companies are having on communities
β: Customers' concerns towards ecological issues
ECO: Ecological value, the positive impact that companies are having on environment
C0: Cost of ordinary products
C1: Cost of sustainable products, resulted from a digital transformation
C2: learning costs
Assuming C0 < C1
S: Subsidies of the government regarding sustainable products
R: Government gain from sustainable companies

Customers' utility function could be expressed as $U = -P + \alpha.SOC + \beta \ ECO$. Using prisoner's' dilemma, the current state of the Moroccan market can be presented as the following simultaneous 3-player game. In each cell, the first, second and third expressions represent the utilities of companies, customers and the government, respectively.

In case of no subsidies from the government (Table 1):

Table 1. Utility functions of the three stakeholders in case of no subsidies from the government

		Customers	
		Ordinary products	Sustainable products
Companies	Ordinary products	P0 − C0; −P0; 0	P0 − C0; −P0; 0
	Sustainable products	P0 − C1 − C2; − P0 + (P1 − P0); R	P1 − C1 − C2; −P1 + α. SOC + β ECO; R

The three stakeholders are looking to maximize their own utility. Companies will not go further on adopting a sustainable mind-set without warranties that their utility functions will be higher. In other words:

$$P0 - C0 < P1 - C1 - C2$$

Same for customers, their utility functions should be higher:

$$-P1 + \alpha.SOC + \beta.ECO > - P0$$

For regulators: $R > 0$

In cases of companies are stuck in situations where $P0 - C0 > P1 - C1 - C2$, the only way out from this unsustainable situation is to beneficiate from a government subsidy S that will ensure a higher utility function:

$$P0 - C0 < P1 - C1 - C2 + S$$

So In case of subsidies from the government, utility functions of all the stakeholders become (Table 2):

Table 2. Utility functions of the three stakeholders in case of subsidies from the government

		Customers	
		Ordinary products	Sustainable products
Companies	Ordinary products	P0 − C0; −P0; 0	P0 − C0; −P0; 0
	Sustainable products	P0 − C1 − C2; −P0 + (P1 − P0); R	**P1 − C1 − C2 + S; −P1 + α.** **SOC + β.ECO; R−S**

In bold, the new and the desirable equilibrium that the stakeholders are looking for.

9 Conclusions

This paper is an attempt to discuss how to reach sustainability while digitally transforming a business. A digital transformation has become a must nowadays to survive to the bloody competition. It is a synonym of a critical review of the whole business model of a given company. The paper answered three main questions related to sustainability' three pillars (economic, social and environmental) from a digital point of view. It discussed how to increase profit and remain competitive in the digital world while increasing the impact on society and avoiding harming environment. It also discussed data from a survey that included major Moroccan companies from different industries. The paper was summed up by a try to model the situation using a game theory approach.

The paper could be reinforced by a case study of a company, from a given industry, in order to illustrate more the recommendations given in this article. It could also be more interesting if the game theory contains multiple iterations.

References

1. Matt, C., Hess, T., Benlian, A.: Digital transformation strategies. Bus. Inform. Syst. Eng. **57**, 339–343 (2015)
2. Herbert, L.: Digital Transformation: Build Your Organization's Future for the Innovation Age. Bloomsbury, USA (2017)
3. Fitzgerald, M., Kruschwitz, N., Bonnet, D., Welch, M.: Embracing digital technology: a new strategic imperative. MIT Sloan Manag. Rev. (55), 1 (2014)
4. Schmarzo, B.: What is digital transformation? http://www.cio.com/article/3199030/analytics/what-is-digital-transformation.html. Last accessed 12 June 2018 (2017)
5. Schallmo, D.R.A., Williams, C.A.: History of digital transformation. In: Digital Transformation Now!: Guiding the Successful Digitalization of Your Business Model. Springer International Publishing, Cham, pp. 3-8. https://doi.org/10.1007/978-3-319-72844-5_2 (2018)
6. Schallmo, DRA.,Williams, C.A.: Digital transformation of business models. In: Digital Transformation Now!: Guiding the Successful Digitalization of Your Business Model. Springer International Publishing, Cham, pp. 9–13. https://doi.org/10.1007/978-3-319-72844-5_3 (2018)
7. Gerasimova, K.: Our Common Future. Taylor & Francis, London, UK (2017)
8. Murray, P.: The Sustainable Self: A Personal Approach to Sustainability Education. Taylor & Francis, London, UK (2012)
9. Mauborgne, R., Kim, W.C.: Blue Ocean Shift. Pan Macmillan UK, UK (2017)
10. Walker, R., Jeffery, M., So, L., Sriram, S., Nathanson, J., Ferreira, J., Feldmeier, J., Merkley, G.: Netflix leading with data: the emergence of data-driven video. Kellogg Sch. Manag. Cases, 1–19 (2017)
11. Aversa, P., Haefliger, S., Reza, D.G.: Building a winning business model portfolio. MIT Sloan Manag. Rev. (58), 49 (2017)
12. Rogers, D.L.: The Digital Transformation Playbook: Rethink your Business for the Digital Age. Columbia University Press, New York, USA (2016)

13. Thakuriah, P., Geers, D.G.: Technology systems for transportation system management and personal use. In: Transportation and Information: Trends in Technology and Policy. Springer New York, New York, NY, pp. 35–71. https://doi.org/10.1007/978-1-4614-7129-5_3 (2013)
14. Pasher, E., Pross, G., Kushnir, U., Neeman, Y.: Tel Aviv: a renaissance revival in the making. In: Formica, P. (ed.) Entrepreneurial Renaissance: Cities Striving Towards an Era of Rebirth and Revival. Springer International Publishing, Cham, pp. 81–88. https://doi.org/10. 1007/978-3-319-52660-7_4 (2017)
15. Châlons, C., Dufft, N.: The role of IT as an enabler of digital transformation. In: Abolhassan, F. (ed.) The Drivers of Digital Transformation: Why There's No Way Around the Cloud. Springer International Publishing, Cham, pp. 13–22. https://doi.org/10.1007/978-3-319-31824-0_2 (2017)
16. Orange: Orange is now also a bank. http://www.orange.com/en/Press-Room/press-releases/press-releases-2017/Orange-is-now-also-a-bank. Last accessed 6 June 2018 (2017)
17. Orange: Orange Bank brings unique customer experience with its virtual advisor powered by IBM Watson. http://www.orange.com/en/Press-Room/press-releases/press-releases-2018/Orange-Bank-brings-unique-customer-experience-with-its-virtual-advisor-powered-by-IBM-Watson. Last accessed 6 June 2018 (2018)
18. Orange: 2017 earnings. http://www.orange.com/en/Press-Room/press-releases/press-releases-2018/2017-earnings. Last accessed 7 June 2018 (2018)
19. Economist, T.: Fuel of the future: data is giving rise to a new economy. http://www.economist.com/briefing/2017/05/06/data-is-giving-rise-to-a-new-economy. Last accessed 8 June 2018 (2017)
20. Zenger, T.: Beyond Competitive Advantage: How to Solve the Puzzle of Sustaining Growth While Creating Value. Harvard Business Review Press, Boston, USA (2016)
21. Porter, M.E., Kramer, M.R.: The big idea: creating shared value. How to reinvent capitalism —and unleash a wave of innovation and growth. Harvard Bus. Rev. (89) (2011)
22. Yekela, O., Thomson, K-L., Van Niekerk, J.: Assessing the effectiveness of the Cisco Networking Academy program in developing countries. In: IFIP World Conference on Information Security Education, pp. 27–38, Springer (2017)
23. Wheatley, S.: Cisco's journey to creating shared value. https://blogs.cisco.com/csr/ciscos-journey-to-creating-shared-value-4. Last accessed 9 June 2018 (2012)
24. Lichtenthaler, U.: Shared value innovation: linking competitiveness and societal goals in the context of digital transformation. Int. J. Innov. Technol. Manag. (14) (2017)
25. Carayannis, E.G., Hanna, N.K.: Mastering Digital Transformation: Towards a Smarter Society, Economy, City and Nation. Emerald Group Publishing Limited (2016)
26. Nielsen: Green Generation: Millennials Say sustainability is a Shopping Priority. http://www.nielsen.com/us/en/insights/news/2015/green-generation-millennials-say-sustainability-is-a-shopping-priority.html. Last accessed 10 June 2018 (2015)
27. Iliad:Iliad Case Study: French Free Mobile to Challenge the French Mobile Incumbents. Datamonitor Plc (2012)
28. Salvatore, R., Carmine, N.: Global sustainability inside and outside the territory. In: Proceedings Of The 1st International Workshop. World Scientific (2014)
29. Magdziarz, T., Mitusińska, K., Gołdowska, S., Płuciennik, A., Stolarczyk, M., Ługowska, M., Góra, A.: AQUA-DUCT: a ligands tracking tool. Bioinformatics (33), 2045–2046 (2017)
30. Srinivasan, S.: Guide to Big Data Applications. Springer International Publishing, Houtson, USA (2017)
31. Kane, G.C., Palmer, D., Phillips, A.N., Kiron, D., Buckley, N.: Strategy, not technology, drives digital transformation. MIT Sloan Manag. Rev. Deloitte University Press (14) (2015)

32. Sebastian, I.M., Ross, J.W., Beath, C., Mocker, M., Moloney, K.G., Fonstad, N.O.: How big old companies navigate digital transformation. MIS Quart. Executive (2017)
33. Müller, A.-L.: Sustainability and customer relationship management: current state of research and future research opportunities. Manag. Rev. Quart. (64), 201–224 (2014)
34. Investopedia: What is the 'Prisoner's Dilemma'. https://www.investopedia.com/terms/p/prisoners-dilemma.asp. Last accessed 18 July 2018
35. Peterson, M.: The Prisoner's Dilemma. Cambridge University Press (2015)

The Effect of Weak Atmospheric Turbulence and Fog on OOK-FSO Communication System

Lamiae Bouanane[1]([⊠]), Fouad Mohamed Abbou[2], Fouad Abdi[1],
Fouad Chaatit[2], and A. Abid[3]

[1] Faculty of Sciences, FST, Fes, Morocco
l.bouanane@aui.ma
[2] Al Akhawayn University, Ifrane, Morocco
[3] Islamic University of Medina Saudi Arabia, Medina, Saudi Arabia

Abstract. In this paper, we investigate the Bit Error Rate (BER) performance for the On-Off keying (OOK) modulated FSO links under weak turbulence condition and atmospheric attenuation. An analytical expression of the average BER in FSO communication system over log normal atmospheric turbulence channels is presented. The BER performance indicators are analyzed for different link characteristics. According to the simulation results, longer propagation distance with lower bit error rates could be achieved by increasing the transmitted power.

Keywords: FSO · Turbulence · Attenuation · BER · SI · Log-Normal

1 Introduction

Recently, The Internet of Things (IOT) is a major challenge for the communications networks. In fact, IOT will allow the communication among devices, which will be massively deployed in smart cars, smart roads, smart cities, smart houses and buildings, security and safety (surveillance, alarm, site networking), and industrial M2M communication. It is anticipated that in 2020, there will be around 50 Billion connections. According to Cisco [1], "Fifty billion things will connect to the Internet of Everything in just a few years". This increasing of number of customers and connections presents a big dilemma for the telecommunications industries. Because of its fast and easy deployment, free space optical communication systems present a potential solution that also provides high bandwidth capacity and high communication security [2].

Although FSO offers such attractive features, it has some major constraints. In FSO, data is mainly transmitted using laser beams through the atmosphere. Hence, the performance of such links is highly dependent on the degradation/turbulence/ attenuations caused by the atmosphere at the level of this link. Weather conditions are one of the factors that highly impact the FSO performance. Each time there is snow, fog, rain or wind, the physical link is affected by these conditions and, thus, the performance is degraded. The atmosphere causes signal degradation and attenuation in a free-space system link in several ways, including absorption and scattering. Absorption is due to gases present in the atmosphere, whereas scattering is caused by big sized rain drops.

© Springer Nature Switzerland AG 2019
M. Ben Ahmed et al. (Eds.): SCA 2018, LNITI, pp. 144–150, 2019.
https://doi.org/10.1007/978-3-030-11196-0_14

The major impacting factor in FSO links is the effect of atmospheric turbulence. Atmospheric turbulence occurs as a result of the variations in the refractive index due to inhomogeneities in temperature and pressure changes. The index-of-refraction fluctuations produce fluctuations in the irradiance (power received per unit area) of the transmitted optical beam, what is known as atmospheric scintillation. These atmospheric turbulences cause rapid fluctuations at the received signal which impairs severely the link performance. The block diagram of a typical terrestrial FSO link is shown in Fig. 1.

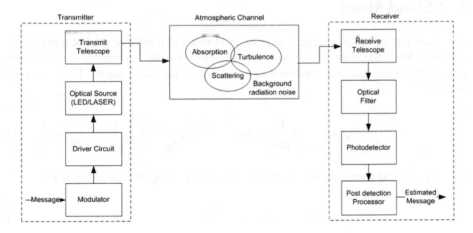

Fig. 1. Block diagram of a terrestrial FSO link

In order to study the effect of the atmospheric turbulence on FSO system, many statistical models have been considered in the literature. The most widely used are log normal distribution for weak turbulence and gamma-gamma for strong turbulence. In this paper, we investigate the effect of weak atmospheric turbulence on the performance of the free space optical system by considering the BER metric. For this purpose, we derive BER expressions for OOK modulated terrestrial FSO links.

The rest of this paper is organized as follows. In Sect. 2, we present some related papers. In Sect. 3, we describe the log normal channel in the atmospheric turbulence conditions followed by BER performance analysis and expressions derivation in Sect. 4. In Sect. 5 we discuss and analyze the obtained results. Section 6 concludes the paper.

2 Related Work

Many papers have investigated the effect of turbulence on the performance of FSO by considering different turbulence modes and hence different statistical models. Kumar [3] carried out a comparative analysis of BER performance for the on-off keying

(OOK) modulated direct detection and coherent detection FSO communication systems in weak turbulence conditions using the Log Normal distribution model. Marvi et al. [4] studied the impairments caused by strong atmospheric turbulence on FSO link by considering the Q factor and the BER. The authors also presented multi-beam technique to increase the achievable link distance of the FSO system. Also, Dhaval et al. [5] analyzed the BER performance of FSO links under strong turbulence mode using the Negative Exponential Model.

3   Channel Model

For weak turbulence modeling, the most used channel model is the log-normal distribution. Since the intensity is log normally distributed, the probability density function (PDF) of the received irradiance I because of turbulence, valid for I > 0, is given by [6]:

$$p(I) = \frac{1}{I\sqrt{2\pi\sigma_{\ln I}^2}} \exp\left\{ -\frac{[\ln(I) - \mu_{\ln I}]^2}{2\sigma_{\ln I}^2} \right\} \tag{1}$$

where $\mu_{\ln I}$ is the mean and $\sigma_{\ln I}^2$ is the variance of the Log-irradiance given by:

$$\mu_{\ln I} = -\frac{\sigma_{\ln I}^2}{2} \tag{2}$$

$$\sigma_{\ln I}^2 = \ln(\sigma_I^2 + 1) \tag{3}$$

where σ_I^2 is the variance of the normalized intensity fluctuations or the scintillation index. In weak turbulence, the scintillation index is expressed as the Rytov variance for plane waves [7]:

$$\sigma_I^2 = \sigma_R^2 = 1.23\, C_n^2 k^{7/6} L^{11/6} \tag{4}$$

where C_n^2 is the refractive-index structure parameter and L is the length of the optical link.

The refractive-index structure parameter is very difficult to measure as it also depends on the temperature, wind strength, altitude, humidity, atmospheric pressure. For a homogenous turbulent field, which can be assumed for near-ground horizontal-path propagation, the refractive-index structure parameter is constant. The scintillation levels are usually divided into three regimes in dependence on the Rytov variance [3]: a weak-fluctuations regime ($\sigma_I^2 < 0.3$), a moderate-fluctuations regime (focusing regime) ($0.3 \leq \sigma_I^2 < 5$), and a strong-fluctuations regime (saturation regime $\sigma_I^2 \geq 5$).

4 BER Analysis

Most terrestrial FSO systems adopt the OOK modulation scheme because of its simplicity and low cost. The probability of error for NRZ-OOK-coded optical data, detected with a photodiode, can be expressed as a function of the Signal-to-Noise Ratio (SNR) [8]:

$$P_e = \frac{1}{2} erfc(\frac{1}{2\sqrt{2}}\sqrt{SNR}) \tag{5}$$

$$P_e = Q\left(\sqrt{\frac{SNR}{8}}\right) \tag{6}$$

where Q (.) is the Gaussian Q-function.

Considering the thermal noise only, the signal-to-noise ratio (SNR) can be written as:

$$SNR = \frac{I^2}{i_{th}^2} \tag{7}$$

where

$$i_{th}^2 = \frac{4KTB}{R} \tag{8}$$

And

$$I^2 = (P_r R_d)^2 \tag{9}$$

where B represents the bandwidth, T is the absolute photodiode temperature in (*Kelvins*), R is the PIN load resistor, k is the Boltzmann's constant and R_d is Responsivity of the detector. In a typical free space optical environment, the FSO transmitter generates optical beams that propagate in free space where weather conditions cause signal losses at the receiver side. The received signal power P_r at a distance L with a transmitter signal power P_t can be written as:

$$P_r = P_t \left(\frac{D}{L\theta}\right) \tau_t \, \tau_r \, 10^{-\gamma(\lambda) \cdot \frac{L}{10}} \tag{10}$$

where D is the receiver diameter, θ is the full transmitting divergence angle, τ_r and τ_t are the optical efficiencies of the transmitter and the receiver respectively and the function of $\gamma(\lambda)$ represents the attenuation of the transmitted light. The atmospheric attenuation consists of absorption and scattering of the light photons by different aerosols and gaseous molecules in the atmosphere. The attenuation is measured as:

$$\gamma(\lambda) = \frac{3.91}{V} \left(\frac{\lambda}{550}\right)^{-\delta} \tag{11}$$

where V is the visibility in (km), λ represent the wavelength in (nm) and the parameter δ is the size distribution of the scattering particles and depends on the visibility distance range [9]. Many models define the value of δ with respect to visibility. The most used models are Kim and Kruse and Naboulssi. For Kim attenuation model, the parameter δ can be found as:

$$\delta = \begin{cases} 1.6, & \text{if } V > 50\,\text{km} \\ 1.3, & \text{if } 6\,\text{km} > V > 50\,\text{km} \\ 0.16V + 0.34, & \text{if } 1\,\text{km} < V < 6\,\text{km} \\ V - 0.5, & \text{if } 0.5\,\text{km} < V < 1\,\text{km} \\ 0, & \text{if } V < 0.5\,\text{km} \end{cases} \tag{12}$$

In the presence of atmospheric turbulence, the BER is considered a conditional probability that is required to be averaged over the probability density function (PDF) of the fluctuating signal. The unconditional BER [3], is given by:

$$BER = \int_0^\infty p(I)\, P_e\, dI \tag{13}$$

Using Eqs. (1) to (13), we can write BER as follows:

$$BER = \frac{1}{\sqrt{\pi}} \int_{-\infty}^{\infty} Q\left(\sqrt{SNR} \ \exp\left(y\sqrt{2}\sigma_{\ln I} - \frac{\sigma_{\ln I}^2}{2}\right)\right) e^{-y^2} dy \tag{14}$$

Further, using the Gauss-Hermite quadrature integration approximation, we get:

$$BER = \frac{1}{\sqrt{\pi}} \sum_{i=1}^{N} w_i \, Q\left(\sqrt{SNR} \ \exp\left(y_i\sqrt{2}\sigma_{\ln I} - \frac{\sigma_{\ln I}^2}{2}\right)\right) \tag{15}$$

where, y_i are the roots and w_i are the corresponding weights of the Hermite polynomial.

5   Results and Discussion

Using the analytical model presented in the previous section, we evaluate the performance of the FSO communication system in the presence of weak turbulence and Kim atmospheric attenuation using Matlab based on numerical values stated in Table 1.

From the computed BER in the presence of Fog using Kim attenuation model as a function of transmitted power for three different distances shown in Fig. 2, it is clear that by increasing the transmitter input power, the effect of fog induced loss is reduced.

Table 1. Simulation parameters

Parameters	Value
Transmission wavelength (λ)	1550 nm
Optical efficiency of transmitter τ_t	0.75
Optical efficiency of receiver τ_r	0.75
Full transmitting divergence angle θ	2×10^{-3} rad
Receiver diameter	1 cm
Electron charge (q)	1.6×10^{-19} C
PIN load resistance (R)	1 kΩ
Boltzmann constant (k)	1.38×10^{-23} J k
Temperature (T)	298 K
Dark current (I_d)	0.05 nA
Responsivity (R_d)	0.6 A/W
Bandwidth (B)	2.5 GHz
Link distance	7000
Planck constant	6.625×10^{-34}

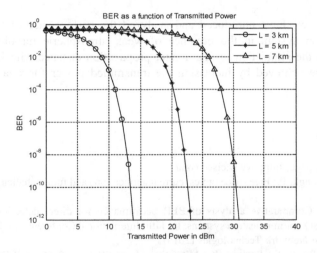

Fig. 2. BER as a function of transmitted power for a visibility of 3 km

Further, the BER in the presence of fog and weak turbulence as a function of Scintillation Index is depicted in Fig. 3. It can be noted that the average BER increases with Scintillation Index and tends to lower values for higher transmitted power.

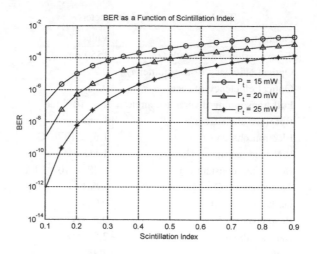

Fig. 3. BER as a function of the Scintillation Index for a distance of 7 km

6 Conclusion

In this paper, we have introduced and analyzed the performance of the OOK-FSO communication system in the presence of weak atmospheric turbulence and fog. According to the simulation results, longer propagation distance with lower bit error rates could be achieved by increasing the transmitted power for weak atmospheric turbulence.

References

1. Cisco Homepage, http://www.cisco.com/
2. Malik, A., Singh, P.: Free space optics: current applications and future challenges. Int. J. Opt., 1–7 (2015)
3. Kumar, P.: Comparative analysis of BER performance for direct detection and coherent detection FSO communication systems. In: Fifth International Conference on Communication Systems and Network Technologies (2015)
4. Marvi, G., Preeti, S., Pardeep, K.: Mitigation of scintillation effects in WDM FSO system using multibeam technique. J. Telecommun. Inform. Technol. (2017)
5. Dhaval, S., Dilip, K.: BER performance of FSO link under strong turbulence with different coding techniques. Int. J. Comput. Sci. Commun. **8**, 4–9 (2017)
6. Vetelino, F.S., Young, C., Andrews, L.: Fade statistics and aperture averaging for Gaussian beam waves in moderate-to-strong turbulence. Appl. Opt. **46**, 3780–3790 (2007)
7. Iniguez, R.R., Idrus, S.M., Sun, Z.: Atmospheric transmission limitations. In: Optical Wireless Communications—IR for Wireless Connectivity, p. 40. Taylor & Francis Group, LLC, London (2008)
8. Senior, J.M., Jamro, M.Y.: Optical Fiber Communications: Principles and practice, 3rd edn. Pearson Education Ltd (2009)
9. Ali, M.A.: Comparison of NRZ, RZ-OOK modulation formats for FSO communications under fog weather condition. Int. J. Comput. Appl. (2014)

MappGuru, A Universal Addressing System for the Unstructured Areas

Valentin Rwerekane[1] and Maurice Ndashimye[2](✉)

[1] University of Rwanda, Kigali, Rwanda
`vrwerekane@gmail.com`
[2] Masters Consulting Company, Nyarugenge, Kigali, Rwanda
`mauricendashimye@gmail.com`

Abstract. An address system is an essential infrastructure which paves the way for social and economic development. It allows people to connect, improves emergency response, increases access to utilities and facilitates postal services and the delivery of goods. On the other hand a digital address infrastructure is undeniably essential for many other processes such as opening bank accounts, the creation of national ID Cards, voting and representing sets of information in data collection and analysis, as well as in marketing and in population census. Designing an addressing system suitable for all countries is challenging. This paper presents a solution by providing a comprehensive universal addressing system—the MappGuru—designed to function as a standardized addressing system with the ability to provide postcode and addresses to any properties. The MappGuru addressing system suggests a unique process to assign addresses based on an existing postcode. The MappGuru addresses are easy for human to understand and can inherently be integrated with information technology systems. To illustrate the MappGuru addressing system, a dedicated section consisting of a case study to assign addresses to an unstructured area in Kigali/Rwanda is described.

Keywords: Address · Postcode · Postal code

1 Introduction

In a connected, globalized world, the need for a comprehensive, universal addressing system has become more pressing.

Addresses serve in wide range of functions [1] including:

- Locating areas where human or business activities take place or areas with economic value and a social role;
- Provision of a physical building location, particularly for cities with a complex structure;
- Properties identification as endpoints of a postal system;
- Location representation in mapping and imagery;
- Parameterization of data for statistics collection (census taking, insurance…);

© Springer Nature Switzerland AG 2019
M. Ben Ahmed et al. (Eds.): SCA 2018, LNITI, pp. 151–164, 2019.
https://doi.org/10.1007/978-3-030-11196-0_15

Many attempts to create a standardized comprehensive addressing system have been made but none succeeded [2, 3]. Geocodes (Geographic object codes) are widely accepted as fundamentally the most suitable systems in regards to providing addresses to the public. However, most geocodes are complex and can't be used without electronic devices. The widely known and most common geocode is the geographic coordinates system. This system can inherently pin point any object on the earth's surface within a range of 1 m; using coordinates (longitude and the latitude) of up to eight digits each; which makes these addresses impractical for human usage.

In the quest for a global addressing system, efforts have been put into simplifying geocodes for human usage. For example, the Mapcode system, an innovative spatial positioning system built specially to make shorter codes for densely populated area; in order to easily be recognized. Mapcode uses rectangular areas called "encoding zone", to define a location and uses letters and digits to represent territories within a certain distance from that location. A 4-letter code can be used to cover a territory of 100 km^2 and a 6-letter code can cover a 250,000 km^2 approximately. Mapcode is very popular in car navigation systems and software that convert coordinates into Mapcode and vice versa are freely available to [4, 5].

Another attempt to create simple geocode systems was the What3words system which aims at making easily memorable codes without using the common grouping of letters and digits. It is based on grids covering the earth of squares of 3×3 m^2 encoded using a format referred to as the 3 m format. The 3 m format is a combination of 3 words to uniquely identify each of the 57 trillion cells of the grid around the globe, for example, the Eiffel Tower: graphics.dads.inched.

The Natural Area Coding (NAC) is another geocode system that can uniquely locate any properties on the earth using an 8–10 alphanumeric code [6]. A NAC consists of three alphanumerical strings separated by spaces representing respectively the longitude, the latitude and the altitude. The NAC system divides the whole surface of the world into 30 divisions longitudinally and 30 divisions latitudinal. Each of these subdivisions forms the outer and biggest NAC cell (level one). NAC cells at level one are once again subdivided into a set of 30×30 cells using the same method to obtain smaller subdivisions (level two). It continues in the same way to get even much smaller subdivisions depending on the sought precision. Blocks at each level are named using numbers between 0 and 9 and 20 capital consonants of the English alphabet.

First level NAC cells are represented by two characters separated by spaces (e.g., NAC: A 2). A second level NAC cell is represented by four characters (e.g., NAC: AC 12). A fourth level NAC (e.g. NAC: 7HGG KJ9L) which covers an area of approximately 50×25 m^2 can represent a block or a building. Therefore, NAC can be used as a universal address system, a global postal code system, a universal property identifier system and universal map grids system.

These geocodes could potentially be used as global addressing systems, however, humans can't use them without resorting to electronic devices; intrinsically, the MappGuru proposed in this research, has the capability to indicate to its users, the orientations and distances with respect to a well known references and/or other users.

Establishing an appropriate addressing system for countries in which year by year a high accumulation of people pours into cities and settle in unstructured fashion without proper plans, can be a long and expensive process. In so far as rapid urbanization is concerned, in all major cities due to informal settlements, an immense proportion of the population lives overcrowded in what is known as slums scattered around cities, whereby properties do not line up and have rarely or no access to street. This phenomenon can be a national tragedy if there is no substantial solution to effectively provide property identification. The MappGuru design features a solution to removing the ambiguity of identifying addresses in unstructured areas or slums.

2 Need for a Comprehensive Address

A comprehensive Addressing system should satisfy following criteria of a well-designed addressing system, including:

1. Human requirements: The addressing system should be composed of meaningful characters and be easy to read (e.g., Initials of cities); addresses size must be short for human to remember. It should also provide the ability to estimates distance and orientation.
2. Digital Representation: It should provide a higher precision and be naturally integrated with mapping systems like Google Maps.
3. Physical Factors: It should coexist with existing addressing systems; each property shall have unique address. It should be easy to implement and the cost of rolling out such a system shall be minimal.
4. Contains a postcode: A postcode is a key constituent of an address, A postcode is added to a postal address to increase the speed and accuracy of identifying places or to be used as area zoning tool; postcodes are essential to remove the ambiguity of addresses where there is no differentiator for addresses;

2.1 Gaps in Postal Addresses

Before the postal addressing era, which dates back towards the end of the 18th century, most houses and buildings had no identification whatsoever. Streets were named after local landmarks or noble people living in the area. Various towns had streets left without a name at all. As mailing services expanded, the good memory of local postmen had become insufficient to facilitate the delivery of packages to thousands of people on a daily basis. The development of postal codes reflected the increasing difficulties of postal delivery.

Postal systems were only implemented in major cities and the creation of the Universal Postal Union in the late 19th century assured the expansion, which became truly global by the 1930s.

Despite its advances and supposed broad coverage, modern postal addressing remains incomplete, even in developed countries. As an example a third of the houses in Ireland had not received proper unique identification until the introduction of the Eircode in 2014 [7].

Even if postal codes systems are adopted and constitute a key element in an addressing system, several addressing systems developed until to date don't profit from the mainstream functionality of the postal code, because are either badly designed or are not embraced and accepted by users [8]; in some cases, postcodes are coupled to administrative boundaries or geographical areas, or are attached to a specific organization. Sometimes they're free-floating and they can't facilitate people to pinpoint their location. Another flaw is that postal code systems in some areas are simply insufficient, requiring address users to further specify the desired address. Other shortcomings include the fact that postcode and addresses' formats are expressed differently across the world.

3   The MappGuru System

MappGuru is a comprehensive digital addressing system built upon a popular geocode, the Natural Area Coding (NAC), to provide addresses to the whole population of the world with minimal effort during the roll-out phase [9, 10].

The MappGuru addressing system suggests a unique process to assign addresses based on defined postcodes; subsequently addresses are connected to the postcode. MappGuru addressing will remove the difficulties of:

- providing addresses to each and every properties with or without access to streets.
- boosting operations of the post with its intelligent zoning
- providing tools for quality management of address data
- Improving the planning for government, private agencies and utility companies.

3.1   MappGuru Conceptual Model

The MappGuru addressing system is structured into three categories of zones centered around a reference point.

These zones are referred to as: (1) Distance zones, (2) Orientation zones and (3) Postal zones.

Reference point: Initially a reference point is chosen, e.g. a city center monument, a landmark place or an emblematic object. A reference point is related to a section of the city or could be related to a whole city. More than one reference points can be used depending on the size of the city. A level 3 NAC code is then assigned to each reference points. E.g., Kigali is a relatively small city and a single reference point such as the main roundabout might be enough; with a NAC: KH4 GN8.

Distance Zones: Distance zones are divisions done by concentric rectangles centered at the reference point. The separation distance between two consecutive rectangles depends on the length and the width of the smallest rectangle that fully covers the city or a section of the city. The separation distance can have three different dimensions: a single level 3 NAC cell, a block of 2×2 or 3×3 level 3 NAC cells. In all three cases, each of the level 3 NAC cells represents a postal zone. Distance zones are named using digits from '0' to '9', therefore, a maximum of 10 orientation zones from the inner rectangle to the outer rectangle is allowed [9].

Orientation zones: Each of the distance zones is divided into blocks of 1×1, 2×2 or 3×3 level 3 NAC cells depending on the width of the first level zones. The obtained number of blocks equals to $2n + 1$, where n is the number of distance zone.

Postal zones: The next step of the design is to locate postal zones inside orientation zones, recalling that each postcode zone has the size of a level 3 NAC cell. Its location depends on the size of the orientation zone.

Quarters: The next step is to define quarters. Postcode zones might suffice depending on the structure and planners preferences. However, In case of congested or unstructured areas (slums), the postcode zone shall be divided into much more refined zones referred to as "Quarters". The postal zone is refined by dividing it into nine Quarters, numbered based on the traffic orientation. Numbers are tracked following the main roads in the ascending order outwards from the reference point.

Address zones: Each Quarter is subsequently subdivided into 10 address blocks from East to West numbered from 0 to 9 and each block is then subdivided into 10 address zones. The results is a 100 Address zones of 25 m x 50 m in each quarter, corresponding to 100 level 4 NAC zones. In the case of structured settlements, addresses can be assigned using Quarters. However, if the area is congested and unstructured, address zones are required.

3.2   The MappGuru Address Format

The format of the MappGuru address is AXPQ NNOO ZZXX,, the last part consisting of four digits serves as a property unique address while the rest of the code can be regarded as a postcode.

The first letter 'A' indicates an alphabetic character, 'N' a numeric character, 'X' an alphanumeric character and 'O' indicates a direction. 'O' can only take values 'N' for North, 'E' for East, 'W' for West and 'S' for South. The postcode is made up of components specified as follows:

AX: Indicates a reference point in a city. Ideally the character A' will be the initial letter, which represents a city, town's name or GIS coordinate, for example K for Kigali. The second character 'X', is intended to distinguish names of cities within a country or different reference points within a city.

P: The third character 'P' identifies the postcodes zone inside the orientation zone.

Q: The Forth character 'Q' identifies the Quarter inside the postcodes zone.

NNOO: Indicates an orientation zone. The first two characters 'NN' indicate distances measured from the reference point in directions OO respectively.

ZZXX: The last part of the postcode is a component required to provide property identity; it can serve as house number, an apartment number or a block number. Only two characters 'ZZ', which indicate the address zone, are sufficient if the property size is an area of 25×50 m^2. However, two characters 'XX' shall be appended in the case the property size is smaller or if the area is unstructured (slum).

4   Illustration of the MappGuru Using a Case Study (Addressing a Slum Located in Kigali—Rwanda)

The current addressing system of Rwanda follows a linear referencing method; this method assumes that addresses vary linearly along a feature which is typically a street [11]. Rwandan address is a six to twelve characters code unique for each address. The structure of the Rwandan addressing system is illustrated in Fig. 1. The addresses are based on streets where address codes are sequential and composed by a coded street name and a house number. For example 1002 KK 204 St, 108 KN 22 Rd.

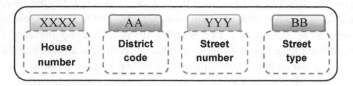

Fig. 1. Structure of the Rwandan Address

Kigali like any other African cities, lately, recognized informal or unplanned settlements which lead to rapid urbanization with a number of slums. These types of settlements impacted development and the provision of improved infrastructure. By 2012, unplanned settlements housed 83 per cent of the city's population (ref World Bank 2012a). Addressing these densely populated areas where multiple household may be living on a plot as small as 10 m² with no access to streets, is a challenge.

In accordance with the City of Kigali's plan to transform existing informal housing, it is expected that each person will have an address identifying them; therefore in the near future, the use of addresses to identify people and properties might be mandatory and each property is expected to have an address.

However, this cannot be achieved using the current addressing system due to the fact that it cannot be extended to properties that are not located alongside streets such as houses in informal settlements neighborhoods and rural areas and also because it is based on streets, and uses district codes KN, KK and KG to name roads or streets, which results in assigning addresses with wrong district initials. Figure 2 shows a location where addresses are based on KN 5 Rd while properties are located in different districts.

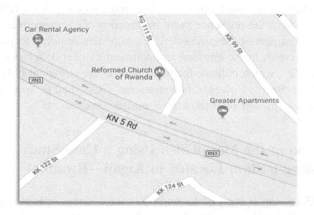

Fig. 2. Different district coding in the same area (Google Maps, 2018)

4.1 MappGuru Case for the City of Kigali (Rwanda)

MappGuru, which is conceptually based on area zoning rather than linear features, has an advantage as it removes the ambiguity of addresses assignment; MappGuru assigns addresses based on an intelligent zoning feature to reflect precisely the locality and to estimate the distance from the property to the reference point chosen from within each district.

Five steps are required to implement MappGuru. (1) Choosing a reference point. (2) Defining distance zones from the reference point. (3) Dividing distance zones into orientation zones. (4) Dividing orientation zones into postcode zones. (5) Defining Quarters. (6) Address zones, (7) Assigning property addresses.

1. Reference point: Kigali has 3 districts with relatively different sizes, in such case, it is reasonable to allocate a reference point for each district. Figure 3 depicts possible reference points in Kigali. Kigali roundabouts could be best candidates for references. Reference points are coded using districts names; to keep the uniformity with the existing addressing. (1) Nyarugenge is KN (Kigali Nyarugenge), (2) Gasabo is KG (Kigali Gasabo), (3)Kicukiro is KK (Kigali Kicukiro).

Fig. 3. Reference point for the City of Kigali (Google Maps, 2018)

2. Distance zones: Since Kigali districts are different in sizes, there 2 possible configurations. Nyarugenge and Kicukiro districts are relatively small, thus utilize concentric rectangles separated by blocks of 2 level 3 NACs whereas Gasabo district which is bigger utilizes blocks of 3 level 3 NACs.

   In principle, these concentric rectangles would go over to adjacent districts but to avoid overlapping rectangles, they are discontinued at the boundaries of each district. Figure 1.5 depicts how distance zones are configured over the City of Kigali (Fig. 4)

3. Orientation zones: Each of the distance zones is divided into orientation zones.

Fig. 4. Distance zones for the City of Kigali

Fig. 5. Orientation zones for the City of Kigali

Nyarugenge and Kicukiro districts orientation zones have groupings of 4 level 3 NAC zones whereas Gasabo has bigger orientation zones of 9 postcodes groupings. Figure 5 illustrates how the orientation zones are configured.

4. **Postcode zones**: The next step is to locate postcode zones inside orientation zones. Two configurations are considered for Kigali. In Nyarugenge and Kicukiro 2 × 2 level 3 NAC cells orientation zone are used. In this case, one orientation zone consists of 4 postcode zones, whereas in Gasabo 3 × 3 level 3 NAC cells orientation zone is used. In this case, one orientation zone encloses 9 postcodes zones named after numbers from '0' to '8' [9] (Fig. 6).

Fig. 6. Postcode zones for the City of Kigali

Postcodes are formed starting from the reference point. Distance zones are assigned number from 0 to 9 in an ascending order moving away from the reference point. Depending on the direction (North, East, West, South), codes for the first level orientation zone are obtained. For example, the first distance zone in the South is coded "1S".

The first level orientation zones are split into two by a middle reference zones, marked by "0" which is added to the fist level orientation zone code to indicate the reference zone; using the same process, second level orientation zones are coded adding two characters indicating the distance and orientation from the reference zone, as an example, the reference zones 1S0 splits the zone 1S into East and West; if we moving two zones in the East, the code becomes 12SE whereas moving two blocks in the west yields the code 12SW [7].

Postcodes inside an orientation zone depend on the size of the district. Since Nyarugenge district use 2 × 2 level 3 NAC cells orientation zones, postcode zones are identified by numbers from 0 to 3. Table 1 illustrates the assignment of a postcode to a locality in Nyarugenge/Kigali.

Table 1. Assignment of a MappGuru postcode

Designation	Value	Code
Province	City of Kigali	K
District	Nyarugenge	N
Reference	Main roundabout	KN
Distance zone	First, South	1S
Orientation zone	Middle	0
Postcode zone	First	0
Quarter	Eighth	8
Postcode	KN1S0	

5. Quarters: The next step is to define quarters. The numbering of Quarters is illustrated in Fig. 7.

Fig. 7. Numbering of Quarters

6. Address zones: The 8th quarter is chosen and divided into address zones as illustrated in Fig. 8 as a yellow rectangle.

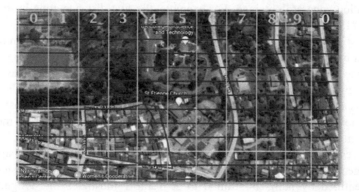

Fig. 8. Address zones

5   Numbering Guidelines

Address zones are 25 m × 50 m blocks. In a structured area, this will fit a maximum of 3 plots with a minimal standard size of 20 m × 15 m. However, since the current project encompasses addressing slums and unstructured cities; following guidelines have been developed to assign addresses to properties within an Address zone.

Each Address zone is virtually divided into two sides (North and South); Numbering begins from West to East using even numbers on right side and odd numbers on left side. Figure 9 illustrates an ideal numbering case.

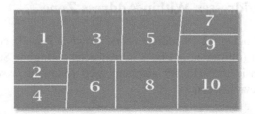

Fig. 9. Typical numbering within an Address zone

Given that in unstructured areas dwellings are randomly arranged. The first part of guidelines describes the internal structure of the Address zone; that is, the way properties are arranged and accessed within the address zones with regard to streets and pathways. The second part of the guidelines defines rules to numbering house.

5.1   Identifying Accesses

This process involves identifying

1. Accesses to properties within the address zone, which in most cases are pathways.
2. Properties' entrances (north, west, south, east)
3. The nature of each property which can be:
   (a)   A compound with a single entrance and a single household
   (b)   A compound with a single entrance and multiple households
   (c)   Habitable houses
   (d)   Small businesses
   (e)   Non habitable houses (toilets, kitchens, …)
   (f)   Public utilities.

5.2   House Numbering Rules

1. Properties are numbered from East to West with even numbers in the Northern side and odd numbers in the Southern side,
2. A property belongs in an address zone if more than a half of the property is in that address zone,

3. A single compound with a single household is assigned a single address,
4. A single compound with multiple households is assigned different addresses; an address for each household,
5. Small businesses are assigned individual addresses,
6. Private non habitable properties are not given addresses (toilets, pet's houses...)
7. Public utilities (water, electrical power station...) are assigned an address each,
8. Properties on streets are given addresses; however they might prefer to keep the existing street addresses.

6   Assigning Addresses Within Address Zone

In this case study, 7 out of the 19 address zones shown in Fig. 10, covering a four-sided unstructured quarter (illustrated in Fig. 8 in red) neighboring the Saint Etienne Church in Kigali were considered. Houses are numbered following a set of guidelines as presented in the previous section.

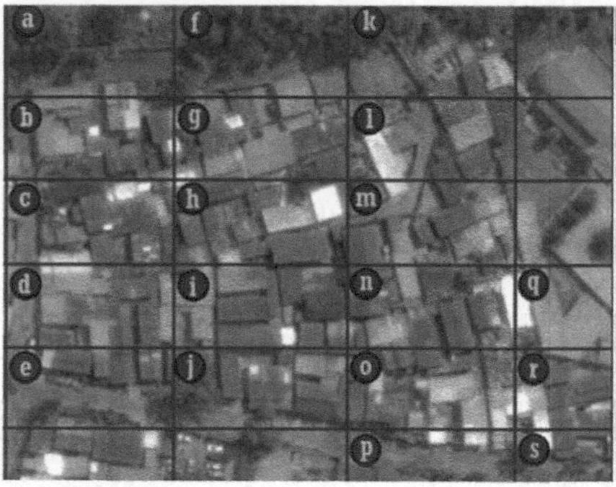

Address zones' codes			
a: KN08 1S 26		k:	KN08 1S 46
b: KN08 1S 27		l:	KN08 1S 47
c: KN08 1S 28		m	KN08 1S 48
d: KN08 1S 29		n:	KN08 1S 49
e: KN09 1S 20		o:	KN09 1S 40
f: KN08 1S 36		p:	KN09 1S 41
g: KN08 1S 37		q:	KN08 1S 59
h: KN08 1S 38		r:	KN09 1S 50
i: KN08 1S 39		s:	KN09 1S 51
j: KN09 1S 30			

Fig. 10. Case study address zones and their codes

6.1   Field Visit Report

The team updated maps, indicated accesses to the location and gained positive response from local authorities, properties' owners and small businesses' owners.

Process: The team visited properties following the updated map. For each of the address block, a form was filled indicating its Address block code, the number of accesses and the number of properties in the block. Then, for each of the properties a form indicating its category, the number of habitable and non habitable houses in the property was filled. After gathering all this information, the research team assigned numbers to properties. The visits produced the following results:

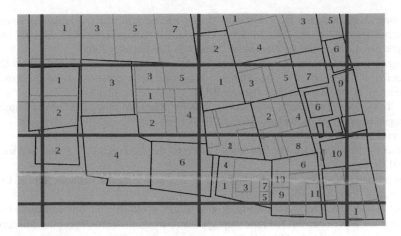

Fig. 11. House numbering map (11)

1. A map of house numbers/addresses,
2. A database of addresses and other information of the sample location (Fig. 11).

7 Conclusion

In general rolling out address system projects take several years to complete; on the other hand updates or changes are inevitable as new or substitutions of city structures occur very often. Considering that there is not a standardized addressing system, the problem becomes difficult and costly.

In this paper we highlighted criteria's of well designed addressing system, which simplifies the task of manually mapping the address to properties and make it easier to define new addresses based on a postcode and automatically adapts to changes without requiring intensive field work. A new solution is presented, which aims at optimizing and revolutionizing the addressing systems—MappGuru a Simple, unified addressing system; with a broader and complete coverage, designed to extends to all places whether structured, unstructured, inhabited or not and be used as a universal addressing standard or standard postcode. In this paper a section is dedicated for a case study to illustrate the MappGuru in Kigali-Rwanda.

References

1. Joint Committee on Communications, Energy and Natural Resources: Annual Report 2007–2009 (2010)
2. Coetzee, S., Cooper, A.K., Lind, M., Wells, M.M., Yurman, S.W., Griffiths, N., Nicholson, M.J.: Towards an international address standard. In: Proceedings of the 10th International Conference for Spatial Data Infrastructure (2008)

3. Cooper, A.K., Coetzee, S.: The South African address standard and initiatives towards an international address standard. In: Proceedings of the Academic Track of the 2008 Free and Open Source Software for Geospatial (FOSS4G) Conference (2008)
4. Potgieter, P.: Mapcodes: a new standard for representing locations. EE Publishers (2014)
5. Geelen, P.: Mapcode: the Netherlands a public location reference standard. Distribution Courtesy of the Mapcode Foundation (2015)
6. N. T. G. Inc.: The natural area coding system. Retrieved July, 2017, from http://www.nacgeo.com/nacsite/documents/nac.asp (2016)
7. Eircode: Code of Practice (2015)
8. Office of Inspector General: The untold story of the zip code. U.S. Postal Service, Tech (2013)
9. Rwerekane, V., Ndashimye, M.: Natural area based postcode scheme. Int. J. Comput. Commun. Eng. 6(3), 161–172 (2017)
10. Post & Parcel: Addressing the world: how geocodes could help billions start using the mail. Retrieved November 2011. From: http://postandparcel.info/43564/in-depth/addressing-the-world-how-geocodes-could-help-billions-start-using-the-mail/
11. Farvacque-Vitkovic, C., et al.: Street Addressing and the Management of Cities. The World Bank (2005)

Toward Mobility Parameter Planning and Enhancement of QoS for 4G Network in Fast Developing Cities_Kigali City

Richard Musabe[✉], Victoire M. Ushindi, Atupenda Mugisha,
Manizabayo Emmanuel, Vienna N. Katambire, Gakwerere Eugene,
and Gaurav Bajpai

School of ICT, College of Science and Technology, University of Rwanda, P.O.
Box: 3900, Kigali, Rwanda
{rmusab10, ushvictoire, atupenda, zabayo92, katavienny72,
gakwerereeugen, gb.bajpai}@gmail.com

Abstract. One of the most engaging challenges for mobile operators in fast developing cities such as Kigali city are switching 3G to 3.9G/4G and to deal with the fast increase of the amount of data traffic . However, there is some overlapping between 3G and Long Term Evolution (LTE). Further, mixed 2G, 3G and LTE network remain in place due to the high cost. So it is needed to measure the four basic radio resource management (RRM) in LTE system in order to ensure the desired quality of services. Moreover the current mobile infrastructure network is unable to handle such massive consumer demand at good quality of service. However, low cost network offloading technique such as relay nodes need to be adopted by operators. The aim of this paper is twofold. First, to collect current network radio coverage data using network drive testing method. Network drive testing data collection was based on the three key performance indicators in mobile parameter as Reference Signal Received Power (RSRP), Reference Signal Received Quality (RSRQ) and Received Signal Strength Indicator (RSSI). A measurement of channel quality represented by Signal to interference plus noise ratio (SINR) is used for measuring link adaptation. Second, study and enhance user QoS based coverage performance by focusing on relay nodes network offloading technique. The work should help to optimize end to end delivery quality for operators in fast developing cities.

Keywords: LTE · Drive test · RSRP · RSRQ · QoS

1 Introduction

Mobile technology which started simply from supporting voice, then data, and now multimedia data; was not enough for human being. Few years back has been proven by the rapid increasing in demand for high data rate applications on mobile device, now in fast developing cities such as Kigali city the volume of mobile data traffic has surpassed the voice traffic and access to mobile internet has become fundamental and integrated part of every day and professional life.

© Springer Nature Switzerland AG 2019
M. Ben Ahmed et al. (Eds.): SCA 2018, LNITI, pp. 165–178, 2019.
https://doi.org/10.1007/978-3-030-11196-0_16

The common marked 4G which is a system developed within the 3rd Generation Partnership Project (3GPP), and 3GPP LTE was the step beyond 3G and towards the 4G [1, 2]. The 3GPP-Standardized Orthogonal Frequency Division Multiple Access in downlink and Single Carrier Frequency Division Multiple Access in Uplink based, which mainly focused on providing high peak data rates, high spectral efficiency and improving system covered. As telecom operator in fast developing cities are struggling to accommodate the existing demand of mobile users and marge 4G network to the existing infrastructure, new data consuming application are emerging in daily routine of mobile users. 4G as the promising technology to achieve better performance such as higher speed, larger capacity and so forth [3], remains the only solution to overcome user demand of high data rate as 4G allows the implementation of heterogeneous mobile network. In LTE radio access network, consists of eNodeBs where all RRM-related functions are performed. The RRM measurements are mainly RSRP, RSRQ and RSSI; ensures that the air interface resources are efficiently utilized so that network efficiency is maximized in one of the main step through network resource optimization. Where network mobility performance has to be reliable and maintain quality without experiencing any interrupt in the service.

Among RRM functions, two measurement criteria such as RSRP and RSRQ are used to make a cell reselection or handover (HO) decision. When the RSRP or RSRQ of the serving cell falls below the RSRP of RSRQ of the neighbor cell by a predefined handover margin for certain period of time, handover occur. Figure 1 show that when a RSRP of the neighbor cell exceeds the single strength of a serving cell by a predefined HO margin for certain duration the handover triggers [4].

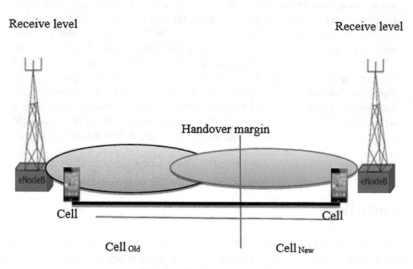

Fig. 1. Handover scenario

In developing country, switching from 3G to 4G in mobile require many efforts. Even if in today's world technology is getting very sophisticated and continues work is going on to make it much better. Rwanda especially Kigali city has proven a great development in ICT, where currently switched to 4G, but still in mobile communication (cellular phone) especial voice and data, 3G is still being optimized before switching to 4G which was evolved after EDGE, UMTS and HSPA Evolution [5].

For fast developing cities, improving mobile system performance compared to existing is one of the main requirements and challenges from network operators, to ensure the competitiveness of LTE. LTE enhanced the Universal Mobile Technology Services (UMTS) in a set of points on account of the future generation cellular technology developing countries needs and growing mobile communication services requirements. First, this paper presents a drive testing report done on Rwandan mobile communication operators' networks. Figure 2 shows measurement location. In blue color indicates path between nodes. It focuses on idle and connected mode mobility parameters design to plan how to switch to 4G in mobile communication. Drive test route was based on three main cluster with high traffic within Kigali City, moreover drive testing was conducted in network pick hours. This gives a lot of flexibility for evaluating network in this pick hours and come up with a low cost offloading techniques to overcome network poor quality of service (QoS).

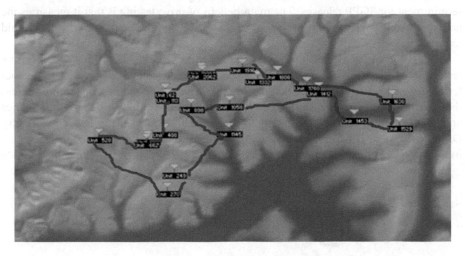

Fig. 2. Drive testing itinerary

This paper is organized as follows, Sect. 2 present related studies. Section 3 present preformed drive testing. In Sect. 4 an overview of network mobile parameter was discussed as main parameter used for data collection and Section analyze drive test result and discuss network performance optimization. Conclusion and area of future work are given in Sect. 6.

2 Related Work

Due to the advance in content sharing, social media application, radio core network has been to meet use demand. On the other side, mobile internet operator has been facing a lot challenge while optimization network coverage. In [6] the author present methods of network optimization by drive testing, where he described the main benefit of drive testing and suggest the actual network coverage measurement and performance that a user on the actual drive route would experience. In [7] the author suggests the method to cover network coverage holes by increasing user quality experience. This paper contributed to the above cited research by investigating 3GPP standard network coverage in fast developing cities by considering Kigali city as a case study.

3 Performed Drive Testing

Drive testing provides an accurate real-world capture of the RF environment under a particular set of network and environment conditions. Drive testing is the best way to analyze and enhance the performance of the network. This CQT Test was performed in the way for characterization of propagation and fading effects in the channel, with the main objective of collected field data and analyses those data according to ITU requirement for 3GPP releases. The main benefit of drive testing is that it measures the actual network coverage and performance that a user on the actual drive route would experience.

The performed drive testing was phone based as seen in Fig. 3; this type of drive testing is useful for evaluating basic network performance and is essential to characterizing the end-user experience while using the network. Phone-based systems address the need to verify network settings such as cell selection and re-selection boundaries and to measure the voice and data application performance in the live network [8]. Drive test system is general built around two measurement components, instrumented mobile phone and measurement receivers. GPS collecting data of latitude and longitude

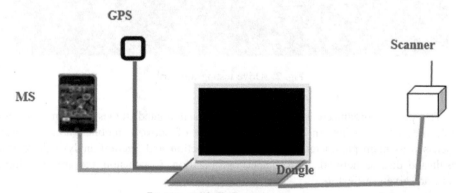

Fig. 3. Drive testing hardware

of each point/measurement data, time and speed. MS collecting mobile data such as signal strength, best server. SCANNER is used to collect data throughout the network, since the mobile radio is a limited and does not handle all the necessary data for a more complete analysis.

3.1 Measurement Setup

A measurement laptop with LTE dongle connected to it, the dongle was used as end user device which operates in both LTE and 3G modes of operation. GPS was connected to the measurement laptop to determine the location coordinates. Initial position to start the measurement was chosen such that UE lie on 3G and 4G dominance area in connected mode. The drive test was performed using two phones. A mobile performing calls for a specific number from time to time, configured in the collecting software. The call test was of long duration and this long call served to verify if the handovers in the network.

4 Key Pefromance Indicator

The types of parameters are measured for 3GPP as for other cellular technologies. Beyond the essential protocol log, which provides visibility of the fundamental interaction with the network, the initial focus was on RF coverage and quality. The key 3GPP LTE items measured in the drive testing are the RSRP, RSSI and RSRQ which comprise the key performance indicators for evaluating the 3GPP LTE physical layer.

4.1 Reference Signal Received Power (RSRP)

RSRP indicates the strength of the received reference signal, it is a cell specific signal strength related metric that is used as an input for cell resection and handover decisions [5].

$$RSRP = \frac{1}{X} \sum_{X=1}^{X} T_{rY}, x \tag{1}$$

where T_{rY}, x is the estimated received power (in watt) of the xth Reference Signal Resource Element transmitted from the first NodeB antenna port. RSRP normally expressed in dBm, is utilized mainly to make ranking among different candidate cells in accordance with their signal strength. RSRP values decreases logarithmically with respect to the distance from the base station based on propagation loss model [see Fig. 1]. As a result, the RSRP is basic parameter for evaluating the level of the radio wave received from the eNodeB and can be used to broadly determine the eNodeB fixed setting conditions [9]. Generally, the reference signals on the first antenna port are used to determine RSRP, however the reference signals sent on the second port can also be used in addition to the RSs on the first port if UE can detect that they are being transmitted.

4.2   Reference Signal Strength Indicator (RSSI)

RSSI is the measured power of all bands, it is the linear average of the total received power observed only in multiplexing methods (OFDM) symbols carrying reference symbols by UE from all sources including co-channel non-serving and serving cells, adjacent channel interference and thermal noise, within the measurement bandwidth [10–12]. RSSI measurement provides information about total received wide-band power; it can be simply computed as the sum of noise, serving cell power and interference power.

RSSI is a parameter which provides information about total received wide-band power including all interference. It is the total power UE observes across the whole band. This includes the main signal and co-channel non-serving cell signal, adjacent channel interference and even the thermal noise within the specified band.

4.3   Reference Signal Received Quality (RSRQ)

RSRQ indicates the quality of the received reference signal for which RSRP is measured. It measurement is a cell specific single quality metric.

RSRQ plays an important role when RSRP value is not sufficient in making cell reselection and handover decision that why It is calculated as the ratio of RSRP and RSSI.

$$RSRQ = N*RSRP/RSSI \tag{2}$$

N: Resource Block Number

Since the RSSI measurement band tends to change with the 3GPP LTE bandwidth, the resource block number is standardized as can be seen in Eq. 2 and as RS (resource block) is fixed, RSSI is influenced by traffic and its value changes also according to the traffic volume [13].

4.4   Relationship Between RSRP, RSSI and RSRQ

From Eq. 2, it is seen that due to the inclusion of RSSI, RSRQ considers the effect of signal strength and interference. The RSSI term in the denominator is the sum of received power over the OFDM symbol that contain own cell power, thermal noise and interference from other cells [14, 15].

$$RSRQ = N \frac{\frac{1}{X}\sum_{X=1}^{X} T_{rY}, x}{\frac{1}{X}\sum_{n=1}^{Nre} Pn} \tag{3}$$

The implementation of the RSRQ measurement is not defined in detail by 3GPP, so by writing down the RSRQ ratio between two cells on the same carrier frequency, we can establish the relationship between those parameters and that is,

$$\frac{RSRQserv}{RSRQneigh} = \frac{N\frac{RSRPserv}{RSSI}}{N\frac{RSRPneigh}{RSSI}} = \frac{RSRPserv}{RSRPneigh} \tag{4}$$

From Eq. 4, it can be well seen that the ratio of RSRQs only depends on the ration of the RSRPS. Although RSRP is an important measure, on its own it gives no indication of signal quality. From the definition where N is the number of RBs over the measurement bandwidth; as it can be seen this is not the direct measurement, it is a kind of derived value from RSRP and RSSI.

$$RSRQ[dB] = 10\log\frac{RSRP}{RSSI} \tag{5}$$

The calculated RSRQ in dB is also a measurement of signal quality, which means the Signal-to-Noise Ratio. This is shown by Eq. 4. By dividing RSRP by RSSI, it could give some information about the interference as well in addition to the strength of the wanted signal. The RSSI parameter represents the entire received power including particularly important near the cell edge when decisions need to be made, regardless of absolute RSRP, to perform a handover to the next cell [16].

5 Performace Result Analysis and Optimization

In mobile communication system, it is very important for end user device to detect and monitor the presence of multiple cells and perform cell reselection to ensure that it is being served by the most suitable cell. The drive test carried out had the main objectives to measure the coverage and performance of 3G and 4G services in Rwanda. Drive test was the first methods to use in determination the quality of the above mentioned services in Rwanda.

For example, for user equipment being supported by a given cell will monitor the cell information and paging of that cell, but it must continue to monitor the quality and strength of other cells to determine if cell reselection is required [17].

Phone based drive test is useful for evaluation basic network performance and are essential to characterizing the end –user experience while using the network [18], that it's why this type of drive testing was the one to be carried out. It shows the really scenario for a cell search, which is performed to determine the detected cells. Reference signal measurements are made (RSRP, RSSI, and RSRQ).

Figure 4 shows the example of basic scenario of cell resection by user Equipment between two cells. The signal powers for both cells in each time are set up accordingly and for three-time period. It can be seen that the measurement of RSRPs are close to the expected ideal values, and that in each period the user equipment selects the expected cell.

5.1 Drive Testing Result Analysis

As mentioned before, a drive test provides measurement about the real condition of the network by showing how the network is performing, the coverage area or the behavior

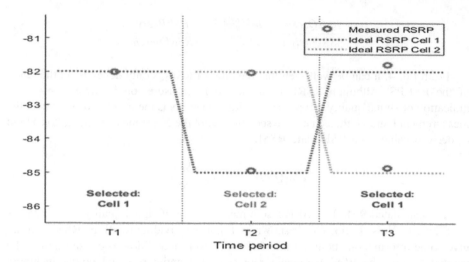

Fig. 4. Basic scenario of cell reselection

of a specific cell or cells. The provided gathered data from the dive test were analyzed. The measured files were analyzed and final data analysis was done by extracting the data into MATLAB. The obtained plots are drawn conclusion for the analysis.

Comparing the RSRP, RSSI and RSRQ graphs in Figs. 5, 6 and 7 respectively; Performed drive tests load environment to obtain the distribution of signals on test routes. The average RSRP shows that many 3G and 4G service drive test route present many coverage holes based on the RSRP measurement reporting range of −140 dBm to −44 dBm and RSRQ of −19.5 dB to −3 dB with 0.5 dB resolution [17, 19].

From Figs. 5 and 6, when examining the relationship between RSRP, RSSI, RSRQ, a match with the identified low coverage areas can be established. So those results reinforce the idea that coverage must be extended, in order to provide a good LTE service in the area. In drive test route, there is some area without dominant cell, where the receive level of the serving cell is similar to the receive levels of its neighboring cells and it can be seen as if the receive levels of downlink signals between different cells are close to cell reselection thresholds. As consequence the SINR of the serving cell becomes unstable because of frequency reuse, and even receive quality becomes unsatisfactory. As a result, frequent handovers or service drops occur on UEs in connected mode because of poor signal quality [see Fig. 6].

It can be observed that both RSRP and RSSI that at every point of measurement slot RSRP is always less than RSSI. From Fig. 7, it can be noticed that lesser the different RSRP and RSSI, better the RSRQ. This is due to the fact that since RSRP is always less than RSSI, lesser the different between RSRP and RSSI implies lesser interference which in turn leads to better received signal quality. Moreover, it can be noticed that there is low coverage level or weak coverage area. The receive level of a UE is less that its minimum access level because downlink receive levels in a weak coverage area are unstable. In this situation, the UE can be disconnected from the network. After entering a weal coverage area, UEs in connected mode cannot be

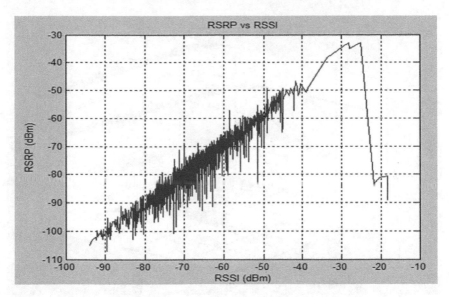

Fig. 5. Average RSRP versus RSSI

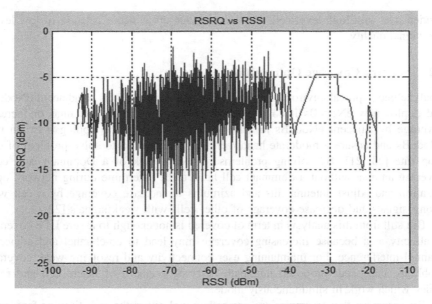

Fig. 6. Average RSRP versus RSSI

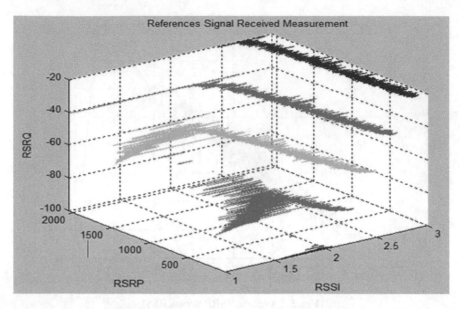

Fig. 7. References signal received measurement

handed over to a high-level cell, and even service drops occur because of low levels and signal quality.

5.2   Network Coverage Optimization

Analyze geographical environments and check the receive levels of adjacent eNodeBs and deploy new eNodeBs if coverage hole problems adjusting antennas or increase coverage by adjacent eNodeBs to achieve large coverage overlapping between two eNodeBs and ensure a moderate handover area cab ne used to solve problem of call drop rate [20, 21]. By solving problems related to lack of a Dominant cell, cells covering an area without a dominant cell have to be determine during network optimization and adjust antenna tilts and azimuths to increase coverage by a cell with strong signals and decrease coverage of other cells with weak signals [22].

But still from this analysis in term of covered is not enough to assure LTE coverage to all city area because Increasing coverage may lead to co-channel and adjacent-channel interference. For maintaining user connectivity and resolving weak coverage problems, this paper proposed multipath diversity scenario of macro cell and relay nodes which work in simultaneously mode.

By assuming that transmissions are orthogonal either through time or frequency division, with all relay nodes participation, the available channel and power resource have been equally distributed between all nodes.

$$\gamma_R = \gamma_{S,D} + \sum_{i=1}^{M} \frac{\gamma_{S,R}\gamma_{R,D}}{1 + \gamma_{s,R} + \gamma_{R,D}} \qquad (6)$$

From Eq. 6, the SNR at the combined output is the sum of the instantaneous end to end SNR γ_R at the destination node with $\gamma_{S,D} = |h_{S,R}|^2 \frac{E_S}{N_0}$; $\gamma_{R,D} = |h_{R,D}|^2 \frac{E_S}{N_0}$ and $\gamma_{S,D} = |h_{S,D}|^2 \frac{E_S}{N_0}$ where $h_{S,D}, h_{S,R}$ *and* $h_{R,D}$ are the mutually independent complex gains form base station user equipment and base station to relay node and relay to user equipment respectively and the average energy per symbol is given by E_S (Fig. 8).

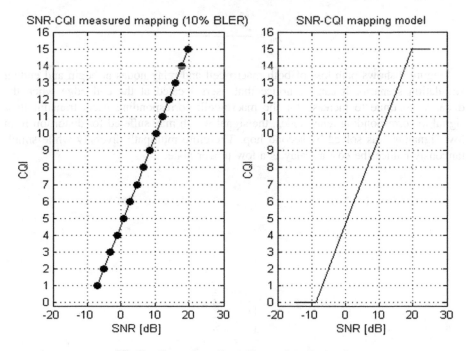

Fig. 8. Channel quality indicator for relay nodes

By focusing on the end to end package delivery and channel quality indicator, the channel quality indicator of second hop consisting of relays, by monitoring its throughput at 10% of Block Error Rate (BLER); the system reflects the efficiency to successfully delivery of traffic.

From drive testing result, it was demonstrated that users of mobile phone usually encounters performance degradation or low data rate due to constraint of distance and low power capture for main cell. By studying the outage channel capacity and probability, source to destination outage channel capacity for all relay participation scheme can be expressed as per Eq. 7:

$$C_{all} = \frac{1}{M+1} \log_2(1 + \gamma_{S,D} + \sum_{R=1}^{M} \frac{\gamma_{S,D}\gamma_{R,D}}{1 + \gamma_{S,R} + \gamma_{R,D}}) \qquad (7)$$

where M indicates the number of relays. From Eq. 7, the outage probability of all relay participation schema can be denoted as:

$$P_{out} = \Pr(\gamma_{S,D} \sum_{R=1}^{M} \frac{\gamma_{S,R}\gamma_{R,D}}{1+\gamma_{S,R}+\gamma_{R,D}} < \gamma_{th}) \tag{8}$$

where γ_{th} is the Signal to Noise Ratio (SNR) threshold value and can be denoted as

$$\gamma_{th} = \frac{2^{(M+1)r} - 1}{\gamma}$$

Figure 9 shows path loss of both macro cell and relay nodes at begin and ending simulation scenario. It can be notice that users located at the cell edge when the distance continue to increase for the macro-cell; corresponding users transmit their signal to corresponding relay nodes, the signal which may suffered for distortion from low signaling was solved by second hop. Therefore multipath diversity while simultaneously reduce the cost as relay is a low power node.

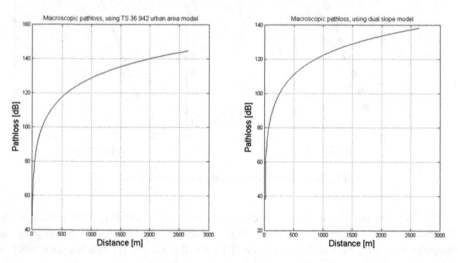

Fig. 9. Outage probability

6 Conclusion

In this work, an analysis of some practical measurement results recorded from a drive testing on a live 3G and 4G network in Kigali city are presented to verify the performance end user quality of service and identify network gap or network coverage holes. So it would be safe to write that, in 3 GPP LTE release, RSSP provides information about signal strength and RSSI helps in determining interference and noise information. This is the reason, RSRQ measurement and calculation is based on both RSRP and RSSI.

It was noticed that when the RSRP and RSRQ of the serving cell drops below the RRM measurement of the neighbor cell, the handover event occurs to maintain the ongoing service. From the data analysis result, the network presents some coverage page gaps and dominant cell which deteriorate quality of service. To ensure good quality of serve performance network coverage was analyzed. This paper studied relay nodes as low cost network offloading technique in fast developing. Relay node simulation was based on multipath diversity scenario and the result showed that signals which may suffered for distortion from low signaling was solved by second hop which relay nodes hop. The future work is intended to study relay node interference of co-channel and adjacent channel.

References

1. Navita, A.: A survey on quality of service in LTE network. Int. J. Sci. Res. Index Copernicus value (2013)
2. Holma, H., Toskala, A.: LTE for UMTS: OFDMA and SC-FDMA Based Radio Access. Wiley (2009)
3. Afroz, F., Subramanian, R.: SINR, RSRP, RSSI and RSRQ measurement in long term evolution networks. Int. J. Wirel. Mob. Netw. (IJWMN) 7 (August 2015)
4. Luan, L., Wu, M., Shen, J., Ye, J., He, X.: Optimization of handover algorithm in LTE high speed railway network. Int. J. Digit. Content Technol. Appl. 5 (2012)
5. Holma, H., Toskala, A. (eds.): WCDMA for UMTS, 3rd edn. Wiley
6. JDSU "Drive Testing LTE" White Paper, Test & Measurement Regional Sales, @2017 JDS Uniphase Corporation 301731160000212 (Feb 2017)
7. The Femto Forum. The Best That LTE Can Be—Why LTE Needs Femtocells. Femto Forum White Paper (May 2016)
8. JDSU, "Drive Testing LTE," Write paper (2011)
9. Sesia, S., Baker, M.: LTE-The UMTS Long Term Evolution: From Theory to Practice. Wiley, Chichester, U.K. (2009)
10. 3GPP TS "LTE; Evolved Universal Terrestrial Radio Access (E-UTRA); Physical layer; Measurements", 36.214 version 11.1.0 Release 11 (2013)
11. Anas, M., et al.: Performance evaluation of received signal strength based hard handover for UTRAN LTE. IEEE VTC (2007)
12. 3GPP TS 36.133, v10.1.0 (March 2011)
13. Legg, P., et al.: A simulation study of LTE intra-frequency handover performance. In: IEEE VTC (2010)
14. Salo, J.: Mobility Parameter Planning for 3GPP LTE: Basic Concept and Intra-Layer Mobility (2010)
15. Afroz, F., Subramanian, R., Heidary, R.: SIRN, RSRP, RSSI and RSRQ measurement in long term evolution network. Int. J. Wireless Mob. Netw. (IJWNM) 7(4) (Aug 2015)
16. Hamao, S., Yoshida, Y.: W-CDMA/LTE Area Optimization using ML8780A/81A (2009)
17. Johansson, B., Sunduin, T.: LTE test bed. Ericcson Rev. 84(2007):1, 9–13 (2010)
18. Döttling, M.: Challenges in mobile network operation: towards self-optimizing networks. In: Nokia Siemens Networks: Nomor Research GmbH, Munich, Germany, Ingo Viering21
19. Stefania, S. Issan, T., Baker, M..: LTE-The UMTS Long Term Evolution from Theory to Practice. Wiley (2009)

20. 3GPP ETSI TS 136 214: LTE; Evolved Universal Terrestrial Radio Access (E-TRA); Physical Layer; Measurements (April 2011)
21. Esswie, A.A.: Intra-Cluster Autonomous Coverage Optimization for Dense LTE-A Network. Network Performance Group, North Africa RSRC, Huawei Technologies (2017)
22. Aldhaibani, J.A., Yahya, A., Salman, M. K.: On coverage analysis for LTE-a cellular networks. Int. J. Eng. Technol. (IJET) (2017)

Towards a Framework for Participatory Strategy Design in Smart Cities

Aroua Taamallah[1(✉)] , Maha Khemaja[2] , and Sami Faiz[3]

[1] ISITCom, GP1 Hammam Sousse 4011, Sousse, Tunisia
aroua_taamallah@yahoo.fr
[2] ISSATS, Taffala 4003, Sousse, Tunisia
maha_khemaja@yahoo.fr
[3] ISAMM, CP 2010, Tunis, Tunisia
sami.faiz@insat.rnu.tn

Abstract. Smart Cities Strategies are designed to mitigate one's city problems and transform it into a smart one. This transformation is repeatedly reused by cities around the world. The complexity of the multi-disciplinary domain of smart cities renders this transformation is very hard. Unless, it is done with the help of experts in the domain. To facilitate this transformational process, this paper relies on the definition of design process of smart cities strategies and on the use of ontologies as a proven formalism to represent knowledge and expertise from experts, more specifically well-designed Ontology Design Patterns (ODP). The use of ODPs seems to be a promising solution that drives the process of cities transformation. The definition of the design process and of the Ontology Design Patterns is guided by a literature review on smart cities strategies. A web-based platform is then used to provide a common space for stakeholders to communicate and propose transformational strategies based on ODPs. The proposed solution is then validated by a use case related to educational domain in the city of Sousse, Tunisia.

Keywords: Smart city · Strategy design · Strategy · Ontology
Ontology design pattern · Design

1 Introduction

Compared to rural areas, cities offer more resources and facilities. Such places are increasingly attracting people for living and working. This phenomenon leads to exponential augmentation of cities inhabitants. According to United Nations [1], this number will reach 70% by 2050. This explosion will cause serious social, environmental and economic problems.

Smart city is a relatively new paradigm that has appeared to solve modern cities problems and increase the quality of life (QoL) using Information and Communication Technologies (ICT). The stakeholders always act to transform cities into smarter ones by making use of innovative strategies. However, thinking and developing strategies is a repetitive and difficult process.

© Springer Nature Switzerland AG 2019
M. Ben Ahmed et al. (Eds.): SCA 2018, LNITI, pp. 179–192, 2019.
https://doi.org/10.1007/978-3-030-11196-0_17

Stakeholders collaboration is a fundamental pillar of smart cities. It can lead to the design of effective strategies. However, stakeholders fail to collaborate in existing smart cities. The emergence of digital tools and web platforms can empower stakeholder's collaboration. The identification of a generalized design process of smart city strategies can assist stakeholders and can help in identifying the tools that may be integrated in the web-based platform.

Another issue is treated in this paper which is the formal design of strategies and knowledge structuring in the web-based platform. The use of an ontology seems to be a promising solution to formally and semantically design of strategies. The formal specification of strategies is important because it allows to manipulate, adapt and simulate them. It also provides a machine and human understandable vocabulary and enables related knowledge organization, management and sharing. The fact that stakeholders have many concerns and many points of view and use various sources to populate the ontology, the use of a modular ontology based on validated Ontology Design Patterns (ODPs) can contribute to the best structuring and the precise semantic of strategies. ODPs are templates that are usually used to model recurrent ontology design problems [2] in many domains such as biology [3], physics [4], gaming [5]. But none of these works have considered the strategy design.

The contributions of this paper are therefore (1) to identify the steps belonging to the design process of smart cities strategies, (2) to propose a Strategy Design Ontology based on set of Ontology Design Patterns (ODPs) that are included in (3) a strategy design platform. The web-based platform can offer a set of tools to increase stakeholder's collaboration in strategies design.

The reminder of the present paper is therefore structured as follows: the Sect. 2 provides works related to strategies definitions, classifications and platforms. Section 3 describes the methodological approach. Section 4 describes the proposed solution including the identification of the design process of strategies and the design of ODPs and their use in a strategy design ontology. Section 5 describes the web platform that ensure stakeholders collaboration during strategies design. Section 6 proposes a use case for strategies design in educational domain. Section 7 incorporates an evaluation of the proposal. Section 8 concludes and gives perspectives of the current work.

2   Related Work

Actually, there is no a clear definition of a smart city strategy, yet a very few works try to give it a definition. The authors of [6] mentions that a smart city strategy is "needed to overcome the existing gap for the actual delivery of smart services to the city's community and to achieve the smart city mission". [7] considers a strategy as a roadmap for stakeholders to solve the city problems. She defines a strategy as "designing and steering a common vision of the city". Moreover, even though many works tried to propose taxonomies or strategies classifications. These taxonomies are quite different and do not tackle the same perspective. Authors in [8] distinguish between government strategy and domain strategy. They define government strategies as "tactic documents with which countries evolve or align to common supranational visions." Compared to government strategies, domain strategies are related to more centralized visions and are designed to evolve city domains related to a specific market or people (transportation, healthcare, economy, etc.).

Author of [9] uses a spatial perspective of smart city strategies and classifies those strategies into categories according to the development level of a city, the level of the strategy, the type of the infrastructure and the application domain.

Author of [7] introduces steps for strategizing a smart city through a model called SMART. The first step consists on the definition of a strategy. The second step is called Multidisciplinary and includes resources and stakeholders coming from the different disciplines that are required for a successful transformation of the city. Appropriation is the third step where actors issued from different backgrounds, collaborate to develop projects. The fourth step is a roadmap that consists on identifying an action plan for definition of projects. The final step consists on implementing the projects by providing services to end-users.

The development of effective and realistic strategies requires stakeholders' inclusion and collaboration as it is a fundamental pillar of the smart city [10]. Before talking about collaborative governance of the city, we should identify the stakeholders that are engaged in that process. In [11], there is a more diversified classification of stakeholders where people, government, companies/industries and universities are considered as the main groups of stakeholders whereas planners, developers, financing organizations and Non-Governmental Organizations (NGOs) are considered as additional groups of stakeholders. Participatory (or collaborative) governance is defined as "a governing arrangement where one or more public agencies directly engage non-state stakeholders in a collective decision-making process that is formal, consensus oriented, and deliberative and that aims to make or implement public policy or manage public programs or assets" [12]. However, stakeholders fail to collaborate in existing smart cities [11]. The emergence of digital platforms may empower that collaboration as it can change radically the existing forms of governance and can strengthen the development process of innovative strategies [13]. STORM Clouds and Online S3 are two digital platforms for collaborative governance. The first platform facilitates citizens participation in collaborative innovation activities. Those activities may include strategies design and implementation activities [14]. This platform is not specialized on strategies development, but its applications may be used in that context. The second platform is designed to support the development of smart specialization strategies [13]. It offers a set of applications that can be used as stand-alone applications or in a sequential manner during the development process of strategies. The disadvantages of Online S3 are that it does not use semantic technologies for the structuring of knowledge coming from various sources and from different stakeholders and for the formal definition of strategies. The semantic technologies also offer a machine and human understandable language that facilitates the communication among stakeholders and the automation of the development process of smart cities strategies. Those issues are treated in this paper.

The proposal of this paper considers therefore the following points (1) the identification of steps included in a strategy design process, (2) the development of ODPs for formal specification of smart city strategies, the integration of the ODPs in addition to other services in (3) a web-based platform to ensure collaborative design of smart cities strategies. The development of the web-based platform and the identification of included services and tools is assisted by strategy design process steps.

3   Methodological Approach

The methodological approach used in this paper is as follows:

First, defining a design process of smart cities strategies as a support for stakeholders. Second, designing a set of ODPs that are used for a formal definition of strategies during the design process. Third, using the ODPs in addition to other tools and services in a web-based platform to implement the design process of smart cities strategies.

The methodological approach for the design of ODPs is inspired from [5] and consists on the following phases.

The first phase consists on the description of the domain for which the ODPs are intended. The second phase aims to identify the key concepts from the domain. The third phase aims to search for ODPs in the literature, or to extract ODPs from ontologies according to the concepts identified. The fourth phase aims to create new required ODPs by following the following steps:

The first step is called specification and consists on the definition of the purpose and the identification of the competency questions (CQs) that the proposed ODPs must answer. Conceptualization is the second step. It consists on structuring the concepts in meaningful models by enumerating relationships, properties and attributes. Formalization is the third step. It consists on writing the appropriate axioms and rules. The fourth step aims to implement the ODPs. Evaluation is the final step. It consists on testing the ODPs against scenario(s) by using a set of criteria (consistency, conciseness, and competency).

4   The Proposed Solution

In this section, the proposed design process and the modules corresponding to each of its steps are described. The design process serves as a guideline for different stakeholders while the formal definition of strategies is important because it allows to manipulate, adapt and simulate them. The use of Ontologies allows to provide a formal representation and description of concepts related to the smart city strategies and to build relationships among them. It also enables to provide a machine and human understandable vocabulary and enables related knowledge organization, management and sharing. In addition to that, ontologies allow reasoning capabilities over knowledge. Stakeholders participating in the design process of strategies have many concerns and use many resources to populate the ontology. The use of a modular ontology (Strategy Design Ontology) based on validated ODPs facilitates the structuring of strategies related knowledge.

4.1 Design Process of Smart City Strategy

The design process of Smart City strategy guides stakeholders all over the world to design strategies in a generalized manner (see Fig. 1). It is identified from the analysis of various smart cities and helps to identify ODPs and digital applications that facilitate strategies design [15].

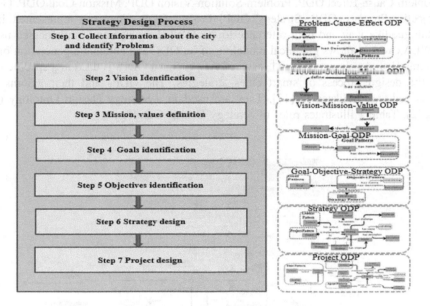

Fig. 1. Strategy Design process and the corresponding ODPs

The Strategy Design Process is composed of the following steps:

The first step is the city problems identification and analysis. The stakeholders identify the city problems from various sources (statistics, interviews, surveys, etc.).

The second step comprises the vision identification where the stakeholders identify what the city will look like in the future. The vision should be clear, powerful and specific to the city. The third step includes mission and values definition. The stakeholders define the strategic directions (mission) sharing various values such as collaboration and cooperation, responsibility and innovation, etc. In the fourth step, they identify the goals to achieve. The goals are determined with reference to identified problems by aligning each problem to each goal as a desired solution to that problem.

During the fifth step, objectives supporting each goal are determined. Conversely to goals which are intended achievements for a long term, objectives are intended achievements for a short term. The sixth step comprises strategies definition. The stakeholders define strategies that might achieve the identified objectives. Two categories of strategies may be defined top-level strategies and sub-strategies. A top-level strategy describes a more abstract plan while sub-strategies (or domain-strategies) are defined under strategic axes. During the seventh step, the defined strategies are implemented by means of scheduled projects (attributing financial and human resources).

4.2   Strategy Design Ontology for Designing Smart City Strategies

Ontology Design Patterns that drive the design of smart city strategies are respectively Problem-Cause-Effect ODP, Problem-Solution-Vision ODP, Mission-Goal ODP, Goal-Objective-Strategy ODP, Strategy ODP, Project ODP. They are represented in a catalogue-like way and with reference to the OWL language and implemented using OWLAx which allows graphical modeling of an ODP and automatic generation of its axioms [16]. As shown in Fig. 1, each of the proposed ODPs is used in a specific step of the design process of smart cities strategies. The ODPs and their axioms are described and validated in [17]. They are used in a Strategy Design Ontology (see Fig. 2). Table 1 illustrates rules of Strategy Design Ontology.

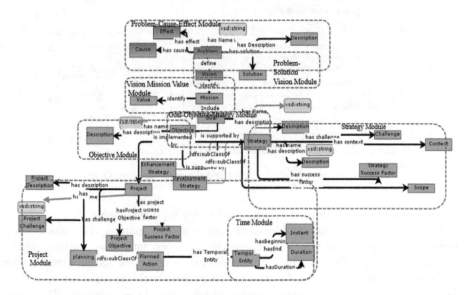

Fig. 2. Strategy design ontology

Stakeholders coming from various domains identify the city problems, then analyze and model them using the Problem-Cause-Effect ODP. The Problem-Cause-Effect ODP is inspired from problem tree [18]. The latter is used in projects planning to represent problems, their causes and effects.

When the stakeholders identify the city problems, they can transform the problems into solutions. In other words, they can reverse the negative statements of the problem into positive ones (e.g. transform the problem "polluted environment" into "a not polluted environment"). They form an idea of the future image of the city in each of strategic axes. Then, they form an idea of the vision of the city. The Problem-Solution-Vision ODP helps stakeholders to define the city. Each of the defined problems has

Table 1. Strategy design ontology rules

Rule	Explanation
Goal (? x) ^ Objective (? y) ^ Strategy (? z) ^is supported by (? x, ?y) ^ is supported by(? y,?z) -> is supported by (?x, ?z)	If a goal is supported by an objective and an objective is supported by a strategy, then the goal is supported by the strategy
Goal (? x) ^ Objective (? y) ^ Strategy (? z) ^ is supported by (?z, ?y) ^ support (?y,?x) -> is support by (?z, ?z)	If a strategy supports an objective and an objective supports a goal, then the strategy supports the goal
Existing City(?x)^ Strategy(?y) ^ has Strategy (?x, ?y) -> Enhancement Strategy(?y)	If an existing city has a strategy then the strategy is an enhancement one
New City(?x)^ Strategy(?y) ^ has Strategy (?x, ?y) -> Development Strategy (?y)	If a new city has a strategy then the strategy is development one
Strategy(?x)^Country(?y)^ is applied to(?x, ?y) -> has scope (?x, local)	If a strategy is applied to a country, the strategy scope is national
Strategy(?x)^City(?y)^ is applied to(?x, ?y) -> has scope (?x, local)	If a strategy is applied to a city, the strategy scope is local

some values of solutions and each solution defines at least one vision. Problem, solution and vision are disjoint classes.

The Vision-Mission-Value ODP allows modelling missions and values with reference to the defined vision. A well-defined and clear vision allows to identify at least one mission that the stakeholders should accomplish. The identified mission(s) allow to identify the participants' values. The classes vision, mission and value are disjoint.

The Mission-Goal ODP models the goals included in a mission. It is used for goals identification. A mission includes at least one goal. Goal, Mission and Description are disjoint classes.

The Goal-Objective-Strategy ODP allows stakeholders to identify Objective(s) supporting each of the defined goal(s) and the strategies(s) supporting each of the identified objective(s).The Objective- ODP is intended to define each of the identified objectives. An objective has au max 1 name and has some descriptions. In the Goal-Objective-Strategy ODP, each goal is supported by some values of objectives, and each objective is supported by some strategies. Goal, Objective, Description and Strategy classes are disjoint.

The strategy ODP is extracted from strategy ontology [19]. Strategies are defined with reference to the identified objectives as described in the previous step. The main concepts to define a strategy are name, strategy description, challenge, some context to mention the contextual information related to the strategy including city context and planning context, strategy success factors that influence the fulfillment of the strategy and the scope of the strategy (local, national and international). A strategy is implemented by one or more projects. Strategy, Challenge, Strategy Success Factor, Context, Scope, Project are disjoint classes.

The project ODP is extracted from strategy ontology [19]. For each of the defined strategies, stakeholders define the project related concepts which are project name, project descriptions, project success factors, project objectives, project challenge, planning including scheduled actions and their temporal entities. The Time Pattern is extracted from Time Ontology [20]. It indicates the beginning and the end or the duration of the scheduled actions.

5   Strategy Design Platform

When they follow the design process of smart city strategies, stakeholders need to use a web-based platform to ensure their collaboration, to use a set of tools that facilitate knowledge construction and strategies formulation. The proposed ontology is integrated in a web-based platform to support stakeholders during strategy design process. The platform is conceived to: (1) structure knowledge coming from various sources and from different stakeholders and have access to it, (2) formally and semantically describe strategies using ontologies, (3) enable co-participation in strategies development and 4) standardize and automate strategy design process.

5.1   Platform Architecture

Figure 3 illustrates the platform architecture. It is mainly composed of: (1) forms for editing and updating data, (2) databases for data storage, (3) online forum for discussions (google docs), (4) services for strategy design and (5) ontologies for knowledge structuring and semantic formulation of strategies.

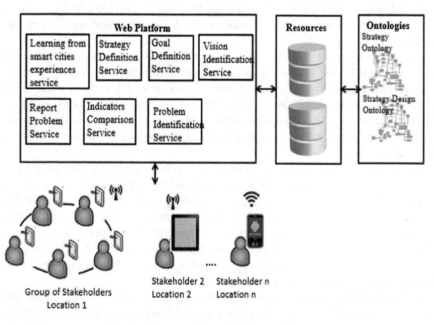

Fig. 3. Strategy design platform architecture

5.2   Platform Implementation

The platform is implemented using Spring Model-View-Controller (MVC) Framework. MVC allows separation of models from their representation. Figure 4 illustrates a screenshot of some web pages of the platform. The platform is based on ontologies as data models. The ontologies facilitate the structuring of knowledge coming from various sources and from different stakeholders and serves as a common language for a formal and semantic description of strategies. The ontologies are implemented with OWL2 (Ontology Web Language version 2). Apache Jena Engine reasoner does inferences and extracts new strategies related knowledge.

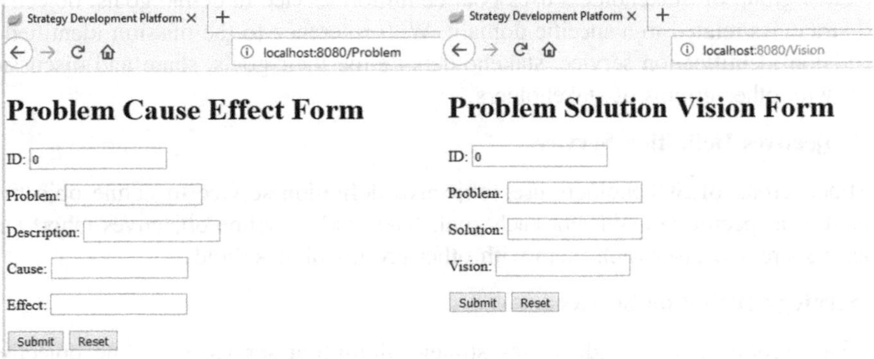

Fig. 4. Screenshot of some web pages of the platform

5.3   Services

Report Problems Service

Report problems is an application where the citizen can send reports about the problems that he observes in his region. He can add his name, his location and choose the type of the problem from the following categories: transport, people, environment, health, safety, education, economy, infrastructure and energy. Then, he describes the observed problem. The information is then stored in a database. The administrative group makes a report ones a week and publish it on the platform. The stakeholders read the report and consider those problems in the strategies design step.

Problem Identification and Analysis Service

Problem Identification and Analysis is a service used for the identification of the city problems in each domain. Stakeholders are divided into groups according to their domain (education, transport, etc.). Each group discusses problems using web forum (e.g. Google Docs) that he identifies from statistics, interviews, surveys. He adds the problems identified in each domain using problem-cause-effect form.

Vision Identification Service

Stakeholders with different backgrounds and disciplines collaborate to identify the city's vision using vision identification service. Each group proposed his vision related to the domain that they are working on (e.g. transport vision) and with reference to the problems identified previously. He shares the vision with other groups using a discussion space (e.g. Googgle docs). They form then a generalized vision on the city. The sub-visions identified for each domain can be considered as the mission that they want to achieve.

Goal Definition Service

Each group of Stakeholders uses goal definition service to define goals, objectives and strategies related to a specific domain. With reference to the mission identified in the vision identification service, stakeholders define their goals, share and discussion them with other groups of stakeholders.

Objectives Definition Service

Each group of Stakeholders uses objective definition service to define objectives related to a specific domain. For each goal, stakeholders define objectives (short term goals); share and discussion them with other groups of stakeholders.

Strategy Definition Service

Each group of stakeholders uses strategy definition service to define objectives related to a specific domain. For each objective, stakeholders define strategies; share and discussion them with other groups of stakeholders. When the strategies are already defined, the stakeholders share them with citizens through the platform and the citizens vote and give their opinions about the strategies.

6   Use Case

To validate the proposed solution, a group of stakeholders, especially researchers and decision-makers, in educational domain use the web-based platform to design strategies and evolve education in the city of Sousse, Tunisia. The education is one of the important domains to increase the intellectual level of citizens, improve their behavior in the city and render them active participants in the transformation of the city to a smart one. For this aim, the stakeholders identify the problems of educational system especially in the primary school to provide transformational strategies. The problems identification is done by re-engineering non-ontological data coming from statistics, interviews, researches, parents and tutors' interviews and many other useful sources of data.

Four problems are identified using the Problem Cause-Effect ODP. The curricula or education programs are outdated. They have not been changed for a long time and suffer from overloaded schedules. Educational approaches are not attractive. The problem of old curricula has many effects including irritated, bored and hating education students. It also does not meet needs of today's society like jobs needs.

The second problem concerns school infrastructure which is old with ruined schools. It has consequences like increasing rates of private tutoring and insufficient schools' numbers in cities.

The third problem concerns the inequality of education between regions and social levels. The causes are that innovative solutions are implemented in northeast regions and government is less interested by southwest regions. Schools are far from students' homes especially in rural regions. Rural regions suffer from poverty and unemployment and students have great dropout rates therefore to help their families.

The fourth problem concerns brilliant students who find integration difficulties in schools. They are bored from learning process and educational programs. They also find difficulties in communicating with other children. Gifted children become isolated or aggressive and suffer from psychological problems. Another cause of the problem is that learning system do not consider other aspects of student's traits like emotional aspects.

The solutions identified from the problems are to seek equality of education between regions and social levels, to build new schools, to offer new curricula and hence generate well educated students. The vision defined from the solutions is new education system with exciting and innovative approaches, new infrastructure and well-educated students. The mission includes two goals. The first goal G1 is to ensure free and quality education for all. It has an objective O1 which is to offer exciting educational programs. The strategy S1 that supports the objective O1 is to include smart technologies in primary schools. P1 is a project that implements S1. P1 includes Internet of Things in learning scenarios to create exciting curricula. Authors of [21] and [22] include preliminary results of the implementation of the strategy. SPARQL queries are used to extract knowledge from the ODPs.

Figure 5 shows the Tunisian educational domain problems with their causes and effects and Fig. 6 illustrates the designed strategies, their planning context (previous projects and strategies) and the projects planned to implement the strategies.

```
PREFIX : <http://www.semanticweb.org/BlemTree#>
PREFIX rdf: <http://www.w3.org/1999/02/22-rdf-syntax-ns#>

SELECT ?problem ?cause ?effect
WHERE {
        ?problem :has_cause ?cause.
        ?problem :has_effect ?effect.
}
```

problem	cause	effect
:inequality_of_education_between_regions	:lack_of_schools_in_rural_areas	:students_coming_from_pover_families
:inequality_of_education_between_regions	:lack_of_schools_in_rural_areas	:high_rate_ol_illitracy
:Old_Curricula	:long_learning_hours	:irritated_students
:Old_Curricula	:long_learning_hours	:education_hate
:Old_Curricula	:long_learning_hours	:boredom_of_students
:bad_school_infrastructure	:schools_are_far_from_students_homes	:ruined_schools
:bad_school_infrastructure	:schools_are_far_from_students_homes	:private_tutoring
:bad_school_infrastructure	:schools_are_far_from_students_homes	:insufficient_schools_in_cities
:inequality_of_education_between_regions	:schools_are_far_from_students_homes	:students_coming_from_pover_families
:inequality_of_education_between_regions	:schools_are_far_from_students_homes	:high_rate_ol_illitracy
:gifted_students_with_difficulties_in_schools	:depression	:decrease_in_school_level
:gifted_students_with_difficulties_in_schools	:depression	:Aggressive_behavior
:Old_Curricula	:no_consideration_of_new_technologies	:irritated_students
:Old_Curricula	:no_consideration_of_new_technologies	:education_hate
:Old_Curricula	:no_consideration_of_new_technologies	:boredom_of_students
:bad_school_infrastructure	:less_maintenance	:ruined_schools
:bad_school_infrastructure	:less_maintenance	:private_tutoring
:bad_school_infrastructure	:less_maintenance	:insufficient_schools_in_cities

Fig. 5. SPARQL query showing educational domain problems

```
PREFIX strategy: <http://www.semanticweb.org/Strategy#>
PREFIX project: <http://www.semanticweb.org/project#>
PREFIX rdf: <http://www.w3.org/1999/02/22-rdf-syntax-ns#>

SELECT ?strategy  ?project ?planningContext
WHERE {    ?strategy strategy:hasplanningContext ?planningContext.
           ?strategy strategy:isImplementedBy ?project.

      }
```

strategy	project	planningContext
strategy:Educating_smart_technologies	strategy:Include_IoT_in_learning_scenarios	strategy:PreviousStrategy_Modernizing_Infrastructure
strategy:Educating_smart_technologies	strategy:Include_IoT_in_learning_scenarios	strategy:PreviousProject_Madrasaty

Fig. 6. SPARQL query illustrating designed strategies

7   Evaluation

This section includes the evaluation of the ODPs and of the web platform.

7.1   ODPs Evaluation

To evaluate the ODPs, a set of criteria are used as described in [23]. Compatibility is the degree to which an ODP is used with other ODPs to construct an ontology. It includes sub-criteria namely reusability, co-existence and interoperability. The ODPs are reusable in use cases that may be related to smart cities strategies development or in other domains. Problem-Cause-Effect ODP can be used for problems analysis of a city at different levels as geographical, social and economic or for the analysis of city domains problems (e.g. transport, energy, education and so on). It can also be used for project planning. Problem-Solution-Vision ODP can be used to identify cities vision(s) at a generic level or at domain level. It may also be used for vision identification in organizational strategic management [24]. Vision-Mission-Value ODP can be used for missions and values identification in various cities domains and for organizational missions and values identification [24]. Goal-Objective-Strategy ODP may be used for cities and for any decision-making projects [25]. Strategy ODP and Project ODP can be used for acquiring expertise from smart cities strategies and projects as the case in [19] and for strategies and projects designs. The competency questions are answered by ODPs as shown by SPARQL query results in Figs. 5 and 6.

7.2   Platform Evaluation

To evaluate the platform, the stakeholders answer to some questions (see Table 2). The preliminary results of the use of the platform shows how the use of Strategy Design Ontology and platform was beneficial in the process of strategies design.

The use of the ontology helps in the structuring of knowledge coming from various sources and from different stakeholders. It is also a data model that the platform is based on and represents a common language understandable by the stakeholders. The web platform ensures stakeholders collaboration in the strategy design process and offers a set of tools that help them during this process.

Table 2. Rules related to the pattern goal-objective-strategy

Question	Response: yes /no/maybe
Do you find the proposed ontology useful for the definition of strategies?	75%: yes 25%: maybe
Do you spend less time in discussions when using the ontology as a common language?	80%: yes 20%: maybe
Does the use of a web-based platform useful?	80%: yes 20%: no (prefer real time discussions)

8   Conclusion

The Smart city concept is developed to mitigate the current cities' problems and to increase the citizens quality of life. Strategies are designed as transformational process. The transformation of cities to smart cities is repeatedly reused by cities stakeholders around the world. The complexity of the multi-disciplinary domain of smart cities renders this transformation hard. Stakeholders collaboration in strategies design can facilitate it. However, this collaboration fails in existing smart cities. The emergence of web platforms can facilitate the collaboration.

In the current paper, a strategy design process was identified, and a Strategy Design Ontology was proposed to support stakeholders during such a difficult process. A web-based platform is also proposed. It includes a set of tools that facilitate smart cities strategies design. The platform represents as a common space for knowledge structuring and strategies definition and sharing among stakeholders.

To validate the proposed framework, we propose a use case of strategies design in the learning domain. Inferences and queries are applied to the Strategy Design Ontology to extract useful knowledge. The future work intends to test and evaluate the designed strategies in a smart city simulator.

References

1. The world cities in 2016, http://www.un.org/en/development/desa/population/publications/pdf/urbanization/the_worlds_cities_in_2016_data_booklet.pdf, last accessed 2017/06/14
2. Gangemi, A., Presutti, V.: Ontology design patterns. In: Handbook on Ontologies, pp. 221–243. Springer, Berlin (2009)
3. Aranguren, M.E., Antezana, E., Kuiper, M., Stevens, R.: Ontology design patterns for bio-ontologies: a case study on the cell cycle ontology. BMC Bioinform. 9(5), S1. BioMed Central, (2008)
4. Vardeman, I.I., Charles, F., Krisnadhi, A.A., Cheatham, M., Janowicz, K., Ferguson, Hitzler, P., Buccellato, A.P.: An ontology design pattern and its use case for modeling material transformation. Semant. Web 8(5), 719–731 (2017)
5. Hitzler, P.: Modeling with ontology design patterns: chess games as a worked example. Ontol. Eng. Ontol. Des. Patterns: Found. Appl. 25, 3 (2016)

6. Maccani, G., Donnellan, B., Helfert, M.: A comprehensive framework for smart cities. In: Proceedings of the 2nd International Conference on Smart Grids and Green IT Systems, pp. 53–63, (2013)
7. Ben Letaifa, S.: How to strategize smart cities: revealing the SMART model. J. Bus. Res. **68** (7), 1414–1419 (2015)
8. Anthopoulos, L., Blanas, N.: Evaluation methods for e-strategic transformation. In: Government E-Strategic Planning and Management, pp. 3–23. Springer, New York (2014)
9. Angelidou, M.: Smart city policies: a spatial approach. Cities **41**, S3–S11 (2014)
10. Angelidou, M.: Smart city planning and development shortcomings. Tema. J. Land Use Mobil. Environ. **10**(1), 77–94 (2017)
11. Mosannenzadeh, F., Vettorato, D.: Defining smart city. A conceptual framework based on keyword analysis. Tema. J. Land Use Mobil. Environ. (2014)
12. Ansell, C., Gash, A.: Collaborative governance in theory and practice. J. Public Adm. Res. Theory **18**, 543–571 (2008)
13. Panori, A., Angelidou, M., Mora, L., Reid, A., Sefertzi, E.: Online platforms for smart specialisation strategies and smart growth. In: The 20th Conference of the Greek Society of Regional Scientists (2018)
14. Kakderi, C., Psaltoglou, A., Fellnhofer, K.: Digital platforms and online applications for user engagement and collaborative innovation. In: The 20th Conference of the Greek Society of Regional Scientists
15. Taamallah, A., Khemaja, M., Faiz, S.: A web-based platform for strategy design in smart cities. Int. J. Web-Based-Communities (in press)
16. Sarker, M.K., Krisnadhi, A.A., Hitzler, P.: OWLAx: a protege plugin to support ontology axiomatization through diagramming. In: International Semantic Web Conference (Posters & Demos) (2016)
17. Taamallah, A., Khemaja, M., Faiz, S.: Toward a framework for smart cities strategies design. In: The 3rd International Conference on Smart Cities Applications, SCA 2018 (2018) (in press)
18. Planning Tools. London: Overseas Development Institute. https://www.odi.org/publications/5258-planning-tools-problem-tree-analysis (2009). Last accessed 15 June 2018
19. Taamallah, A., Khemaja, M., Faiz, S.: Strategy ontology construction and learning: insights from smart city strategies. Int. J. Knowledge-Based Dev. **8**(3), 206–228 (2017)
20. Hobbs, J.R., Pan, F.: Time ontology in OWL. W3C working draft 27, 133 (2006)
21. Taamallah, A., Khemaja, M.: Designing and eXperiencing smart objects based learning scenarios: an approach combining IMS LD, XAPI and IoT. In: Proceedings of the Second International Conference on Technological Ecosystems for Enhancing Multiculturality, pp. 373–379. ACM (2014)
22. Taamallah, A., Khemaja, M.: Providing pervasive learning eXperiences by combining internet of things and e-learning standards/Proporcionar experiencias de aprendizaje ubicuo mediante la combinación de Internet de las Cosas y los estándares de e-Learning. Educ. Knowl. Soc. **16**(4), 98 (2015)
23. Hammar, K.: Content Ontology Design Patterns: Qualities, Methods, and Tools, vol. 1879. Linköping University Electronic Press (2017)
24. Dalmau Espert, J.L., Llorens Largo, F., Molina-Carmona, R.: An ontology for formalizing and automating the strategic planning process (2015)
25. Paradies, S., Zillner, S., Skubacz, M.: Towards collaborative strategy content management using ontologies. In: International Semantic Web Conference. Washington, DC (2009)

Urban Traffic Flow Management Based on Air Quality Measurement by IoT Using LabVIEW

Mohamed El Khaili[(✉)], Abdelkarim Alloubane, Loubna Terrada, and Azeddine Khiat

SSDIA Laboratory, ENSET Mohammedia, Hassan II University of Casablanca, Casablanca, Morocco
{elkhailimed,a.alloubane,loubna.terrada,azeddine. khiat}@gmail.com

Abstract. Our challenge has two dimensions: social and technological. We want to solve a serious problem that is assessing the air quality in cities and traffic management according to the air quality indices. To inform and sensitize people to the air pollution problem, our project will bring the locals in participatory situation and actor for the improvement of air quality by managing the urban traffic. We hear more and more talk about the Internet of Things, connected objects, or even connected world, or even intelligent home; new concepts that invade the world and enhance our way of life. Internet of Things called the third industrial revolution will profoundly change the lives of people with home automation, health and recreation, energy, distribution and our environment with intelligent cities or transport connected. The collection of information remains a major challenge without the participation of a large group of people or partners. The Crowdsourcing allows obtaining information due to a large group of people by the internet.

Keywords: Smart city · Crowdsourcing · Internet of things · Air quality
Urban traffic management · LabView

1   Introduction

In recent years, the world has experienced an impressive development of the multimedia world. This is due to the technical and technological progress and major innovations that have revolutionized the world of telecommunication, IT cloud (Cloud Computing), social media, Internet of Things....

This development is shown as the extension and the invasion of the internet of things in the physical world. While the Internet usually does not extend beyond the electronic world, the Internet of Things represents the exchange of information and data from devices present in the real world to the Internet. Internet of Things is regarded as the third evolution of the Internet, known as Web 3.0. The objects constituting the "Internet of objects" called "connected", "communicating" or "smart". Currently Connected or intelligent objects are everywhere; they invaded the world and impact our personal and professional lives. They generate billions of information that must be processed and analyzed to make them usable.

© Springer Nature Switzerland AG 2019
M. Ben Ahmed et al. (Eds.): SCA 2018, LNITI, pp. 193–207, 2019.
https://doi.org/10.1007/978-3-030-11196-0_18

Urban Traffic Management is a very complex problem. Millions of people every day use urban road networks and suffer from congestion. Researchers are led to solve whether is possible traffic flow management problem in order to reduce the congestion and to improve the air quality. A part from long-term structural solutions, it is possible in principle to control the urban traffic flow either acting on the individual cars or acting on the traffic signalization.

Our project is to demonstrate that the connected objects can consist of a participatory information collection (Crowdsourcing) for the control, supervision or conduct of a system. We use LabView graphical programming for its capabilities and interfacing with electronic devices. We treat the case of index measurement of air quality in the city of Casablanca. To what extent can we consider the Internet of Things and LabVIEW features as a key technology for decision support for the Smart City?

In this communication, we give an overview of the Internet of Things (IoT) and urban traffic control. Also, we give a brief introduction to LabVIEW features applied to the connected objects. We present our project of crowdsourcing based on Connected Objects for measuring air pollution using the features and benefits of LabView. We propose our approach to control urban traffic. We show and discuss at the end experimental results. A pilot experiment well be done in Casablanca which has about 4.2 million inhabitants.

2 Internet of Things (IoT)

The Internet of Things (IoT), also called the Internet of Everything or the Industrial Internet, is a new technology paradigm illustrated as a global network of machines and devices able to interact with each other [1]. Recently, the world has experienced an impressive development of the multimedia world. This is due to the technical and technological progress and major innovations that have revolutionized the world of telecommunication, IT cloud (i.e. Cloud Computing), social media, Internet of Things....

Connected or intelligent objects are used everywhere, while the Internet usually is not extended beyond the electronic world, the Internet of Things represents the exchange of information and data from devices present in the real world to the Internet [1]. They invaded the world and affected our personal and professional lives. They generate billions of information that must be processed and analyzed then stored to make them usable [2]. Today, connected objects begin to take part in our daily lives and are translated into several and different objects in multiple fields of application [3, 4].

2.1 Architecture of IoT

Different models with various supports IoT technologies can illustrate the Internet of Things architecture [5]. It serves to illustrate how they are interconnected in different scenarios. Figure 1 illustrates the role of the various processes of the architecture of IoT:

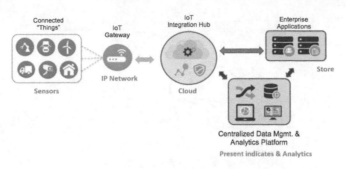

Fig. 1. Architecture modeling of IoT

- Sensors to transform a physical quantity analog to a digital signal.
- Connect allows interfacing a specialized object network to a standard IP network (LAN) or consumer devices.
- Store calls made to aggregate raw data produced in real time, Meta tagged, arriving in unpredictable ways.
- Present indicates the ability to return the information in a comprehensible way by humans, while providing a means to do it and/or interact.

Five IoT technologies are used in industrial firms for the deployment of successful IoT-based products and services [1], namely: Radio frequency identification (RFID), Wireless Sensor Networks (WSN), Middleware, Cloud computing and IoT application software.

2.2 IoT's Functions

The Internet of Things (IoT) concept allows objects to connect between what happens online and the physical world. Christened in 1999 by researcher Kevin Ashton center at MIT (Massachusetts Institute of Technology), his team launched the promotion of open connectivity of all objects using RFID (Radio Frequency IDentification). The first solutions in this area seemed simple calling itself Machine To Machine (M2M), they are based on the fact that a single type of hardware/sensor connects via a service gateway (Internet) to a single application. Then these solutions have become more complex while using several sensor/display/actuator using a multi serving gateway (Internet TCP/IP) to a single application. Since the work get more complicated there are other standards taken into consideration as the geographic dispersion and the increase of device data consumers [5]. The functioning of the Internet of Things via cloud is shown in the illustration below (Fig. 2) [6]:

Today, the Internet of Things has simplified processes with Cloud [7], from which the collected data is centralized in a single space. The Internet of Things can be used in plenty and different domains, such as: Healthcare, Industry, sport, Smart cities, Smart grids, logistics....

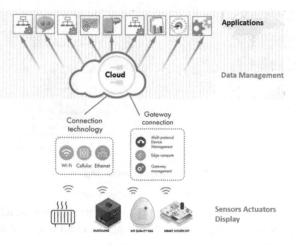

Fig. 2. IoT functionning via cloud

With the IoT, the whole world is upset and entering a new era. IoT brings many benefits for individuals, businesses and the business community and for any other user. However, this new innovation is still limited by sensitive security issues. So, what are these different advantages and disadvantages [5, 7] (Table 1).

Table 1. Advantages and inconvenient of IoT [6]

Advantages	Inconvenient
• Time • Perfection • Accessibility and mobility: "anytime, anywhere, any device" • Connected objects will be an integral part of our everyday life in all areas (health, car, lifestyle) • New wave of internet that comes in the continuity of cloud	• Security (locks connected) • Data piracy • Geolocation laws • Data security • Privacy or Image Rights (Google glass) • Bandwidth • Health Influence

3 Overview of Urban Traffic Management UTM

3.1 Urban Traffic Management

Traffic Operators use traffic control systems in large urban areas to perform the crucial role of tackling road congestion and minimizing traffic related environmental effects. Conventional road traffic signal management techniques, such as traffic-responsive systems or fixed time light strategies optimized using historical data, work reasonably well in normal or expected conditions. Currently, software systems in the urban traffic management area tend to be based around a syntactic, product specific integration of data, which at best share data externally at a relational database level. They have a vertical systems design, and though eminently configurable within the range of their function, they are not integrated at a horizontal level with the overall function of the urban management centre where they operate. Within urban traffic management and control (UTMC) operations, this perpetuates the status quo of recurrent system replacement, rather than system evolution [8].

The context of the novel application of planning that we describe in this paper is developing semantic technology in order to better capture and exploit real-time and historical urban data sources, while pursuing a higher level of data integration. We aim to make UTMC systems less brittle and more adaptable by raising the level of traffic control software integration via semantic component interoperability. In doing this, we have the longer-time aim of utilizing an autonomic approach to UTMC in particular, and road transport support in general. Results of the Network supported the idea of the construction of a semantic systems level for UTMC, consistent with previous work on integrating decision support within semantic technologies. Among the benefits of a higher level of information integration are a more joined up UTMC capability, where the flexibility of a knowledge level representation gives the opportunity to use general AI techniques such as automated planning to provide a more intelligent approach to tackle UTMC issues.

Within this context, we present a novel Planning application addressing a well known functional drawback of established UTMC tools referred to above. In these cases, Transport Operators may struggle to find a strategy intervention tailored to solve the unexpected situation. Creating such strategies is a manual task that may last several days or weeks, and it is therefore infeasible to hand craft one in real-time. For example, transport operators may want to reduce traffic concentrations in a targeted urban zone to ameliorate effects of predicted road traffic pollution; or optimise the flow of saturated road links due to an emergency road closure; or produce a strategy to deal with a forthcoming complex situation such as optimizing the light timings to deal with the combination of a concert, a football match and some emergency roadworks.

3.2   Urban Traffic Management Approaches

Generating a detailed strategy of interventions, such as changes to traffic signal timings over a period of time, to manage an emergency situation in real time is considered to be beyond the capacity of human operators. Transport operators need the ability to produce local strategies in real time which will deal with abnormal or unexpected events such as road closures. These cause huge delays and increased air pollution because of excessive congestion and stationary traffic. The existing conditions and set of corrective goals required to deal with these events are so varied that detailed strategies are impossible to draw up a priori in a large, dense urban area. Approaches to local or regional traffic control has been trialed using Model Predictive Control (MPC) strategies and optimization [8, 9].

This topic of research uses a control theory approach which, given an adequate dynamical model, can be used to derive a solution that can offer continuous responses to changing situations. Under changing state conditions, researchers have designed MPC algorithms which can continuously adjust the controlled signal timings to optimize some goal in real time. This approach tends to be less flexible than a goal directed

AI approach, however, as a solution needs to be designed, implemented and tuned using a specific model of traffic flow and a specific objective function. Additionally, it is less scrutable, as it generates strategies over a restricted time horizon. For instance, there is a need in UTM to be able to achieve more focused or detailed goals than simply minimizing delay in particular operators may want to generate strategies that avoid problems with air quality caused by traffic congestion. While there are several examples of the application of general AI techniques to road traffic monitoring and management, the trial of UTM systems embodying an AI planning engine within a real urban traffic management centre, with an evaluation performed by transport operators and technology developers, is novel. On the scheduling side, however, the SURTRAC project uses a real time distributed scheduling system which controls traffic signals in urban areas [9]. In SURTRAC, each intersection is controlled by a scheduling agent that communicates with connected neighbors to predict future traffic demand, and to minimize predicted vehicles waiting time at the traffic signal.

Two recent lines of research showed the feasibility of using AI planning to generate actions to deal with unexpected circumstances in complex urban traffic control situations. Gulić et al's system involves joining together a SUMO simulator [10] to an AI Planner, via a monitoring and execution module called the "Intelligent Autonomic System". The planning representation was done using PDDL [11], with no explicit representation of vehicles in the planner. Instead, traffic concentrations on road links are represented by relative density descriptors, such as very-low, low, medium and high. Traffic light change actions are enumerated to cover all the ways that a particular configuration would affect the arrangements of road links. By abstracting away from explicit counts of vehicles, the system can deal with regions containing thousands of vehicles. Also, the close coupling with SUMO demonstrates the use of monitoring and replanning very effectively, and allows exhaustive testing of the system under sets of disturbances (vehicle influx, road closures).The second work was inspired by works such as, where traffic is modeled using "flows", and then analyzed through model-predictive controllers. Vallati et al. exploit PDDL+ for encoding a flow model of vehicles through traffic-light controlled junctions. The length of traffic light phases are under the control of the planner, that can decide to prioritize some traffic flows, in order to reach specified goals. Goals are specified in terms of numbers of vehicles desired on some critical road links. Encoded problems are then solved using the UPMurphi solver, extended with domain-specific heuristics. Their experimental analysis demonstrated that UPMurphi could solve traffic problems containing thousands of vehicles, in response to exceptional conditions. They showed the efficacy of the resulting strategy by comparing its execution against fixed time and reactive approaches, using SUMO [8].

4   LabVIEW and Connected Objects

LabVIEW 2015 introduces key innovations to address the challenges faced by engineers who create, design and test the IoT. These productivity tool that will shorten the development time of systems (Time To Market) and increase the size and complexity at an unprecedented pace [12].

Users can extend the functionality of the LabVIEW development environment and reduce the development time of new plug-ins on single click to perform common programming tasks more quickly, such as the conversion of a scalar object into an array. Using part of plug-in included with LabVIEW 2015, the developer can set their own button to edit articles on the panels and block diagram objects before editing the diagram and objects at runtime. Users can also add links to websites for panels intuitively and full documentation of block diagram with the new hyperlink support.

5   Crowdsourcing Based on IoT for Measuring Air Pollution

5.1   Air Pollution

The air consists of several gases and solid particles. We talk about pollution in the case of unexpected occurrences of new bodies or exceeding the normal concentration of an air constituent. Studies have shown adverse and even fatal effects of some element of pollution [13]. We will focus on four of the most dangerous classified types namely SO_2, NO_2, O_3 and PM_{10}.

(a) Sulfur Dioxide SO_2

Sulfur Dioxide gas SO_2 is a toxic gas with a strong irritating smell. It is present at very low concentrations in the atmosphere and is naturally emitted during volcanic eruptions. Inhaling sulfur dioxide has been associated with respiratory disease and difficulty breathing. The most Sulfur Dioxide gas sensors operate using proven fuel cell technology.

The SO_2 range provides OEMs with reliable sensors for use in a variety of applications. Strong signal levels combined with low zero current allows resolution below 1 ppm and an operating range for safety applications up to 2000 ppm. The sensors are designed for use in both fixed site and portable instrumentation.

(b) Nitrogen dioxide NO_2

Nitrogen dioxide is a byproduct of the burning of hydrocarbons. It is a found primarily as a toxic component of vehicle exhaust in any space in which engines are burning gasoline, propane, diesel fuel, propane, natural gas, and kerosene or jet fuel. Areas include enclosed parking structures, ambulance bays, fire halls, warehouses, loading docks, ice arenas, maintenance facilities and municipal works garages.

For vehicle exhaust detection applications, there are a wide array of hazardous chemicals, including particulates, sulfur compounds, carbon monoxide, carbon dioxide and oxides of nitrogen. As it is not practical to monitor for all of these compounds, a combination of nitrogen dioxide and carbon monoxide sensors will give the best indication of overall air quality.

(c) Ozone O_3

Ozone or trioxygen, is an inorganic molecule with the chemical formula O_3. It is a pale blue gas with a distinctively pungent smell. It is an allotrope of oxygen that is much less stable than the diatomic allotrope O_2, breaking down in the lower atmosphere to O_2 or dioxygen. Ozone is formed from dioxygen by the action of ultraviolet light and also atmospheric electrical discharges, and is present in very low concentrations throughout the Earth's atmosphere (stratosphere).

Ozone is a powerful oxidant (far more so than dioxygen) and has many industrial and consumer applications related to oxidation. This same high oxidising potential, however, causes ozone to damage mucous and respiratory tissues in animals, and also tissues in plants, above concentrations of about 100 ppb.

(d) Particulate Matter PM

Atmospheric aerosol particles, also known as atmospheric particulate matter, particulate matter PM, particulates, or suspended particulate matter SPM are microscopic solid or liquid matter suspended in Earth's atmosphere. Sources of particulate matter can be natural or anthropogenic. They have impacts on climate and precipitation that adversely affect human health. Subtypes of atmospheric particles include suspended particulate matter SPM, thoracic and respirable particles, inhalable coarse particles, which are coarse particles with a diameter between 2.5 and 10 micrometers PM_{10}, fine particles with a diameter of 2.5 µm or less $PM_{2.5}$, ultrafine particles, and soot. The International Agency for Research on Cancer IARC and The World Health Organization WHO designate airborne particulates a Group1 carcinogen. Particulates are the deadliest form of air pollution due to their ability to penetrate deep into the lungs and blood streams unfiltered, causing permanent DNA mutations, heart attacks, and premature death. The smaller $PM_{2.5}$ were particularly deadly, with a 36% increase in lung cancer per 10 $µg/m^3$ as it can penetrate deeper into the lungs.

5.2 Our Project Presentation

The measurement of pollution in the air is traditionally based on the deployment by the local authority of fixed sensors in the city (Fig. 3). Several initiatives have emerged at the intersection of the citizen sensors and the Open Hardware, to design low cost sensors increase coupled smartphones to enable a collaborative data collection (crowdsourcing) and so multiply the observations and measurements (Fig. 4).

Fig. 3. Casablanca MAP

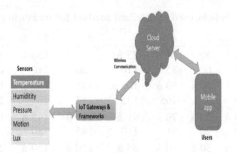

Fig. 4. Deploying our project

Our project now aims to design objects connected miniaturized and portable ("wearables") capable of informing users about the quality of the air around them. They would be the equivalent, for the measurement of the environment, "tracer activities" which measure the efforts and physical performance. This information will be collected and processed in a GIS to help authorities in traffic management. Our project calls this new generation of connected objects "enviro-trackers" ("environmental tracers"). A pilot experiment is planned on the urban area of Greater Casablanca, a city of 4.2 million inhabitants.

Our project takes place in three phases:

1. Design of connected objects and embedded applications.
2. Design of the pilot platform with GIS and web portal.
3. Operations of results (Decision support based on the availability of information via multi-modal facilities).

6 Implementation for Data Collection

6.1 Indicators: AirParif Method

The index is calculated taking into account the pollutants SO_2, NO_2, O_3 and PM_{10} which are recognized by their health effects. Table indices according to European standards is based on the measurement of the concentration of each pollutant in $\mu g/m^3$ for a given time interval. Each city sets tolerance levels. From Table 2, the maximum of the four sub-indices is the value of the general index of air quality.

6.2 Crowdsourcing Architecture

The proposed system is based on a communication between a Treatment Unit and multiple Tracers. The network shares the same concept of habitual network such as addresses for planning and routing system. Each Tracer can transmit data to the Treatment Unit frequently or after a request of the Treatment Unit. Figure 5 illustrates data which are exchanged on the network.

Table 2. Indices used by AirParif method for measuring air quality

	1	2	3	4	5	6	7	8	9	10
SO_2 max 24 h	0 à 15	15 à 30	30 à 60	60 à 85	85 à 110	110 à 150	150 à 210	210 à 270	270 à 350	>350
NO_2 max 1 h	0 à 30	30 à 60	60 à 80	80 à 105	105 à 135	135 à 155	155 à 180	180 à 270	270 à 400	>400
O_3 max 1 h	0 à 30	30 à 50	50 à 70	70 à 91	91 à 110	110 à 145	145 à 180	180 à 250	250 à 360	>360
PM_{10} max 24 h	0 à 15	15 à 30	30 à 60	60 à 85	85 à 110	110 à 150	150 à 210	210 à 270	270 à 350	>350

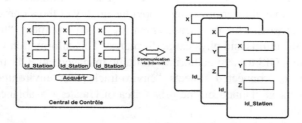

Fig. 5. Platform of simulation

(a) **Sensors connectivity**

A sensor for each pollutant (SO_2, NO_2, O_3 and PM_{10}) to be designed in a chip containing analog and digital blocks to measure the density and the presence of the pollutants. Figure 6 illustrates the sensors to be connected to the analog port of the data acquisition devices (DAQ device) plus the temperature acquisition.

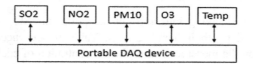

Fig. 6. Sensors and DAQ connection

(b) **Wireless technology**

Communications means naturally rely on wireless communication technologies [8–11, 14–16]. Optimized to consume as little energy as possible and designed to be produced at very low cost. Wireless technology and wireless networks are widely used today, but it is quite new in industrial automation systems. There are different technologies and wireless standards available: Bluetooth, Wireless USB, ZigBee (IEEE 802.15.4) and Wi-Fi (IEEE 802.11).

Wi-Fi and ZigBee are the primary wireless technologies for measurement and control systems.

(c) **Wireless data acquisition**

Wi-Fi DAQ is [12]:

- Simple—Direct sensors connectivity and graphical programming
- Secure—Highest commercially available data encryption and authentication
- Wireless—802.11 g.

Wi-Fi data acquisition is an extension of PC-based data acquisition to measurement applications where cables are inconvenient or uneconomical. NI Wi-Fi data acquisition (DAQ) devices combine IEEE 802.11 g wireless or Ethernet communication; direct sensor connectivity; and the flexibility of NI-DAQmx driver software for remote monitoring of electrical, physical, mechanical, and acoustical signals. NI Wi-Fi DAQ devices can stream data on each channel at up to 250 kS/s. In addition, built-in NIST-approved 128-bit AES encryption and advanced network authentication methods offer the highest commercially available network security [13]. Unlike most wireless sensors or wireless sensor networks, wireless data acquisition devices are meant to stream data continuously back to a host PC or laptop. A wireless sensor node is typically a low-power, autonomous battery-operated device intended for long-term deployment in applications where measurements are needed only every few minutes, hours, or even days (Fig. 7).

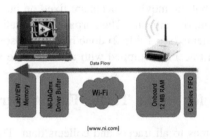

Fig. 7. The data flow in WIFI-DAQ

Common components in Wi-Fi DAQ are as follows:

- DAQ Devices
- Wireless Access Points (WAP) or a Wireless Router
- Network Switches.

(d) **Wireless Sensor Network (WSN) devices**

A Wireless Sensor Network (WSN) [17–19] is a wireless network consisting of spatially distributed autonomous devices that use sensors to monitor physical or environmental conditions. These autonomous devices, or nodes, combine with routers and a gateway to create a typical WSN system. The distributed measurement nodes communicate wirelessly to a central gateway, which provides a connection to the wired world where you can collect, process, analyze, and present your measurement data. To extend distance and reliability in a WSN, you can use routers to gain an additional communication link between end nodes and the gateway (Fig. 8).

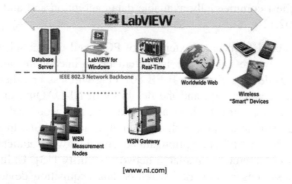

[www.ni.com]

Fig. 8. Wireless sensor network components

6.3   Tracers

We propose two type of tracers: static and moving tracers. The static tracers are fixed in facilities like buildings but the moving tracers are fixed on buses or used as application on smartphones of voluntaries citizens. The Airparif Air Quality Assessment System is based on four measurements (see Table 2) done by SO_2 sensor, NO_2 sensor, O_3 sensor and PM_{10} sensor. Each tracer is equipped with at least three sensors.

7   Our Approach for Urban Traffic Flow Control

The Treatment Unit listens to all tracers and collects data. The index of air quality is computed per each region who correspond to it tracer. The indices can be displayed on a map of the city or used for regulating traffics and other way of management for the city's facilities (urban logistics).

The initial phase of our collaborative project concentrated on the semantic enrichment of the data. The raw data was taken from transport and environment sources and integrated into a Data Hub. The method was to take real time feeds and process them until they produced logical facts about a traffic scenario, which could serve as part of an initial sate of a AI planner. The abstract system architecture is used to test the generation and operation of the control strategies. To work towards that, however, the data enrichment and strategy generation had to be tested in a real scenario, hence rather than taking in real time current data, we adjusted the system so that what would be translated into the current state would be from historical data. This would allow checking the performance of the system against the observed performance from historical data, in order to evaluate it off line.

As a basis for exploring exceptional or emergency traffic conditions, we chose to use historically averaged traffic data from a time/day when the road links were most congested: morning rush hour on a non holiday weekday. The main data source was the probabilistic processes. From this and other transport engineer documentation records our partners extracted, for the selected region within the urban area, the following:

1. the topology of the road links (direction of each link and junctions between roads).
2. The vehicle capacity of all the road links taking into account the differing size of vehicles.
3. The average traffic flows between links in number of vehicle per second. This number represents the number of vehicles existing in a particular junction at a fixed time of day, when the corresponding traffic signal phase is green.
4. The traffic signal features: position, phases of signals, minimum and maximum time that a signal phase can be set for.
5. Inter-green timings between each of the phases of the signals. Inter-green intervals are used between two traffic light phases for clearing the intersection from vehicles, and allowing pedestrian crossings.

These data items made up the initial state of a data file in planning terms. The goal language of the planner is what the actions in the domain model can do. In this case the goals are made up of numerical constraints denoting predicates on the occupancy levels of each road link.

Considering the simulation, the traffic models were run independently by analytic functions to estimate real traffic flow. In the first test, after validating that the simulations were fairly consistent, the reduction in time to clear a junction using the planner strategy was remarkable using the off-line simulations. We will compare in future with similar results offered by a known simulator. Our simulations tended to produce slightly long times to clear congestion.

As well as estimating the emissions reduction in the link referred to in the goal, the emissions reduction from the overall effect of applying the strategy to the model given that certain links carry more weight was calculated. The emissions around the link to be cleared were calculated to drop by 5%, whereas the overall drop over the region was 2.5% defined as objective target. It is worth stressing that these results are preliminary, however, with more testing to be done to accurately determine the effect on air quality levels.

8 Conclusions

Internet of Things and the connected objects have shown many benefits to users, including in areas such as health, transportation, home automation. Using LabVIEW is a quality contribution to our work. Indeed, its interfaces and libraries have allowed us to improve our approaches to work on the project. We have confirmed the efficiency of such technology by its implementation to measure the air quality in the city of Casablanca.

In this paper we presented an approach to traffic control using IoT. Traffic signalization control has been presented meanwhile the mechanism to control cars is also needed. While the reported tests are encouraging, a more thorough evaluation is required to better assess real world potential.

References

1. LeMag Web4. 10 Juillet 2015.: https://lemag.agenceweb4.ch/site/fr/lemag-web4?tag=1
2. EVANS, Dave. Cisco. L'internet des objets. Avril, 2011 http://www.cisco.com/web/CA/solutions/executive/assets/pdf/internet-of-things-fr.pdf
3. COULON, Alain. L'internet des objets un gisement à exploiter. Hiver 2010 http://www.adeli.org/document/23-l78p26pdf
4. Vineela, A., Sudha, L.: Internet of things—overview. Sudha Rani Int. J. Res. Sci. Technol. **2** (4) (April 2015). https://ia800501.us.archive.org/7/items/2.IJRST020402812/2.IJRST020402(8-12).pdf
5. https://vision.cloudera.com/an-end-to-end-open-source-architecture-for-iot/. Last visited: 29 July 2018
6. Terrada, L., Bakkoury, J., El Khaili, M., Khiat, A.: Collaborative and communicative logistics flows Management using Internet of Things. In: Mizera-Pietraszko J., Pichappan P., Mohamed L. (eds.) Lecture Notes in Real-Time Intelligent Systems. RTIS 2017. Advances in Intelligent Systems and Computing, vol. 756, pp. 216–224. Springer, Cham (2019)
7. Farahzadi, A., Shams, P., Rezazadeh, J., Farahbakhsh, R.: Middleware technologies for cloud of things-a survey. Digit. Commun. Netw. (2017). ISSN 2352-8648. https://doi.org/10.1016/j.dcan.2017.04.005
8. McCluskey, T.L., Vallati, M.: Embedding automated planning within urban traffic management operations. In: Proceedings of the Twenty-Seventh International Conference on Automated Planning and Scheduling (ICAPS 2017), Pittsburgh, USA, 18–23 June 2017
9. Gulić, M., Olivares, R., Borrajo, D.: Using automated planning for traffic signals control. PROMET Traffic Transp. **28**(4), 383–391 (2016)
10. Krajzewicz, D., Erdmann, J., Behrisch, M., Bieker, L.: Recent development and applications of SUMO—Simulation of urban MObility. Int. J. Adv. Syst. Measur. **5**(3 & 4), 128–138 (2012)
11. Vallati, M., Magazzeni, D., De Schutter, B., Chrpa, L., McCluskey, T.L.: Efficient macroscopic urban traffic models for reducing congestion: a PDDL+ planning approach. In: The Thirtieth AAAI Conference on Artificial Intelligence (AAAI), pp. 3188–3194 (2016)
12. National Instrument, WirelessData Acquisition in LabVIEW tutorial (www.ni.com)
13. Huang, J., Pan, X., Guo, X., Li, G.: Impacts of air pollution wave on years of life lost: a crucial way to communicate the health risks of air pollution to the public. Environ. Int. **113**, 42–49 (2018)
14. Khiat, A., Bahnasse, A., El Khaili, M., Bakkoury, J.: Study and evaluation of vertical and horizontal handover's scalability using OPNET modeler. Int. J. Comput. Sci. Inform. Secur. (USA) **14**(11) November 2016
15. Khiat, A., Bahnasse, A., El Khaili, M., Bakkoury, J.: Study, evaluation and measurement of IEEE 802.16e secured by dynamic and multipoint VPN IPsec. Int. J. Comput. Sci. Inform. Secur. (USA) **15**(1) (January 2017)
16. Khiat, A., Bahnasse, A., El Khaili, M., Bakkoury, J.: Study, evaluation and measurement of 802.11e and 802.16e quality of service mechanisms in the context of a vertical handover

case of real time applications. Int. J. Comput. Sci. Netw. Secur. (KOREA) **17**(2) (February 2017)

17. Jin, M., Gu, X., He, Y., et al.: Wireless sensor networks. In: Conformal Geometry, pp. 253–296. Springer, Cham (2018)

18. Chandana, L.S., Sekhar, A.R.: Weather monitoring using wireless sensor networks based on IOT (2018)

19. Kuo, Y.W., Li, C.L., Jhang, J.H., Lin, S.: Design of a wireless sensor network-based IoT platform for wide area and heterogeneous applications. IEEE Sens. J. **18**(12), 5187–5197 (2018)

20. International Telecommunication.: Sudha. The Internet of Things. ITU (2005) https://www.itu.int/net/wsis/tunis/newsroom/stats/The-Internet-of-Things-2005.pdf

21. BUSINESS.: Objets Connectés Et Machines Communicantes (M2M). 27 November 2015 http://pme.sfrbusinessteam.fr/machine-to-machine/

22. Nemri, M.: France Stratégie, Janvier 2015 N°22 Source: Source: IDATE (2013). www.strategie.gouv.fr/sites/strategie.gouv.fr/files/atoms/files/notesdanalyse22.pdf

23. Caelen, J.: ObjetsConnectes. 2011. http://www-clips.imag.fr/geod/User/jean.caelen/Publis_fichiers/ObjetsConnectes.pptx. Last visited 29 July 2018

24. Schmidt, J.F., Neuhold, D., Klaue, J., Schupke, D., Bettstetter, C.: Experimental study of UWB connectivity in industrial environments. In: European Wireless 2018; 24th European Wireless Conference, pp. 1–4. VDE (2018 May)

Big Data for Smart Cities

An MDA Approach Based on UML and ODM Standards to Support Big Data Analytics Regarding Ontology Development

Naziha Laaz[(✉)] and Samir Mbarki

MISC Laboratory, Faculty of Science, Ibn Tofail University, Kenitra, Morocco
laaznaziha@gmail.com, mbarkisamir@hotmail.com

Abstract. Today, the large increase in the amount of data produced by different sources and the development of technologies to store and analyze them offer many perspectives for the ontology modeling. The creation of domain ontologies will form the basis for application developers to target business professional contexts, however the future of big data will depend on the use of technologies to model ontologies. With that said, many researches recommend the combination of ontologies and big data approaches as the most efficient way to store, extract and analyze data. In this paper, we present a new methodology supporting ontology modeling for the automatic generation of domain ontologies. We propose a transformation from UML class diagrams to ODM models in agreement with the MDA approach. MDA provides opportunities to present ontology artifacts in an intuitive way by defining them in a high level of abstraction using the UML graphical syntax. With the MDA process, the ontology represented as a class diagram will automatically be generated through an ODM metamodel. In this proposal, we founded on an analytical survey. To validate our proposal, we applied it to an e-learning domain ontology.

Keywords: Ontology modeling · Big data · E-learning · Unified modeling language (UML) · Ontology definition metamodel (ODM) · Model-driven architecture (MDA) · Ontologies

1 Introduction

The big data is an abstract concept [1], it is the accumulation of a significant amount of data so that its treatment becomes difficult and requires much technical knowledge. Nothing seems to stop the Big Data revolution. One of the main tasks of analysts and professionals in the field, however, goes well beyond simple data and their universal availability: it is to give meaning to these data. For this reason, it is strongly recommended to integrate ontologies in the big data analytics. Effective ontology modeling efforts have been conducted with respect to big data [2–5]. They are confident that the combination of big data and ontologies provide new solutions for many problems notably ontology can provide semantics to add raw data, generalized concepts in ontology can connect data in various concept levels across domains. Furthermore, we can use ontology as given (and authorized) knowledge to analysis big data [3].

© Springer Nature Switzerland AG 2019
M. Ben Ahmed et al. (Eds.): SCA 2018, LNITI, pp. 211–225, 2019.
https://doi.org/10.1007/978-3-030-11196-0_19

Despite such progress, researchers are just beginning to address the fact that the treatment of these tasks require tools that will allow them to automate the modeling and development of ontologies. The integration time will be shorter, projects will be implemented more easily, which will allow us to develop and diversify the use of user needs. On the other hand, we notice the great effort of the OMG group which has conducted several researches on the implication of ontologies in the model-driven engineering to exploit it later in application domains of different areas. In this context, we propose a new approach that use MDA technologies based on the two OMG standards; UML and ODM as a basis to model ontologies. The idea is to begin with UML representation of classes and then using the concept of model to model transformation Language, to generate a target owl model and use it to get the source code of owl file. Our methodology ensures the recommendations of MDA approach by presenting domain ontology in PSM level through ODM which is OMG standard for ontology modeling. However, we founded on UML to model the platform independent level, because UML is advisable by MDA in PIM. As an illustration, we present along this paper the model driven development process applied for e-learning UML class diagram to generate ontology documents respecting an RDF/XML format.

The remainder of the paper is organized as follows. Section 2 presents the backgrounds of our research work. In Sect. 3, we review previous efforts in ontology development regarding the generation of domain ontologies from UML class diagrams. Section 4 describes our contribution and gives an overview of model driven development process describing its different phases. In Sect. 5, we show the results of our proposal by applying the model driven development process to an e-learning domain ontology. Finally, Sect. 6 concludes the paper.

2   Background

In the followings, we describe the background of this research work and its contribution. The foundations of this study are presented under different subsections. Each of these subsections discuss the backgrounds of our proposal.

2.1   MDA

Model-Driven Engineering (MDE) is an engineering paradigm that uses models as the primary artifact in the software development process. The aim of MDE is to raise the level of abstraction and increase automation in software development.

Model-driven Architecture(MDA) is an OMG's (Object Management Group) vision on MDE and thus relies on the use of OMG standards. The MDA focuses on the technical variability in software and provides architecture for the design of software systems [7]. Furthermore, this approach provides the ability to understand complex systems and the real world through the definition of three different levels representing the abstraction layers of the application [8]. Figure 1 shows the three MDA layers:

- **CIM**: The Computation Independent Model describes the application and its environment and hides the details related to its implementation.

- **PIM**: The Platform Independent Model presents a vision of analysis and design of the system. Models in this level are independent of any technological details.
- **PSM**: The Platform Specific Model is the model that is closest to the final application code.

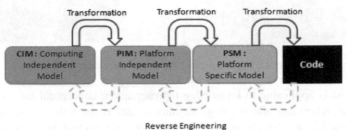

Fig. 1. Overview of the MDA approach [8]

The MDA approach advocates the use of OMG standards, among them UML [10], MOF [11] and XMI [12]. In this work, we considered the two standards: ODM (Ontology Definition Metamodel), and UML (Unified Modeling Language) as central elements for design, transformation of models and generation of final code.

UML

The OMG publishes a variety of standard specifications, the best known is UML (Unified Modeling Language) [13]. UML is the standard notation in the object-oriented modeling world advocating its use by the MDA [14]. Its Metamodel integrates all the concepts necessary for the development of UML diagrams (use case, classes, sequences diagram …), in addition, it provides a mechanism for extending UML modeling entities; notably classes, associations, properties etc., for building domain specific metamodel elements.

On the other hand, class modeling concepts are the most widely used UML concepts [15]. UML class model is a key ingredient of MDE. According to the UML specification, a class model allows to express the conceptual model of information systems, even non-computer scientists use it for its simple graphical syntax that helps to understand the models easily. Moreover, several tools are available to model class diagrams like UML Designer [16], etc.

ODM

The ODM specification was appeared in 2007. It defines a set of ontology metamodels; conforming to the MOF, and associated transformation methods (profiles and mappings) [17]. ODM reflects an abstract syntax of different knowledge representation languages such as RDF, OWL or Topic Maps [18], etc. ODM defines five metamodels; RDFS, OWL, Topic Maps, Common Logic and Description Logic, two UML Profiles (RDFS/OWL Profile, Topic Maps Profile) and a set of QVT mappings from UML to OWL, Topic Maps to OWL and RDFS/OWL to Common Logic. This work uses two of technical artifacts defined by the present specification, which are the OWL and RDFS metamodels as depicted in Fig. 2.

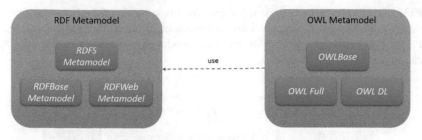

Fig. 2. Dependencies between ODM metamodels

The ODM is applicable to knowledge representation, conceptual modeling, formal taxonomy development and ontology definition, and enables the use of a variety of enterprise models as starting points for ontology development through mappings to UML and MOF [17]. The scope is to define semantics of several knowledge defined according to different formalisms, and thus, facilitate the exchange of knowledge models. By using these metamodels, it's possible to transform models expressed with UML into ontologies [19].

With ODM, we can use MDA's capabilities for ontology development. There are two methods mentioned in ODM for the passage of a UML model to a model RDF Schema, either by using UML profiles or by transformation rules using language processing such as QVT, ATL, etc. We adopt the second method in order to generate our domain ontology linked to the graphical interfaces.

2.2 Big Data

Along with other technology trends like Artificial Intelligence and Semantic web, Big data has become a buzzword around the world. Big Data is a very large data sets that no classic database management or information management tool can really manage. This massive volume is categorized into three categories specifically: Structured data, Semi structured and Unstructured data [20]. Web giants, first and foremost Yahoo but also Facebook and Google, were the first to deploy this type of technology [21].

Figure 3 mentions the 5 V's essential characteristics of Big data. They refer to five key elements to consider and optimize as part of a process to optimize the big data management [22]. Table 1 give an overview of each of these characteristics.

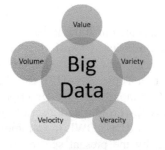

Fig. 3. The five versus of big data

Table 1. 5 V's Big data definition

Big data characteristic	Description
Volume	the volumes of data to be collected and analyzed are considerable and constantly increasing
Variety	a wide variety of information from various sources, unstructured, organized, open, etc. The data can take very heterogeneous forms (texts, web analytics, images, etc.)
Velocity	a certain level of Velocity to reach, in other words frequency of creation, collection and sharing of these data
Value	Be able to focus on data with real value. It is not easy to make sense of Big data analytics tools
Veracity	It's a reliability of data. It raises issues of the meaning in the data itself such as word variation

We can identify five key technology families for the big data industry: text-mining, graph-mining, machine learning, data-visualization and knowledge description(ontologies). Efforts are made for one purpose, which is to simplify the analysis of big data sets by giving them meaning. Big data refers to a large scale that is used to represent a huge collection of datasets. The retrieval of data from these structures benefits from semantic knowledge. For this purpose, it is desirable to provide an ontology for data. The use of ontology is essential to integrate thinking and intelligent recovery of big data. Moreover, generalized concepts in an ontology can connect data in various concept levels across domains, they allow to add metadata ontology for annotating data, and they bring benefits for search and retrieval information [3]. Since, there are other advantages that ontologies offer to solve the difficulties faced in big data analytics.

2.3   Domain Ontologies

Ontological engineering is born from the needs of representations of knowledge in various fields especially Web technologies, database integration, artificial intelligence, big datasets [23]. This utility is motivated by the fact that ontologies are an effective way to manage and share knowledge in a particular area, as well as, allowing systems to touch the semantic part of information and manipulate it by establishing representations and models. Roche et al. [24] define an ontology as a conceptualization of a domain to which one or more vocabularies of terms are associated. The concepts are structured in a system and participate in the meaning of the terms. In particular, a domain.

Domain ontology is defined for a given purpose and expresses a point of view shared by a group of people. It's a structured set of concepts and relationships between them intended to represent objects of the world in a form that can be understood by both human and machines [25]. In fact, it is formalized using five kinds of elements: classes, properties, datatypes, individuals and axioms. To clarify, classes have the same meaning of classes of objects modeled that exist in the modeling area, and individuals represent instances of classes. Properties represent attributes defined for characterizing classes. There are two kinds of properties; a property that link two classes or properties between a class and range of values. Datatypes are sets of literals such as strings or integers. All these declarations are grouped by axioms, in order to form complex descriptions from the basic entities. Besides, semantic technologies proposed by the W3C [26–28] are used to standardize the terms and concepts of each domain ontology, which facilitates communication and knowledge sharing. Domain ontologies have already shown their application in many domains, such as education, government, commerce, health, etc. This work focuses on the e-learning domain.

3   Related Work

In this Paper, we propose an enhanced MDA approach to facilitate the modeling of ontologies and to generate them automatically. The main purpose of our proposal is to encourage software analysts, developers, or designers that expect to easily use ontologies for improving Big data analytics and management, either for data integration and the implementation of big data management or data quality measurement, etc. On the other hand, several works based on MDA for ontology modeling exist in the literature. The most relevant are: [19, 29–36].

A transformation from UML class diagrams to owl ontologies was presented by Belghiat et al. [29]. Authors implement an application of the transformation based on graph transformation and by using the tool AToM3 [37]. This approach has the advantage of generating owl file. But authors don't use transformation language like QVT or ATL. Nevertheless, they don't use ODM to generate ontologies and it's a great inconvenience because this meta model is considered one of standards defined by the object management group (OMG) for modeling ontologies.

Aßmann et al. [30] describe the role of ontology in model-driven engineering. Authors present a scheme combining descriptive ontologies and prescriptive models in the meta-pyramid, the multi-level modelling approach of MDE. In this scheme, MDE starts from ontologies, refines, and augments them towards system models, respecting their relationships to prescriptive models on all metalevels.

Musumbu et al. [19] Propose an approach that use the benefice and advanced researches in Semantics Web, to combine it to Model Driven Architecture in the goal of making automatic business rules generation. This approach is based on merging MDA technologies and Semantics Web, but it does not allow an ontology generation from UML class diagrams while passing through the ODM metamodel. The authors of [31] presented an UML profile for OWL DL and OWL Full respecting ODM Metamodel. However, they did not achieve a real transformation based on the ODM standard while respecting the MDA architecture.

The last approach cited has some similarities with the work proposed by Gašević et al. [36]; They define an approach based on MDA concepts This architecture consists of the ODM defined using the OWL, as well as the related Ontology UML Profile (OUP). The model to model transformation was based on two phases, the first one consisted on an M2M transformation from Uml model to ODM model after that it executes another M2M transformation from ODM model to Xml Model and finally comes the phase of using an XML extractor they generate the owl file. This work done by these researchers has supported OMG's efforts in ontology development, and it is the only one that has covered the maximum of uml to odm transformation rules. Nevertheless, this approach has some lacks; use of two M2M transformations, no transformation language recommended by omg such as QVT. ODM implementation presented in [35] used the ATL language to develop the same approach presented by Gašević et al. [36].

In [32], an OWL translation of UML diagrams has been defined. Bahaj et al. [34] presented a method to convert a UML class diagram to an ontology using OWL/XML language but they don't use an MDA approach and its standards. On the other hand, Zedlitz et al. [33] elaborate a QVT transformation between UML class diagrams and OWL 2 ontologies. They specify the transformation rules on the M2 level and the meta-models of UML and OWL 2. however, this conversion has some limitations; doesn't support ABOX part of ontology and it would have been better if they used the ODM, furthermore, the authors talk about a QVT transformation while they have not used the MDA approach.

In this section, we highlighted interesting researches about ontology modeling in MDA. In brief, we can say that all these approaches did not propose an MDA-based ontology generation process that exhibits the ODM standard, and the only one which generate owl file from ontology UML profile has gaps. This work represent an enhanced approach proposed for the generation of ontologies from UML class diagrams and ODM models using the MDA approach and QVT transformation to simplify their use in big data analytics.

4 Our Proposal

This work is based on the integration OMG efforts with the various works done regarding the ontology development. The main goal of our approach is to provide an improved and optimized solution to represent and generate domain ontologies in a standard format automatically using OMG standards. So, the idea is to represent ontologies in a simple way, which will allow its use by a large community of different domains especially the Big data area.

We propose a model-driven development process that allows the development of domain ontologies to support ontology modeling for Big data analytics. The process is based on the MDA-based standards approach. It highlights two OMG standards: UML and ODM. One of the open source tools which support MDA and follows the standards of the Object Management Group (OMG) for both UML and MDA is Eclipse providing an implementation of MDA through Eclipse Modeling Framework (EMF). Accordingly, we implemented this process using eclipse modeling tools.

The process is divided into three steps: The definition of PIM Model and PSM MetaModel, the definition of M2M Transformation and the definition of M2T transformation. Figure 4 shows the Model-driven process to generate domain ontologies. Indeed, the model driven process starts with an abstract Model conformed to UML metamodel, in order to produce an ODM Model as a target model. We imported PSM metamodel, then we established the transformation rules that gives us the model result

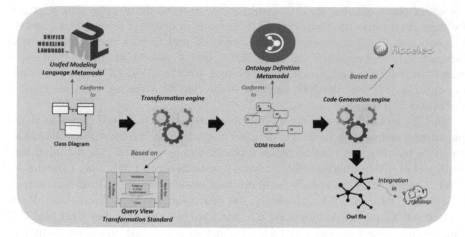

Fig. 4. The model driven development process

which describe sufficiently eLearning domain. Finally, we transform the PSM model in the code generation phase to generate an ontology document respecting an RDF/XML serialization format.

4.1 Construction Rules of PIM and PSM Levels

UML was used in the PIM level to model the class diagram. The necessary rules and steps for constructing Class diagram are very easy. With that said, we need only to define the construction rules of ODM metamodel. Custom Ontology metamodels and ODM Metamodel have many differences that made the creation of the dynamic instance of the last one a bit ambiguous. In other metamodels, we observe a hierarchical architecture which define the elements and the components of this elements in a different level, unlike the dynamic instance of ODM Metamodel which has the characteristic of having all its components in the same level (OWLGraph, OWLOntology, OWLClass ...etc.) and the dependencies between them reside in references (see Fig. 5).

As mentioned above in Sect. 2, we use just the two metamodels of ODM metamodel; OWL and RDFS, to design our domain ontology. OWL metamodel includes all concepts of RDFS and RDF. The main and most important construction rules that is valid for all the components in ODM Model, is the fact that all of them have an URI that identify each one of them. These URIs are referenced when needed in all the components with URIReference which has a "LocalName" and the referenced URI.

Having said that, we proceed to other construction rules:

- "OWLGraph" is the root element. "OWLOntology" elements are referenced by "OWLClass" components.
- "RDFSClass", "OWLObjectProperty" and "OWLDatatypeProperty" are elements that define the Properties for Domain and Range.

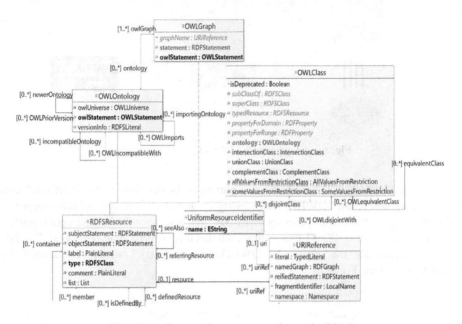

Fig. 5. References between metaclasses in ODM

- "OWLClass" elements reference Properties for Domain and Range and contains a Plain Literal that defines its name. OWL Classes are RDFS Classes and OWL Properties are properties of RDF.
- RDFS defines two properties, rdfs:domain and rdfs:range that connect an rdf: Property with an rdfs:Class, making it possible to distinguish various types of relations between various types of resources.
- "ClassExpression" Specifies the different extensions and restrictions of classes, such as, *ComplementClass, DisjointClass, EquivalentClass and IntersectionClass*.

4.2 M2M Transformation Rules

Miller and Mukerji [14] propose three ways to define mapping: It can be in natural language, an algorithm in an action language, or a model in a mapping language. We were based on the third concept using QVT operational transformation language defined by the OMG [38]. So, Fig. 6 shows the adopted method to elaborate a model to model transformation in our process.

The UML to ODM mapping covered the most used UML components needed for representing UML models (class diagram) and representing them in the ODM standard.

The source model in our mapping is a Class Model, so we need to build an URI for the target OWLGraph and we do the same thing to transform Package to OWLOntology. UML and OWL support the concept of namespace represented by "Package" in UML and "Ontology" in OWL. UML Class is transformed into OWL-Class. The "ownedAttribute" relationship defines the attributes of each class. It is an "OrdredSet" of "Property" that can be mapped to either "OWLDatatypeProperty" or

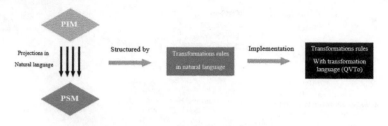

Fig. 6. Adopted method for M2M transformation

"OWLObjectProperty". If a property is a part of an association's "memberEnds" then the mapping will result "OWLObjectProperty", otherwise if the type of the property is "PrimitveType" then the property will be mapped to "OWLDatatypeProperty". "Domain" is the set of "OWLClass" that contains this "OWLDatatypeProperty". while "range" represents the type of "OWLDatatyoeProperty" which is defined by "RDFSClass" (see Fig. 7).

Fig. 7. UML property to OWLDatatypeProperty

A Binary Association specifies a relationship between typed instances. It has exactly two ends represented by properties, each is connected to the end type. This association is mapped to OWLObjectProperty in a target model. The multiplicity of an UML class is mapped to (minimum and maximum) cardinality of the OWL object property.

4.3   Transformation Engine

Our process derive directly the ontology file from a PSM model using Acceleo template [39] without the need to use an XML extractor tool. We opted for the use of RDF/XML type to generate the owl file, and defined generation rules that bordered the construction of Ontology file. Our plugin allows us to generate ontologies not only in RDF/XML format, but in other standard formats such as, JSON, OWL format, etc.

5   Results and Evaluation

In order to validate the utility and applicability of our proposal, this section shows an evaluation of its application on the case study presented in Fig. 8. The main goal of this evaluation is to analyze the Model-driven development process that can be functional in real projects. The case study designed represents a domain of an e-learning. ELearning helps users from all around the world to take courses online using electronic devices. It offers multiple advantages in the education process for different parties. For learners, eLearning allows them to directly interact with lessons for an engaging learning experience, and they can learn at their own pace. For teachers, eLearning permits to present a consistent course material that can be accessed anytime and anywhere, etc.

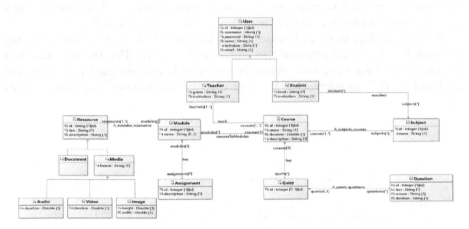

Fig. 8. E-learning class diagram

Big data has a significant impact on eLearning by bringing more benefits, in fact it allows by a data-driven analytics to enhance the eLearning process, for example teachers and institutions can identify the most popular strategies that provide a better experience for individuals which will help improve the efficiency of eLearning by advertising the most popular courses, updating the platform, etc. With that said, big data analytics provide much of help for institutions in such a way that with less effort and less cost, can provide much better and effective user experience.

Once the e-learning system has been sufficiently modeled using the Class diagram, the model is used as an input for the transformation engine developed for the approach. Indeed, we will first generate a PSM model for the e-learning domain that respects the ODM metamodel. Then, it will become the input for the code generation part; that was developed using Acceleo. The resulting file is an owl file representing an e-learning domain ontology. the result is very long either in the ODM model or the ontology generated. Using SPARQL queries, we obtained the following result: Forty-six *Subjects*, 6 *Predicates* and thirty-seven *Objects* distributed as: fifteen classes, 5 object

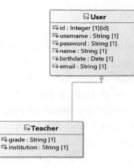

Fig. 9. Class diagram excerpt

properties, seventeen data properties. So it is difficult to display all the results. that's why, we chose to present a part of the obtained results depicted in Fig. 9.

As shown in Fig. 10 the owl classes (User, Teacher) are defined by "owl:class" tag and identified by an uri embedded in "rdf:about" attribute. The Teacher class is a subclass of the User class, this relation is presented in the first one by an contained "owl:subClassOf" tag referencing the User class in "rdf:resource" tag containing its Uri.

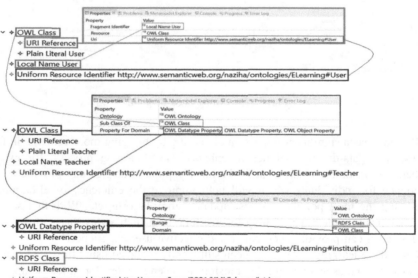

Fig. 10. An excerpt of the ODM model obtained from the QVTo transformation

The "institution" data type property is represented by "owl:DatatypeProperty" tag and identified by an Uri in the "rdf:about" attribute, this data type property belongs to the Teacher class presented in the "rdfs:domain" tag by a reference of its Uri in the "rdf:resource" attribute, the data type of this data type property is defined in the "rdfs: range" tag and referencing the type by an Uri in the "rdf:resource" attribute; This data type property is valid for all the other ones.

Figure 11 presents a part of the output file in RDF/XML format, it describes the result of the M2T transformation. Many XML Namespaces are used to present our ontology, from which we mension: xmlns:owl="http://www.w3.org/2002/07/owl#".

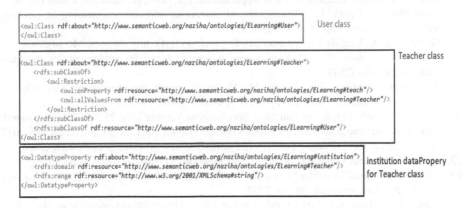

Fig. 11. An owl file excerpt of e-learning domain

6 Conclusion

In this paper, we presented an improved solution for generating domain ontologies from UML class diagrams automatically using the MDA approach. The main goal of this work is to stress the overwhelming importance of ontology development in big data analytics. Our model driven development process starts with UML Class diagram, then we transform it to ODM model. Finally, we used generation engine developed to obtain owl file respecting a standard syntax. As a result, our ontology generated files can be shared with ontological engineering tools notably protégé. Our solution is a best useful development technique for all software engineers who participates in ontology modeling and development process. It will improve everyday work productivity in Big data. The ontology generated can easily be stored in one of big data technologies such as the open source Hadoop Distributed File System. As a future work, we can derive ontologies in other serialization formats like JSON, OWL, etc.

References

1. Chen, M., Mao, S., Liu, Y.: Big data: a survey. Mob. Netw. Appl. **19**(2), 171–209 (2014)
2. Kozaki, K.: Ontology engineering for big data. In: Ontology and Semantic Web for Big Data (ONSD2013) Workshop in the 2013 International Computer Science and Engineering Conference (ICSEC2013), Bangkok, Thailand (2013)

3. Konys, A.: Ontology-based approaches to big data analytics. In: International Multi-Conference on Advanced Computer Systems, pp. 355–365. Springer, Cham (2016)
4. Nadal, S., Romero, O., Abelló, A., Vassiliadis, P., Vansummeren, S.: An integration-oriented ontology to govern evolution in big data ecosystems. ArXiv preprint arXiv:1801.05161 (2018)
5. Saber, A., Al-Zoghby, A.M., Elmougy, S.: Big-data aggregating, linking, integrating and representing using semantic web technologies. In: International Conference on Advanced Machine Learning Technologies and Applications, pp. 331–342. Springer, Cham (2018)
6. Brambilla, M., Cabot, J., Wimmer, M.: Model-driven software engineering in practice. Synth. Lect. Softw. Eng. 3(1), 1–207 (2017)
7. Guide, M.D.A. Version 1.0, Version 2.0, Document OMG. http://www.OMG.org/mda
8. Blanc, X., Salvatori, O.: MDA en action: ingénierie logicielle guidée par les modèles. Editions Eyrolles (2011)
9. Raibulet, C., Fontana, F.A., Zanoni, M.: Model-driven reverse engineering approaches: a systematic literature review. IEEE Access 5, 14516–14542 (2017)
10. «About the Unified Modeling Language Specification Version 2.0», 25 Nov 2017 [En ligne]. Disponible sur: http://www.omg.org/spec/UML/2.0/About-UML/. Consulté le: 25 Nov 2017
11. Object Management Group. Meta Object Facility (MOF) 2.0 Core Specification (2003)
12. OMG, X.: Metadata Interchange (XMI) Specification. Version, 1, 02-01 (2000)
13. «About the Unified Modeling Language Specification Version 2.5.1». https://www.omg.org/spec/UML/. Last accessed: 08 Feb 2018
14. Miller, J., & Mukerji, J.: MDA Guide Version 1.0. 1, Object Management Group. Inc., June 2003
15. Shaikh, A., Wiil, U.K.: Overview of slicing and feedback techniques for efficient verification of UML/OCL class diagrams. IEEE Access (2018)
16. «UML Designer Documentation». http://www.umldesigner.org/. Last accessed: 08 Feb 2018
17. ODM, O.: Ontology definition Metamodel–OMG adopted specification. Object Management Group, En Ligne. http://www.omg.org/spec/ODM/1.0/Beta2/PDF/ (2007)
18. «TopicMaps.org—Topic Maps». http://www.topicmaps.org/. Last accessed: 10 Feb 2018
19. Musumbu, K.: Towards a model driven semantics web using the ontology. In: The 2013 International Conference on Advanced ICT for Business and Management (ICAICTBM2013), p. 700 (2013)
20. Taneja, R., Gaur, D.: Robust fuzzy neuro system for big data analytics. In: Big Data Analytics, pp. 543–552. Springer, Singapore (2018)
21. Liu, J., Pacitti, E., Valduriez, P.: A survey of scheduling frameworks in big data systems. Int. J. Cloud Comput. 1–27 (2018)
22. Storey, V.C., Song, I.Y.: Big data technologies and management: what conceptual modeling can do. Data Knowl. Eng. 108, 50–67 (2017)
23. De Giacomo, G., Lembo, D., Lenzerini, M., Poggi, A., Rosati, R.: Using ontologies for semantic data integration. In: A comprehensive guide through the italian database research over the last 25 years, pp. 187–202. Springer, Cham (2018)
24. Roche, C.: Terminologie et ontologie. Langages 1, 48–62 (2005)
25. Berners-Lee, T., Hendler, J., Lassila, O.: The semantic web. Sci. Am. 284(5), 34–43 (2001)
26. «RDF—Semantic Web Standards», https://www.w3.org/RDF/. Last accessed: 18 Feb 2018
27. «OWL—Semantic Web Standards», https://www.w3.org/OWL/. Last accessed: 18 Feb 2018
28. «Extensible Markup Language (XML)», https://www.w3.org/XML/. Last accessed: 26 Nov 2017
29. Belghiat, A., Bourahla, M.: Automatic generation of OWL ontologies from UML class diagrams based on meta-modelling and graph grammars. World Acad. Sci. Eng. Technol. 6 (8), 380–385 (2012)

30. Aßmann, U., Zschaler, S., Wagner, G.: Ontologies, meta-models, and the model-driven paradigm. In: Ontologies for software engineering and software technology, pp. 249–273. Springer, Berlin, Heidelberg (2006)
31. Brockmans, S., Colomb, R.M., Haase, P., Kendall, E.F., Wallace, E.K., Welty, C., Xie, G.T.: A model driven approach for building OWL DL and OWL full ontologies. In: International Semantic Web Conference, pp. 187–200. Springer, Berlin, Heidelberg (2006)
32. Saripalle, R.K., Demurjian, S.A., De la Rosa Algarín, A., Blechner, M.: A software modeling approach to ontology design via extensions to ODM and OWL. Int. J. Semant. Web Inf. Syst. (IJSWIS) 9(2), 62–97 (2013)
33. Zedlitz, J., Jörke, J., Luttenberger, N.: From UML to OWL 2. In: Knowledge Technology, pp. 154–163. Springer, Berlin, Heidelberg (2012)
34. Bahaj, M., Bakkas, J.: Automatic conversion method of class diagrams to ontologies maintaining their semantic features. Int. J. Soft Comput. Eng. (IJSCE) 2 (2013)
35. Hillairet, G.: ATL use case-ODM implementation (Bridging UML and OWL). http://www.eclipse.org/M2M/atl/usecases/ODMImplementation (2007)
36. Gašević, D., Djurić, D., Devedžić, V.: MDA-based automatic OWL ontology development. Int. J. Softw. Tools Technol. Transfer 9(2), 103 (2007)
37. De Lara, J., Vangheluwe, H.: AToM 3: a tool for multi-formalism and meta-modelling. In: International Conference on Fundamental Approaches to Software Engineering, pp. 174–188. Springer, Berlin, Heidelberg (2002)
38. OMG, Q.: Meta Object Facility 2.0, Query/View/Transformation Specification (2011)
39. Musset, J., Juliot, É., Lacrampe, S., Piers, W., Brun, C., Goubet, L., Lussaud, Y., Allilaire, F.: Acceleo user guide, vol. 2. See also http://acceleo.org/doc/obeo/en/acceleo-2.6-user-guide.pdf (2006)

Combining CRM Strength and Big Data Tools for Customers Profile Analysis

Z. Elyusufi, Y. Elyusufi, and M. Aitkbir$^{(\boxtimes)}$

LIST Laboratory, Doctoral Studies Center STI, Abdelmalek Essaadi University,
Tétouan, Morocco
{zelyusufi, elyusufiyasyn}@gmail.com, m.aitkbir@fstt.ac.
ma

Abstract. Today Big Data tools are not just a phenomenon of the massive information collection; they are also the best way to approach a customer target. These technologies allow the profiling of the customers of an organization thanks to the histories of purchases, the products that they consult; the data that they share through the social networks. They also make it possible to anticipate the purchase of actions via behavioral analysis. Therefore, the combination of the power of CRM and the performance of BIG DATA tools brings a great added value for customers profile analysis, especially if it is about events triggered in real time. It is in this context that the present work is positioned. Our goal is to intercept events (customer behaviors) and analyze them in real time. We will use the Complex Events Process (CEP) architecture that perfectly meets this need. In order to successfully implement our CEP architecture, we will use the ontology approach.

Keywords: Profiling · CEP · CRM · Ontology · Big data

1 Introduction

With the democratization of the internet and especially social networks, we are now faced with a huge flow of information with its variety (Images, sound, text), that's why the investigation of events has become more and more important. Large companies want to generate new knowledge from existing knowledge bases and analyze the mass data in because individual events often have little or no meaning. Currently several works deal with systems that can analyze events. Among the most important are CEP systems (Complex Event Processing). A CEP system essentially aims to detect, in an infinite stream of events, predefined sequences of events [1]. Once a predefined sequence is detected by the CEP system, it produces a combination of events that can possibly trigger certain associated actions (generation of an alert, storage of the combination detected in a database, etc.). A CEP system therefore consumes an input stream of events, and detects sequences described by detection rules, and outputs complex events. Furthermore, ontologies play a very important role in combination with event processing systems [2–4]. But ontologies make it possible to create strategies for describing semantic models [5]. In addition ontologies can model any knowledge domain (possibly several domains). They can also determine the properties,

© Springer Nature Switzerland AG 2019
M. Ben Ahmed et al. (Eds.): SCA 2018, LNITI, pp. 226–237, 2019.
https://doi.org/10.1007/978-3-030-11196-0_20

restrictions and axioms of any field of knowledge [6]. The peculiarity of our work lies in the combination between the CEP system and the ontology approach to take advantage of the synergy between the two approaches. In order to successfully implement this architecture, this study will focus our work on the detection rules engine. These rules are derived from the rules of management of a specific area of knowledge, including the credit industry and the customer relationship management. Several works have been directed in this direction. Astrova I. described five ontologies that deal with the approach of events [7]. They presented a comparison of the existing event models. The main criteria of this comparison were to model events taking into account the concept of time, location and participation of objects. However, other factors can also be taken into account. Some ontologies have been developed to model events taking into account the main components of CEP architecture such as time, causality .... Also the concepts of space and geography have been object of development of many ontologies. These ontologies include: Event—Model-F, Event Ontology model, Timeline in the Timeline ontology and LODE Ontology. In our work we will develop a new ontology called CTOP (Customer to Profile Ontology). Thanks to the combination of CEP and CTOP, we will classify the customers of a financial institution into profiles that will be defined later. Also the system will be able to detect these profiles in real time thanks to CEP, and to be able to react in real-time and in an efficient and intelligent way towards these customers.

2 State of the Art

Several research works have addressed the problem of complex event flow detection. In this work, we are interested in the model of detection and specification of rules. Demers, A. propose a CEP system named Cayuga Event Language (CEL) [8]. The Cayuga system is used to represent events that last in time and not just instant events. CEL adopts syntax close to SQL. Among the main limitations of CEL is that it does not propose to specify negations in the rules. Furthermore, Suhothayan, S. present the language Siddhi Query Language [9], the syntax of this language is based on SQL, from which the name Siddhi Query Language (SiddhiQL). A very important aspect addressed by the Siddhi system is the partitioning of data, as well as the partitioning of flows. Finally Cugola, G. propose a T-Rex CEP system based on the TESLA rule specification language [10]. However, the main limitations of TESLA [11] are that it does not propose logical operators of conjunction and disjunction, nor does it propose an iteration operator. A comparative study of these architectures (CAYUGA, SIDDHI and T-Rex) shows that Siddhi is more complete than the other systems, it contains the criteria of a perfect CEP system, it proposes in particular the logical operators, a sequence operator, an iteration operator as well as negations. The CEP architecture that we propose must satisfy all these requirements, by introducing the ontological paradigm. The latter aims to fill up the lack of logical manipulation of the rules, since the rules of ontology are based on the logical operations (negation, conjunction, transitivity ...). The role of CEP will be limited in our approach to event management while the main role of ontologies will be the qualification and profiling of clients.

3   Our Architecture

The purpose of our architecture is to set up a system for the detection of complex events in real time and at the same time to take advantage of the existing CRM database already in business, a database that contains enough detail about customers and prospects of the company. The combination of the power of the CEP and a CRM database is proving to be a very good result in terms of profiling customers and to follow their behavior.

The architecture that is proposed will be used in a context of profiling clients of a financial institution. The rules engine must then combine between the credit business and customer relationship management. Our goal is to integrate the solution with an existing Customer Relationship Management (CRM) system in order to take advantage of the maturity of this system in terms of rules and data. Figure 1 gives a global vision of our target architecture.

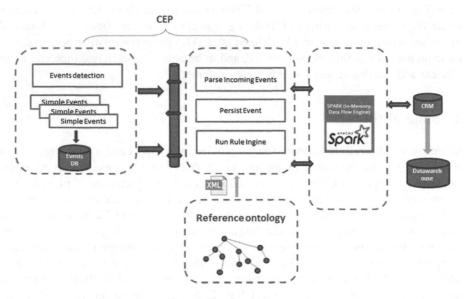

Fig. 1. Target architecture

Events intercepted through an event detection layer are parsed and analyzed, inference rules and business rules are handled at a reference ontology level. The output of the CEP Layer is a complex event that will be used to enrich the CRM database.

3.1   Event Management

Our system will be equipped with a set of event sensors that will be represented by classes. Our approach must combine multi-source data (data from social networks and data collected at the level of the organization's digital platform). According to the reference architecture proposed by Tim Bass a pre-treatment step is essential, it allows formatting the data of an event in a unitary way before proceeding to the next steps. Figure 2 shows the CEP reference model proposed by Bass [12].

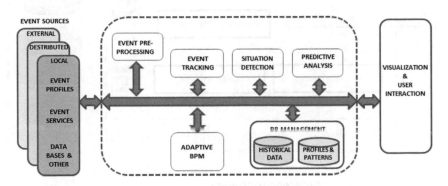

Fig. 2. The CEP reference model proposed by Tim Bass

Tim Bass model is a multi-level inference model in distributed event decision architecture; it is based on five levels:

- Level 1: Event pre-processing
- Level 2: Event Tracking
- Level 3: Situation detection
- Level 4: Predictive analysis
- Level 5: Adaptive BPM.

In this work we focus on the first three levels of the Tim Bass model, event pre-processing, event tracking and event detection.

3.2   Event Pre-processing

Our architecture will be able to identify profiles and recognize if it is a new user or it is a customer or prospect in a preliminary stage, it will be done with a simple verification of existence in our CRM database.

This verification of the customers existence in the CRM database will also be generalized on other repositories of the company (Example: bases of the incidents of payments or base of the persons who belong to specific lists), this will make it possible to refine further our inference engine rules and make the event detection step easier and more relevant.

3.3   Event Tracking

The management rules and inference rules that will be implemented at the level of our ontology will then be personalized. The first rule engine will be dedicated to new customers, its ultimate goal is to follow the behavior of the profile in order to transform it into a customer while the second engine will be dedicated to the customers and prospects of the company, and its role is the supervision of their behavior in real time (Fig. 3).

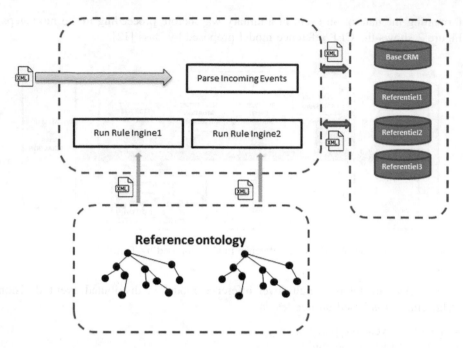

Fig. 3. The choice of the reference otology specific to each profile

3.4 Situation Detection

Once the profile has been identified, our CTOP ontology will use the part of the profile-specific inference engine as well as the rules for detecting appropriated events and subsequently trigger the rules engine for decision-making (Behavioral reaction logic)

$$\text{Complex event} + \text{Decisioning conditions} \rightarrow \text{response actions}$$

Finally, the CRM database will be enriched by the output information of our CEP to build what is called the social CRM.

4 The CEP System

The operation of the CEP system involves queries capable of obtaining data from events that arrive in the system over a period of time. CEP also allows having algorithms able to analyze this set of data (statistics, algebraic, rules …) producing precise decisions and actions. Our architecture must meet a specific need, mainly related to the recognition and monitoring requirements of customers and business prospects. If we take into account the size of the enterprise digital platform as well as the Customer Relationship Management System, future requirements in terms of volume and in terms of data variety are a real challenge. In order to meet these needs, we have chosen to move towards a BIG DATA context. From a technological point of view, the correlation between continuous events in large numbers over time (spatial and temporal) inevitably raises a problem of scaling up. The solution then lies in in-memory technologies [13]. We chose Apache Spark for the treatment of massive and heterogeneous data, first because it is an engine that favors In-memory processing, second it is suitable for the case of a CEP architecture that requires a real-time processing. Apache spark streaming is an extension of the Apache Spark API, it allows the processing of high throughput flows with fault tolerance. Data can come from multiple sources and be processed in parallel. The streaming mode used allows the detection of events in a real-time event cloud. Figure 4 illustrates the DStream model proposed by Apache Spark.

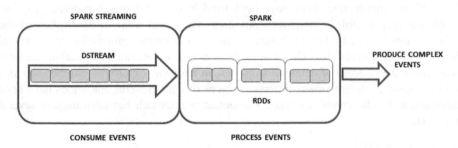

Fig. 4. The DStream event detection model offered by Apache Spark

Apache Spark DStream event detection is performed as follow:

- Client nodes inject continuous data streams into caches using Data Streamers.
- The data is automatically partitioned between the nodes and each node receives the same amount of data.
- The streaming data can be processed simultaneously on the data nodes in a co-located manner.
- Clients can also perform concurrent queries on the data being broadcast.

The Spark Streaming module allows you to see the event flow as an infinite DStream. In fact, the DStream is a sequence of event batches representing each period of the event flow (Example 1 second). Spark then associates each batch of events with a data structure called Resilient Distributed Dataset (RDD). In our approach, the customer event management task will be accomplished by the DStream model of Spark. While the profile qualification task will be handled by the CTOP ontology we are developing.

5   Our Approach

We have already worked in our series of researches on building profiles based on ontology approach. Among the major limitations of our researches is that the system will not be able to follow the evolution of profiles over time and on real time. In our approach we focused our research on the criteria "**Velocity**", since we use CEP architecure. This later will analyze user events in real time, meaning a high frequency during main activity hours.

In order to implement an architecture meeting the requirements mentioned above, we chose to combine the two technological paradigms (Social CRM, and CEP). The first paradigm will make it possible to build a mine of data through the collection of data from several data sources (social networks, mobile application …). These will produce a giant mass of data. The second CEP paradigm (Complex Event Processing) will be needed to process events in real time and manage them using a set of rules.

In this work, we used the ontology approach. The latter will be able to model business rules in the area of customer relationship management. Ontologies are the heart of our architecture. They have been used in several research projects [14, 15]. They are very useful for heterogeneous data. They also allow a description of the relationships between different domains. They can combine information from several heterogeneous sources [16]. They can facilitate access to relevant information and even respond to the concern of temporal windows. In addition, ontologies can be used as a tool to extract data from many different sources of data. Several ontologies have been designed to handle complex events. The ontology approach has been used in several projects.

- Event—Model-F
- Event Ontology
- Timeline in the Timeline ontology
- LODE Ontology.

One of these projects is EU project WeKnowlt. A case study in this project used Event-Model-F ontology. The objective was to recognize incidents of a company in real time, in order to better react in critical situations as soon as possible.

The basic tasks in our proposed approach for detecting events from a data stream are described as follows (Fig. 5):

- **Collection phase**: In practice, we manually limit the scope of the tweets collected from social networks "Twitter, Facebook …" by using Social CRM. The data are then cleaned into a normalized form and all unnecessary elements are removed before storing them in database.
- **Extracting terms**: Normalized tweets are analyzed to extract all of the potential terms. We can improve this step by using ontology approach as shown in Sect. 7, in order to recognition with predefined knowledge data (ontology rules).
- **Inference system**: The system will be able to group customers in clusters, i e instances of ontology classes (individuals) that meet the rules described in the reference ontology will allow to qualify (infer) in real time the customers carrying a set of properties (see Sect. 5.2).

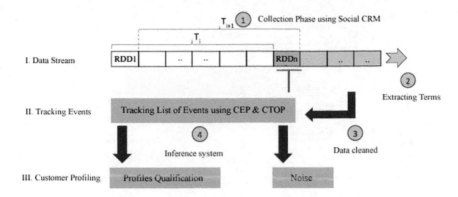

Fig. 5. Workflow of detecting and tracking events

5.1 Using the Ontology Approach

The ontology we propose is a domain ontology of customer relationship management in a financial context (which will be designed by CRM and credit business experts). This ontology will help us structure the data and semantic links that exist between these varied, multi-source and unstructured data. This ontology will allow the system to categorize costumers into well-identified profiles. Thanks to the integration of this ontology into the CEP architecture, the system will be able on the one hand to identify the customers in their profile and on the other hand to react in real time to alert the detection of a customer who just belonged to such a profile. This will allow answering in real time the queries related to customers and their behavior. The CTOP (Customer to Profile) ontology is composed of several classes and properties. The following figures illustrate an application prototype and show some examples of implementations. For the sake of clarity, a minimal version of the CTOP ontology has been adapted, as shown in Fig. 6.

In order to implement the CTOP ontology we defined a set of business rules, among which:

1. A **Customer** *is intersted in a* **Domain** (Thing)
2. A **Domain** *interests one or more* **Customer** (Thing)
3. A **Customer** *has a* **OnlineAccount** (Thing)
4. A **Customer** *has a account on* **TheDigitalPlatForm** (Thing)
5. A **Customer** *is located in a* **Geographic Location**
6. A **GeographicLocation** *has a* longitude (float)
7. A **GeographicLocation** *has a* latitude (float)
8. A **Customer** *Benefits from* **Opportunity (Thing)**
9. A **Suspect** *posts_Like* **on the portal** (Thing)
10. A **Prospect** *has a* **Credit_simulation** (Thing)
11. A **PaperForm** *is completed by* **Prospect** (Thing)
12. A **Qualified Prospect** *is a* **Prospect** *has a* **(OnlineAccount or** *goesto* **Agency)** (Thing).

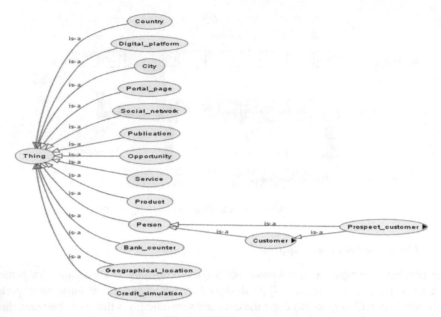

Fig. 6. CTOP ontology

Defining class relationships and business rules will help identify customers with well-defined properties. Figures 7 and 8 show the relationships that exist between classes: *Person, Customer, Silent_Customer, Suspicious_customer, Prospect_customer* and *Prospect_customer_qualified.*

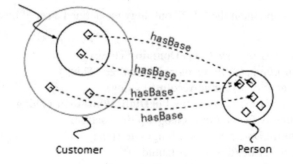

Fig. 7. Relationship between the customer and person classes

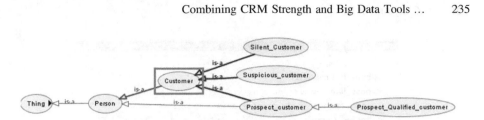

Fig. 8. Definition of the "Customer" class and these subclasses

5.2 Customer Profiling Results

Thanks to the combination of the CEP and the CTOP ontology, the system will be able to recognize a costumer as soon as possible based on rules described into CTOP ontology (Fig. 8). If we take the example of **Customer1** which is an instance of Customer class (individual) as shown in (Fig. 9). It is a customer who has benefited from a product **Product1**, he has already posted a like on the site www.credit.org, he has already connected to the bank account Bank_counter_anfa. The system infers that **Custolmer1** knows **Cusomer2** (Figure 10). This will allow the system to recommend as soon as possible to **Cusomer2** to benefit from **Product1** since they have a knowledge link see Fig. 11.

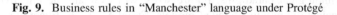

SubClass Of
- Customer
- Person and (post_like some Social_network)

SubClass Of
- interested_in only (Product or Service)
- Prospect_customer

SubClass Of (Anonymous Ancestor)
- Person and (knows some Person)
- benefit min 1 Product
- Person and (post_like some Portal_page)
- Person and (connects_via some Bank_counter)

SubClass Of (Anonymous Ancestor)
- Person and (knows some Person)
- benefit min 1 Product
- Person and (post_like some Portal_page)
- Person and (connects_via some Bank_counter)
- Person and (post_like some Social_network)

Fig. 9. Business rules in "Manchester" language under Protégé

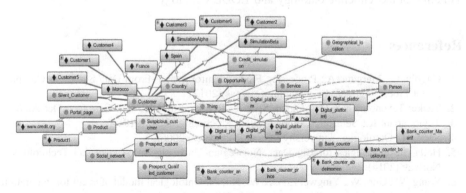

Fig. 10. Ontology CTOP after creation of instances (individuals)

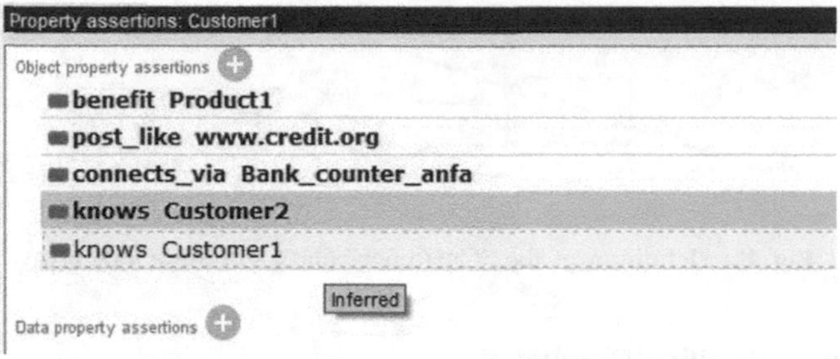

Fig. 11. Example of inference offered by our approach

As we can see in the example above, the system allows inferences about profiles and their behavior towards different products. By following the logic of inference. We can enrich our system by adding other constraints. Among these constraints we can mention, time, place, action and category. These will be translated into rules and integrated into the PTOP ontology and will allow to improve the quality of customer profiling.

6 Conclusion

The added value of this work lies in the qualification of customer profiles in real time in a Big Data context. In order to implement a seemly qualification of the customers, we combined the social CRM and the CEP architecture for the collection of the data by means of a fine expression of the scenarios to be identified. We used the ontology approach first for knowledge domain structuring, and then for the qualification of customers in profiles. The combination of CEP (Complex Event Processing) and CTOP (Customer to Profile Ontology) allowed us to take advantage of the synergy between the two technological paradigms. Future work will evaluate the effectiveness of our approach against the previously cited models, Event-Model-F, Event Ontology model, Timeline in the Timeline ontology and LODE Ontology.

References

1. Cugola, G., Margara, A.: Processing Flows of Information: From Data Stream to Complex Event Processing. ACM Computing Surveys (2012)
2. Tacke: Taxonomien & Ontologien. Soziales Retrieval im Web 2.0. Germany, Oct 2008
3. Teymourian, K.: Semantic Complex Event Processing, Sept 2010
4. Sack, H.: Ontologien. University of Jena, Germany (2006)
5. Hella, L.: Ontologies for Big Data. Norwegian University of Science and Technologies, Norway (2014)
6. Wang, W., Guo, W., Yingwei Luo, X.W., Xu, Z.: Ontological model of event for integration of inter-organization applications, pp. 301–310 (2005)

7. Astrova, A., Koschel, A., Lukanowski, J., Martinez, J.L.M., Procenko, V., Schaaf, M: Ontologies for complex event processing. World Acad. Sci. Eng. Technol. Int. J. Comput. Inf. Eng. **8**(5) (2014)

8. Demers, A., Gehrke, J., Hong, M., Riedewald, M., White, W.: Towards Expressive Publish/Subscribe Systems EDBT 2006, numéro 3896 de Lecture Notes in Computer Science. Springer, Berlin, Heidelberg. https://doi.org/10.1007/11687238_38

9. Suhothayan, S., Gajasinghe, K., Loku Narangoda, I., Chaturanga, S., Perera, S., Nanayakkara, V.: Siddhi: A second look at complex event processing architectures. In: Proceedings of the 2011 ACM Workshop on Gateway Computing Environments, GCE'11. ACM, New York, NY, USA (2011)

10. Cugola, G., Margara, A.: Complex event processing with T-REX. J. Syst. Softw. **85**(8), 1709–1728 (2012)

11. Cugola, G., Margara, A.: TESLA: a formally defined event specification language. In: Proceedings of the Fourth ACM International Conference on Distributed Event-Based Systems, DEBS'10 (2010)

12. Bass, T.: CISSP, Next Generation Security Event Management (SEM) with Complex Event Processing (CEP), CDIC (2007)

13. Zaharia, M., Chowdhury, M., Das, T., Dave, A., Ma, J., McCauley, M., Franklin, M.J., Shenker, S., Stoica, I.: Resilient distributed datasets: a fault-tolerant abstraction for in-memory cluster computing. In: Proceedings of the 9th USENIX Conference on Networked Systems Design and Implementation, NSDI'12 (2012)

14. Anett, H., Ana, R., Christophe, N.: Ontology-based user profile learning from heterogeneous web resources in a big data context. CheckSem Research Group, Laboratoire Electronique, Informatique et Image (LE2I), Université de Bourgogne

15. Hoppe: Automatic ontology based user profile learning from heterogeneous web resources in a big data context, 26–30 Aug 2013, Riva del Garda, Trento, Italy. Proc. VLDB Endowm. **6** (12) (2013)

16. Twardowski, B., Ryzko, D.: Multi-agent architecture for real-time big data processing. In: International Joint Conferences on Web Intelligence (WI) and Intelligent Agent Technologies (IAT), 2014 IEEE/WIC/ACM

Framework Architecture for Querying Distributed RDF Data

Lamrani Kaoutar[✉], Ghadi Abderrahim,
and Florent Kunalè Kudagba

IT Department Faculty of Science and Technology, Tangier, Morocco
{kaoutar.lamrani1,ghadi05,kkunale}@gmail.com

Abstract. Today, the Web knows a rapid increase in data level that makes their processing and storage limited in traditional technologies. That is why future technology tries to exploit the notion of semantics and ontology by adapting them to big data technology to allow a fundamental change in the access to voluminous information in the web. That Intended to have a complete and relevant response to the user request. Our research work focuses on the semantic web. Focus exactly on the semantic search on many data expressed by RDF (Resource Description Framework) in distributed system. The semantic language proposed by W3C (World Wide Web Consortium) provides the formalism necessary for the representation of data for the Semantic Web. However, only a knowledge representation format is insufficient and we need powerful response mechanisms to manage effectively global and distributed queries across a set of stand-alone and heterogeneous RDF resources marked by the dynamic and scalable nature of their content.

Keywords: RDF · RDFS/OWL · SPARQL · MapReduce · HDFS Ontology

1 Introduction

Today the resources and services available in the web are growing at a fast way day by day, this massive data considered unstructured, distributed and heterogeneous presents a problem in the time of data analysis, access and maintain relevant resources for a user query. Thanks to this increase, the notion of big data appears as a new term described by 6 V (volume, velocity, variety, value, variability, veracity), which to face the need for filtering, distribution, analysis and storage of data.

This big data phenomenon presents a challenge for Data Architect and data scientist, who are faced with the question of how to exploit big data technologies to run an algorithm that deals with the problem of storing and processing this massive data [1]. In this massive data another factor that must be taken into consideration is the possibility that sources describing the same resource may be conflict with their descriptions. In this case, there is the question of the reliability of the data sources, and the precautions should take to avoid inconsistencies that may arise from a combination of contradictory resources. In this context, the W3C recommends the RDF formalism to describe the

© Springer Nature Switzerland AG 2019
M. Ben Ahmed et al. (Eds.): SCA 2018, LNITI, pp. 238–246, 2019.
https://doi.org/10.1007/978-3-030-11196-0_21

resources semantically, however, only a format of resource representation is not enough and we need RDF resource query languages called SPARQL.

When a user sends a request to have an information that satisfies certain conditions of an object, its purpose is to search for a coherent and complete answer, without having the worry of the sources of information that participates in the combination of information to describe the same object from two different points of view depending on the context in which the descriptions are considered. Because today's web or infrastructure interacts with humans does not always respond to the completeness and relevance of the responses to the user's request, the web gets ready to interact with intelligent agents capable of reasoning about resources collected and communicate their results to humans, this generation of web is the semantic web considered as a step in the evaluation of the sharing of human knowledge. That is why having an information retrieval mechanism to access the semantic content of the sources is necessary and the use of language to express the semantics of data is more than ever in the attention of organizations holding big data.

To deal with this problem of massive and distributed RDF data, finding data in this flow of information to have a relevant and complete answer and increase the quality of response service we will exploit the technologies of big data and semantic web. The researchers are investing their efforts to design RDF data storage architecture and an analysis system for processing massive data [2]. On the one hand the data must be interrogated by the SPARQL query to guarantee a quality of service in the semantic search of information [3], on the other hand to provide users requirement who demand a fast access to the data and a high data quality. So to design big data architecture in a distribute, intelligent and efficient way, the storage of the global graph built by coherent combination of the origin graphs must be efficient and the treatment of these data should be in distributed computation way of a set of sub query decomposed from SPARQL query. Therefore, it will exploit the MapReduce for the parallels of the calculation of the SPARQL query, and the storage maintenance of the global RDF graph in HDFS.

So the rest of this paper is organized as follow, in Sect. 2 we introduce the related work to our topic Framework architecture for querying big and distributed RDF data, Sect. 3 we mention technologies to integrate big RDF data, our architecture to access big and distributed RDF data are in Sect. 4, finally in Sect. 5 we concluded our paper.

2 Related Work

The main idea provides by data integration is the combination of resources coming from different sources for giving a user a unified view of data. For that many investigator are focused there searches in this field by exploiting semantic techniques like using Karma 5 to integrate Big Data in here summit [4], using a semi-automatic solution to mapping the structured sources to the ontology to have sources model proposed in [5]. In [6] propose a WETSUIT to integrate and query a web data coming from different sources by parallel process.

When the user request this unified view of data the answers are limited because the current search engine use a traditional techniques to search the information, among these search engine we finds Jena6 and Sesame7. They are limited because they do not support the big data on the one hand, on the other hand the current technologies provide solution to overcome the lack of this traditional engine, as the existence of Cliquesquare [7], H2RDF [8], Inferray [9], Triad [10], TripleStore and other similar solution that solves RDF storage problems.

3 Technology to Integrate Big RDF Data

3.1 Big Data Technologies

Storage in HDFS (Hadoop Distributed File System). HDFS is considered a distributed file system of Hadoop proposed by the Apache Foundation, written in Java. HDFS is used to store in a distributed manner very large files in a directory. The files are physically decomposed into blocks of large bytes; these blocks are broadcast over several machines to process these files in parallel way. The blocks in each file are intelligently replicated to multiple machines in the event of a failure.

Processing by MapReduce. MapReduce proposed by Google as a generic model and a framework for processing in parallel way the treatment. Use two functions to process data Map and Reduce. The Map function provides a pair of inputs and provides a set of intermediate key/value pairs. The Reduce function accepts the intermediate key with their set of values sent by the MapReduce library. In the context of Big Data, MapReduce shows its interest in the processing of large volumes of data, associated with a dedicated software infrastructure to run the MapReduce schema massively distributed on a cluster of machines while taking into account the stakes [11].

3.2 Web Semantic Technologies

RDF. The RDF is considered as a standard data model proposed by W3C to describe things in a semantic, simple and unambiguous way. Presented as a triplet ⟨s, p, o⟩ subject, predicate and object, such that the subject and the predicate can be Uniform Resource Identifier (URI) resources and the object can be a resource or a literal value. This abstract presentation can concretely modeled as an oriented graph with vertices and arcs labeled with the arcs are always directing resources to other resources or literals (Fig. 1).

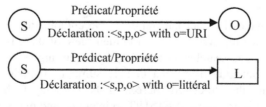

Fig. 1. Example of RDF graph data

RDFS/OWL. These languages proposed by W3C make it possible to share a common domain to exchange between the users and the applications. The RDFS seen as a framework of construction of ontology describes the vocabulary and the concepts of a particular domain. this semantics description can be partially supported by RDFS, on the other hand OWL associates the basic concepts of a specific domain and the relationships between these concepts encodes the knowledge of a particular domain allows a richer integration and guarantees a capacity to be distributed as he adds the notion of disjunction, cardinality, equivalence, typography and symmetry.

SPARQL. SPARQL recommended by the W3C as the standard RDF query language. Identified by two clauses: SELECT called the header of the query, which presents the variables that will appear in the results, the clause WHERE called body of the query that presents the conditions of selection. These conditions determine the sub graph that will selected from the global RDF graph. This is why the result of the SPARQL query is considers as a sub graph of RDF. Mapping SPARQL query considered as pattern model consisting of a set of triples models of RDF model. This is an example of SPARQL Query:

$$
\begin{aligned}
&\text{SELECT ?nom ?prenom ?description WHERE} \\
&\{ \\
&\qquad \text{?personne rdf:type foaf:Person.} \\
&\qquad \text{?personne foaf:img ?image.} \\
&\qquad \text{?image dc:description ?description} \\
&\}
\end{aligned} \tag{1}
$$

In this example of SPARQL query we have three conditions with Q is a global query and Q'i a sub query:

$$
Q = UQ'i/i = \{1\ldots3\} \tag{2}
$$

4 Architecture to Access Big and Distributed RDF Data

This architecture gives to the user the possibility to query a homogeneous and centralized data in order to give the user a complete and pertinent answer that can combined from different sources. The following architecture present steps carry out such supports (Fig. 2).

4.1 Formalisms for the Representation of Big RDF Data

The first step present the data source that are consider heterogeneous because the existed of different models of data and schemas. Multiple objects, entities, concepts and the presence of different interpretation of the same data used to present the semantic descriptions of resources. Autonomous because each source data described independently from other source, distributed because the sources of the data are stored in a distributed manner And this resource are mapped by ontology to give a structured data

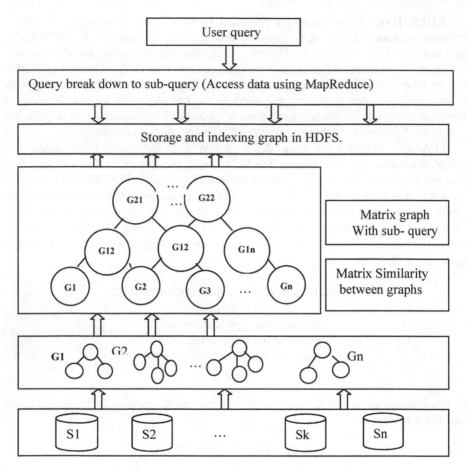

Fig. 2. Architecture to query big and distributed RDF data

presented by a graph this graph can transformed to a declaration that will be exploited by mechanisms of inference to do the combination between graphs.

4.2   Similarity Matrix Between RDF Graphs

To generate a complete and pertinent answer of the user query, the combination between the RDF graphs must be consistent. To calculate this index of consistency we should detect if the graphs can be combined or not. For that it is necessary to calculates for each pair of RDF graphs the measurement of similarity to indicate to which point the graph can be combined or not. To make this comparison between two graphs we use the ontology [12] to have an approximation between the resources by the use of the mechanism of inference on the ontology, which accompany the RDF description. This approximation can translate either with a synonymic relation, that expressed by:

$$\text{OWL: equivalentClass}$$
$$\text{OWL: equivalentProperty} \tag{3}$$
$$\text{OWL: SameAs}$$

On the other hand, with a relation Subsumption between equivalent resources presented by the following ontology:

$$\text{RDFS: subClassOf}$$
$$\text{RDFS: subPropertOf} \tag{4}$$

In addition, to have an approximation of the literal ones we uses the comparison of texts to say to which point are close enough based on a threshold. This matrix makes it possible to detect if the triplets coming from a graph will be able or not to combine with the triplet coming from another RDF graph. The similarity matrix can present in this way:

$$\begin{array}{cccccc} & \mathbf{G1} & \mathbf{G2} & \mathbf{G3} & \dots\dots & \mathbf{Gn} \\ \mathbf{G1} & * & Sim(G1,G2) & Sim(G1,G2) & \dots & Sim(G1,Gn) \\ \mathbf{G2} & Sim(G1,G2) & * & Sim(G1,G2) & \dots & Sim(G1,Gn) \\ \mathbf{G3} & Sim(G1,G2) & Sim(G1,G2) & * & \dots & Sim(G1,Gn) \\ \\ \mathbf{Gn} & Sim(Gn,G2) & Sim(Gn,G2) & * & \dots & Sim(Gn,Gn) \end{array} \tag{5}$$

That mean if the similarity between two couple of graphs is null $SIM(Gi, Gj) = 0$ the two graphs can't be combined. Also if $SIM (Gi, Gj) = 1$ the two graphs can be combined. For the other values to determinate if two couple of graphs can combined or not depends on similarity threshold.

4.3   Construction of the RDF Global Graph

The processes of searching the answer in each graph RDF of source contained in the knowledge can take a lot of time to have a response. So for optimize the time of response it is necessary to construct an abstract graph that plays the role of intermediary between the user and the set of RDF graphs.

The global graph can constructed from the graphs contained in the knowledge base by an intersection operation of these graphs in an iterative manner using the OWL (Web Ontology Language) and RDFS formalism languages.

The first level of global graph constitutes the original graphs note (G1...Gn) can be linked between as by using the condition of combination given by similarity matrix. The virtual graph is built by calculating intersection between original graphs presented in the knowledge base or between two virtual graphs. The graphs on the top level are the virtual graphs Gij. i level of graph global and j the number of graphs already created.

The global graph can used to determine which RDF graphs can respond to a sub query without going through a check of all knowledge base graphs. This allows a faster build of the graph-sub query matrix. Compare each sub query by the highest level of abstraction graph if a graph Gmk responds to the sub query bi then the graphs that build this graph also responds and so on. The graph expanded until either its highest level contains a single node or several nodes whose graphs correspond have no other intersection than the empty set.

The graphs that satisfy the query Q′i and denoted by 1 in the matrix graph sub query. The propagation of the sub query continues to all the marked nodes until the graph traversed entirely.

4.4   Storage of Graph in HDFS

Global RDF graph are distributes and indexes in HDFS. They partitioned into files that are stored in the Hadoop node to reduce processing by exploitation the parallelism of Map Reduce when sub queries are matching with the RDF graph.

4.5   Decomposition of SPARQL Query

The request considered like a conjunction of conditions decomposed into a minimal query called a sub query that contains a single condition of the original request. The query admits two part header specified by the 'SELECT' clause and the body part which specify in the 'WHERE' clause content. The decomposition of the query is in fact the separation of conditions in the body of the query gives the following set:

$$Q = UQ'i/i = \{1\ldots n\} \tag{6}$$

4.6   Construction of Matrix Graph with Sub Query Using MapReduce

The query considered like a conjunction of conditions. This global query SPARL (Q) is decomposing into a set of sub queries (Q′). Each sub query contains a single condition of the original query. The matrix graph with sub query is a Boolean matrix.

To optimize the time of construction of this matrix we will use the MapReduce processing. The Boolean matrix can present in this way:

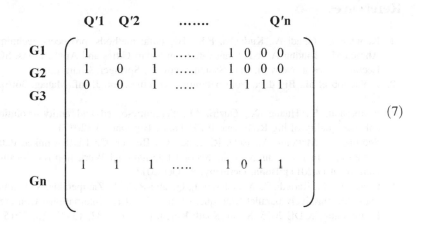

$$(7)$$

4.7 Constructing a Complete Query Answer

This step provide the sorting of the graphs by the number of sub queries to which they answers also allows to know the graphs that answer to the large number of sub-queries. The main idea of sorting is reducing the number of combinations needed to generate a complete response to the user query and make optimization by ranking in the first plan the graphs that take the most of sub query.

The graphs that do not answer all the sub queries are completed by other graph responses with checking the possibility to combine these graphs by using the similarity matrix. The construction of the answer takes place by complete a partial answer provided by the answers of another graph.

5   Conclusion

In this article, we present some aspect of traditional web that influences the responses sent to the user's query that is considered irrelevant and incomplete. For that we thought to exploit the new technologies and algorithms like big data technologies and semantic web technologies, which take into account these traditional problems, gives a high quality of service and allows to have a relevant and complete access to information. On the one hand, the Big Data technology provides a parallel processing of the SPARQL query that interrogates the data stored in HDFS to have relevant and complete answers to the user query. On the other hand, the semantic Web allows us to present data in a structured way using RDF and to apply the ontology language inference mechanism (RDFS/OWL) in order to achieve a coherent combination between these graphs taking into account: their degree of contradiction based on the similarity matrix.

References

1. Kaoutar, L., Ghadi A., Kudagba, F.K.: Big data: methods, prospects, techniques. In: Ben Ahmed M., Boudhir A. (eds.) Innovations in Smart Cities and Applications. SCAMS 2017. Lecture Notes in Networks and Systems, vol 37. Springer, Cham (2018)
2. I. Yaqoob et al., Big data: from beginning to future. Int. J. Inf. Manag. **36**(6), 1231–1247 (2016)
3. Benbernou, S., Huang, X., Ouziri, M.: Semantic-based and entity-resolution fusion to enhance quality of big RDF data. IEEE Trans. Big Data, 1 (2017)
4. Schultz, A., Matteini, A., Isele, R., Bizer, C., Becker, C.: LDIF—linked data integration framework. In Proceedings of the Second International Workshop on Consuming Linked Data (COLD 2011), Bonn, Germany, 23 Oct 2011
5. Goasdoué, F., Kaoudi, Z., Manolescu, I., Quiané-Ruiz, J., Zampetakis, S.: Cliquesquare: Flat plans for massively parallel RDF queries. In: 31st IEEE International Conference on Data Engineering, ICDE 2015, Seoul, South Korea, pp. 771–782, 13–17 Apr 2015
6. Koziris, N.: H2rdf+: an efficient data management system for big RDF graphs. In: International Conference on Management of Data, SIGMOD 2014, Snowbird, UT, USA, pp. 909–912, 22–27 June 2014. Jacobs, I.S., Bean, C.P.: Fine particles, thin films and exchange anisotropy. In: Rado, G.T., Suhl, H. (eds.) Magnetism, vol. III, pp. 271–350. Academic, New York (1963)
7. Subercaze, J., Gravier, C., Chevalier, J., Laforest, F.: Inferray: fast in-memory RDF inference. PVLDB **9**(6), 468–479 (2016)
8. Gurajada, S., Seufert, S., Miliaraki, I., Theobald, M.: Triad: a distributed shared-nothing RDF engine based on asynchronous message passing. In: International Conference on Management of Data, SIGMOD 2014, Snowbird, UT, USA, pp. 289–300, 22–27 June 2014
9. Knoblock, C.A., Szekely, P.: Semantics for big data integration and analysis. In: Proceedings of the AAAI Fall Symposium on Semantics for Big Data (2013)
10. Knoblock, C.A., Szekely, P., Ambite, J.L., Gupta, S., Goel, A., Muslea, M., Lerman, K., Taheriyan, M., Mallick, P.: Semi-automatically mapping structured sources into the semantic web. In: Proceedings of the Extended SemanticWeb Conference (2012). Young, M.: The Technical Writer's Handbook. University Science, Mill Valley, CA (1989)
11. Endrullis, S., Thor, A., Rahm, E.: WETSUIT: an efficient mashup tool for searching and fusing web entities. PVLDB **5**(12), 1970–1973 (2012)
12. Calvanese, D., Giacomo, G.D., Lembo, D., Lenzerini, M., Rosati, R.: Tractable reasoning and efficient query answering in description logics: The dl-lite family. J. Autom. Reason. **39**, 385 (2007)

Hirbalink: CAM Collection & Tracking System

Ouissam El Andaloussi[1]([⊠]) [iD], Mhamed Ait Kbir[2],
and BD Rossi Hassani[1]

[1] LABIPHABE Laboratory, FSTT-UAE, Tangier, Morocco
ouissam.and@gmail.com
[2] LIST Laboratory, FSTT-UAE, Tangier, Morocco

Abstract. The web contains huge volume of information related to complementary and alternative medicine. However, healthcare recommendation with medicinal plants has become complicated because precious information about medicinal resources are available now. Moreover, the existing scientific search engines are not quite efficient and require excessive manual processing. As a result, the search for accurate and reliable data about herbal plants has become a highly difficult and time-consuming task for scientists. Till date, a wide mapping of the already available data concerning herbal plants hasn't been carried out. In this regard, the complementary and alternative medicine collection tracking system (Hirbalink) introduced in this work was created for the purpose of organizing and storing related data.

Keywords: Complementary and alternative medicine · Herbal medicine
Library · Database · Bioinformatics · Tracking system · Biological data

1 Introduction

Recent studies of National Health Interview Survey (NHIS) have reported that approximately 38% of the studied populations are using complementary and alternative medicine (CAM) [1]. The CAM component of the NHIS, developed by the National Center for Complementary and Integrative Health (NCCIH) and the National Center for Health Statistics (NCHS), also collected data about CAM costs, including cost of CAM use, frequency of visits made to CAM practitioners, and frequency of purchases of self-care CAM therapies [2, 3]. In this context, our project was initiated to create a database for CAM's work. The aim is to collect and organize scientific articles published in numerous databases (e.g. PubMed), and keep track of the most recent works by a constant update and synchronizing of the library. The main objectives of the Herbalink are the establishment of an information structure to support scientists in their research, while ensuring the accuracy of the up-to-date data. The present work describes the rationale, content, methods, and implementation of Herbalink.

© Springer Nature Switzerland AG 2019
M. Ben Ahmed et al. (Eds.): SCA 2018, LNITI, pp. 247–256, 2019.
https://doi.org/10.1007/978-3-030-11196-0_22

2   Related Work

Eisenberg and his collaborators are published first population-based to assess the use of alternative medicine in the United States, which found that 34% of respondents had used at least one non-conventional therapy in the past year [4, 5]. Based on a 1997 follow-up survey, alternative medicine use increased from 34 to 42% [6].

The growing use of complementary and alternative medicine therapies by people encouraged scientists and government to research and develop this section [7], in mission to "to explore complementary and alternative healing practices in the context of rigorous science; to educate and training CAM researchers; and to disseminate authoritative information to the public and professionals" [8].

In 2014, More than 11,000 articles lauding alternative medicine appear in the PubMed database, but there are only a few articles describing the complications of such care [9]. Currently, we find more than 21,000 articles about complementary and alternative medicine.

CAM became used in with all health problems, A majority of cancer patients use complementary and alternative medicine (CAM) [10, 11]. The National Center for Complementary and Integrative Health defines Complementary Medicine as non-mainstream practice used together with conventional medical treatment [12]. In contrast, alternative medicine describes the use of non-mainstream medicine instead of conventional medicine [13].

Complementary health approaches can be divided into several main categories as, for example, natural products (such as herbs, vitamins, minerals and probiotics), mind and body practices (including yoga, meditation, massage, acupuncture, etc.) and so-called holistic approaches (e.g., Traditional Chinese medicine, homeopathy, Ayurvedic medicine, etc.) [14, 15].

Currently, with the evolution of IT tools, we can refer back to specific database dedicated to herbal medicine and molecular interaction …, but we cannot find tools for tracking CAM's literature. For these reasons, we are interested in the subject, this research will be helpful as we notified earlier to scientist to focus their research on a specific topic with an easy system, in addition for functions available, we develop an analysis tools in order of detailing this data.

3   Database Content

To collect the targeted scientific publications, a literature survey must be carried out using many available sources such as scientific publications databases (e.g. PubMed, Science direct, NCBI [16] …). As a first step, we restricted our focus to PubMed only due to its positioning in scientific research fields and the ease of interrogation using PubMed utilities [17]. In fact, PubMed is one of the leading search engines for bibliographic data across all research fields in biology and medicine. It was developed by the American Center for Biotechnology Information (NCBI), and is hosted by the American Library of Medicine of the US Institutes of Health. PubMed is a free search engine providing access to the MEDLINE bibliographic database, gathering citations and abstracts of biomedical research articles, and it is the reference database for the biomedical sciences (Fig. 1).

Fig. 1. Data storage process

After the identification of the subject and the concepts as well as defining the terms related to the search, a search query replacing the concepts by the terms found should be written.

In order to build a search query (to link the terms between them, exclude others, etc.), it is necessary to use different logical tools such as the AND and OR operators and limitations (which makes it possible to combine, include or exclude one or several terms).

PubMed offers an Advanced Search Builder. It is necessary to note that those queries were well targeted on the subject to have a certain number of results thus allowing a good analysis (Fig. 2).

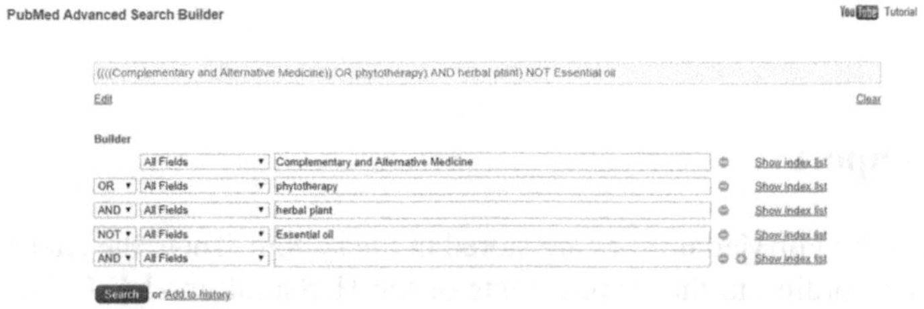

Fig. 2. PubMed advanced search builder

4   Software Design

The web is a constant changing environment. Many technologies and tools are emerging and are offering more and more interactivities in our web pages on a daily basis. All these technologies transform our browsers into real platforms. Symfony is a PHP web application framework and a set of reusable PHP components/libraries [18]. It makes heavy use of existing PHP open-source projects as part of the framework.

5   Management Rules

Herbalink contains:

An "Article" serves as the primary entity in the model and identifies particular keywords that are to be collected and tracked. An "Article" is imported by "Title",

"Authors", "abstract", "Publication Date" and a "DOI". The track can be linked to any entity (title, keyword, authors). A "Keyword" entity is used to save the keywords recovered, also an "Authors" entity to save author's names of articles.

A "Queries" entity is created to store all queries introduced, serves at the time of the update.

An Administrator can:

- Add users, modify their data or delete them.
- Grant the actions of the various partners, or the canceled ones.
- Manage (Add, modify and delete) the different data (queries, update, articles) of our database.

   The users are same rules as administrator except "user management".
   A visitor can:

- Make research in Herbalink
- Show charts of publication progress
- Access to the result page
- Access to information of a chosen article.

6   Conceptual Data Model

See Fig. 3.

7   Data Import

Data imported by PubMed utilities are stored in the system. Then, the data are analyzed and verified according to the requirements of the Herbalink model. Using the request shown in Fig. 2 the targeted data are imported afterwards and stored in a relational database. The admin may review and correct any possible errors or inconsistencies, then resubmit the input data once again.

8   Reporting

After importing targeted data from PubMed, and recording in Herbalink database, the User will thereafter have the possibility to review not only the latest articles but also all the related information. With google charts it also possible to check the progress of publication in some fields of research during last years, and how much a keyword is cited in articles.

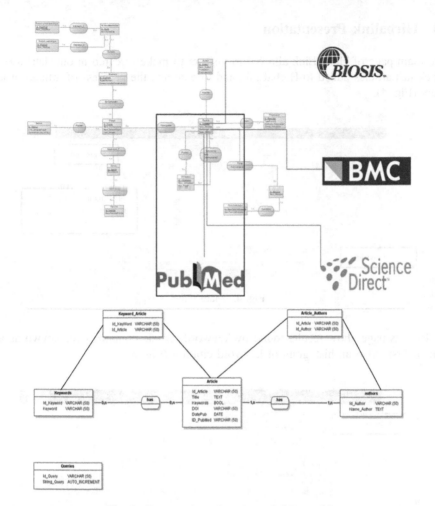

Fig. 3. Presentation of conceptual data model

9 Discussion and Results

The results obtained using Hirbalink are interesting for any scientist, because with this system one can keep track and be updated about all the publications in his field of research just on one click. We have presented the interest of the Herbalink as a database which allows storing and retrieving a set of information of different natures as well as the links that exist between them.

Herbalink presents a high potential to be an asset in the field of phytotherapy. The proposed functionalities introduce a participatory approach recommending active involvement of the various partners in order to share the knowledge of any actor in the field of CAM.

10   Hirbalink Presentation

The main page of HerbaLink allowed to visitors to make research in our database, to check last articles added to HerbaLink, and to examine the progress of articles at last years (Fig. 4).

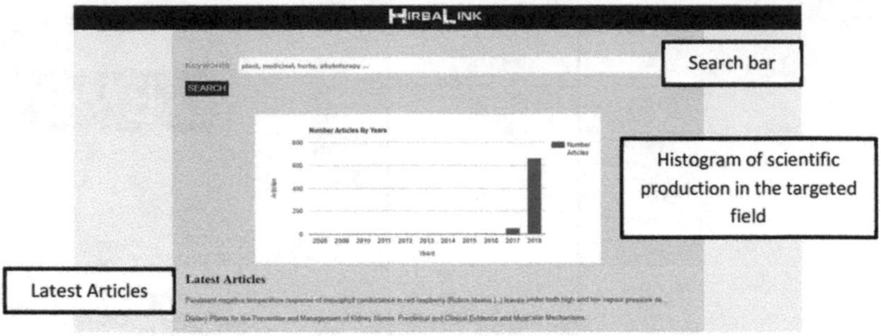

Fig. 4. Main page

Results page show results found by keyword, article contains over keywords are ranked first. Also an histogram of keyword citation (Fig. 5).

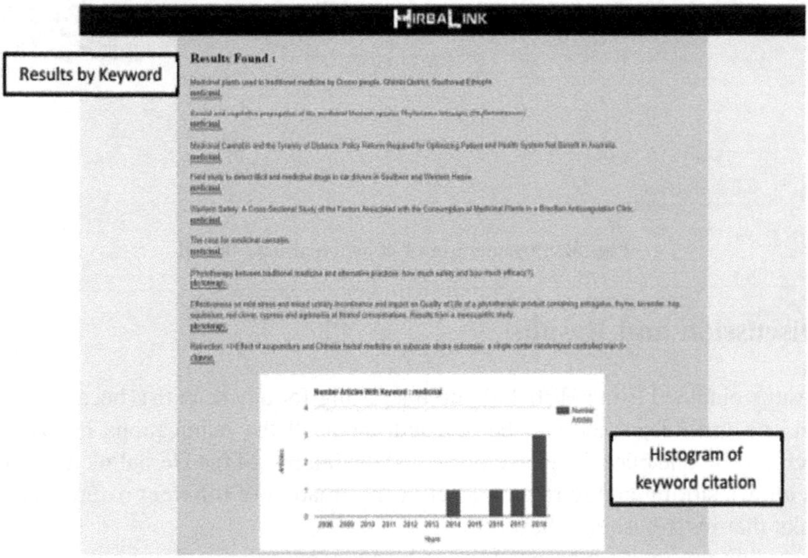

Fig. 5. Results page

To access the administration page, an account with login and password is required (Fig. 6).

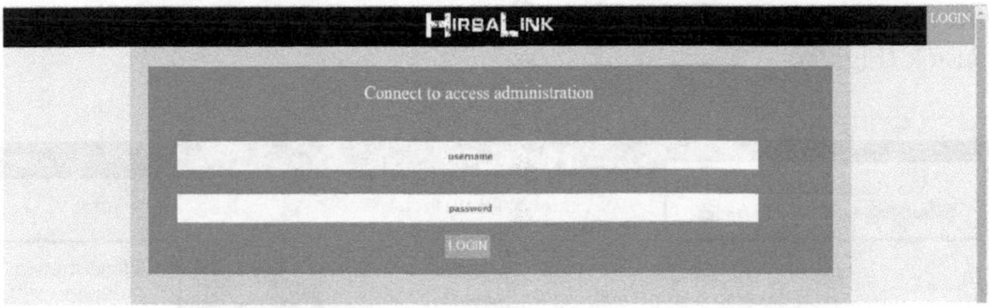

Fig. 6. Login page

This section allowed to administrators to add new queries to Hirbalink (Fig. 7).

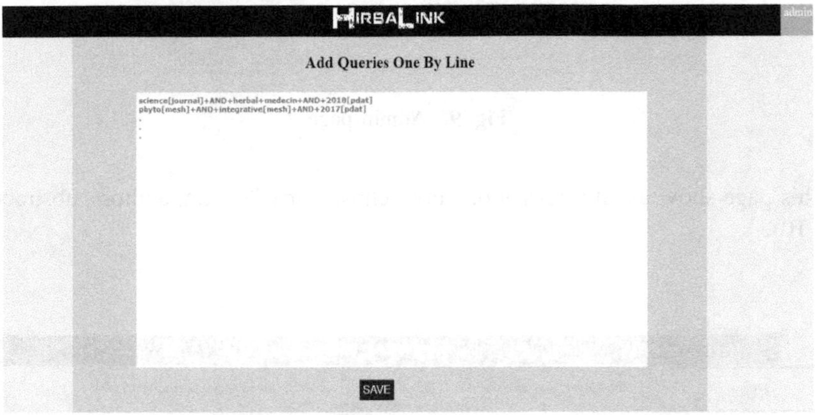

Fig. 7. Add queries page

We can consult the number of existing articles, and choice queries updated (Fig. 8).

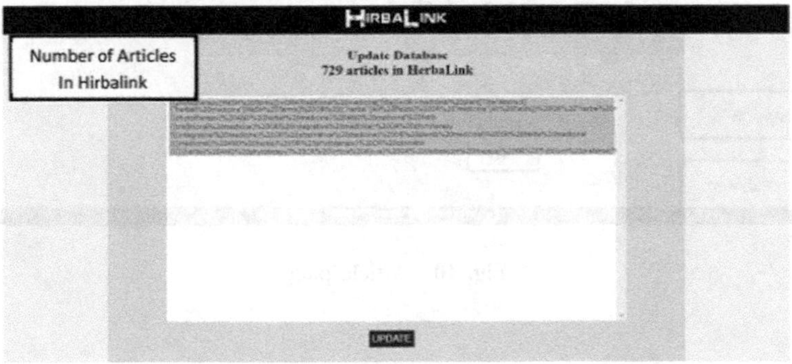

Fig. 8. Update page

We can consult the new number of articles (after updating), and manage data in HerbaLink (Fig. 9).

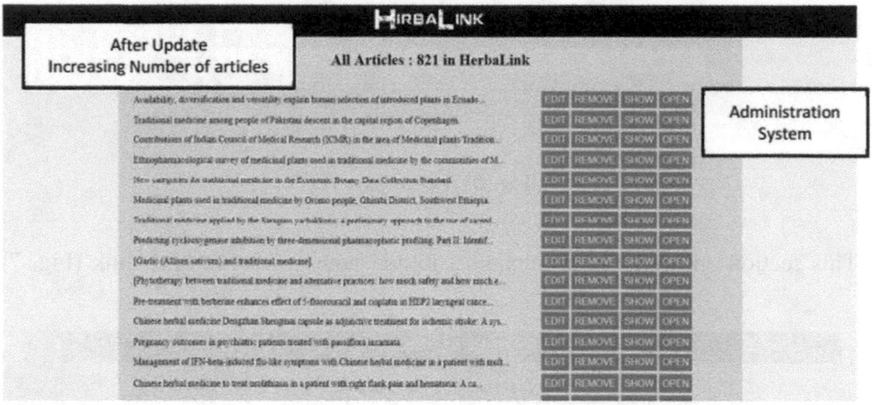

Fig. 9. Admin page

This page show as all information about chosen article (title, authors, abstract, …) (Fig. 10).

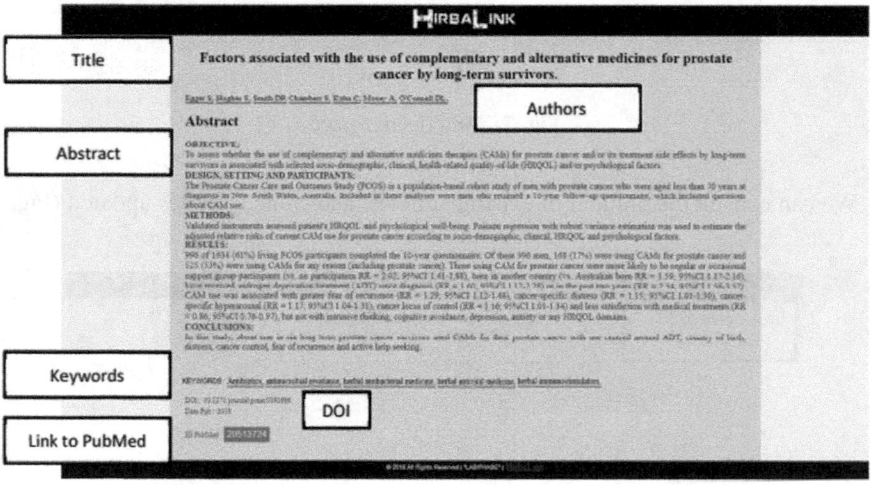

Fig. 10. Article page

11 Conclusion

We have already implemented HerbaLink in a local server, and the collection of data tracked is so far well performing.

Herbalink offers access to content that can be difficult to locate through other databases. The inclusion of CAM topics and hard-to-find articles make Herbalink a helpful resource for scientists researching complementary and alternative medicine topics. Scientists, doctors or practitioners conducting research in the areas of CAM (herbalism, acupuncture, homeopathy, chiropractic, physiotherapy, podiatry, and rehabilitation ...) will want to consider using Herbalink in their searches.

The next step of our work is the analysis and the links between the collected data by HerbaLink, and also interrogates other database to offers a large choice to scientist.

References

1. Vadivu, G., et al.: Ontology mapping of indian medicinal plants with standardized medical terms. J. Comput. Sci. **8**(9), 1576–1584 (2012)
2. Nahin, R.L., et al.: Costs of complementary and alternative medicine (CAM) and frequency of visits to CAM practitioners. National Health Statistics Reports No. 18 (2009)
3. Harris, E.S.J., et al.: Traditional medicine collection tracking system (TM-CTS): a database for ethnobotanically-driven drug-discovery programs. J. Ethnopharm. **135**(2), 590–593 (2011)
4. Eisenberg, D.M., et al.: Unconventional medicine in the United States. N. Engl. J. Med. **328**, 246–252 (1993)
5. Eisenberg, D.M., et al.: Trends in alternative medicine use in the United States, 1990–1997. Results of a follow-up national survey. JAMA **280**, 1569–1575 (1998)
6. Patterson, Ruth E., et al.: Types of alternative medicine used by patients with breast, colon, or prostate cancer: predictors, motives, and costs. J. Altern. Complement. Med. **8**(4), 477–485 (2002)
7. Nahin, R.L., et al.: Research into complementary and alternative medicine: problems and potential. BMJ **322**, 161–164 (2001)
8. National Center for Complementary and Alternative Medicine: Five year strategic plan. http://nccam.nih.gov. Accessed 13 Dec 2000
9. Bayme, M.J., et al.: The perils of complementary alternative medicine. Rambam Maimonides Med. J. **53**, e0019 (2014)
10. Berretta, M., et al.: Use of complementary and alternative medicine (CAM) in cancer patients: an Italian multicenter survey. Oncotarget **8**(15), 24401–24414 (2016)
11. Horneber, M., et al.: How many cancer patients use complementary and alternative medicine: a systematic review and metaanalysis. Integr. Cancer Ther. **11**(3), 187–203 (2012)
12. Huebner, J., et al.: Online survey of patients with breast cancer on complementary and alternative medicine. Breast Care (Basel, Switz.) **9**(1), 60–63 (2014)
13. Huebner, J., et al.: User rate of complementary and alternative medicine (CAM) of patients visiting a counseling facility for CAM of a German comprehensive cancer center. Anticancer Res. **34**(2), 943–948 (2014)
14. Firkins, R., et al.: The use of complementary and alternative medicine by patients in routine care and the risk of interactions. J. Cancer Res. Clin. Oncol. **144**(3), 551–557 (2018)

15. Complementary, alternative, or integrative health: what's in a name? NCCIH: national center for complementary and integrative health. https://nccih.nih.gov/health/integrative-health. Accessed Nov 2017
16. National Center for Biotechnology Information. http://www.ncbi.nlm.nih.gov
17. Entrez Programming Utilities. National Center for Biotechnology Information (2018)
18. Peltier, M.: Développement d'applications Web avec le framework PHP (2011)

Identifying the Centers of Interests of User Profiles in a Big Data Context

Ismail Bensassi, Yasyn Elyusufi, and El Mokhtar En-Naimi[✉]

LIST Laboratory, Faculty of Sciences and Technologies, Tangier, Morocco
ennaimi@gmail.com

Abstract. The Idea we propose in this article is a follow up to our research series in the ontology-based profiling framework. The approach relies on tracking user profile changes for better user connection within a Big Data context. We have worked in our series of research on the identification and qualification of profiles in web 2.0 context based on the ontological approach and multi agent system. Among the limitations of our research is the fact that changing interests over time does not affect the relationships between profiles. The goal of our approach is to follow the change of the interests of internet users and to propose afterwards new relations having changed activities in the same direction. In order to implement this approach, we will first use the ontology approach. The ontology we propose will generate a set of domains and sub domains of activities, used to identify user profiles. On the other hand, we will use the Multi Agents approach to process users' activities before classifying them in their profiles.

Keywords: Profiling · Ontology · MAS · Big data

1 Introduction

Today, with the evolution of information systems, the number of users using these systems increases exponentially. The users have by default different cultures, different geographical region, different school levels, and different activity centers. The goal of our approach is to delegate the search task of profiles to the system and make it automatable. It is in this context that this article is positioned, which aims to design and build a prototype that makes it possible to intelligently recommend to users relationships that have a number of common properties at a percentage that we define later. In this framework, we will analyze the text flow exchanged and stored at the database servers, then the calculation of the priorities of each profile to several application domains. The calculation of priorities will use a domain ontology that we develop. Thanks to a synergy between the latter and the text-mining process applied on the existing masses of data by software agents, the system will allow a classification of the profiles, subsequently serving to improve the process of user inter-connection by smart recommendation between similar profiles in Big Data context.

© Springer Nature Switzerland AG 2019
M. Ben Ahmed et al. (Eds.): SCA 2018, LNITI, pp. 257–268, 2019.
https://doi.org/10.1007/978-3-030-11196-0_23

2 State of the Art

With the expansion of new informations and communication technologies and more particularly the internet, social networks and the web, there is a proliferation of sources and documents made available to users. This gives rise to complex and heterogeneous information systems. Also research and exploitation of information in this context proves to be a very difficult task. Several information access tools (search engines, recommendation systems) have been developed to assist the user in this job. Through these tools, the question of the relevance of the results presented to the users was the subject of a very thorough reflection. However, another question that has been much less detailed is that of evaluating whether these results are really adapted to the user, relative to a certain number of criteria. This is to ensure among other things that the results obtained are understandable by the user and that they correspond well to the goals and preferences of the latter. Among the most used approaches to recommend information relevant to users is the profiling approach. The latter is based on the description of a user's characteristics (demographics, interests, preferences, etc.). Several approaches to acquiring the elements of a user profile exist and can be grouped into manual approaches and automatic or semi-automatic approaches. In several works, we have discussed profiling techniques based on ontologies [1, 2]. On the other hand the profiling application context is often associated with massive data. These last ones concern the environments where the users are lost before the masses of data which are presented to them (Social networks, and Background Big Data). This later was introduced by the giants of the web. These are presented as a solution to allow everyone to access real-time massive and giant databases. They aim to provide alternative solutions to existing database and analytics solutions. Many approaches have been proposed in the profiling framework in a Big Data context. In this very context Silva and Ma propose in [3] a Big Data based approach that harnesses collective intelligence of crowd in (research) social networking platforms and scientific databases for expert profiling. In the same way Ramalingam and Chinnaiah present in [4] a survey of the existing and latest technical work on fake profile detection in Big Data context. Also Anett et al. present in [5] a new approach based on ontologies in order to apply a profiling system. The system is based on online browsing traces, with the ability to distinguish the interests of users based on the implicit information described in cookies and log files. Along the same lines Hoppe present in [6] a combination of data analysis, based on ontological engineering and the processing of large data provided by an industrial partner. The end goal is to automatically build a user profile based on ontologies. Contrary to our previous research, we chose in this approach to apply the tests on a few million users. It is within this general framework that the present article is positioned, which aims to offer a relevant recommendation of user profiles to users of Big Data contexts.

3 Our Approach

We have already worked in our work [1] to build user profiles. In this approach we have proposed to represent each user according to a vector model V which will be used later for the semantic classification of the profiles. In order to set up a precise classification, we have chosen to present the choices of the answers proposed by questionnaires in matrix form where the weight ωij of the choice belongs to a matrix of choices. The classification of the users according to profiles is based on a vector similarity measurement between the representative vector Vi of a user and the other vectors of the users having the weights of the identical terms, term by term. Among the limitations of this method is that if users refuse to answer the questionnaires either out of ignorance or refusal. The system will not be able to follow the evolution of profiles over time. For this reason we are going to define a new model based on a text-mining process on the text stream generated by profiles (searches, paper title posted, keywords …). This model is based on a coupling between the ontology approach and the multi-agents approach. Thanks to the synergy between the two approaches, the system will be able to first extract, qualify, and calculate the frequency of appearance of keywords in user text streams, then the calculation of the interests' priorities of user profiles to different domains of application. The vocabulary and structure of these domains will be described using an ontology that we detail later (see Fig. 1).

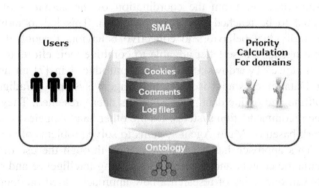

Fig. 1. Framework for users' inter-connexion

In order to better grasp our approach, Table 1 shows the frequency of appearance of keywords in an environment (Social Network). For example we find that the user $U1$ is first interested in the field of technology by a frequency of appearance of 331 keywords, then by the field of exact sciences by 122 keywords, by sports and a little bit 12 political domain keywords. This will categorize users into profiles, based on the calculation of priorities for different domains.

Table 1. Appearance frequency example of keywords in users' text feeds

	Sciences			Social and political			Leisure		
	Exact sciences	Technologies	Health	Sociology	Political	Human	Art	Cinema	Sport
U1	122	331	0	0	12	0	0	0	120
U2	512	0	133	0	0	0	0	2	12
U3	110	0	0	11	0	0	0	0	0
U4	11	312	4	0	0	6	0	12	0
U5	13	511	0	2	0	0	2	0	300
U6	16	0	111	12	3	2	15	0	3
U7	500	312	0	25	13	2	5	0	0
U8	12	145	0	60	0	0	0	0	15
U9	0	0	3	212	412	16	15	17	33
U10	2	14	0	190	120	114	66	16	2

4 Multi-agent Systems and Big Data

Multi-agent systems are part of distributed intelligence (DAI). The basic idea is to distribute the intelligence on several agents where each one is associated with a sub-problem or sub-objective. From the coordination of the activities of these agents emerges the goal to be reached (solve the problem). This led to moving from the description of individual behaviors to collective behaviors through the cooperation between several agents. These latter are able to combine their efforts to increase their collective intelligence. Twardowski and Ryzko introduced in [7] an architecture for real-time Big Data processing based on multi-agent systems paradigms. A global approach to offline and online data processing has been presented. They validated the approach in the recommendation systems. On the other hand Angeles et al. presented in [8] a Framework based on Multi-Agent systems to solve problems of performance and data security. This approach has been implemented through the use of ontologies to incorporate semantic content and improve data mining intelligence and efficiency. We already adopted in our series of research a new approach based on coupling between the ontology approach and the Multi-Agents approach [9]. Thanks to a synergy between the two approaches, we managed to automate the primordial task of our research which aims to automate the text-mining process which allows on the one hand to extract keywords from the generated text stream by the users, and on the other hand to qualify these keywords by interacting with a reference ontology that is called PTOP ontology. The role of the multi agent concept is therefore essential in our approach in order to benefit from the specific characteristics of these systems which are: Cooperation, coordination and communication. Whereas the domains of user profiling and resource analysis have been studied for decades, Big Data is a domain that emerged during the last years, mainly motivated by the immense amount of data produced by web applications. Indeed, interpreting incoming events along with the dynamic creation of the user profile and its exploitation for content suggestion have to scale to the number of partner sites and the number of users that surf those pages every day.

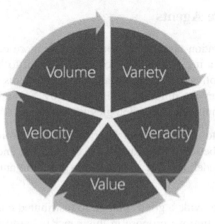

Fig. 2. Characteristics of big data

Thus, the profiling task at hand qualifies in all criteria defined by the IEEE Company as characteristics of Big Data (see Fig. 2), notably:

Volume: The amount of data assembled by the company each month reaches about 150 million user events per month for an average partner site, in sum 2,4 billion events each month, leading to a high volume of data to analyze.

Variety: The analysis includes web data in all its formats and orientations, leading to a highly variable spectrum.

Velocity: User events arrive for analysis in real time, meaning a high frequency during main activity hours.

Value: As the analysis is performed on real-life data, generated by the activity of users, it bears information about their interests and habits that is commercially exploitable.

Veracity: In fast arriving, variably shaped data ambiguities and erroneous entries are not avoidable. The trustworthiness of the source and uncertainty are to be considered in the analysis.

In our approach we focused our research on the criteria *"Variety"*, since we use several sources of heterogeneous data, and of different formats, particularly (Cookies, log files, comments, etc.). In this very context Trusov et al. used the cookies for user profiling in customer-base analysis and behavioral targeting [10]. In the same way Kazai et al. proposed in [11] an approach based on social-network which aims to analyzing users' behavior in social networks (e.g. activities such as comments as well as likes posted to social networks, and shared publications). However, social-network based methods are weak in capturing subjective expertise knowledge claimed by experts and verifying objective expertise. In our approach and in order to make efficient use of different heterogeneous data (cookies, logs, comments ...), we used software agents.

5   Using Software Agents

Modeling and computation tasks are becoming much more complex as the size continues to increase. As a result, it is laborious and difficult to handle using centralized methods. Although motivations to apply multi-agent systems (MASs) for researchers from various disciplines are different, as indicated by Yu and Liu in [12], the major advantages of using multi-agent technologies include:

– Individuals take into account the application-specific nature and environment;
– Local interactions between individuals can be modeled and investigated;
– Difficulties in modeling and computation are organized as sub-layers or components.

Therefore, MASs provide a good solution to distributed control as a computational paradigm. The architecture we propose in this paper is based on cognitive agents that allow interaction with the user. These agents are in permanent communication thanks to a managing agent (MA Manager Agent). Their ability to memorize messages and analyze them will subsequently categorize users into profiles. This categorization will allow the system to decide which relationships (profiles) will be recommended for each type of profile. The proposed architecture is based on the following agents (see Fig. 3):

1. Interface Agent (IA);
2. Manager Agent (MA);
3. Profile Manager Agent (PMA);
4. Keyword Ontology Agent (KOA);
5. User Ontology Agent (UOA).

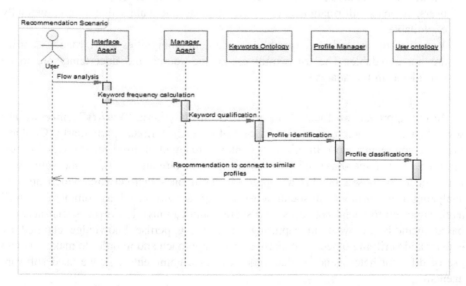

Fig. 3. Communication scenario between agent and PTOP ontology

5.1   Interface Agent (IA)

This agent acts as an interface between the user and other agents through Agent Manager. The agent uses the KQML language interface for communication with other agents. It receives the message flow generated by users (cookies, log files), then proceeds with the detection of keywords (KW), marking of the KW and then the calculation of the frequency of each KW. Next, it reformulates the result of the analysis and returns the result of treatment to the agent Manager in KQML language.

5.2   Manager Agent (MA)

This agent controls all the operations performed by the system and assigns tasks to agents according to their roles. The agent MA plays an orchestral role between the various agents. Once the agent (MA) receives the request of the agent (IA), it determines the nature of the application to select agents that can handle this request. Requests exchanged between the IA and MA are identified by an identification number.

5.3   Profile Manager Agent (PMA)

The role of Profile Manager Agent (PMA) is to create, initialize, and treat the user profile after a request from (IA). It can also add a new profile and initialize resources (attributes and values). It also allows changing the profile setting, and then returning the result of modifications to (IA).

5.4   Keyword Ontology Agent (KOA)

Keyword Ontology Agent (KOA) is an agent associated with a human expert and communicates with other agents. Its main role is to qualify keywords (KW)s with instances of the PTOP ontology (for more explanation, see Sect. 6). Through communication with the agents Manager Agent (MA) and Profile Manager Agent (PMA), it allows the profile identification by communication with (UOA).

5.5   User Ontology Agent (UOA)

User Ontology Agent (UOA) is a profile filtering agent according to profile interests. When (UOA) receives responses from (KOA) and (PMA), it will be able to categorize each user in a profile, and then return the filtering result to (IA) through (MA).

6   Use of Ontology Approach

Hella [13] states that ontologies are very useful for heterogeneous data. They also allow a description of the relationships between different domains. They can combine informations from multiple heterogeneous sources. They can facilitate access to relevant information. Also ontologies can be used as a tool to extract data from many different sources of data. Calegari and Pasi [14] suggest in their work to use the basic profile terms from a user's collection as seed terms to then collect additional informations from reference ontology. This concepts included in the keywords are identified and their neighboring concepts integrated to the user profile, depending on the distance to the original concept and the relation types that exist between them. In our work we will take advantage of the following characteristics:

1. Semantic description of domains and sub domains.
2. Design and develop our ontology (PTOP ontology) to describe domains, sub-domains and their keywords.
3. Describing rules in the form of user ontology that we develop.

The proposed architecture is based on the use on the one hand of multi-agent systems for user-generated text flow analysis (cookies, traces, text, logs, etc.). On the other hand the system is designed so that each of the actors (sociologists, terminologists

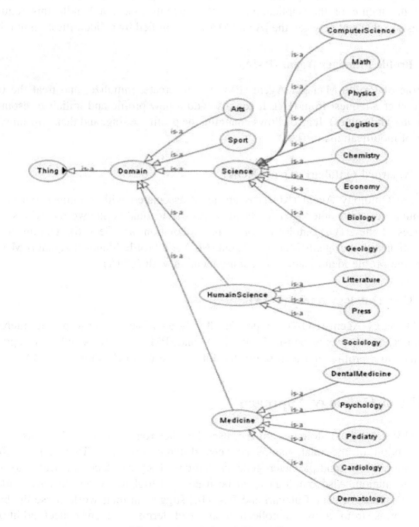

Fig. 4. Part of the PTOP ontology

...), has a cognitive agent allowing to compare the centers of interest of the users with the instances of the PTOP ontology. This ontology must be designed by a community specialized in terminology. Among the main domains described at the ontology: Exact sciences (mathematics, physics, chemistry, geology, logistics ...), Humanities (Sociology, Literature ...), Medicine (cardiology, pediatrics, dermatology, gastronomy ...), Sport and Art (see Fig. 4).

We associate to each ontology subject (class) a set of keywords. These keywords are provided by the ontology designers according to domain of users, where each keyword accepts many instances *"individuals"*. This structure will help the system to automatically qualify the extracted words from users' text feeds (see Fig. 5).

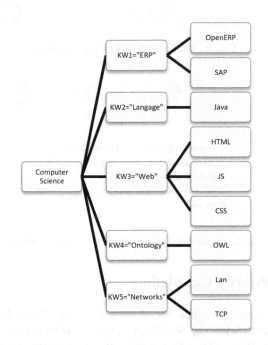

Fig. 5. The section of the class "Computer Science" in the reference ontology

The results of priority calculation will be displayed in the form of user ontology which will be presented using the ontology editor Protégé 4.3 (cf. Fig. 6). The class *"Profil"* is created as a sub-class of the *"Person"* class, the class which assembles all concepts that are combinations of base attributes to a higher abstraction level.

For example, we chose the class *"Profil"*, designating a person that is connected from certain Hotspot, lives in a city, interested with same specific domains, and no interested with others. All those rules attributes and rules (see Fig. 7) belong to the basic profile proposition defined by user ontology. Thanks to the meta-data presented at the level of user ontology, and inferences offered by this later. The system recognizes the priorities of interests of each profile, and will be able to offer an automatic interconnection between users.

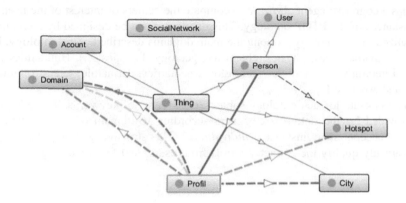

Fig. 6. Minimal version of user ontology

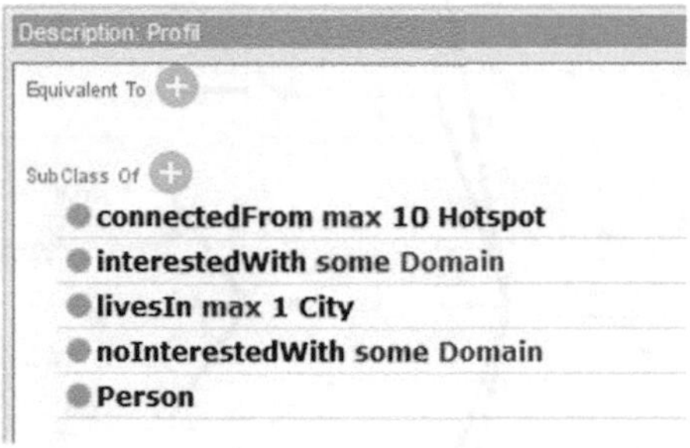

Fig. 7. Some rules of "*Profil*" class in user ontology

7 Results of Priority Calculation

After calculating users' priorities for several domains (Table 2): Exact Sciences, Technologies, Health Sciences, Sociology, Politics, Humanities, Art, Cinema and Sport, the system will be able to find users with first interests. Common in the areas mentioned above and having an order of priority similar to a defined percentage.

From the Table of priorities above (Table 2), we can see that user priorities for domains change from one user to another, for example *U1* is a profile that is primarily interested in the technology domain, then the field of exact sciences, after the field of sport and finally politics. Also we can see that users *U2, U3,* and *U7* are interested in the field of exact sciences in the first row, this allows the system to categorize them in the same profile (profiling of order 1, i.e. taking into account only priority 1), and therefore make them an inter-profile recommendation. These results will be directed to user ontology. The latter will be charged by the classification of profiles that have similar priorities to a percentage. With a very large sample that exceeds a few miles of users (Big Data context), the system will be able to find users with similar priorities who reach the order 7, 8 and 9, i.e. similarities of important order of priorities. And will be able to make a relevant inter-user recommendation.

Table 2. Table of priorities for a minimal sample of users

	Sciences			Social and political			Leisure		
	Sciences exactes	Tech	Health	Sociology	Politics	Human	Art	Cinema	Sport
U1	2	1	0	0	4	0	0	0	3
U2	1	0	2	0	0	0	0	4	3
U3	1	0	0	2	0	0	0	0	0
U4	3	1	4	0	0	4	0	2	0
U5	3	1	0	4	0	0	5	0	2
U6	2	0	1	4	5	7	3	0	6
U7	1	2	0	3	4	6	5	0	0
U8	4	1	0	2	0	0	0	0	3
U9	0	0	3	2	1	4	6	5	3
U10	7	6	0	1	2	3	4	5	8

8   Conclusion and Future Work

We presented in this paper a new profile recommendation approach in a Big Data context, to implement this approach; we introduced a coupling between the ontological approach and the multi-agent approach. This model mainly allows efficient communication between the various system actors in the recommendation process, while taking advantage of the flow analysis (texts, traces, logs, cookies, etc.) introduced by users of Big Data environments. This method allows a recommendation of profiles between users. This recommendation is based on profiles with the same interest priorities in several areas. Future development of this method will predicts the future needs of users in Big Data contexts by using the priorities calculation method introduced in this paper.

References

1. Elyusufi, Y., Seghiouer, H., Alimam, M.A.: Building profiles based on ontology for recommendation custom interfaces. In: 2014 International Conference on Multimedia Computing and Systems (ICMCS), Anonymous IEEE, pp. 558–562 (2014)
2. Elyusufi, Y., Seghiouer, H., Benkaddour, A.: Customization of human computer interface guided by ontological approach in web 2.0. Int. J. Comput. Appl. **81**(5) (2013) (0975 8887). ISBN: 973-93-80878-45-9 (Foundation of Computer Science, New York, USA)
3. Silva, T., Ma, J.: Expert profiling for collaborative innovation: Big data perspective. Inf. Discov. Deliv. **45**(4), 169–180 (2017)
4. Ramalingam, D., Chinnaiah, V.: Fake profile detection techniques in large-scale online social networks: a comprehensive review. Comput. Electr. Eng. **65**, 165–177 (2017)
5. Anett, H., Ana, R., Christophe, N.: Ontology-based user profile learning from heterogeneous web resources in a big data context. In: Check Sem Research Group, Laboratoire Electronique, Informatique et Image (LE2I), Université de Bourgogne (2013)

6. Hoppe, A.: Automatic ontology based user profile learning from heterogeneous web resources in a big data context, 26–30 Aug 2013, Riva del Garda, Trento, Italy. Proc. VLDB Endow. **6**(12), 1428–1433 (2013)
7. Twardowski, B., Ryzko, D.: Multi-agent architecture for real-time big data processing. In: International Joint Conferences on Web Intelligence (WI) and Intelligent Agent Technologies (IAT), IEEE/WIC/ACM (2014)
8. Angeles, M.D., Córdoba-Luna, J.: Multi-agent distributed data mining by ontologies. Int. J. Adv. Softw. **6**(3&4) (2013)
9. Elyusufi, Y., Alimam, M.A., Seghiouer, H.: Recommendation of personalized RSS feeds based on ontology approach and multi-agent system in web 2.0. J. Theoret. Appl. Inf. Technol. **70**(2), 324–332 (2014) (9p.)
10. Trusov, M., Ma, L., Jamal, Z.: Crumbs of the cookie: user profiling in customer-base analysis and behavioral targeting. Mark. Sci. **35**(3), 405–426 (2016)
11. Kazai, G., Yusof, I., Clarke, D.: Personalised news and blog recommendations based on user location, Facebook and Twitter user profiling. In: Proceedings of the 39th International ACM SIGIR Conference on Research and Development in Information Retrieval (SIGIR'16), pp 1129–1132. ACM, New York, NY, USA (2016)
12. Yu, N.-P., Liu, C.-C.: Multi-agent systems. In: Advanced Solutions in Power Systems: HVDC, FACTS, and Artificial Intelligence. Wiley, Hoboken, NJ (2016)
13. Hella, L.: Ontologies for big data. Department of Computer and Information Science, Faculty of Information Technology, Mathematics and Electrical Engineering, 10 June 2014
14. Calegari, S., Pasi, G.: Personal ontologies: generation of user profiles based on the yago ontology. Inf. Process. Manag. **49**(3), 640–658 (2012)
15. Chen, Z., Zhu, F., Guo, G., Liu, H.: User profiling via affinity-aware friendship network. In: Aiello, L., McFarland, D. (eds.) Social Informatics, vol. 8851, pp. 151–165. Springer International Publishing (2014)
16. Li, Y., Zhang, C., Swan, J.R.: An information filtering model on the web and its application in job agent. Knowl.-Based Syst. **13**(5), 285–296 (2000)

Integration Methods for Biological Data Sources

H. Hanafi[1(✉)], F. Rafii[1], B. D. Rossi Hassani[2], and M. Aït Kbir[1]

[1] LIST Laboratory, CED STI, University, UAE, Tangier, Morocco
hamzahanafil@gmail.com
[2] LABIPHABE Laboratory, CED STI, University, UAE, Tangier, Morocco

Abstract. Nowadays, recent technologies in biology has gained a lot of attention, because of the massive data they produced with different types, very complex structures and various interaction categories. They allowed to perform deep analysis on cell structure and it's sub-system. Moreover, They enabled construction of complex networks that represent the extracted data and the mutual interactions between biological entities of diverse types. However, most of users, especially researchers and biologists, find it difficult to do their experiments on a set of data of various types stored in multiple databases. In this paper, we present the state of the art for data integration based on collective mining, using various types of networked biological data. Moreover, we propose a new approach to make it possible to integrate heterogeneous data in the MicroCancer platform, recently developed by our laboratory, to deal with micro-array data.

Keywords: Biological networks · MicroCancer generalization
Data integration

1 Introduction

A cell is a complex system, it contains several structures of different molecules forming a complex machinery, which can be naturally represented as a system of different types of interconnected molecular and functional networks. Recent research in the field of biology has generated a large mass of data that can describe different aspects of the functional system of a cell also known as omics layers (see Fig. 1) [1].

Several technologies and experiments make it possible to obtain various layers of data of a cell, such as next-generation sequencing [2–5], microarrays [6, 7] and RNA-sequencing technologies [8–10] have enabled construction and analyses of diverse omics layers. These layers are linked to each other to illustrate the interactions that occur within the cell.

Figure 1 illustrates these layers and their constituents: genes in the genome, mRNA in the transcriptome, proteins in the proteome, metabolites in the metabolome and phenotypes in the phenome. It describes the procedures by which genes (in the genome layer) lead to complex phenotypes (in the phenome layer) depend on all intermediate layers and their mutual relations (e.g. protein–DNA interactions).

© Springer Nature Switzerland AG 2019
M. Ben Ahmed et al. (Eds.): SCA 2018, LNITI, pp. 269–274, 2019.
https://doi.org/10.1007/978-3-030-11196-0_24

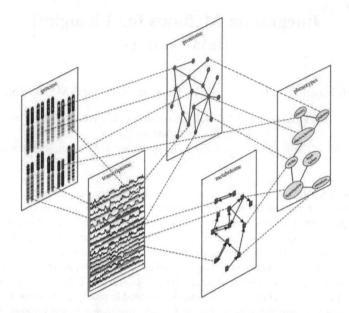

Fig. 1. Molecular information layers of a cell (omics layers) [10]

2 Data Integration

2.1 Biological Data and Graph Representation

Data integration makes it possible to define a general model that represents all the data of the layers, it starts by representing the layers in graph.

Biological networks governed the way of perceiving biological systems, as they enabled studies and experiments from a system level perspective. In graph theory a network is composed of a set of nodes and a set of edges [11], based on the type of data the network represents, the edges might have several nature, edges might be directed or undirected and weighted or unweighted [12].

2.2 Biological Networks

In this section, we are going to give a brief description of commonly used networks and data integration to construct biological networks.

Data integration can be divided into three categories based on the structure and the nature of edges of the graph [1]:

1. Molecular interaction networks:

   - PPI network (protein-protein interactions)
   - MI network (metabolic interaction)
   - DTI network (drug target interaction)

2. Functional association network:

- GI network (gene interaction)
- GDA network (gene diseases association)
- ONs (Ontology networks)

3. Functional/Structural similarity network:

- Gene Co-Ex network
- DCS (drug chemical similarity)
- Protein similarity networks.

2.3 Data Integration Methods

Data integration methods are divided into two categories based on their data type, especially homogeneous and heterogeneous [1].

A brief description for some methods is given below:

- Network-based methods: the most of these methods are using simple ways to integrate diverse types of network data, to generate a model of a set of networks.
- Bayesian networks: these methods are based on concepts of probability, and graph theory.
- Kernel based methods: these methods belong to the class of statistical machine learning methods, for example learning tasks, such as clustering, classification, regression, correlation, feature selection.

3 MicroCancer Platform Generalization

MicroCancer [13] is a web application that aims to manage the experiments of different users, interested in microarrays, by offering them services to facilitate their access, analysis and manipulation results. It provides some roles for the end user with to different profiles. MicroCancer is targeting the experiments designed for cancer studies; it also allows users to integrate experiments from two databases: GEO and ArrayExpress based on semantic web concepts. The plate-form aims to implement data analysis, import and export as Web services. In this paper, we are taking advantages of network data integration to make it to deal with diverse types of biological data (see Fig. 2) within the extended MicroCancer platform. In addition, we want to reinforce the Data Mining methods already developed within the platform, and the most commonly used data integration methods, discussed in the second section.

On the recent version, MicroCancer has been concentrated on investigating the noteworthy information produced by the Microarray technology [14]. Microarray experiments are producing multiple datasets, which are publicly accessible in the two repositories: ArrayExpress and Gene Expression Omnibus (GEO).

The access to Microarray data was a challenging task by the majority of the researchers. For this reason, this paper represents a generalization to those presented in [15–17].

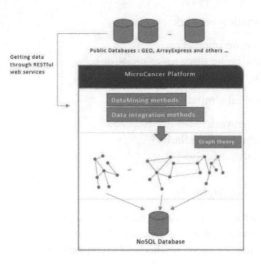

Fig. 2. MicroCancer global structure

Microarray datasets are very large; however, the analysis are affected by a number of variables. Consequently, it is not easy to generate efficient decisions about the produced data. To extract gainful knowledge from the Microarray data, two methods have been used for depth analysis: Clustering and Classification. Clustering has allowed to grouping genes that have same patterns that characterize the problem under study.

Classification was implemented to assign genes to the adequate class in a supervised context.

Various techniques have been implemented into Microarray data and there have shown important results:

- Two versions of Adaptive Resonance theory (ART) [18, 19]
- Kohonen Self Organizing Map
- K means
- Artificial neural networks
- MLP network [20].

The integration of several data type into the MicroCancer platform can be performed through the RESTful web services that are available for each public database. Then we perform Data Mining methods into the collected data in order to make useful decisions, with taking into consideration data integration methods to generate graphs for each omics layers or the ones we are interested into. In results, we have a set of inter-connected graphs, of various types of nodes and links, these graphs represent valuable information to the end user. Each graph records data can have a specific format, and creating a general model for the set of graphs to store the data in a relational database is a very hard task. Consequently, we can take advantages of a no relational database and store the entire graphs into a NoSQL database, because it requires no data structure or data type.

4 Conclusion

Biological networks represent a very important step to understand the functional system of a cell, in a specific way, and in general to understand the biological system, the combination of multiple biological networks is therefore required to improve and reinforce the results of experiments. This work aims to expend the domain of Micro-Cancer platform by making it extensible to diverse types of biological data. We look forward to implement data integration methods in the MicroCancer platform alongside the Data Mining methods that will be available as Web services.

References

1. Gligorijevic, V., Przulj, N.: Methods for biological data integration: perspectives and challenges. J. R. Soc. Interface **12**, 20150571 (2015)
2. Hawkins, R., Hon, G., Ren, B.: Next-generation genomics: an integrative approach. Nat. Rev. Genet. **11**, 476–486 (2010)
3. Nielsen, R., Paul, J., Albrechtsen, A., Song, Y.: Genotype and SNP calling from next-generation sequencing data. Nat. Rev. Genet. **12**, 443–451 (2011)
4. Hirschhorn, J., Daly, M.: Genome-wide association studies for common diseases and complex traits. Nat. Rev. Genet. **6**, 95–108 (2005)
5. Duerr, R., et al.: A genome-wide association study identifies IL23R as an inflammatory bowel disease gene. Science **314**, 1461–1463 (2006)
6. Quackenbush, J.: Computational analysis of microarray data. Nat. Rev. Genet. **2**, 418–427 (2001)
7. Dahlquist, K., Salomonis, N., Vranizan, K., Lawlor, S., Conklin, B.: GenMAPP, a new tool for viewing and analyzing microarray data on biological pathways. Nat. Genet. **31**, 19–20 (2002)
8. Marioni, J., Mason, C., Mane, S., Stephens, M., Gilad, Y.: RNA-seq: an assessment of technical reproducibility and comparison with gene expression arrays. Genome Res. **18**, 1509–1517 (2008)
9. Mortazavi, A., Williams, B., McCue, K., Schaeffer, L., Wold, B.: Mapping and quantifying mammalian transcriptomes by RNA-seq. Nat. Methods **5**(7), 621–628 (2008)
10. Wang, Z., Gerstein, M., Snyder, M.: RNA-seq: are volutionary tool for transcriptomics. Nat. Rev. Genet. **10**, 57–63 (2009)
11. West, D.: Introduction to Graph Theory, 2nd edn. Prentice Hall, New York, NY (2000)
12. Newman, M.: Networks: An Introduction. Oxford University Press, Inc., New York, NY (2010)
13. Fadoua, R.: Techniques de data mining pour la prise de décision sur les données despuces à ADN. Ph.D. dissertation. Department Informatique, UAE University (2017)
14. Rafii, F., Hassani, B.D.R., Kbir, M.A.: New approach for microarray data decision making with respect to multiple source. In: Proceeding BDCA'17 Proceedings of the 2nd International Conference on Big Data, Cloud and Applications. Article No. 106, 29–30 Mar 2017. Tetouan, Morocco (2016)
15. Rafii, F., Hassani, B.D.R., Kbir, M.A.: Exploring semantic web technologies to integrate microarray experiments for cancer studies. Int. J. Emerg. Trends Eng. Dev. **6**(6), 251–265 (2016)

16. Rafii, F., Kbir, M.A. Hassani, B.D.R.; Microarray data integration to explore the wealth of sources generated by modern molecular biology. In: Veille Stratégique Scientifique et Technologique, pp. 11–13. Granada, Spain (2015)
17. Rafii, F., Kbir M.A., Hassani, B.D.R.: Microarray data integration for efficient decision making. In: Conférence sur les Avancées des Systèmes Décisionnels, pp. 10–12. Tangier, Morocco (2015)
18. Rafii, F., Hassani, B.D.R., Kbir, M.A.: Lung cancer diagnosis based on microarray data by using ART2 network. Int. J. Comput. Sci. Trends Technol. **4**(3), 129–136 (2016)
19. Rafii, F., Hassani, B.D.R., Kbir, M.A.: Automatic clustering of microarray data using ART2 neural network. J. Theoret. Appl. Inf. Technol **90**(1), 175–184 (2016)
20. Rafii, F. Hassani, B.D.R., Kbir, M.A.: MLP network for lung cancer presence prediction based on microarray data. In: Third World Conference on Complex Systems, pp. 23–25. IEEE, Marrakech, Morocco (2015)

Processing Unstructured Databases Using a Quantum Approach

H. Amellal[1(✉)], A. Meslouhi[1], and A. El Allati[2]

[1] University Mohammed V Faculty of Sciences, Av.Ibn Battouta,
B.P 1014, Rabat, Morocco
amellal@yandex.ru
[2] Laboratory of Engineering Sciences, Faculty of Sciences
and Techniques, B.P.3, C.P. 32003 Ajdir, Al-Hoceima, Morocco

Abstract. One of the most fundamental choices to store the big data it's the use of unstructured databases. However, the classical algorithms used in NoSQL databases suffer from slow execution of orders, especially in search operations. In order to decrease the data time processing in general and more particularly the search period in unstructured databases, we suggest in this work the use of a quantum approach based on Grover's algorithm.

Keywords: Data mining · Unstructured databases · Relational databases
Quantum algorithms

1 Introduction

Databases have taken an important place in computer science, and particularly in the field of management. The study of databases led to the development of specific concepts, methods and algorithms, particularly for managing data in secondary memory (hard disk). Indeed, from the beginning of the discipline, computer scientists have observed that the size of the RAM does not allow loading the entire database in memory. This assumption is always verified, because the volume of data continues to grow under the pressure of new technologies. Accordingly, the big explosion of digital data has forced the scientists to find new methods to study, search, share, storage, analysis and presentation of data. Thus was born the "Big Data" [1]. It is a concept for storing an unspeakable number of information on a digital basis. The Big data represent literally a massive data, they refer to a very voluminous set of information that no conventional database management or information management tool can really work. Therefore, we can generate trillions bytes of data every day. Generally, the information coming from everywhere: Emails, social media, phone messages, GPS signals and many more. In fact, the Big Data represents a promising solution designed to allow everyone real-time access to big databases. It aims to propose a solution to the traditional databases such as: Platform of Business Intelligence in server SQL and much other. Accordingly, to optimize the processing times on Big Data, there are different methods that developed, such as NoSQL databases. Classically, big data based on several data meaning algorithms for the data processing, however this solutions was founded on principles of classical computing, which make the operations process take a

long time. Therefore, we propose in this study to profit advantages of quantum computing to reduce processing time in big data by using Grover's algorithm [2, 3].

The paper is organized as follows: In Sect. 2, classical processing of databases. In
Sect. 3 proposition of quantum processing of NoSQL databases. Finally, conclusion is
drawn in the last section.

2   Classical Processing of Databases

2.1   Relational Databases Review

In simplest terms, a database is a means of storing data in such a way that information
can be retrieved from it. A relational database is one that presents data in tables with
rows and columns. Relational databases are known as a relational database management system (RDBMS). Moreover, virtually all relational database systems use
structured query language (SQL) as a language for querying and maintaining the
database. Accordingly, when we talk about high-traffic sites and databases, we rarely
hear about relational databases (see Fig. 1). Indeed, guaranteeing the consistency of
data costs expensive in time and is often incompatible with performance. Since, the
relational model does not seem appropriate in environments requiring large architectures. In fact, using RDBMS it has many advantages this includes: the technology is
mature, so that today SQL is a standardized language, the ability to implement complex
queries and wide support is available and also strong communities [4, 5].

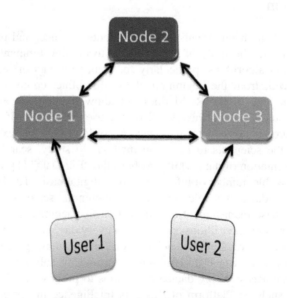

Fig. 1. Relational database principle

The success of relational databases has several origins and relies notably on SQL:

- SQL is a standard since 1986, which is enriched over time.
- SQL can interface with third generation languages like C or COBOL, but also with more advanced languages like C++, Java or C#.
- Modifying a data structure does not necessarily lead to a major overhaul of the application programs.
- These systems are well suited to large IT management applications (client-server and Internet architectures).
- They integrate development tools such as precompiles, code generators, reports, forms.

The model of relational databases is based on a rigorous theory while adopting simple principles this includes: relational tables and rows. A table is made up of columns that describe the records [6–8]. For example to create table by SQL we use the next code:

```
CREATE TABLE author
(
idArticle INT not null AUTO_INCREMENT,
Titre VARCHAR (255) not null,
Institution VARCHAR (255) not null,
PRIMARY KEY (idArticle)
)
```

Unfortunately, the relational database management system (RDBMS) suffers from a combination of drawbacks such as: the modification of the established model can be costly, on a very large volume of data the model performance can be limited, for some companies, the license price is very high. In order to achieve the requirements of customers, a new movement was born from the initiative of cloud computing architects and social sites: NoSQL (Not Only SQL) was going to enter the world of data representation [1, 9, 10].

2.2 Unstructured Databases

NoSQL refers a new family of database management systems that deviates from the classical paradigm of relational databases (RDBMS). It consists of many databases, which are characterized by non-relational logic in data representation (see Fig. 2). Therefore no query interfaces in SQL. Unstructured database is designed to solve the problems of volume, multi-source and multi-format data processing in big data environments. However, no distinction is made in terms of volume or diversity of data when talking about NoSQL technologies. Accordingly, nonreltional storage encompasses a wide variety of different database technologies have evolved in order to overcome those scalability limits in building modern applications [9, 10]. In general there are different types of NoSQL databases:

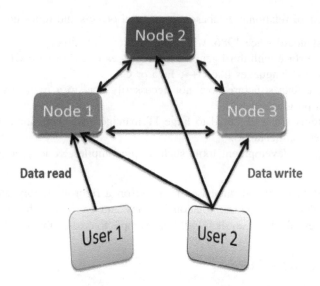

Fig. 2. NoSQL database principle

- Key-value stores: The data was organized in tables, there is no schema and the value of the data is opaque. Values are identified and accessed using a key. Each key has to be unique to provide non-ambiguous identification of values. The stored values can be numbers, strings, or counters. The key-value bases are for example adapted to the collection of events (online games), the management of traces (measurement of audience) and user profiles management of high audience sites (e-commerce).
- Wide column store: The wide-column store is a type of key-value database, has some passing similarity to principle of RDBMS in its use of rows, columns and tables. The important difference here is the names and format of the columns can change from row to row in the same table.
- Graph databases: is a type of database that uses graph theory to store, map, and query relations. It consists essentially of a set of nodes and edges. Each node represents an entity and each edge, connection, or relation between two nodes. Each edge is defined in turn by a unique identifier, a starting node and / or an arrival node, as well as a set of properties. Graph databases are particularly suited to the analysis of interconnections, which explains their great interest in exploring data from social networks. They are also useful for manipulating data in disciplines involving complex relation and dynamic patterns, such as supply chain management, identification of the origin of an IP telephony problem.
- Document stores: is a database for applications that manage documents. It is a compilation of documents or other information presented in the form of a database in which the full text of each referenced document can be viewed online, printed or downloaded. For example, this type of databases is used by university libraries and is intended for students and faculty. They are particularly suitable for online courses that can be downloaded from the Internet for distance learning.

Managing and processing the big data is seen as a new challenge for IT, and relational database traditional engines, highly transactional, seem totally outdated. In fact, a number of solutions emerged, the most important of which are the following:

– Google solution:

Google is one of the most companies concerned with manipulating the big data, which is why it has sought and developed a solution to address these challenges. Accordingly, to enable storage of these large volumes of data, Google has developed in the first a distributed file system named GoogleFS, or GFS, proprietary system used only at Google. The goal of GFS was to provide a redundant and resilient storage environment running on a cluster of medium-sized. Moreover, Google has developed the technical base to have the possibility of storing the data on GFS, to be able to perform processing on this data in a distributed way and to be capable to render the results. To realize this goal, the engineers inspired by the functional languages and extracted two primitives, the Map and Reduce functions. Basically, Map allows processing on a list of values. By implementing it on GFS and consolidating the results with the Reduce function, Google had managed to build a distributed processing environment that allowed it to solve a large number of problems [11, 12].

– Hadoop distributed file system:

Hadoop Distributed FileSystem is an open source framework that relies on Java. Hadoop supports Big Data processing in distributed computing environments. Hadoop is an integral part of the Apache project sponsored by the Apache Software Foundation. Hadoop is inspired by Google MapReduce, a software model that consists of breaking up an application into many small components. Each of these components (called fragment or block) can run on any node of the cluster. The Apache Hadoop ecosystem consists of the Hadoop kernel, MapReduce, the Hadoop Distributed File System (HDFS), and a number of related projects, including Apache Hive, HBase, and Zookeeper [13–15].

– BigTable

BigTable is a distributed storage system for structured Data developed by Google to support a large volume of structured data associated with the group's search tools and web services. BigTable looks like a gigantic, distributed hash table that incorporates mechanisms to manage the consistency and distribution of data on GFS. It is based on a simple data model. The data are assembled in line order and indexing is performed according to the rows, column keys, and the timestamp. Compression algorithms achieve high levels of capacity. Google Big Table serves as a database for applications like Google App Engine Datastore, Custom Search, Google Earth and Google Analytics. Google retains software maintenance as a proprietary home technology. However, BigTable has had a huge impact on the design of NoSQL databases [10].

– DynamoDB

Amazon DynamoDB is a non-relational database providing reliable performance, regardless of the scale. Dynamo consists of a warehouse of key-value pairs destined to be also totally distributed, in a masterless architecture. With DynamoDB, we can create database tables that can store and retrieve any amount of data, as well as handle any level of request traffic. You can increase or decrease the throughput capacity of your tables without downtime or performance degradation, and use the AWS Management Console to monitor performance metrics and resource utilization. Moreover, Amazon DynamoDB offers on-demand backup capabilities. It allows you to create full backups of your tables for archiving and long-term retention, for regulatory compliance purposes. Accordingly, DynamoDB allows automatic removal of expired items from tables to help the user to reduce storage usage and the cost of storing data that is no longer relevant. In the same context, DynamoDB automatically distributes data and traffic from your tables to a sufficient number of servers to manage your throughput and storage requirements, while ensuring consistent, fast performance.

From the above we conclude that the importance of NoSQL databases is indisputable. Therefore, increasing data processing speed is inevitable specially that the quantity of information circulating increases every day. Accordingly, we propose to profit quantum computing to reduce the data time processing in NoSQL databases via a quantum process.

3   Proposition of Quantum Processing of NoSQL Databases

Quantum computing takes advantage of the strange ability of the quantum superposition which makes the subatomic particles exist in more than one state at any time. Due to the way the tiniest of particles behave, operations can be done much more quickly and use less energy than classical computers. Moreover, the superposition makes the quantum algorithms faster than the classical ones. Accordingly, the following we study the effectiveness of Grover's algorithm on NoSQL databases processing [20–25].

3.1   Grover's Algorithm

Grover's algorithm proposes a smart search through a massive data in order to find the intended information. We consider a search space of size N, and no prior knowledge about the data presentation in it, where we want to find an element of that search space satisfying a known property. Classically, this problem requires approximately N operations. Grover's database search algorithm enables a dramatic reduction in the computational complexity of seeking in NoSQL database, where this algorithm allows it to be solved using approximately $\sqrt{N}$ operations [2, 3, 24, 25].

We consider a function $f(y)$ on bits with n inputs where $y \in \{0, 1\}^n$. The output of the function is a single bit, so we can have $f(y) = 1$ if y is solution to the search problem and $f(y) = 0$ otherwise (no such y exists). Classically, to find the value of y that realize $f(y) = 1$ in the generating outputs we should have $2 - 1$ tries to solve the problem (see Fig. 3). By using Grover's algorithm we can solve the problem with only $\sqrt{2}$ tries. For example, if we consider 4 bits, the classical computer will be need $2^4 - 1$ operation to find the correct y. By using Grover's algorithm we need $\sqrt{2^4} = 4$

operation to find the correct y. Therefore, Grover's algorithm can be described as a smart quantum database searching algorithm, which reduces the number of operations necessary to solve the problem as compared to a classical algorithms.

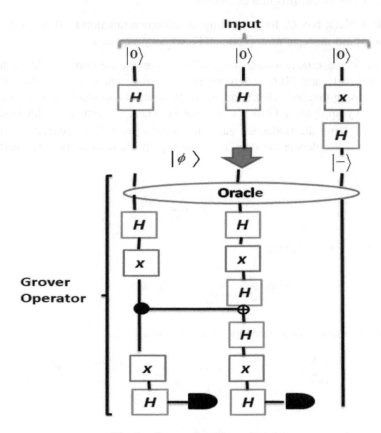

Fig. 3. Grover algorithm principle

We can resume Grover's algorithm by the following steps:

- Initial state $|0^{\otimes n}\rangle$ it is required to prepare the equally probable superposition of states of all input qubits. We can do it by applying a corresponding Hadamard gate that is equal to the tensor product of n unary Hadamard gates multiplied by each other
- Apply the Hadamard transform to all qubits: $H^{\otimes n}|0^{\otimes n}\rangle$
- Apply the Grover iteration: The given iteration lies in the consistent application of two gates, the oracle and the so-called Grover diffusion operator. We will review both of them below. We perform this iteration $\sqrt{2^n}$ times
- Measure the register: it is required to measure the input register of qubits. With a really high probability, the measured value will point to the sought parameter.

3.2 Application: Web Site Database

Considering a user's site web database UD with N different elements where $id \in \{0, N-1\}$ we are searching for id_0 where $UD[id_0] = Gr$. We give this search problem a general mathematical structure as follows:

- **Input:** A black box U_f for computing an unknown function $f : 0, n^n \rightarrow 0, n$
- **Output:** Find an input $id \in \{0, N-1\}^n$, where $f(id_0) = 1$.

We consider a quantum state $|id_0\rangle$, which represent the correct value searching, where $f(id_0) = 1$, and $f(id) = 0$ represent the error values of $|id\rangle$. To solve the problem by using Grover's algorithm we create an input superposition state and rotate it into $|id_0\rangle$ by applying a Grover's operator G. Then, we consider n bit input state $|0\rangle^{\otimes n}$, and applying the Hadamard gate $|0\rangle^{\otimes n}$ on the state $|0\rangle$ to generate a superposition of states. We denote the state $|\phi\rangle$ is the superposition of all possible states $|id\rangle$, where:

$$|\phi\rangle = \frac{1}{\sqrt{2}} \sum_{id \in \{0,1\}^n} |id\rangle \tag{1}$$

We have $|id_0\rangle \in |\phi\rangle$, then:

$$\langle id \mid \phi\rangle \frac{1}{\sqrt{2}} \sum_{id \in \{0,1\}^n} |id\rangle\langle id_0 \mid id\rangle = \frac{1}{\sqrt{2^2}} \tag{2}$$

Then, we consider two operators. The first given by:

$$U_f \sum_{id \in \{0,1\}^n} (-1)^{f(id)} \langle id \mid id\rangle = \sum_{id \in \{0,1\}^n} (-1)^{\lambda_{id,id_0}} \langle id \mid id\rangle \tag{3}$$

- If $\lambda_{id,id_0} = 1$

$$id = id_0 \tag{4}$$

- If $\lambda_{id,id_0} = 0$

$$id \neq id_0 \tag{5}$$

We define the second operator by:

$$U_s = 2|\phi\rangle\langle\phi| \tag{6}$$

if we divide $|\phi\rangle$ into two parts. The first contain $|\phi'\rangle$ and the second part containing the rest. Where:

$$|\phi\rangle = \sqrt{\frac{2^n - 1}{2^n}}|\phi'\rangle + \frac{1}{\sqrt{2^n}}|id_0\rangle \qquad (7)$$

Then:

$$\langle\phi \mid \phi'\rangle = \sqrt{\frac{2^n - 1}{2^n}} \qquad (8)$$

After some calculus we found:

$$|id_0\rangle = \sqrt{2^n}|\phi\rangle - \sqrt{2^n - 1}|\phi'\rangle \qquad (9)$$

Therefor:

$$U_s|id_0\rangle = (2|\phi\rangle\langle\phi - I|)(\sqrt{2^n}|\phi\rangle - \sqrt{2^n - 1}|\phi'\rangle) \qquad (10)$$

Then:

$$U_s|id_0\rangle = \sqrt{2^n}|\phi\rangle - \sqrt{2^2 - 1}\frac{2\sqrt{2^n - 1}}{2^n}|\phi\rangle + \sqrt{2^n - 1}|\phi'\rangle \qquad (11)$$

Using Eq. 9 to substitute $|\phi\rangle$ as following:

$$U_s|id_0\rangle = \frac{2\sqrt{2^n - 1}}{2^n}|\phi'\rangle + (\frac{2}{2^n} - 1)|id_0\rangle \qquad (12)$$

Therefore:

$$U_s|\phi'\rangle = \frac{2\sqrt{2^n - 1}}{2^n} \qquad (13)$$

Accordingly, we consider an angle θ, where: $\sin\theta = \frac{2\sqrt{2^n-1}}{2^n}$. And we notice that Eqs. 12 and 13 are a rotation, therefore:

$$U_s|id_0\rangle = \cos\theta|id_0\rangle - \sin\theta|\phi'\rangle \qquad (14)$$

And

$$U_s|\phi'\rangle = \sin\theta|id_0\rangle - \cos\theta|\phi'\rangle \qquad (15)$$

We apply a Grover operator defined by: $G = U_s U_f$. Then, we obtain a more familiar form of a rotation, namely:

$$G|id_0\rangle = \cos\theta|id_0\rangle - \sin\theta|\phi'\rangle \qquad (16)$$

And

$$G|\phi'\rangle = \sin\theta|id_0\rangle - \cos\theta|\phi'\rangle \qquad (17)$$

The principle is that the Grover operator rotates the state $|\phi'\rangle$ into the state we are searching for $|\,|id_0\rangle$. We have applied this rotation multiple times t. Then:

$$G^t|id_0\rangle = \cos\theta|id_0\rangle - \sin\theta|\phi'\rangle \qquad (18)$$

And

$$G^t|\phi'\rangle = \sin\theta|id_0\rangle - \cos\theta|\phi'\rangle \qquad (19)$$

4 Conclusion

In this paper, we have suggested the use of Grover's algorithm for unstructured database processing. Accordingly, we proved by a theoretical study on web site database that the results are very promising to be implemented in big data. In fact, we found that the quantum process decreases the search duration at last 4 times compared with the classical algorithms. However, the implementation of quantum algorithms with the current technologies, it still very complicated which makes the use of Grover's algorithm is linked to the emergence of quantum computer.

References

1. Zadrozny, P., Kodali, R.: Big data analytics using Splunk: deriving operational intelligence from social media, machine data, existing data warehouses, and other real-time streaming sources (2013). ISBN: 143025761X, 9781430257615
2. Grover, L.: Fast quantum mechanical algorithm for database search. In: Proceedings of the 28th Annual ACM Symposium on Theory of Computing (STOC_96), pp. 212–219 (1996)
3. Grover: Quantum mechanics helps in searching for a needle in a haystack. Phys. Rev. Lett. **79**(2), 325–328 (1997)
4. Grof, J., Weinberg, P.: SQL the Complete Reference, 3rd edn. McGraw-Hill Inc., New York (2010)
5. Cur, O., Blin, G.: RDF Database Systems: Triples Storage and SPARQL Query Processing, 1st edn. Morgan Kaufmann Publishers Inc., San Francisco (2014)
6. Gulutzan, P., Pelzer, T.: SQL Performance Turning. Addison-Wesley Longman Publishing Co., Inc., Boston (2002)
7. Wood, P.T.: Query languages for graph databases. SIGMOD Rec. **41**(1), 50–60 (2012)
8. Ohlhorst, F.J.: Big Data Analytics: Turning Big Data Into Big Money, p. 21. Wiley, New York (2012)

9. Demchenko, Y., Zhao, Z., Grosso, P., Wibisono, A., De Laat, C.: Addressing big data challenges for scientific data infrastructure. In: 2012 IEEE 4th International Conference on Cloud Computing Technology and Science (CloudCom), pp. 614–617 (2012)
10. John, Walker S.: Big Data: A Revolution That Will Transform How We Live, Work, and Think. Taylor & Francis, New York (2014)
11. Dean, J., Ghemawat, S.: MapReduce: simplifed data processing on large clusters. Commun. ACM **51**(1), 107–113 (2008)
12. Riondato, M., DeBrabant, J.A., Fonseca, R., Upfal, E.: PARMA: a parallel randomized algorithm for approximate association rules mining in MapReduce. In: Proceedings of the 21st ACM International Conference on Information and Knowledge Management, pp. 85–94. ACM, New York (2012)
13. Oruganti, S., Ding, Q., Tabrizi, N.: Exploring Hadoop as a platform for distributed association rule mining. In: Future Computing 2013 the fifth International Conference on Future Computational Technologies and Applications, pp. 62–67 (2013)
14. Kovacs, F., Illés, J.: Frequent itemset mining on hadoop. In: 2013 IEEE 9th International Conference on Computational Cybernetics (ICCC), p. 241–245. IEEE, New York (2013)
15. White, T.: Hadoop: The Definitive Guide. O'Reilly Media Inc., Farnham (2009)
16. Khan, M., Jin, Y., Li, M., Xiang, Y., Jiang, C.: Hadoop performance modeling for job estimation and resource provisioning. IEEE Trans. Parallel Distrib. Syst. **27**(2), 441–454 (2016). https://doi.org/10.1109/TPDS.2015.2405552
17. Hadoop, A.: Welcome to Apache Hadoop. http://hadoop.apache.org/. Accessed 10 Mar 2017
18. Plimpton, S.J., Devine, K.D.: Mapreduce in mpi for large-scale graph algorithms. Parallel Comput. **37**(9), 610–632 (2011)
19. Kollmitzer, C., Pivk, M.: Applied Quantum Cryptogtraphy, Lecture Notes in Physics, vol. 797. Springer (2010), ISBN 978-3-642-04829-6
20. McMahon, D.: Quantum computing explained. Wile Interscience A John Wiley Sons, Inc., Publication, Computer society IEEE (2007)
21. Kaye, P., Laflamme, R., Mosca, M.: An Introduction to Quantum Computing. Oxford University Press, Oxford (2007)
22. Imre, S., Balazs, F.: Quantum Computing and Communications an Engineering Approach. Wiley (2005)
23. Aharonov, D.: Quantum computation a review. Annual Review of Computational Physics VI, pp. 259–346. World Scientific (1998)
24. Ambainis, A.: Quantum walk algorithm for element distinctness. SIAM J. Comput. **37**, 210239 (2007)
25. Dirac, P.A.M.: The Principles of Quantum Mechanics, 3rd cdn. Clarendon Press, Oxford (1947)

Towards Remote Sensing Datasets Collection and Processing

Boudriki Semlali Badr-eddine[✉] and Chaker El Amrani

Abdelmalek Essaâdi University, Tétouan, Morocco
Badreddine.boudrikisemlali@uae.ac.ma,
ch.elamrani@fstt.ac.ma

Abstract. The world is witnessing important increases in industrial, transport and agriculture activities. This leads to economic growth, but, on the other hand, causes substantial damage in urban air, due to emissions of harmful gases, mainly CO, SO_2, NO_2 and PM. The World Health Organization (WHO) confirms that daily exposure to pollutants causes approximately three millions of deaths (Smith et al. in Annu Rev Public Health 35(1):185–206, [3]). It is therefore necessary to assess continuously the air quality. In this context, a Java-based application was developed to acquire data from EUMETSAT geostationary and Polar Orbit satellites, through the Mediterranean Dialogue Earth Observatory (MDEO) terrestrial station (El Amrani et al. in Development of a real-time urban remote sensing initiative in the mediterranean region for early warning and mitigation of disasters. IEEE, pp. 2782–2785, [2]). This application filters, subsets, processes and visualizes products covering Morocco zone. Thanks to this program, significant correlations were found between emissions and industrial activities related to power thermal plants, factories, transportation and ports.

Keywords: Remote sensing · Air pollution · Data processing
Data visualization · JAVA application

1 Introduction

Satellite data of atmospheric and earth components are becoming widely used by some large scientific communities. Remote sensing techniques aim to collect information about a remote object without making a physical contact with it [4]. These techniques employ satellites helping to monitor air quality and supervising climate changes. Recently, the word is suffering of air pollution and the increase of natural disasters. That is why the continuous monitoring of air quality is highly significant, by preventing or reducing eventual hazards or damages, and helping to decision makers. For this purpose, we developed a JAVA-based application that acquire, filter, and process satellite data in near-real-time. We had collecting data from MDEO platform represented in Fig. 1, enabling access to various polar and geostationary satellites of EUMETSAT organization. MDEO, includes a lots of channels, we were interested only in some specific data thus, VCD (Vertical Column Density) of some gases and some environmental variables including temperature, humidity, wind Speed and Pressure.

© Springer Nature Switzerland AG 2019
M. Ben Ahmed et al. (Eds.): SCA 2018, LNITI, pp. 286–294, 2019.
https://doi.org/10.1007/978-3-030-11196-0_26

This JAVA application illustrated in Fig. 4, could represents also final results into interactive maps of Morocco. Therefore, we found out a correlation between the density of traces gases, and industrial emitting sources.

Fig. 1. MDEO ground station of UAE

2 Collection of Pollution Data from MDEO

In this study, we make use of MeTop A and B satellites to acquire environmental data, in near-real-time [5], Fig. 2 explain the different stage of EUMETSAT data acquisition, so that we can monitors pollution patterns at regional scales and following the rapid change in atmospheric composition by IASI (Infrared Atmospheric Sounding Interferometer) sensor. IASI is an active sensor measuring the upwelling radiances in the thermal infrared spectral range with a swath of 2200 km and more than 1 million spectra collected per day. This instrument helps also to derive the total column of some traces gases from space, additionally, it measures the temperature, humidity and ozone profile on 101 levels pressure but it is affected by cloud conditions. Accordingly, we find some gaps during scans. Among retrieved variables from IASI in this project, there are the VCD of (CO_2, N_2O, CO and Ozone) in atmosphere with Kg/m^2 unit, and some climatic variables like (Temperature, humidity, pressure and wind speed) at the low tropospheric layer [1] (Fig. 3).

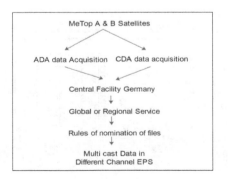

Fig. 2. EUMETSAT data acquisition

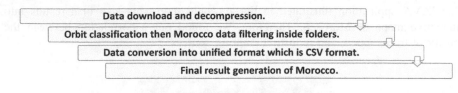

| Data download and decompression. |
| Orbit classification then Morocco data filtering inside folders. |
| Data conversion into unified format which is CSV format. |
| Final result generation of Morocco. |

Fig. 3. MDEO data processing steps

3 Filtering MDEO Data

Data were amassed from the EUMETCast data access service, an FTP server makes several channels available, containing various data files hailing from numerous satellites and sensors. As a result, atmospheric, land, forecasting, pollution and environmental datasets are intensely huge. We need only some interest variables, such as temperature, humidity and the VCD of gases. We chose appropriate channels and files to be processed. EUMETSAT indicates the nature of each file in her nomination. Channels most used in this work, are EPS-Africa and EUMETSAT-Channel-3, the first one contains files providing data of temperature, humidity and Vertical Column Density of traces gases measured by IASI, AMSU and MHS Sensors. The second one, accommodates datasets of Ozone profile, wind speed, measured by the SEVIRI Instruments.

4 Storing Morocco Datasets

MeTop A and B are flying at about 817 km on a sun synchronous orbit making about 14 orbits daily. The Two satellites provide global scale data, besides, in this study, we need only data of Morocco zone. Filtering these data, is based into an analyze of orbits, and the specific times of scan of this region. Fortunately, the sun synchronous orbit across the same location in the same local time after 29 days. Furthermore, EUMETSAT structure's names of file is uniform, containing several useful information, particularly, orbit number. As a result, to refine Morocco data we just have knowledge of orbits number and time of scans flying on Morocco area. This idea makes the filtering, more fast and easy. Ultimately, the JAVA application executes automatically the aforementioned tasks (Fig. 4).

Fig. 4. GUI of JAVA application of acquisition and processing of MDEO data

5   Extraction and Conversion of MDEO Data

Meteorological data are usually stored into scientific files like HDF5, NETcdf and BUFR that make data access, described in Fig. 6, search and subset more quiet. EUMETCast is a Near-real-time service offering data from satellites, with definite latency. In general data will be in level 2 of processing with BUFR and NETcdf formats. EPS-Africa contains a lot of BUFR and Binary files, to decrypt data we need some libraries such as BUFRDisplyay. In some channels like SAF-Africa we found NETcdf files more accurate, corrected, calibrated, and interpolated [6]. To explore this type of data we exploited, Panoply software. In order to have an automatic access of datasets we converted the cited formats into CSV format, because it is more simple and heavy to manipulate. Consequently, we used a Python library to achieve this task. Finally, we developed a Java code that enable extracting, removing noised and blank value, converting units and saving datasets into another output file more refined. Figure 5, shows the average processing time of MDEO data, we notice that the processing time of MSG channel take more than 12 min since, data are coming from a geostationary satellite scanning some area all the day.

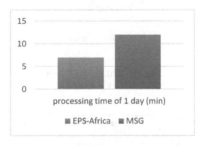

Fig. 5.   MDEO processing time

6   Near-Real Time Visualization

Visualization of data is the final feature of the JAVA application, it represents data into interactive map and charts of Morocco. Maps was developed with Folium library, it allows geo-temporal plots of different environmental and pollution data [7]. Furthermore, the map shows the location of several factories, ports, thermals power plants and dumps of Morocco in order to apply a correlation between emitting sources and concentration of pollutant gases. Moreover, the application plots NETcdf files exploiting MatploLib Library [8] (Figs. 6 and 7). Validation of these data refers to comparing the experimental data with ground sensors in our project we retrieve values of Temperature, Humidity, Pressure, and Wind speed from MeTop and METEOSAT satellite, Figs. 8 and 9 show respectively the Temperature and Humidity in Morocco at 04/19/2018. Some websites gives as ground sensors data, distributed over 28 Morocco. We validated data of Tangier. Hence, we remark that Remote sensing and ground sensors values are much closed.

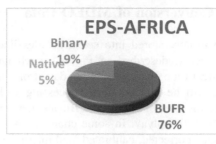

Fig. 6. MDEO data formats

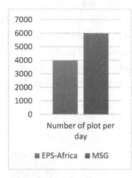

Fig. 7. Number of plots

Fig. 8. Temperature

7   Results and Discussion

In Morocco we have many industrial factories, power thermal plants, dumps detailed respectively in Tables 1, 2 and 3 with their longitude, latitude in intended gases [9].

These interactive maps help to monitor and visualize data of Morocco. We notice that climate was cool in Morocco on March, temperature Reached minus five Celsius in

Fig. 9. Humidity

Table 1. Factories of Morocco

Name of the factory	Activity	City	Longitude	Latitude	Altitude	Intended gases
Dairy central its	Processing of milk and tendencies	Casablanca	33.60	−7.62	16.25	CO_2 CH_4
SEGAPLAST	Processing plastic material	Casablanca	33.61	−7.53	17.17	CO_2 CO
Lafarge	Production of cement	Casablanca	33.55	−7.53	17	
SANACOB	Production of painting	Agadir	30.41	−9.45	17	CFC SO_2
METALFER	Metallic construction	Tangier	35.71	−5.90	17	CO SO_2 PM2,5/10
SOTAFER	Mechanical manufacturing	Tangier	35.76	−5.83	17	CO_2 NO_2 CO

Table 2. Thermals power plants of Morocco

Name	Type	City	Longitude	Latitude	Altitude	Intended gas
Appy-be	Central Thermal	Mohamadia	33.86	−7.43	15	CO_2
JORF-Al Asfar	Central Thermal	Safi	33.10	−8.63	17	SO_2
Tkc-one	Central Thermal	Kenitra	33.20	−6.56	17	NO_2
CTC	Central Thermal	Casablanaca	33.60	−7.57	16	CH_4
OCP	Phosphate mine	Khouribga	33.11	−8.60	15	PM 2.5 PM 10 CO_2

Table 3. Dumps in Morocco

City	Longitude	Latitude	Altitude	Intended gas
Tangier	35.75	−5.95	17	CO_2
Tetouan	35.56	−5.41	17	NO_2
Casablanca	33.43	−7.58	15	CH_4
Rabat	33.86	−6.77	17	N_2O
Safi	32.28	−9.21	17	
Marakech	31.69	−8.31	15	
Dakhela	23.71	−15.95	16	

middle Rif area. Also, humidity was highest in Kenitra, Rabat and Casablanca cities. In pollution maps, Figs. 10, 11 and 12, we perceive important Emission of Dioxide of Carbon (CO_2) around factories and thermal power plants area. Furthermore, we remark a significant concentration of Methane (CH_4) around agricultural extent such as Fes and Larache cities. So we could deduce a correlation between density of pollution gases and emitting sources.

Fig. 10. VCD of CO_2

Fig. 11. VCD of CH$_4$

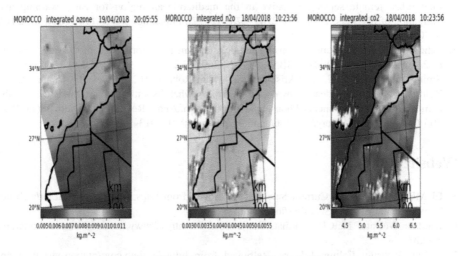

Fig. 12. VCD of Ozone, NO$_2$ and CO$_2$

8 Conclusion

There are several environmental variables like temperature, pressure and pollutants like CO$_2$, NO$_2$, etc. that should continuously be monitored by satellite sensors, to prevent natural disasters and health problems. In this study we developed a remote sensing application that enable the data processing of polar satellites (MeTop) and geostationary satellites METEOSAT acquired from EUMETCast service. Filtered and refined datasets were plotted in near-real time into geo-temporal interactives map of Morocco. Moreover, we found out a correlation between emitting sources of morocco and trace gases concentration Among perspectives, we would like collect remote sensing data from other satellites and sensors in order to get various and more detailed data. However, these data are regarded as Big Data so we are looking forward to manage, store and process these data with Big data solutions. Besides we aim to include Cloud technologies to enhance Remote sensing data processing Validation of received satellite data with ground sensors would also be an interesting work to conduct in the future.

References

Journal article

1. Clerbaux, C., Hadji-Lazaro, J., Turquety, S., George, M., Boynard, A., Pommier, M., … Van Damme, M.: Tracking pollutants from space: eight years of IASI satellite observation. C. R. Geosci. **347**(3), 134–144 (2015). https://doi.org/10.1016/j.crte.2015.06.001
2. El Amrani, C., Rochon, G.L., El-Ghazawi, T., Altay, G., Rachidi, T.: Development of a real-time urban remote sensing initiative in the mediterranean region for early warning and mitigation of disasters. IEEE, pp. 2782–2785 (2012). https://doi.org/10.1109/IGARSS.2012.6350855
3. Schowengerdt, R.A.: Remote sensing: models and methods for image processing. 3rd ed., p. 2. Academic Press (2007). ISBN 978-0-12-369407-2
4. Smith, K.R., Bruce, N., Balakrishnan, K., Adair-Rohani, H., Balmes, J., Chafe, Z., … Rehfuess, E.: Millions dead: how do we know and what does it mean? Methods used in the comparative risk assessment of household air pollution. Annu. Rev. Public Health **35**(1), 185–206 (2014). https://doi.org/10.1146/annurev-publhealth-032013-182356

Website

5. EUMETSAT Website: Current Satellites. Retrieved from https://www.eumetsat.int/website/home/Satellites/CurrentSatellites/index.html
6. Unidata Website: NetCDF format. Retrieved from https://www.unidata.ucar.edu/software/netcdf/
7. Github Website: Folium Library. Retrieved from https://github.com/python-visualization/folium
8. Matplotlib Website: Basemap Library. Retrieved from https://matplotlib.org/basemap/users/examples.html
9. Maroc-address Website: factories and thermal power plants of Morocco. Retrieved from http://www.maroc-adresses.com/

Tracking Luggage System in Aerial Transport via RFID Technology

Achraf Haibi[1]([✉]), Kenza Oufaska[2], and Khalid El Yassini[1]

[1] Faculty of Sciences Meknès, IA Laboratory, Moulay Ismail University,
Meknes, Morocco
{achraf.haibi,Khalid.ElYassini}@gmail.com
[2] Faculty of Computer Science and Logistics, TICLab Laboratory, International
University of Rabat, Salé, Morocco
Kenza.Oufaska@uir.ac.ma

Abstract. The aim of this study is to enhance baggage traceability by inte-grating Radio Frequency Identification (RFID) systems into the baggage tracking process in airports. Considering current preoccupations, we present the general concepts and realization & implementation. In the beginning, we present different definitions, Radio Frequency Identification components, operating principle, architecture and advantages. Secondly, we discuss the design and implementation of a traceability system that consists on the one hand of a middleware called BTMiddleware designed for real-time collecting and filtering of a large volume of data transmitted by RFID tags placed on the objects to be identified, on the other hand a mobile application named BagTrac that offers users the ability to follow their luggage from their smartphone.

Keywords: RFID · Tag · Middleware · Reader · BTMiddleware
BagTrac

1 Introduction

Radio Frequency Identification (RFID) is a form of Automatic Identification and Data Capture (AIDC) [1] that uses radio waves to automatically identify people or other objects on condition that an RFID tag has been placed upon them. This technology has recently seen growing interest from a wide range of industries such as retail, phar-maceutical, and logistics [2]. RFID technology can be used to track objects in a manner similar to using barcode based systems and Optical Character Recognition (OCR) systems [3], but RFID also brings additional benefits. RFID technology does not require line of sight readings, can read multiple tags simultaneously, and store large amounts of data in addition to the ID of the object tracked [4].

The outline of the paper is as follows, first, we introduce Radio Frequency Iden-tification systems and the operating principle. Secondly, we present the Middleware components and some existing RFID middleware solutions. Finally, we discuss BTMiddleware, defining its architecture and implementation, followed by an example of application describing luggage-tracking system that use the proposed BTMiddleware and the BacTrac application, to sum up the work with a conclusion and perspectives.

© Springer Nature Switzerland AG 2019
M. Ben Ahmed et al. (Eds.): SCA 2018, LNITI, pp. 295–306, 2019.
https://doi.org/10.1007/978-3-030-11196-0_27

2   RFID Systems

The RFID systems typically composed of two or three elements: a tag/transponder and a reader for a reduced RFID system, or a tag/transponder a reader and a middleware deployed at a host computer. The RFID tag is a data carrier part of the RFID system, which is attached on the objects to be uniquely identified. The RFID reader is a device that transmits and receives data through radio waves using the connected antennas. Among its features are the powering of RFID tags, and read and/ or write data to the RFID label. As shown in Fig. 1, the antennas emitted signals, which subsequently creates a zone of interrogation consisting of an electromagnetic field, when an object containing an RFID tag enters this zone, it will be activated to exchange data with the reader [4]. Later, the RFID tag data collected by the RFID readers then transmits to a software system called Middleware for processing. The RFID middleware manages RFID readers i.e. it is responsible for filtering, the consolidation and formatting of RFID tag data so that they can be accessible by various enterprise applications [5].

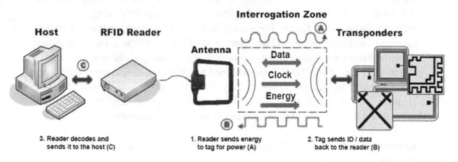

Fig. 1. RFID system components [3]

2.1   RFID System Components

RFID Transponder/Tag. A RFID transponder, or tag, consists of a chip and an antenna [4]. A chip contains a serial number that identifies the object in a unique way around the world and other information based on the tag's type of memory. The tag's type of memory can be read-only, read-write, or write-once and read-many [1]. Read-only tags are much cheaper to produce and are used in most current applications. The antenna is used to transmit information from the chip to the reader, and the larger the antenna the longer the read range. The RFID tag can be either attached or embedded in an object to be identified, and can be scanned by mobile or stationary using radio waves [6].

   RFID Reader. An RFID reader is a scanning device that agitates with both tags and middleware, it reads the tags reliably and communicates the results to the middleware. Readers use their antennas to transmit radio waves so as to all the tags within range will answer. Readers can acquire data from multiple tag RFID at the same time, allowing for increased read processing times. They can be either mobile or stationary that we can find their optimal deployment in the area [5], and they are differentiated by their storage capacity, processing capability, and the frequency they can read [4] (Fig. 2).

Fig. 2. Example of RFID reader

RFID Middleware. Middleware, is a combination of two English words middle and software [7], it represents a software intermediate between the hardware part of the RFID system and the information system of the company. The middleware manages the Automatic Identification and Data Capture (AIDC) technology equipment, receives the traceability events and transfers the formatted data to the information system [7].

Since the middleware is an interface between the readers and the information system, it plays a key role in the process of managing assembled data flows from RFID readers.

The main functions of an RFID middleware are:

- It hides the entire hardware part from back-end applications
- It applies filtering: Any redundant, meaningless or useless information will be filtered
- It is responsible for the raw data processing part before sending it to the relevant applications. It offers the possibility of the management of the readers.

3   Middleware RFID

As shown in Fig. 3, An RFID middleware generally consists of four layers:

1. Reader Interface
2. Data Processor and Storage
3. Application Interface
4. Middleware Management.

3.1   Reader Interface

Represents the lowest layer, it is responsible for interactions of the middleware with the various hardware components of the RFID system. This layer maintains the device drivers for all the devices supported by the system.

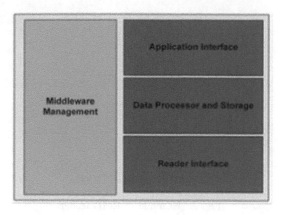

Fig. 3. Middleware architecture [6]

3.2 Data Processor and Storage

It is the layer responsible for many services that an RFID middleware is supposed to provide, it process the raw data stream sent by RFID readers, it filters and regrows the information of the RFID tags.

This component also manages the data level events associated with the application.

3.3 Application Interface

This layer is responsible for interfacing heterogeneous client applications with the middleware by giving them access to the different services provided by the middleware.

It provides the application with an API to communicate and configure the RFID middleware.

3.4 Middleware Management

Middleware Management: it provides information about all the processes running in the middleware. It allows to:

1. Add, remove, and modify the RFID readers connected to the system.
2. Change various settings by applications.
3. Enable and disable various functions supported by the middleware.

4 Examples of RFID Middleware

4.1 WinRFID

WinRFID is a middleware developed using the Microsoft .NET Framework. WinRFID demonstrates the ease of integration of RFID technology into existing IT infrastructures. It is presented as a scalable and extensible infrastructure (ease of integration with the applications of the company and those of its partners, thanks in particular to the numerous API16 that it has).

4.2 Fosstrak

Fosstrak is an open source platform (formerly known as the "Accada platform") that is designed to meet the needs of track & trace applications. This platform implements the specifications of EPCglobal Inc. (including the EPC Network standard) as three separate modules [8, 9]: a reader module, a middleware and the EPCIS service.

4.3 AspireRFID

AspireRFID is (Advanced Sensors and lightweight Programmable middleware for Innovative Rfid Enterprise applications) an open-source European project launched in the second half of 2008 by the OW221 consortium for the development and promotion of a reliable open source tool whose role is to facilitate deployment (with minimal cost) and management of RFID applications [10, 11].

4.4 SUN Java System RFID

SUN Java System RFID is an RFID platform developed by Sun Microsystems, which supports in its design the big standards accepted by the industrialists (like those developed by EPCglobal Inc.) [12]. It is designed to provide a high level of reliability and scalability of the EPC network by simplifying its integration into existing enterprise systems.

5 Comparison of Identification Technologies

RFID technology has a number of advantages compared to other identification technologies. It does not require line-of-sight alignment, several labels can be identified almost simultaneously, and labels do not destroy the integrity or aesthetics of the original object. The location of marked objects can be monitored automatically and continuously (Table 1).

Table 1. Comparison of identification technologies [14]

	Code-barres	RFID
Line of sight	Required	Not required (in most cases)
Memory	No memory	Small
Cost	Low	Medium
Range	Inches to feet	Inches to 100 s of feet
Reusability	No	Yes
Read rate	One at a time	Multiple simultaneously
Security	Very low (coding)	Medium (authentication)

6 BTMiddleware

6.1 Introduction: Why BTMiddleware?

There are a variety of Middleware each of them has its particularity, some of them use standard functions, others are classified among multi-layered middleware, for our case we designed and built a lightweight middleware named BTmiddleware, Our Middleware it is differentiated by its architecture and also characterized by a robustness from the point of view of the large volume of data that can be collected thanks to the integration of the MongoDB NoSQL database that can underpin a real-time Big Data system..

6.2 BTMiddleware Architecture

BTMiddleware aims to meet the requirements of the large data volume without forgetting to eliminate redundant data, i.e. it represents a Middleware capable to receive a large flow of data that can be transmitted by the objects containing RFID tags circulating in the interrogation area (Fig. 4).

The middleware collaborate with four external actors.

- With RFID readers: RFID readers have the possibility to open and close the connection with the middleware.
- With the database: distance applications able to consult the data recorded in the database by middleware.

It interact also with the user: this user is the main actor of the system; it comes into play in all the parameters. It has the right to update the data, enable and add the readers.

Figure 5 shows working process of our Middleware BTMiddleware.

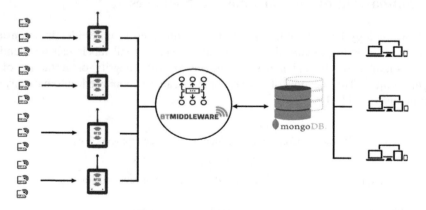

Fig. 4. BTMiddleware architecture

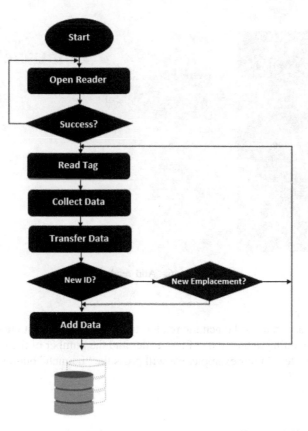

Fig. 5. The working process

6.3   BTMiddleware Implementation

Regarding the implementation part, we chose the Netbeans IDE development environment because it offers several tools, for the programming language, we chose the Java language and for storage we used the MongoDB NoSQL database.

Add a reader

The user is responsible for filling these fields for each reader:

- ReaderID: the reader ID
- ReaderName: represents the name of the reader
- ReaderIPAddr: The IP address of the reader.

When the user completes the filling of the fields, he must finish the operation by clicking on the "add" button to add the new reader (Fig. 6).

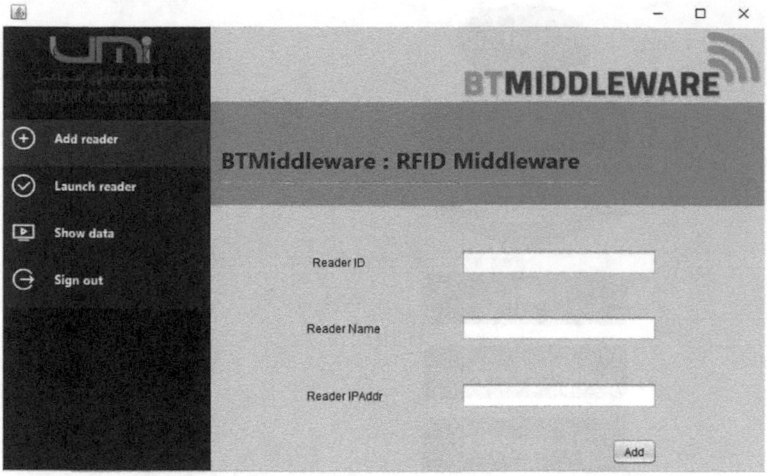

Fig. 6. Add reader

Launch reader

Subsequently user have to launch the reader, to do this the he must click on the "Start" button corresponding to the name of the reader and the number of the port to which the reader is connected. In our example, we will press the "Launch" button associated with "reader1" (Fig. 7).

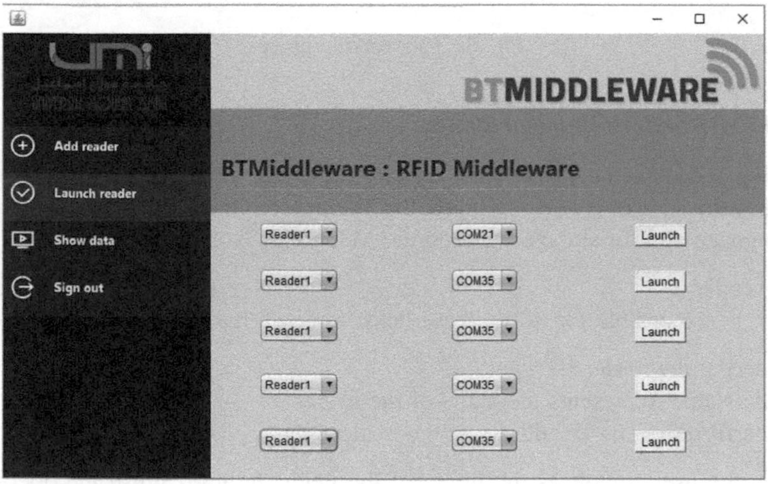

Fig. 7. Launch reader

Show Data

To view the data collected by the RFID readers, simply choose the interface named "show data" (Fig. 8).

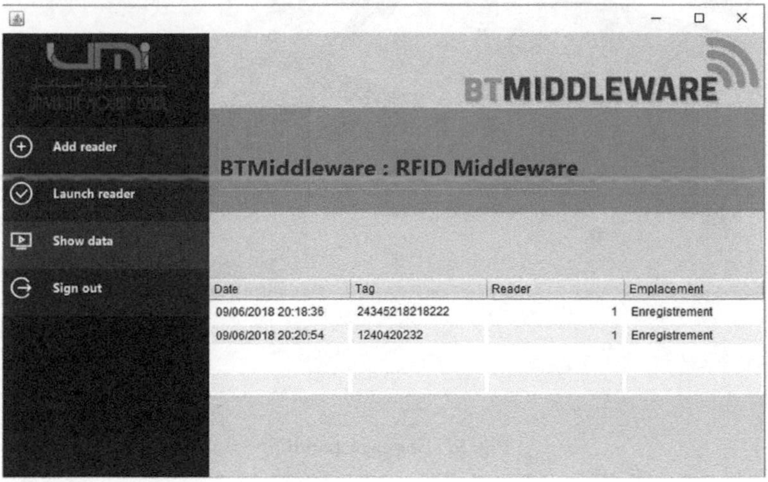

Fig. 8. Show data

7 Application Example

SITA 2017 (IT specialist for air transportation and communication) reported that during 2007 until 2016 there is 48% increasing of airline passenger, from 2,5 Billion at 2007 to 3,7 Billion at 2016 [13], so the number of luggage is increasing. As shown in the Fig. 9, bar code requires a line of sight for the reading of the labels so a baggage tracking system with bar code technology has a number of limits, taking as an example a label can be hidden and then the suitcase will be unidentifiable.

According to a Liligo/Next Content survey, luggage collection is the number one source of stress for French travelers at an airport, at 73%. Travelers are more stressed than travelers, and surprisingly, traveling regularly and being a regular at the airport does not lessen the level of stress.

According to the liligo/Next Content study, the 6 main sources of stress at the airport are:

- Baggage recovery (73%)
- Registration (62%)
- Access to the airport terminal (59%)
- Security checks (51%)
- Customs controls (43%)
- Boarding (40%).

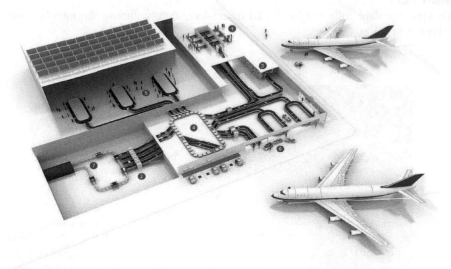

Fig. 9. Baggage handling

The purpose of this study is to create a baggage tracking system with three parts, first, RFID tags, which guarantees a location of bags through the airports, secondly, the middleware BTMiddleware, which collects, filters and formats data, finally, the Bag-Trac mobile application that allows users to know the location of their luggage in real time.

1. Registration of luggage
2. Check and scan luggage
3. Storage of luggage
4. Sorting the luggage
5. Withdrawal of luggage
6. Loading and unloading planes
7. Transfer of baggage between terminals.

The use of RFID systems allow a more precise follow-up through all the phases of the baggage tracking process at an airport, namely: Registration of luggage, Check and scan luggage, Storage of luggage, Sorting the luggage, Withdrawal of luggage, Loading and unloading planes, transfer of baggage between terminals.

RFID systems at the airport help also reducing baggage loss, which reduces costs and increases customer satisfaction, and without the need for line of sight and the ability to read RFID tags while moving, RFID offers significant advantages over bar codes.

To benefit from the advantages offered by this technology we have proposed luggage tracking systems via RFID technology, it is the Middleware BTMiddleware and the BagTrac mobile application (Fig. 10).

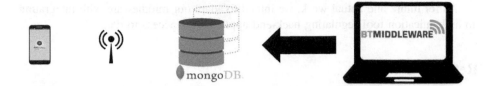

Fig. 10. Communication between BTMiddleware and BagTrac

To install this traceability system, on the bag or suitcase is affixed an RFID tag; the readers disposed on the path of the bag follows its position throughout their journey by sending the id of the tag retrieved to BTMiddleware.

BagTrac communicates with the database to display to the user the location of his suitcase. The position is then collected, centralized in real time and transmitted to the corresponding user (Fig. 11).

Fig. 11. BagTrac application

8   Conclusion

This article presents the different components of RFID technology plus the role of RFID middleware. It also describes the design and realization of our BTMidleware middleware and the mobile application BagTrac.

With such a system of baggage traceability, the airport can benefit from several advantages taking as an example:

- Baggage tracking in real time
- Reduction of manpower
- Name reduction number of lost luggage
- Higher luggage reading rate
- Offers travelers the ability to track their luggage via their smartphone.

As for future and actual work, we intend to adapt our middleware with integrating an authentication tool regulating back-end application's access to data.

References

1. Ishikawa, T., Yumoto, Y., Kurata, M., Endo, M., Kinoshita, S., Hoshino, F., Yagi S., Nomachi, M. Applying Auto-ID to the Japanese Publication Business to Deliver Advanced Supply Chain Management, Innovative Retail Applications, and Convenient and Safe Reader Services, Auto-ID Center, Keio University (2003)
2. Gaukler, G.M.: Item-level RFID in a retail supply chain with stock-out based substitution. IEEE Trans. Ind. Inf. 7(2), 362–370 (2011)
3. Ajana, M.E., Boulmalf, M., Harroud, H., Hamam H.: A policy based event management middleware for implementing RFID applications. In: Proceedings of WiMOB 2009 5th International Conference on Wireless and Mobile Computing, Networking and Communications, Marrakesh, Morocco, 12–14 Oct 2009. ISBN 978-0-7695-3841-9
4. Burnell, J.: What is RFID middleware and where is it needed? RFID Update (2008)
5. Rouchdi, Y., El Yassini, K., Oufaska, K.: Resolving security and privacy issues in radio frequency identification middleware. Int. J. Innov. Sci. Eng. Technol. (IJISET) 5(2), 2348–7968 (2018)
6. Rouchdi, Y., El Yassini, K., Oufaska, K.: Complex event processing and role-based access control implementation in ESN middleware innovations in smart cities and applications. LNNS 37, pp. 966–975. Springer (2018)
7. Venot, E.: Middleware RFID: traçabilité et objets connectés, Editions T.I (2015)
8. Floerkemeier, C., Roduner, C., Lampe, M.: RFID application development with the Accada middleware platform. IEEE Syst. J. 1–13 (2007)
9. Floerkemeier, C., Lampe M., Roduner, C.: Facilitating RFID development with the Accada prototyping platform. In: Fifth Annual IEEE International Conference on Pervasive Computing and Communications Workshops. PerCom Workshops'07, Zurich, Switzerland, 2007
10. Soldatos, J.: AspireRFID can lower deployment costs. 16 Mars 2009. Available: http://www.rfidjournal.com/article/view/4661
11. ASPIRE Project: ASPIRE—the EU funded project that brings RFID to SMEs (2009). Available: http://www.fp7-aspire.eu
12. Gupta, A., Srivastava, M.: Developing auto-ID solutions using sun java system RFID software. October 2004. Available: http://java.sun.com/developer/technicalArticles/Ecommerce/rfid/sjsrfid/RFID.html
13. SITA: Air transport industry insights, the baggage report (2017)
14. Aqeel-ur-Rehman, Abbasi, A.Z., Shaikh, Z.A.: Building a smart university using RFID technology. In: International Conference on Computer Science and Software Engineering, 2008

Using Feature Selection Techniques to Improve the Accuracy of Breast Cancer Classification

Hajar Saoud[1(✉)], Abderrahim Ghadi[1], Mohamed Ghailani[2],
and Boudhir Anouar Abdelhakim[1]

[1] LIST Laboratory, University of Abdelmalek Essaadi (UAE), Tangier, Morocco
{saoudhajar1994,ghadi05,boudhir.anouar}@gmail.com
[2] LabTIC Laboratory, University of Abdelmalek Essaadi (UAE), Tangier,
Morocco
ghalamed@gmail.com

Abstract. Classification is a data mining process that aims to divide data into classes to facilitate decision-making; it is therefore an important task in medical field. In this paper we will try to improve the accuracy of the classification of six machines learning algorithms: Bayes Network (BN), Support Vector Machine (SVM), k-nearest neighbors algorithm (Knn), Artificial Neural Network (ANN), Decision Tree (C4.5) and Logistic Regression using feature selection techniques, for breast cancer classification and diagnosis. We examined those methods of classification and techniques of feature selection in WEKA Tool (The Waikato Environment for Knowledge Analysis) using two databases, Wisconsin breast cancer datasets original (WBC) and diagnostic (WBCD) available in UCI machine learning repository.

Keywords: Breast cancer · Diagnostic · Machines learning algorithms
Feature selection · Classification · WEKA

1 Introduction

Breast cancer is one of the diseases that make higher number of incidence and mortality in the word. It represents the second cause of death for women after lung cancer [1]. Early diagnosis can reduce the breast cancer mortality rate by 40% [2].

Breast cancer can be defined as dangerous disease where cancer cells form in the tissue of the breast of the women and can spread to the others organs of the body. Machines learning algorithms and feature selection techniques will be interesting tools to predict and diagnose breast cancer also to classify it into its two categories either benign or malignant tumor.

We examined accuracy of six machines learning algorithms in the classification and diagnosis of the breast cancer: Bayes Network (BN), Support Vector Machine (SVM), k-Nearest Neighbors Algorithm (Knn), Artificial Neural Network (ANN), Decision Tree (C4.5) and Logistic Regression.

After that we want to improve the accuracy of those classifier using feature selection techniques that can be defined as the process of eliminating irrelevant and redundant features to improve the accuracy of classification.

© Springer Nature Switzerland AG 2019
M. Ben Ahmed et al. (Eds.): SCA 2018, LNITI, pp. 307–315, 2019.
https://doi.org/10.1007/978-3-030-11196-0_28

The rest of this paper is structured as follows. Part two is a presentation of breast cancer. Part three gives a vision about similar research. Part four is a theoretic presentation of machine learning algorithms. Part five give the definition of feature selection techniques. Part six describes the database used. Part seven explain confusion matrix. Part height shows the experiments performed by WEKA software on Wisconsin breast cancer dataset and results of these experiments and finally conclusion and perspectives in part nine.

2 Breast Cancer

Breast cancer is an abnormal production of cells in the breast that grow in an anarchic way. The masses of cells formed in the breast are called tumors. The cancer cells can stay in the breast or spread to other organs of the body. This is called metastasis.

2.1 Types of Breast Cancer

Breast cancer is decomposed into two types benign and malignant tumors:

- Benign tumors are non-dangerous tumors, they have well-defined contours. They develop slowly in the organ where they appeared without producing metastatic cases. Benign tumors are composed of cells that resemble to normal cells of the breast tissue.
- Malignant tumors are dangerous tumors, because they spread to other organs of the body and can produce metastatic cases. Cancer cells of malignant tumors have several abnormalities compared with normal cells in shape, size and contours where cells lose their original characteristics.

2.2 Causes of Breast Cancer

The first risk factor that can increase the probability of breast cancer is the age factor, the risk of breast cancer increases with age. Other factors that can intervene like:

Family or genetic factors. Gender: women are the most infected with breast cancer.

- A woman history: The woman that had already breast cancer in one breast, she has an increased risk to have cancer in the other breast.
- A family history: If several parents of the woman has been diagnosed with breast cancer, especially at a younger age, the risk to develop breast cancer increases.
- Genetic factors: Some genetic mutations increase the risk of breast cancer.

Characteristics of the individual. Obesity: The obesity increases the risk of breast cancer.

- Having period in early age: Having the period before the age of 12 increases the risk of breast cancer.
- Late menopause: Woman that started menopause at a later age, she is more likely to develop breast cancer.

- Having the first child in old age: Women who give birth to their first child after the age of 30 may have an increased risk of breast cancer.
- Women who have never been pregnant: The fact of not having a child increases the risk of developing breast cancer.
- Hormone replacement therapy: (Estrogen and progesterone) increases the risk of having breast cancer after 5 years of treatment.
- Drinking Alcohol: Drinking alcohol increases the risk of breast cancer.

3 Related Works

Several researches have been carried out in this field, some of them used only machine learning algorithms to classify and diagnose cancers others tried to improve the accuracy of classification using feature selection techniques.

Aalaei et al. [3] in this research they have chosen Genetic Algorithm for selecting the best subset of feature, they evaluate it using three classifier ANN, PS-classifier and GA-classifier on Wisconsin breast cancer original, diagnostic and prognosis. The GA improved the accuracy of the classifier, for WBC PS-classifier achieved the beast accuracy for WBCD and WBCP the ANN algorithm that achieved higher accuracy.

Saabith et al. [4] examined the accuracy of three classifiers Decision Tree, Neural Network and Rough set with and without feature selection techniques for breast cancer effective prediction.

Gowri and Ramar [5] they tried to propose a hybrid approach that combines between the best feature selection technique and best machine learning algorithm. The feature selection techniques examined are Correlation based Feature Selection (CFS), Information Gain (IG), Relief (R), Principle Components Analysis (PCA), Consistency based Subset Evaluation (CSE), and symmetrical uncertainty (SU) and techniques of classification examined are Naïve Bayes, K-Nearest Neighbor and Decision Tree on WBC, WBCD and WBCP.

Ahmed Abd El-Hafeez Ibrahim et al. [6] in this research they tried to present multi-classifiers fusion approach that fusion between classifiers to get the best multi-classifier fusion approach for each dataset. Dataset used are WBC, WBCD, WBCP and BCD, in WBCP and WBC 4th fusion level is better than other, for BCD the 3rd is the best and for WDBC the 2nd level is the beast.

Hamsagayathri and Sampath [7] in this research they analyzed the performance of four decision three algorithms in breast cancer dataset original, diagnostic and prognosis, they concluded that Priority based decision tree classifier is the beast classifier it classify data with 93.63% of accuracy.

Lavanya and Usha Rani [8] in this paper they analyzed the performance of decision tree classifier-CART with feature selection techniques and without them, they concluded that classification with feature selection is better than classification without feature selection because feature selection technique enhance the accuracy of classification.

Kaur et al. [9] in this research they compared the existing feature selection methods and compared their performance by calculating TPR, FPR, Classification accuracy, ROC Area, Precision, Kappa Statistic and Training Time.

4 Machine Learning Algorithms

4.1 Bayes Network

Bayes Networks [10], also called (Bayesian belief networks), are methods that are widely used for modeling and presenting knowledge of uncertain domains. Bayes Network is a directed acyclic graph (DAG), consisting of several nodes that represent variables and arcs that represent the probabilistic dependencies between those variables.

4.2 Support Vector Machines (SVM)

Support Vector Machines [11] are supervised learning models that can be used in prediction also in the classification of the linear and nonlinear data. The principle of the SVM algorithm is to use a non-linear mapping to transform the original learning data into a larger dimension. In this new dimension, it looks for the linear hyperplane of optimal separation. SVM algorithm aims to find a hyperplane with the largest margin named maximum marginal hyperplane (MMH) to be more accurate in the classification of future data i.e. it looks the shortest distance between the MMH and the closest training tuple of each class.

4.3 k-Nearest Neighbors Algorithm (Knn)

The k-nearest neighbors classifiers [11] are based on analog learning, i.e. they compare a given test tuple with similar learning tuples. They classify the tuples using more than one nearest neighbor. The principle of the k-nearest neighbors classifier is that it looks in the space model for the K test tuples closest to the unknown tuple. These tuples are named (k nearest neighbors) of the unknown tuple.

4.4 Decision Tree (C4.5)

The decision tree algorithm [12] is a classification algorithm that is similar to an organizational chart where the internal nodes (not- leaf) of a decision tree represent the tests on the attributes, the branches represent the results of the test and the external nodes (leaves) represent the predicted results. At each node, the algorithm chooses the best attribute to partition the data into individual classes. At the end a tree will be built when selecting subsets from the data provided.

4.5 Logistic Regression

Logistic regression [13] is one of the generalized linear models much used in machine learning. Logistic regression predicts the probability of a result that can take two values from a set of predictor variables. Logistic regression is mainly used for prediction and also to calculate the probability of success.

4.6 Artificial Neural Network (ANN)

Neural Network [14] can be defined as a reasoning model based on the human brain. An Artificial Neural Network is a set of processors (or neurons) very simple, very interconnected by weighted connections to pass signals from one neuron to another and they operate in a parallel manner. These neurons are similar to the biological neurons of the human brain, Artificial Neural Network consists of three layers: input layer, output layer and between them they are extra layers called hidden layers.

5 Feature Selection

Feature selection also called variable selection is the process of choosing the most relevant features to improve the process of the classification. They are three types of feature selection:

5.1 Filter Method

Filter methods select the variables independently of the chosen classification model; they select the variables by correlating the predictors and the answer variable. They take the variables that are relevant for the classification and delete the others.

5.2 Wrapper Method

Wrapper methods choose the variables by doing a combination between them, unlike the filter methods, wrapper methods try to find the interactions between the variables.

5.3 Embedded Methods

Embedded methods have been recently proposed, they try to combine the advantages of wrapper and filter methods. The choice of feature is done at the same time with the execution of the algorithm.

6 Description of the Dataset

In this paper we used two databases: Wisconsin breast cancer original (WBC) [15] and Wisconsin breast cancer diagnostic (WBCD) [16] available in UCI machine learning repository. WBC contains 699 records (458 benign tumors and 241 malignant tumors). It is composed of 10 variables 9 predictor variables and one result variable that shows whether the tumor is benign or malignant. The predictive attributes vary between 0 and 10. The value 0 corresponds to the normal state and the value 10 corresponds to the most abnormal state.

WBCD contains 569 records (357 benign tumors and 212 malignant tumors). It is composed of 33 variables 32 predictor variables and one result variable that shows also whether the tumor is benign or malignant.

7   Confusion Matrix

Confusion matrix gives the possibility to evaluate the performance of each classifier by calculating its accuracy, Sensitivity, Specificity. It contains information about real classifications or (current) and predicted (Table 1):

Table 1. Confusion matrix

	Predicted benign	Predicted malignant
Actual benign	TP (true positives)	FN (false negatives)
Actual malignant	FP (false positives)	TN (true negatives)

TP the cases predicted as benign tumors, they are in fact benign tumors
TN the cases predicted as malignant tumors, they are in fact malignant tumors
FP the cases predicted as benign tumors but in the reality they are malignant tumors
FN the cases predicted as malignant tumors but in the reality they are benign tumors

From the confusion matrix we can calculate:

- Accuracy $= \frac{TP+TN}{TP+FP+TN+FN}$
- Sensitivity $= \frac{TP}{TP+FN}$
- Specificity $= \frac{TN}{TN+FP}$

8   Experimentations and Results

8.1   WEKA Tool

The platform that we used to apply the machine learning algorithms on the breast cancer database is WEKA [17], because WEKA is a collection of open source machine learning algorithms, which allows realizing the tasks of data mining to solve real world problems. It contains tools for data preprocessing, classification, regression, grouping, and association rules. Also it offers an environment to develop new models.

8.2   K-Fold Cross-Validation

To evaluate the performance of machine learning algorithms based on breast cancer data we used the K-fold cross validation test method. This method aims to divide the database in two sets, the training data to run the model and the test data to evaluate the performance of the model. This is the most used method in the evaluation of machine learning techniques.

8.3   Feature Selected

Wrapper methods are used for feature selecting using Best first as search method with classifier Subset Evaluator technique to improve the accuracy of the classification:

- Bayes Net:

	Feature selected
WBC	1,2,3,4,5,6,7,8
WBCD	1,9,22,23,28

- Support Vector Machines:

	Feature selected
WBC	1,2,3,4,6,7,8
WBCD	4,5,22,23,24,25,26,28,30

- k-Nearest Neighbors Algorithm

	Feature selected
WBC	1,2,6,7
WBCD	1

- Decision Tree

	Feature selected
WBC	1,2,3,4,6
WBCD	1,22,23,29,31

- Logistic Regression

	Feature selected
WBC	1,2,3,5,6,8,9
WBCD	3,21,24,26

- Artificial Neural Network

	Feature selected
WBC	1,2,3,4,5,6,7,9
WBCD	3,6,10,12,16,17,21,25,26,28,31

8.4 Results

- WBC (Table 2 and 3):
- WBCD:

Table 2. Classification accuracy in WBC

	Without FS (%)	With FS (%)
BN	97.2818	97.4249
SVM	97.2818	95.279
KNN	95.279	95.8512
DT	95.1359	95.7082
LR	96.5665	96.7096
ANN	95.422	95.8512

Table 3. Classification accuracy in WBCD

	Without FS (%)	With FS (%)
BN	95.2548	96.1336
SVM	97.891	97.3638
KNN	96.1336	96.1336
DT	92.9701	94.9033
LR	94.2004	95.6063
ANN	96.1336	95.6063

9 Conclusion

To conclude, in this paper we tried to improve the accuracy of the classification of breast cancer using feature selection techniques. We use to databases Wisconsin breast cancer dataset original (WBC) and diagnostic (WBCD). The feature selection technique improved the accuracy of some classifier like Bayes net in both WBC and WBCD but we see the opposite for some classifier like SVM the feature selection technique has reduced the accuracy of classification. The best model to classify breast cancer in WBC is Bayes Network with feature selection and the beast one for WBCD is Support Vector Machines without feature selection. In the feature work we will try to propose methods that can improve more the accuracy of the classification of breast cancer.

References

1. «Breast cancer statistics»: World Cancer Research Fund, 22 Aug 2018. Available on: https://www.wcrf.org/dietandcancer/cancer-trends/breast-cancer-statistics
2. Ganesan, K., Acharya, U.R., Chua, C.K., Min, L.C., Abraham, K.T., Ng, K.-H.: Computer-aided breast cancer detection using mammograms: a review. IEEE Rev. Biomed. Eng. **6**, 77–98 (2013)
3. Aalaei, S., Shahraki, H., Rowhanimanesh, A., Eslami, S.: «Feature selection using genetic algorithm for breast cancer diagnosis: experiment on three different datasets». Iran J. Basic Med. Sci. **19**(5), 7 (2016)
4. Saabith, A.L.S., Sundararajan, E., Bakar, A.A.: «Comparative Study on Different Classification Techniques for Breast Cancer Dataset», p. 8 (2014)
5. Gowri A.S. Ramar, D.K.: «A novel approach of feature selection techniques for image dataset». **3**(2), 5
6. Abd El-Hafeez Ibrahim, A., Hashad, A.I., El-Deen Mohamed Shawky, N. Maher, A., Arab Academy for Science, Technology Maritime Transport, Cairo, Egypt: «Robust breast cancer diagnosis on four different datasets using multi-classifiers fusion». Int. J. Eng. Res., **V4**(03) (Mars 2015)
7. Hamsagayathri P., Sampath, P.: «Performance analysis of breast cancer classification using decision tree classifiers». Int. J. Curr. Pharm. Res. **9**(2), 19 (Mars 2017)
8. Lavanya, D., Usha Rani K.: «Analysis of feature selection with classification: breast cancer datasets». Indian J. Comput. Sci. Eng. (IJCSE) **2**(5), 9 (2011)
9. Kaur, R.: «Study and comparison of feature selection approaches for intrusion detection». Int. J. Comput. Appl. **7**
10. Mahmood, A.: «Structure Learning of Causal Bayesian Networks: A Survey», p. 6
11. Han, J., Kamber, M.: Data mining: concepts and techniques, 2nd ed., [Nachdr.]. Elsevier/Morgan Kaufmann, Amsterdam (2010)
12. Han, J., Kamber, M.: Data mining: concepts and techniques, 3rd edn. Elsevier, Burlington, MA (2011)
13. Yusuff, H., Mohamad, N., Ngah, U., Yahaya, A.: «Breast cancer analysis using logistic regression». Int. J. Res. Appl. Stud. **11** (2012)
14. Negnevitsky, M.: Artificial intelligence: a guide to intelligent systems. 2nd ed. Addison-Wesley, Harlow, England; New York: (2005)
15. «UCI Machine Learning Repository: Breast Cancer Wisconsin (Original) Data Set»: Available on: https://archive.ics.uci.edu/ml/datasets/breast+cancer+wisconsin+(original)
16. «UCI Machine Learning Repository: Breast Cancer Wisconsin (Diagnostic) Data Set»: Available on: https://archive.ics.uci.edu/ml/datasets/breast+cancer+wisconsin+(Diagnostic)
17. «Machine Learning Project at the University of Waikato in New Zealand»: Available in: https://www.cs.waikato.ac.nz/ml/index.html
18. Saoud, H., Ghadi, A., Ghailani, M.: Analysis of evolutionary trends of incidence and mortality by cancers. In: Ben Ahmed M., Boudhir A. (eds.) Innovations in Smart Cities and Applications. SCAMS 2017. Lecture Notes in Networks and Systems, vol 37. Springer, Cham (2018)

Smart Education

Application of Cloud Computing in E-learning: A Basic Architecture of Cloud-Based E-learning Systems for Higher Education

Abderrahim El Mhouti[✉], Mohamed Erradi, and Azeddine Nasseh

SIIPU Lab, Abdelmalek Essaadi University, Tetouan, Morocco
abderrahim.elmhouti@gmail.com

Abstract. Today, there is a growing trend regarding the use of e-learning approach in higher education. However, e-learning systems require huge investments in infrastructure and hardware and software resources of which many academics institutions cannot afford. The coming of cloud computing has brought new development ideas to overcome this problem and can be adopted to cut-down such investments. The purpose of this study is to put forward an overview on what is the current state of the application of cloud computing in e-learning in higher education context, where the use of computers is increasingly intensive. The paper introduces e-learning and cloud computing concepts, by analyzing e-learning systems challenges and trends, the convenience of cloud computing for e-learning and the key benefits of e-learning on the cloud. The study shows that cloud computing has tremendous effects on e-learning modes and it is the core technology of the next generation of e-learning in higher education. Cloud-based e-learning systems are emerging as an attractive method for providing flexible and scalable e-learning services, that can be accessed anytime, anywhere and from any device. The paper exposes also some applications' solutions using cloud computing in e-learning for higher education, by presenting the most common architecture that have been adopted. Finally, the paper discusses issues related to the implementation of cloud-based e-learning systems for higher education and presents some potential ways to overcome them.

Keywords: E-learning · Higher education · Cloud computing Cloud-based e-learning

1 Introduction

In recent years, e-learning technology offers a wide range of new opportunities for the development of higher education practices. Due to the rapid growth of internet technology, universities are investing heavily in e-learning systems to improve their students' learning experience and performance [1]. Recently, with the evolution of Web 2.0 technologies, providing rich and simple collaboration tools (wiki, blog, forum, social networks ...), one of the most important benefits of e-learning is that it enables collaboration and interaction between various actors involved.

Nevertheless, with these new orientations of e-learning practices, featured by social interaction and collaboration and with the daily rising trend on requirement's dynamic

© Springer Nature Switzerland AG 2019
M. Ben Ahmed et al. (Eds.): SCA 2018, LNITI, pp. 319–333, 2019.
https://doi.org/10.1007/978-3-030-11196-0_29

changes in service, e-learning systems deployed in universities are facing many challenges of optimizing large-scale resources management and provisioning, according to the huge growth of the number of users, services, courses and resources [2]. Indeed, the evolution of e-learning systems has affected theirs different dimensions (courses, users, hardware and software resources, etc.). Firstly, the infrastructure provisions those are necessary to provide a competitor service for a large amount of users clearly exceed the capabilities of a simple web server. On the other hand, the demands of software and hardware resources usually vary in a dynamic and very quick way, and presents high peaks of activity for e-learning systems. To attend flexible and dynamic requests of resources, it will be necessary to prepare a quite superior infrastructure than that required for the regular working of higher education institutions [3].

To confront these issues, an alternative solution for higher education institutions would be to offer those services depending on the demand and only paying for the resources that are actually used. With these characteristics, cloud computing arises as an accurate alternative to traditional e-learning systems. Much research work has been done on analyzing how cloud computing influence learning experience and effectiveness in e-learning systems [4–6].

This work is a contribution to the research efforts in cloud computing used in e-learning systems deployed in universities, where the use of computers is increasingly intensive (online labs, computing centers, data centers, …). The objective of this paper is to bring forward an overview of the current status of the forms of cloud computing technology and e-learning deployment solutions in higher educational context. Thus, the paper consists to explore and discuss the potential of cloud computing services to design open and dynamic e-learning systems for higher education purposes. The paper gives details of the architectures that have been developed for cloud-based e-learning systems and provides some examples of cloud-based e-learning systems that can be found in the specialized literature. The paper discusses also issues related to the implementation of cloud-based e-learning systems for higher education and presents some potential ways to overcome them. By leading this review study, it is concluded that cloud-based e-learning systems are emerging as an attractive method for providing flexible and scalable e-learning services that can be accessed anytime, anywhere and from any device.

The remainder of this paper is arranged as follows: e-learning concept and e-learning systems trends and challenges will be discussed in the next section (Sect. 2). Therefore, Sect. 3 gives a review of cloud computing model within its characteristics, services and deployment modes. Section 4 introduces the reasons for the move towards e-learning based on cloud computing and presents the benefits of e-learning on the cloud. This section discusses also the applications' solutions of cloud computing in e-learning and describes the basic cloud-based e-learning architecture. Issues facing the implementation of cloud-based e-learning systems and recommendations to overcome them are discussed in Sect. 5. Conclusions and perspectives can be found in the last section of this paper.

2 E-learning Systems for Higher Education: Trends and Challeges

2.1 E-learning Concept: Overview

E-learning refers to the use of information technologies in education to deliver information, where students and instructors are separated by distance, time, or both. The aim is to enhance student's learning experience and performance [7]. E-learning is as an Internet-based learning process that uses computer technologies and Internet to design, implement, manage, support and extend learning, which will not replace traditional education techniques, but will greatly enhance the efficiency of education [8].

In higher education, e-learning is widely enclosure of all kinds of educational technology in learning and teaching. The term e-learning is widely synonymous with online education, technology-enhanced learning, computer-assisted instruction, web-based training, and virtual education. These other alternative names dwell on a specific component, aspect or delivery method [9].

The evolution of the Web to Web 2.0 and the influence of new practices on the Web have resulted in a new array of services, which can be collectively termed "e-learning 2.0". Today, and with the current digital age, e-learning has become a necessary instrument in higher educational environment, allowing the creation of student-centered learning and offering new more flexible learning methods.

2.2 E-learning Systems

E-learning systems are a technological development designed in order to reform and restructure the delivery and interaction of instructors and students with course materials and pedagogical resources. An e-learning system is a software application, system or platform for flexible learning. Its aim is the realization of learning process theory: organization of contents and resources, delivery of educational courses and training programs, tracking, documentation and administration tasks.

In academic field, the first e-learning systems were really only set up to deliver information to students but as we entered the 70s e-learning started to become more interactive. In the last years, many higher educational institutions have used and deployed e-learning systems as a tool to ensure the delivery of course's contents and improve the access of the courses and subjects by both teachers and students [10].

In higher education, the key features of an e-learning system includes learning management, courses management, assessment, reporting, content management, content creation and delivery. Also, these tolls provide facilities for managing the learning experience, communicating the intended learning experience and facilitating tutors' and learners' involvement in that experience. The learning experience needs to be communicated via syllabi, course content or copies of visual aids/handouts, additional resources, links to resources in libraries and Internet. The learning experience is facilitated typically via self-assessment quizzes and communications tools such as e-mail, forum, discussions and chat rooms. All the various functions and resources need to be capable of being hyperlinked together within a consistent interface [11, 12].

Figure 1 shows the typical structure of an e-learning system used in higher education context.

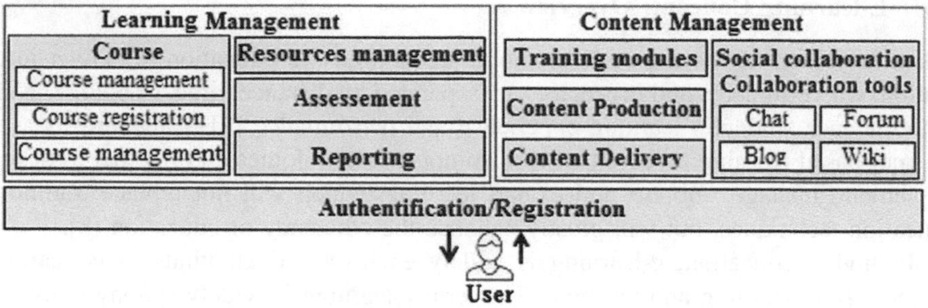

Learning Management		Content Management	
Course	**Resources management**	**Training modules**	**Social collaboration**
Course management	**Assessement**	**Content Production**	**Collaboration tools**
Course registration			Chat — Forum
Course management	**Reporting**	**Content Delivery**	Blog — Wiki
Authentification/Registration			

User

Fig. 1. Typical structure of e-learning systems

From the technical point of view, the majority of e-learning systems for universities are usually developed as distributed applications deployed using a three-tiered architecture that includes software components and the necessary hardware components [13]. The most used systems are VLE (Virtual Learning Environment), LCMS (Learning Content Management System), LMS (Learning Management Systems), etc.

2.3 E-learning Systems Trends

The trend of using e-leaning systems as teaching and learning tools is now rapidly expanding into higher education. However, with the rapid evolution of web technologies, these systems are opening up new trends to improve learning experiences. Indeed, rapid evolutions of web technologies are introducing new opportunities in the development of e-learning systems. If conventional e-learning systems have been used to organize and publish learning materials, the tendency for the new emerging e-learning systems focuses on how technology can facilitate the sharing and creation of knowledge and expertise through peer interaction and group learning processes [14].

In higher education institutions, e-learning featured by collaboration has become more and more advocated and the nature of e-learning systems was constantly changing from static to a highly dynamic media and more collaborative learning environments. Current e-learning systems are based on a learning strategy that embodies the application of new technologies and where several students interact with each other in order to achieve their common goals. They use the collaborative environment supported by the computer network to carry out the collaborative learning, in the form of group work, between teachers, tutors and students, based on their discussion, cooperation and communication [15], using the various interaction tools.

Indeed, e-learning systems for higher education are based on Web 2.0 and emerging trends in e-learning that are built around collaboration, which assumes that knowledge is socially constructed. They are based on creating and sharing of information and academic knowledge with others using social media tools like blogs, wikis and social networks to support collaborative approach to learning [16].

2.4   Current E-learning Systems Challenges

In higher education, it is obvious that e-learning systems offer several benefits over conventional classroom-based learning. Nevertheless, there are several problems and challenges facing the implementation of e-learning systems providing active learning. These challenges concern pedagogical, technical and organizational aspects. Thus, it has become imperative for practitioners to understand these challenges, which must be addressed prior to the full integration of e-learning into the academic field.

Several studies conducted in this context [17, 18] identify that the most important challenges facing higher education institutions are related to the efficient utilization of e-learning systems software and hardware resources. These challenges concern also the keeping pace of the rapid increase in the size and variety of data in these systems. Thus, e-learning systems are still weak on scalability at the level of their infrastructure. In an e-learning system, several resources are deployed and assigned just for specific tasks, which implies to add and configure new resources of the same type when receiving high workloads.

On the other hand, it is important to understand that, in an e-learning system, there is a cost related to the hardware resources maintenance, but also software resources. In that case, the institution must pay for the site licensing, installation and technical support for the individual software packages [19].

Indeed, with great growth of the number of services offered, users, contents and resources made available by higher education institutions, e-learning systems dimensions grow at an exponential rate. This emerging evolution concerns the computing resources optimization and storage and communication requirements. The challenges regarding this evolution highlight the necessity of the use of infrastructures that meet scalable demands and cost control.

Challenges for e-learning systems include also some aspects related to the pedagogical, technical and cost implications of e-learning technologies availed. The challenge of storage facilities is a hindrance for hosting e-learning systems that support multimedia content as pointed by Gamundani et al. [20].

3   Cloud Computing

3.1   The Concept of Cloud Computing

Cloud computing is a promising technology that refers to the use of applications and services that run on a distributed network using virtualized resources and accessed by common Internet protocols and networking standards [21]. Cloud computing has been recently emerging as a compelling paradigm of the present century, for managing and delivering services over the Internet.

Reference [22] defines cloud computing as a model for enabling ubiquitous, convenient, on-demand network access to a shared pool of configurable computing resources (networks, servers, storage, applications) that can be rapidly provisioned and released with minimal management effort or service provider interaction.

The cloud computing model is composed of five essential characteristics, three service models, and four deployment models. In the following, we describe the principles of these three cloud computing dimensions.

3.2   Cloud Characteristics

The five essential characteristics that distinguish cloud computing paradigm from other computing approaches are synthesized in the following aspects [23]:

- *On-demand self-service*: users can request and manage their own computing resources on demand as well as personalize their computing environments later on;
- *Broad network access*: cloud services are offered over the Internet or private networks and accessed through standard mechanisms;
- *Resource pooling*: computing resources (processing, storage, memory …) are pooled together to serve multiple consumers according to theirs demands;
- *Scalability and rapid elasticity*: cloud services and computing platforms can be rapidly and elastically provisioned and can be scaled across various concerns;
- *Measured service*: cloud computing resource usage can be measured, controlled, and reported providing transparency for both the provider and consumer.

3.3   Cloud Services

In terms of services, there are many services provided by cloud computing. The three basic service models (or layers) are Software as a Service (SaaS), Platform as a Service (PaaS) and Infrastructure as a Service (IaaS) (Fig. 2).

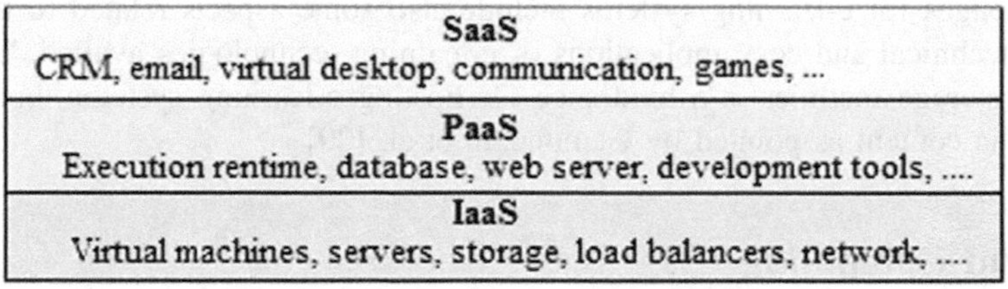

SaaS
CRM, email, virtual desktop, communication, games, ...
PaaS
Execution rentime, database, web server, development tools, ...
IaaS
Virtual machines, servers, storage, load balancers, network, ...

Fig. 2.   Cloud computing layers

In a SaaS model, software applications are offered as services on the Internet rather than as software packages to be purchased by individual customers.

In PaaS, an operating system, hardware and network are provided to support the entire application development lifecycle (design, implementation, debugging, testing …). Internet browsers are used as the development environment.

The IaaS model provides just the hardware resources (such as storage), computing power (such as CPU and memory) and network as services. This enables customers to rent these resources rather than spending money to buy dedicated servers and networking equipment [13].

3.4   Deployment Models

Deployment models define the nature of how the cloud is located. From this perspective, there are four ways used to deploy a cloud environment [21]:

- Public cloud: it is provisioned for public and open use alternatively for a large industry group. It is owned by an organization selling cloud services;
- Private cloud: it is provisioned for the exclusive use by a single organization. The cloud may be managed by that organization or a third party;
- Hybrid cloud: it combines multiple clouds (private, community of public) where those clouds retain their unique identities, but are bound together as a unit;
- Community cloud: the community cloud is one where the cloud has been organized to serve a common function or purpose.

4   Cloud-Based E-learning Systems

The challenges facing e-learning systems implementation, and which concern optimization of resources, scalability and storage requirements, request to point out the necessity to migrate towards a solution that alleviate these issues and which meets scalable demands and cost control. Cloud computing has brought opportunities for this. It is the future e-learning architecture, including all the necessary software and hardware resources engaging in e-learning.

Thus, following the challenges enumerated above, this section gives the implications and reasons for the migration towards cloud-based e-learning. The section stresses the benefits of cloud computing in e-learning and reviews some of the cloud-based applications of e-learning that have been already developed. Also, the common architecture of the cloud-based e-learning environments will be addressed.

4.1   Why Cloud Computing Is Convenient for E-learning

With its dynamic scalability and usage of virtualized resources as a service through the Internet, cloud computing has become an adoptable technology for many fields. E-learning for higher education is one of the technologies of interest in cloud computing. Today, cloud computing has become an alternative and attractive model for delivering IT services with which most universities would like to be incorporated with for various reasons.

Indeed, cloud computing plays a big role in the higher education industry and has a great potential to make significant changes in e-learning systems used by universities. Cloud computing has been adopted in e-learning to increase the efficiency and availability of such e-learning systems. After the computing resources are virtualized, they can be afforded in the form of services for higher educational institutions and students to rent computing resources. On the other hand, due to the scalability and cost reduction, cloud computing services allow implementing easier, faster and less expensive e-learning solutions.

In addition, the use of computers in higher education institutions is increasingly intensive (online labs, computing and data centers ...). This means that these institutions are seeking to provide free or low-cost alternatives to expensive and exclusive tools. In this sense, cloud computing paradigm has promoted the growth of e-learning systems in higher education with its pay as you go model: users can use computer resources anywhere, anytime, simply on demand and only pay for the usage thereof. This model is adapted to all scales of budgets and requirements of universities. By developing usage of Internet and networks, cloud computing was introduced as one of the best and economical option to the needs of higher education institutions.

Thus, cloud computing offers for higher education institutions a new type of business model where the services that are delivered become computer resources. It allows to universities to develop their services and use resources in a flexible manner. When academics need more resources to their e-learning system, it is no need to purchase hardware or install software, but these resources are automatically transferred to user, which constitutes a cost-effective platform to respond the educational needs.

Also, the adoption of cloud computing in e-learning supports the creation of a new generation of e-learning environments which are able to run on a number of hardware devices, while the data is stored in the cloud. Also, Cloud computing offers a natural platform to support e-learning systems and this by enabling the implementation of data mining techniques that becomes important when large databases are being used so that meaning can be extracted from data [3].

Moreover, adopting cloud-based e-learning allows managing educational and technical tasks better. By adopting cloud computing, universities become responsible for the content creation, management and delivery, while the cloud services providers are responsible for the construction as well as management of the e-learning system [24].

Finally, we can say that cloud computing responds to the aims for which e-learning environments were deployed in higher education institutions: collaboration learning and exchange between students and teachers. Cloud solutions can be used to support collaborative learning by creating virtualized resources that can be made available to users for collaboration purposes.

4.2   Cloud-Based E-learning Benefits in Higher Education

Because of the costs increase, institutional performance and competition, cloud computing has become an important requirement for many higher educational institutions. The appropriate use of cloud computing tools can enhance engagement among educators, students, and researchers in a cost effective manner.

By using virtualization, centralized data storage and facilities for data access monitoring, cloud computing offers many benefits to e-learning solutions by providing three types of supports: 1/Infrastructure: deploy an e-learning solution on the provider's infrastructure; 2/Platform: implement an e-learning solution based on the provider's development interface; 3/Services: use the e-learning solution given by the provider [25]. By detailing these three supports, we conclude that, by using cloud computing, universities can concentrate more on teaching and research activities rather than on complex IT configuration and software systems management.

Thus, a lot of benefits covering the three aspects mentioned above (infrastructure, services and platform) can be achieved when implementing the e-learning systems in the cloud. These benefits include [26, 27]:

- Cut-down cost of e-learning technologies investment: the need for higher education institutions is increasing constantly. E-learning systems require sophisticated resources of which educational institutions cannot afford the huge investments.
- Unlimited and centralized storage of data: in cloud-based e-learning, there is a large part of data, learning resources and applications that are stored into the cloud, which offers an almost unlimited storage capacity.
- Scalability: cloud computing is highly scalable and creates virtualized resources that can be made available to users. Cloud computing allows to educational institutions to scale theirs services as according to their demand.
- High availability: the integration of high-performance computing power and mass storage in e-learning systems can provide a higher quality of service.
- Accessibility: cloud computing services can be accessed through heterogeneous systems. On the other hand, the data access is easy since anywhere, any time and any student/teacher can access the application.
- Data security: in the cloud computing model, data is storied intensively. Based on one or more data center, the managers allocate the resources, manage the unified data, deploy the software, control security, and do the reliable real time monitoring, thus guarantee the users' data security to the greatest possible degree [28].
- Virtualization: it is one of the most important characteristics of the cloud architecture. Each application deployment environment and physical platform is not related. It is managed, expensed, migrated, and backup through virtualization platform.
- Easy monitoring: only one place should be supervised, not thousands of computers distributed over the world, which make the monitoring of data access easier.
- No user-side software needed: there is no installation, software maintenance, deployment and server administration costs. This allows reducing costs for higher educational institutions, as, which leads to a lower total cost, and a fewer IT staff.

4.3 Examples of Cloud Computing Applications in E-learning

In recent years, cloud computing applications in e-learning environments for higher education have been scarcely explored. Various examples of cloud-based learning solutions in higher education have been discussed.

As a first example, reference [13] have proposed a cloud based e-learning framework for Ladoke Akintola University of Technology Open Distance Learning using computer science and related courses as a case study. The proposed system addresses the cloud services in a new dimension and each layer in the cloud-based architecture specifies the essential components needed to construct an academic cloud in an open distance learning environment. Also, a way of implementing the framework has been described.

Also, reference [20] have designed a system architecture based on distributed resources which provided by users' computers. In addition to the elastic characteristics of cloud computing, the system provides the high scalability by supporting the coordination of distributed resources of the node to the central computer system.

On its side, reference [29] has developed a new service model that improves the effectiveness within a virtual personalized learning environment. The presented framework is devoted for the subscription of the selected learning resources as well as the creation of a personalized virtual classroom. It allows the educational content providers to registry their applications in the server and the students integrate other resources to their learning application pools.

Reference [30] has proposed an open source software for e-learning based on cloud computing technology. Authors have implemented the EduCloud platform to deploy their e-learning environment on a public cloud, based on IaaS and SaaS in order to overcome resource limitation and lack of scalability.

Reference [31] has oriented their work to manage a virtual Cloud lab's resources allocation, with the ability to deploy the proposed virtual lab on a public or private cloud. The implemented virtual cloud lab allows enhancing resource utilization and sharing. The framework manages PaaS in virtual computing labs.

Reference [32] have developed BlueSky, an e-learning framework embracing cloud computing. It's about a cloud-based architecture that has several components devoted to the efficient provision and management of the e-learning services. The architecture is able to preprogram the necessary resources for the demanding contents and applications before they are actually needed.

Another example of cloud-based learning is CloudIA system developed by reference [33]. CloudIA is a framework that delivers on-demand creation and configuring of Virtual Machines images so that students are able to have their own Java servlet environment for experimentation. The framework allows students to focus more on the development, deployment and test of their applications in a servlet container.

Finally, reference [34] has developed StartHPC, a system for teaching parallel programming at MIT. StartHPC is designed on the basis of a virtual image of Amazon EC2 which is used to create the class cluster. In this cloud-based system, students and teachers were allowed to focus on the parallel programming concepts in OpenMPI and OpenMP without being distracted by non-related.

4.4 Basic Architecture of the Cloud-Based E-learning Systems

According to the objectives set, cloud computing services can be pooled to e-learning systems in different ways. In most cases, and in the higher education context, this combination takes into account several universities' demands, such as centralized data storage, resources virtualization, low cost of running, flexibility, availability and scalability of e-learning systems.

Indeed, the cloud-based e-learning architecture is usually common to the most e-learning approaches on the cloud [3]. This architecture includes, in addition to the cloud management system, all hardware and software computing resources and services offered by the cloud to engage in e-learning.

The basic architecture of cloud-based e-learning systems is shown in Fig. 3.

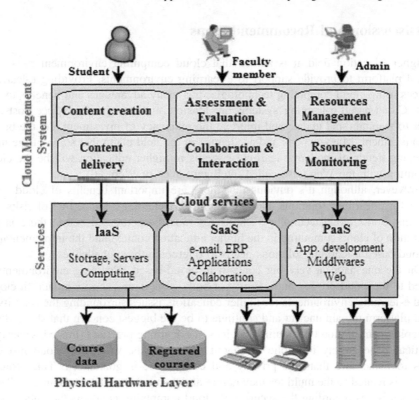

Fig. 3. Basic architecture of cloud-based e-learning systems

This cloud-based architecture is divided into three main layers. The first layer is the Cloud Management System Layer, which represents the interface of the e-learning system with the cloud environment. This layer includes many management subsystems permitting the integration of e-learning practices in the cloud computing model. Through this layer, there is no user-side software needed. Rather than having to install course design and management software, academics can simply use their Internet browsers to upload content, create new courses and collaborate between them.

The second layer represents cloud services delivered by the virtual machines implemented within the system. This layer delivers three types of services: SaaS, Paas and IaaS. Users use services delivered via the Internet. They do not need to make installation, purchase software and hardware nor to maintain or upgrade them and server administration costs. This allows reducing costs for universities.

The third and lowest layer is the Physical Hardware Layer. It includes all the physical architecture of the cloud-based e-learning system. This layer represents the information infrastructure and all resources used. It represents also for students the basic computing power like physical memory, CPU ... The physical host pool is dynamic and scalable. This means that new physical host can be added in order to enhance physical computing power for cloud middleware services.

5 Discussion and Recommendations

In higher education field, it is agreed that cloud computing environment rises as a natural platform to provide support to e-learning environments. Providing e-learning services using cloud computing technologies has many advantages and benefits to end users. Cloud-based e-learning systems are feasible and can greatly reduce costs, be easier to maintain and update and improve the efficiency of investment and the power of management. Many experts and academics in the field claim that some of the most promising trends of modern e-learning systems in higher education will be the cloud computing with the Web (also called intelligent Web or Web 3.0).

However, although it's obvious that there are important benefits of cloud computing for e-learning solutions, cloud computing technology is not free of risks and concerns. Some cloud technology challenges must be addressed before there is full integration of cloud computing in the higher education context and the implementation of cloud-based e-learning solutions needs to overcome these challenges.

On the one hand, a very big concern in cloud-based e-learning environments is related to the cloud privacy and security of both services and sensitive data. In cloud-based e-learning environments for higher education, issues surrounding the security of cloud platform remain unclear and continue to be the biggest concern that slows down the deployment of cloud computing services in e-learning practices. Indeed, security is a critical issue largely in public or shared environments, where the cloud provider needs to make sure that data privacy and compliance is guaranteed. This issue of security is related to the multi tenancy nature and resource and data outsourcing. Thus, the mains issues regarding the security of cloud computing concerns the data protection, insecure or incomplete data deletion, compliance risks, protection of intellectual property and of the data in cloud, loss of governance, etc.

In practice, the security issues are still not convinced by many higher education institutions. Academics are always concerned about the processing and storage of their data and critical applications deployed on the cloud. Academics continue to raise many questions about risks related to data protection and security, such as: who manages their data, who uses their personal data, where are their data located, but also, what will be the fate of their data in case of cloud-based e-learning system failure.

Thus, so that the cloud security concerns do not overshadow its benefits for e-learning, the future work in cloud computing and its applications in higher education, must focus on security and data protection risks by developing approaches and mechanisms that are able to address such issues. In this context, strict norms and rules are to be enforced so that higher education institution and actors will feel secured to use cloud computing services.

Another concern that must be resolved when higher educational institutions are brought to deploy cloud-based e-learning solutions is related to the connectivity weakness. Indeed, e-learning systems based on cloud computing architecture require reliable and fast Internet access because cloud-based applications require a lot of bandwidth. The low speed connections limit accessibility and reduce the efficiency of the provision of e-learning services. Therefore, cloud computing requires a stable and fast connection to the internet at all times.

Thus, to ensure consistency and availability for universities that have campuses dispersed in various sides, the low Internet throughput is one handle to tackle before a practical implementation of the cloud-based e-learning solutions. Indeed, as the stability and speed of the Internet continue to enhance, the deployment of cloud computing solutions for e-learning will increase.

6 Conclusion and Future Works

In conclusion, e-learning systems are being used increasingly by higher education institutions to provide efficient learning services. However, most of the conventional e-learning systems are becoming not being suitable for requirements of educational development and not being able to catch up with the changes of learning demand in time. Due to its dynamic scalability and effective usage of resources, cloud computing is becoming an attractive technology that rises swiftly as a natural platform to provide support to e-learning systems.

The aim of this work was to explore the current state of the application of cloud computing in e-learning in higher education context, where the use of computers is increasingly intensive. Thus, the review study has outlined an overview of the concepts of e-learning and cloud computing. More specifically, this paper has presented an insight on the characteristics, structure, trends as well as the challenges of e-learning systems. Likewise, the paper has presented an insight on cloud computing characteristics by focusing on the impact of using cloud computing services for e-learning in higher education and theirs key benefits.

The study shows that the use of cloud computing becomes a necessity and not an option for higher education institutions. This is due to a various factors such as the payment per use model and the management policies of risks and security purposes. By including the cloud services, that can be accessed anytime, anywhere and from any device, academic institutions achieve a substantially decreasing of expenses with software licensing and at the same time to reduce the campus IT staff. Thus, cloud-based e-learning systems can reduce costs due to lower requirements of hardware and software and offer more powerful functional capabilities to end users. Cloud-based e-learning systems are also easier to deploy across multiple locations as they are centrally administered.

Besides, to the potential advantages of using cloud computing in higher education, this review study has addressed some cloud-based e-learning solutions in higher education and the common and basic architecture of cloud-based e-learning systems. The study shows that the cloud computing economic strategy has forced different educational institutions and organizations to consider adopting a cloud solution. Universities have begun to adhere to this initiative.

At the end, the review study has discussed the cloud computing limitations that should be considered in cloud-based e-learning solutions. The study shows that the main risks in cloud computing are security and data protection risks. On the other hand, these limitations are related to issues surrounding the low speed of Internet connections that reduce the efficiency of the provision of e-learning services.

In the future research, we are working on the design and implementation of a cloud-based e-learning environment for enhancing collaborative learning in higher education context. In addition to the cloud services, the proposed e-learning environment integrates a multi-agents system to ensure students' tracking and improve their collaboration.

References

1. Kattoua, T., Al-Lozi, M., Alrowwad, A.: A review of literature on E-learning systems in higher education. Int. J. Bus. Manag. Econ. Res. **7**(5), 754–762 (2016)
2. Paul, C.J., Santhi, R.: A study of E-learning in cloud computing. Int. J. Adv. Res. Comput. Sci. Softw. Eng. **4**(4), 729–734 (2014)
3. Fernández, A., Peralta, D., Herrera, F., Benítez, J.M.: An overview of E-learning in cloud computing. In: Workshop on learning technology for education in cloud (LTEC'12), pp. 35–46. Springer, Berlin (2012)
4. El Mhouti, A., Nasseh, N., Erradi, M.: Using cloud computing and a multi-agents system to improve collaborative e-learning in LMS. In: proceedings of the 11th International Conference on Intelligent Systems: Theories and Applications. Mohammedia, Morocco (2016)
5. Li, J., Peng, J., Zhang, W., Han, F., Yuan, Q.: A Computer-supported collaborative learning platform based on clouds. J. Comput. Inf. Syst. **7**(11), 3811–3818 (2011)
6. Huang, L., Liu, C.: Construction of collaborative learning environment supported by cloud computing. Appl. Mech. Mater. **543–547**, 3581–3585 (2014)
7. Keller, C., Hrastinski, S., Carlsson. S.A.: Students' acceptanc of e-learning environments: a comparative study in Sweden and Lithuania. Int. Bus. 395–406 (2007)
8. Jain, A., Chawla, S.: E-learning in the cloud. Int. J. Latest Res. Sci. Technol. **2**(1), 478–481 (2013)
9. Sneha, J.M., Nagaraja, G.S.: Virtual learning environments: a survey. Int. J. Comput. Trends Technol. **4**(6), 1705–1709 (2013)
10. Alkhalaf, S., Drew, S., AlGhamdi, R., Alfarraj, O.: E-learning system on higher education institutions in KSA: attitudes and perceptions of faculty members. Proc.—Soc. Behav. Sci. **47**, 1199–1205 (2012)
11. Gaeta, M., Ritrovato, P., Talia, D.: Grid enabled virtual organizations for next-generation learning environments. IEEE Trans. Syst. Man, Cybern.—Part A: Syst. Humans **41**(4), 784–797 (2011)
12. Blas, N.D., Bucciero, A., Mainetti, L., Paolini, P.: Multi-user virtual environments for learning: experience and technology design. IEEE Trans. Learn. Technol. **5**(4) (2012)
13. Oladimeji, I.W., Folashade I.M.: Design of cloud-based e-learning system for virtual classroom. Int. J. Sci. Appl. Inf. Technol. **5**(1), 1–6 (2016)
14. Resta, P., Laferrière, T.: Technology in support of collaborative learning. Educ. Psychol. Rev. **19**(1), 65–83 (2007)
15. Wang, J., Sun, Y.H., Fan, Z.P., Liu, Y.: A collaborative e-learning system based on multi-agent. In: 1st International Workshop on Internet and Network Economics, China (2005)
16. Rupesh, K.A.: E-learning 2.0: learning redefined. Libr. Philos. Pract. **284**, 1–5 (2009)
17. Karim, F., Goodwin, R.: Using cloud computing in e-learning systems. Int. J. Adv. Res. Comput. Sci. Technol. **1**(1), 65–69 (2013)

18. Guoli, Z., Wanjun, L.: The applied research of cloud computing platform architecture in the e-learning area. In Computer and Automation Engineering. In: Proceedings of the 2nd International Conference on Computer and Automation Engineering, Singapore (2010)

19. Kwan, R., Fox, R., Chan, F., Tsang, P.: Enhancing Learning Through Technology: Research on Emerging Technologies and Pedagogies. World Scientific Publishing Co. Pty. Ltd. (2008)

20. Gamundani, A.M., Rupere, T., Nyambo, B.M.: A cloud computing architecture for e-leaning platform, supporting multimedia content. Int. J. Comput. Sci. Inf. Secur. 11(3), 92–99 (2013)

21. Bhure, G.C., Bansod, S.M.: E-learning using cloud computing. Int. J. Inf. Comput. Technol. 4(1), 41–46 (2014)

22. Mell, P., Grance, T.: The NIST Definition of Cloud Computing. In: NIST Special Publication 800-145, (2011)

23. Wang, L., Laszewski, G.: Scientific cloud computing: early definition and experience, In: Proceedings of 10th IEEE International Conference on High Performance Computing and Communications, China (2008)

24. Sharma, M.K., Rana, S.: G-cloud (e-Governance in cloud). In: 5th National Conference; INDIACom (2011)

25. Bora, U.J., Ahmed, M.: E-learning using cloud computing. Int. J. Sci. Mod. Eng. 1(2), 9–12 (2013)

26. Masud, M.A.H., Huang, X.: ESaaS: a new education software model in e-learning systems. In: Zhu M. (ed.) Information and Management Engineering. ICCIC 2011. Communications in Computer and Information Science, vol. 235. Springer, Berlin, Heidelberg (2011)

27. Ouf, S., Nasr, M.: Business intelligence in the cloud. In: Proceedings of the IEEE 3rd International Conference on Communication Software and Networks, Xi'an, China (2011)

28. Masud, M.A.H., Huang, X.: An e-learning system architecture based on cloud computing. Int. Sch. Sci. Res. Innov. 6(2), 736–740 (2012)

29. Liang, P.-H., Yang, J.-M.: Virtual personalized learning environment (VPLE) on the cloud. In: Gong, Z., Luo, X., Chen, J., Lei, J., Wang, F.L. (eds.) WISM 2011, Part II. LNCS, vol. 6988, pp. 403–411. Springer, Heidelberg (2011)

30. Yang, Z., Zhu, Z.: Construction of OSSBased e-learning cloud in China. In: Proceedings of the 2nd International Conference on Education Technology & Computer, Shanghai, China (2010)

31. Tian, W., Su, S., Lu, G.: A framework for implementing and managing platform as a service in a virtual cloud computing Lab. In: Proceedings of 2nd International Workshop on Education Technology and Computer Science, China (2010)

32. Dong, B., Zheng, Q., Qiao, M., Shu, J., Yang, J.: BlueSky Cloud Framework: an e learning framework embracing cloud computing. In: Jaatun, M.G., Zhao, G., Rong C. (eds.) Cloud Computing. LNCS, vol. 5931, pp. 577–582. Springer, Heidelberg (2009)

33. Sulistio, A., Reich, C., Doelitzscher, F.: Cloud infrastructure & applications—CloudIA, In: Jaatun, M.G. Zhao, G. Rong C. (eds.) Cloud Computing. LNCS, vol. 5931, pp. 583–588. Springer, Heidelberg (2009)

34. Ivica, C., Riley, J.T., Shubert, C.: StarHPC-teaching parallel programming within elastic compute cloud. In: Proceedings of the 31st International Conference on Information Technology Interfaces, Cavtat/Dubrovnik, Croatia (2009)

Designing an IMS-LD Model for Sharing Space of Learning Management System

Mohammed Ouadoud[1][(✉)] and Mohamed Yassin Chkouri[2]

[1] LIROSA Lab, UAE, Faculty of Sciences, Tetouan, Morocco
mohammed.ouadoud@gmail.com
[2] SIGL Lab, UAE, National School of Applied Sciences, Tetouan, Morocco

Abstract. The context of this work is that of designing an IMS-LD model for sharing space of Learning Management System (LMS). Our work is specifically in the field or seeking to promote, by means of information technology from a distance. Our approach is to first think about the conditions for creating a real sharing space for LMS between learners, and designing the IT environment that supports this space. In this paper, we try to adapt the IMS-LD model with a sharing model for LMS based on the traditional pedagogy and the social constructivism. This adaptation will go through three stages. Firstly, the development of an LMS model. Secondly, the study of correspondence between the developed model and IMS-LD model and their transformation to IMS-LD model.

Keywords: Sharing space · IMS-LD · e-Learning platform · Designing an IMS-LD model

1 Introduction

In the 20th century, there was an international movement in favor of e-learning integration in higher education. This movement has been operationalized due to the variety of the educational offer by universities, which most have opted to diversify knowledge dissemination platforms to meet the needs of their target public.

E-learning is promoted through educational platforms: integrated systems offer a wide range of activities in the learning process. Teachers use the platforms to monitor or evaluate the work of students. They use learning management systems (LMS) to create courses, tests, etc. However, the LMS e-learning platforms do not offer personalized services and therefore do not take into account the aspects of personalization such as the level of knowledge, interest, motivation and goals of learners. They access the same resource sets in the same way.

In fact, we present an easy sharing model of LMS to create and administer the educational content online. This tool allows the generation and editing structures of websites through database rather than pedagogical models, with a variety of choices that ensures better adaptation to the teaching of the course and learning style.

© Springer Nature Switzerland AG 2019
M. Ben Ahmed et al. (Eds.): SCA 2018, LNITI, pp. 334–347, 2019.
https://doi.org/10.1007/978-3-030-11196-0_30

Therefore, it is necessary to find a method to model all LMS types. In order to modeling the sharing space of LMS we have based ourselves upon the IMS-LD[1] specification focusing on learning theory that was judged the most important and relevant to our modeling, namely the social constructivism. Then, this learning theory which have inspired for a long time the design of computer applications are combined and put into perspective with several emergent pedagogical functionalities to build an original modeling for our sharing space of LMS. This reveals that this proposed modeling that is presented to readers here looks for ways to leverage technology for learning by considering users as being human actors and not human factors [12].

The IMS-LD specification or instructional design engineering uses pedagogical concepts, allowing to model learning units. IMS-LD takes into account a wide variety of teaching models it is there its flexibility. A course plan extract of a general or specific database can be modeled with IMS-LD, through the description of the different roles, activities, environments, methods,[2] properties, conditions and notifications. It is used to transform the course plans into formal learning units (UOL) that can be performed with an IMS-LD editor based on an engine such as Coppercore [18]. These executable units can be designed from the beginning using an editor such as Reload [19].

During the last decade, the LMS e-learning platforms have evolved considerably. However, several modeling of LMS platforms have been developed previously [13–17], but they have been abandoned because platforms' life cycle is changing apace. Therefore, we have conducted a comparative and analytic study on free e-learning platforms based on an approach of evaluating the e-learning platforms quality [1–3, 22]. Based on these various research works, which seemed to us incomplete, we proposed a modeling portrait of a designing an IMS-LD sharing model for LMS platform. This latter is anthropocentric and relies on a learning conception that is located at the intersection of the most used learning theory. Indeed, the idea is to orient the design work research towards a great and optimal compatibility between the services offered by e-learning platforms and the needs of all users, particularly learners, for better optimization of online learning.

To concretize our modeling work, we present in the section "Theoretical approach and concepts" the different theories of learning used for the modeling of this space, namely the traditional pedagogy and the social constructivism. We also present the online learning specification: Instructional Management Systems—Learning Design (IMS-LD). Then, we present in the section "Designing an IMS-LD model of sharing space", the use case diagram of the sharing space. Thus, we try to adapt the class diagram model of the sharing space to the equivalent IMS-LD model.

---

[1] **IMS-LD**: Instructional Management Systems-Learning Design. Available at https://www.imsglobal.org/learningdesign.

[2] The word "method" used by IMS-LD means the unfolding of the scenario.

2 LMS and Activity Spaces

2.1 Activity Spaces

The learning management system consists of different activity spaces for activities of teaching and learning [28, 29, 24]. Each model represents a space, in these spaces, both teachers and learners can have a:

- Disciplinary information space. In this space, teachers or tutors can export content from the LMS as IMS/SCORM conformant Content Packages that can be viewed offline, or imported into the LMS.
- Communication space. Communication is the act of conveying intended meanings from one entity or group to another using mutually understood signs and semiotic rules. Learners can communicate with others through their Inbox using LMS's private mail, through the discussion forums or the chat rooms.
- Collaboration space. Collaboration is the process of two or more people or organizations working together to realize or achieve something successfully [8]. Teams that work collaboratively can obtain greater resources, recognition and reward when facing competition for finite resources.
- Sharing space. Sharing space, it allows discussing, to define and to follow the implementation of collaborative projects of one or several courses.
- Evaluation space. Evaluation is a process where learners are assessed to determine their suitability to take a course. The learners' aptitudes are determined by a variety of techniques including interviews, group exercises, presentations, examinations and psychometric testing.
- Production space. In production space, the teacher or tutor can create notifications, and assess the productions undertaken by learners.
- Self-management space is the management of an LMS by the administrators themselves. The customizations are decentralized as much as possible and the sharing of tasks between all the users is done equitably.
- Assistance space. Assistance designates the action of bringing help or relief. It is about providing aid, support or relief in all matters.

2.2 Benefits of LMS

The LMS on which we increasingly rely as a means of learning have a considerable potential in the construction of knowledge and competence development. Thanks to the different services offered by these e-learning platforms, individuals can access and use interactively the multiple sources of information available to them everywhere, at all times. They can also compose customized training programs and thus develop their abilities to the highest level of their potential, according to their needs [25].

Based on the work of De Vries [26], the main pedagogical functions that may be assigned to the LMS as computer applications for learning are:

- Presenting information,
- Providing exercises,
- Really teaching,

- Providing a space of exploration,
- Providing a space of exchange between educational actors (learners, teachers, tutors…).

These different pedagogical functions, that correspond to one or many learning theories, allow the learner to acquire individual and collective knowledge according to the type of interaction that takes place between him/her and the sources of information made at his/her disposal. In practice, each individual has a set of tasks to deal with such as:

- Consulting and reading the pedagogical resources,
- Realizing the interactive exercises,
- Exploring the learning environment,
- Solving the problem situations,
- Discussing via synchronous and asynchronous tools of communication.

3   Theoretical Approach and Concepts

3.1   LMS and the Social Constructivism

The social constructivism is the fruit of the development of learning theories under the influence of some researchers, particularly Lev Vygotski in 1934 [21, 11], who wanted to depart from the behaviorism by integrating other factors that are able to positively influencing the knowledge acquisition. Thus, new ideas emerged in connection with the possible interaction of individuals with the environment.

The social constructivism outlines learning by construction in a community of learners. In this light, learners are expected to interact with the available human resources (teachers, tutors, other learners…) in the proposed learning environment. In this way, the learners' psychological functions increase through social cognitive conflicts that occur between them. These conflicts lead to the development of the zone of proximal development[3] [10] and thus facilitate the acquisition of knowledge.

Learning is seen as the process of acquisition of knowledge through the exchange between teachers and learners or between learners. These latter learn not only through the transmission of knowledge by their teacher but also through interactions [9]. According to this model, learning is a matter of the development of the zone of proximal development: this zone includes the tasks that learners can achieve under the guidance of an adult; they are not very tough or so easy. The development of this zone is a sign that the learners' level of potential development increases efficiently [27].

The teacher's role is to define precisely this zone in order to design suitable exercises for learners. Furthermore, designing collaborative tasks, which involve discussions and exchange (socio-cognitive conflicts) between learners is so important in

---

[3] "The distance between actual development level as determined by independent problem solving and the level of potential development as determined through problem solving under adult guidance or in collaboration with more capable peers [27, p. 86]".

this model. Errors are considered as a point of support for the construction of new knowledge.

Based on the social constructivism approach (cf. Fig. 1), the design of the LMS was oriented towards integrating online communication and collaboration tools. In practice, a wide range of platforms, particularly the social constructivist ones, proposes a set of tools, which allow sharing, exchanging and interacting in synchronous and asynchronous mode such as blogs, wikis, forums....

In summary, the ideas of social constructivist authors have highlighted the social nature of learning. Other authors have taken one-step further by emphasizing the distribution of intelligence between individuals and the environment. Furthermore, considering that learning occur in a social context is no longer enough to ensure deep learning. Indeed, working in groups can affect negatively the quality of learning if these following conditions are not taken into consideration: Learning styles, the way groups are formed, interaction modality, and the characteristics of tasks.

3.2 Instructional Management Systems—Learning Design

There are several e-learning specifications, for example: SCORM,[4] DCMI,[5] IMS-SS[6] or IMS-CP.[7] Among these, IMS-LD, one of the last publications, seeks to incorporate pedagogical flexibility and complements certain aspects treated by others.

IMS-LD was published in 2003 by the IMS/GLC. (Instructional Management Systems Global Learning Consortium: Consortium for global learning management systems with training, the original name when IMS was started in 1997 Instructional Management Systems project) [4, 7]. Reminds us of its origins: the source (EML[8]) of the proposed language was assessed by the European Committee for Standardization (CEN) in a comparative study of different SRMS [5, 7], as best suited to satisfy the criteria definition of an EML. EML is defined by CEN/ ISS as "*an information aggregation and semantic model describing the content and processes involved in a unit of learning from an educational perspective and in order to ensure the reusability and interoperability.*" [23] In this context, the North American IMS consortium undertook a study and provided a specification of such a language, giving birth in February 2003, the Learning Design specification V1.0 (IMS-LD). She adds that proposal, largely inspired EML developed by [6, 7] (OUNL) provides a conceptual framework for modeling a Learning Unit and claims to offer a good compromise between on the one hand to the generic implement a variety of instructional approaches and secondly, the power of expression that allows an accurate description of each learning unit.

---

[4] **SCORM**: Sharable Content Object Reference Model. Available at https://www.adlnet.gov/adl-research/scorm/.

[5] **DCMI**: Dublin Core Metadata Initiative. Available at http://dublincore.org/.

[6] **IMS-SS**: Instructional Management Systems-Simple Sequencing. Available at https://www.imsglobal.org/simplesequencing.

[7] **IMS-CP**: Instructional Management Systems-Content Packaging. Available at https://www.imsglobal.org/content/packaging.

[8] **EML**: Educational Modelling Language.

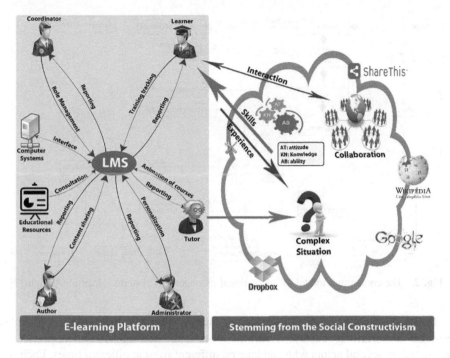

Fig. 1. LMS and underlying social constructivism model

This specification allows us to represent and encode learning structures for learners both alone and in groups, compiled by roles, such as "learners" and "Team"(cf. Fig. 2) [7]. We can model a lesson plan in IMS-LD, defining roles, learning activities, services and many other elements and building learning units. The syllabus is modeled and built with resources assembled in a compressed Zip file then started by an executable ("player"). It coordinates the teachers, students and activities as long as the respective learning process progresses. A user takes a "role" to play and execute the activities related to in order to achieve a satisfactory learning unit. In all, the unit structure, roles and activities build the learning scenario to be executed in a system compatible with IMS-LD.

IMS-LD does not impose a particular pedagogical model but can be used with a large number of scenarios and pedagogic models, demonstrating its flexibility. Therefore, IMS-LD is often called a meta-pedagogic model. Previous e-learning initiatives claim to be pedagogically neutral, IMS-LS is not intended to pedagogical neutrality but seeks to raise awareness of e-learning on the need for a flexible approach.

IMS-LD has been developed for e-learning and virtual classes, but a course face to face can be done and integrated into a structure created with this specification, as an activity of learning or support activity. If the ultimate goal to create rich learning units, with support to achieve the learning objectives by providing the best possible experience, face-to-face meetings, and any other learning resource are permitted such as video conferencing, collaborative table or any field action research.

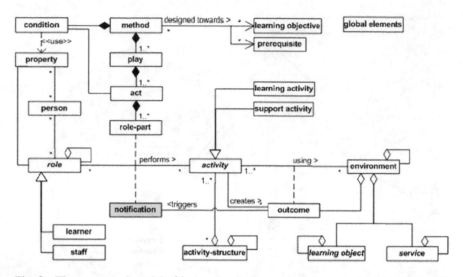

Fig. 2. The conceptual model of instructional management systems—learning design [7]

IMS-LD uses the theatrical metaphor, which implies the existence of roles, resources and learning scenario itself: one room is divided into one or more acts and conducted by several actors who can take on different roles at different times. Each role is to carry out a number of activities to complete the learning process. In addition, all roles must be synchronized at the end of each act before processing the next act.

4   Designing an IMS-LD Model of Sharing Space

In our work, we tried to design an LMS sharing model of learning online, beginning with the study of the IDM approach, (Model based Engineering) based on four stages of implementation:

- The development of a model without IT preoccupation (CIM: Computer Independent Model);
- Its manual transformation into a model in a particular technological environment (PIM: Platform Independent Model);
- The automatic transformation into a model associated with the target implementation platform (PSM: Platform Specific Model) model must be refined;
- Its implementation in the target platform.

In this section, we will talk about the IT design of our LMS sharing model without using the same approach that we have adopted in previous works, because we have detected the real problems of semantic loss during the transformation of the model.

This led us to develop our model through the outline of the diagram in which we will eventually identify the features of the constituent entities of our model and the class diagrams in which we will specify the different classes' constituents our LMS sharing model.

4.1   Use Case Diagram

The use case diagrams identify the functionality provided by the model (use case); users interact with the system (actors), and the interactions between them. The main objectives of the use case diagrams are:

- Provide high-level view of the model.
- Identify users ("actors") of the model.
- Define the roles of the actors in the model.

Table 1 describes the service function for each actor.

Here are the use case diagram (cf. Fig. 3) of the model representing the external actors who will interact with the system and how they will use it:

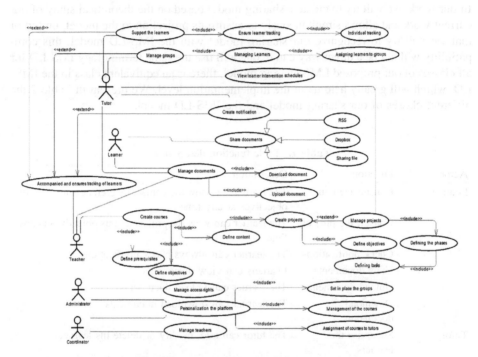

Fig. 3. Use case diagram of sharing space

4.2   Correspondence Between the Terminology of IMS-LD and that of the Sharing Model

In this research, we try to adapt the model IMS-LD with a sharing model supporting LMS. This adaptation will go through three stages, first the development of an LMS sharing model, secondly, the study of correspondence between the developed model and IMS-LD model and their transformation to IMS-LD model that reduced the MDA approach in a transformation that is based on the rules implemented in the ATL language. However, we will talk about the IT design of our LMS sharing model without using the transformation that is based on the rules implemented in the ATL language, because we have detected the same problems as in several works [20, 23, 30–32]. In IMS-LD, we do not have the opportunity to build a course, which consists of several chapters. For this, there is at this level a semantic loss.

Consequently, the majority of classes designed in our LMS sharing model correspond perfectly with the IMS-LD model, which makes possible their transformation to it. The transformation of model is a technique aims to put links between models in order to avoid unnecessary reproductions. In the table II, we have tried to collect all the classes of LMS sharing model and their equivalent at the IMS-LD.

4.3   Class Diagram IMS-LD

In our work, we will try to create a sharing model based on the theoretical study of our current work and allows simultaneously to ensure its projection to the model, on top of that we will try to recognize our sharing model with the IMS-LD model, this compatibility will not be a direct way c to d, one will use the same terminology IMS-LD for all classes of our proposed LMS sharing model, there is an equivalent class in the IMS-LD, which will greatly help us in the implementation level. We present in Table 2 the different classes of our sharing model and the IMS-LD model.

Table 1. Table function description

Actor	Function	Description
Learner	Consult the course	The learner can view the course, its prerequisites, and its objectives at any time
	Consult the project	The learner can view the project and its objectives at any time
	Check notifications	The learner can always check notifications
	View documents	Learners can view documents
	Upload documents	The learner can upload documents
	Download documents	The learner can download documents
Tutor	Supervise the learners	The tutor can add, modify or delete his learners
	Manage groups	The tutor can add, modify or delete groups
	Assign learners to groups	The tutor can assign learners to groups
	Create notifications	The tutor can create notifications
	Set phases	The tutor can set phases of project
	Set tasks	The tutor can set tasks of project

(continued)

Table 1. (*continued*)

Actor	Function	Description
Teacher	Create courses	The teacher can create courses
	Create projects	The teacher can create projects
	Set prerequisites	The teacher can set prerequisites for course
	Set objectives	The teacher can set objectives for course and projects
	Upload documents	The teacher can upload documents
	Download documents	Teachers can download documents
Administrator	Set in place groups	The administrator can set in place the classes' groups
	Assignment of courses to tutors	The administrator can assign courses to tutors according to the specialty of each one
	Management of the courses	The administrator can manage the courses
	Manage access rights	The administrator can manage the access rights of teachers and learners
Coordinator	Manage teachers	The coordinator can add, modify, or delete teachers
	Set in place groups	The coordinator can set in place the classes' groups
	Assignment of courses to tutors	The coordinator can assign courses to tutors according to the specialty of each one
	Management of the courses	The coordinator can manage the courses

Table 2. Correspondence between the terminology of IMS-LD and that of the sharing model

Sharing model	IMS-LD
Activity	Activity
Document, Dropbox, and project	Learning activity
RSS and notification	Support activity
Task, and subtask of project	Activity structure
Prerequisite	Prerequisite
Project phase	Play
Members, and team	Person
Coordinator, teacher, and tutor	Staff
Learner	Learner
Notification	Notification
Project objective	Learning objective
Course	Learning object
Sharing space	Services

In the following, we propose sharing model to achieve the needs of LMS platform, and the needs of teachers and learners. Therefore, we establish the following class diagram (cf. Fig. 4) as a proposal of sharing model for LMS platform.

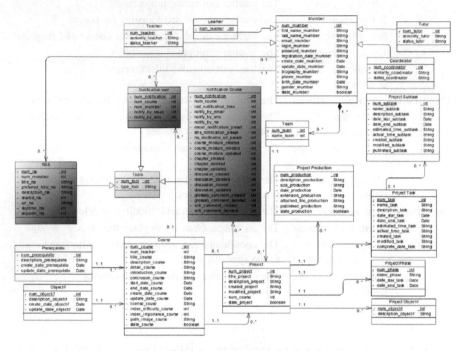

Fig. 4. Class diagram of sharing space

Teachers and tutors can create sharing through the following technological advances:

- File sharing. Use file-sharing tool to gather and distribute course documents. Learners can gather their own files in a personal workspace, and hand in assignments from there. Create group workspaces for private file sharing within workgroups.
- Notification. This notification block implements a solution that periodically notifies learners about new content or activities included in an LMS course. This solution will act as a sentinel that detects whether new contents or new activities have been included in the LMS course, and notifies learners, the teachers and/or the tutors about that.
- RSS. The RSS Plugin offers you the possibility to load multiple feeds into a block. The RSS block allows authorized users to configure multiple RSS feeds per instance.
- Assignment DropBox. Extending the File Storage. Teachers, tutors, or learners can create file folders for collecting assignment submissions, collected from all course members, from group members, or from individuals.

5 Conclusion and Perspectives

In our work, we are on the way to the design and modeling a sharing Model of Learning Management System compatible with IMS-LD. This design is based on active teachings learner-centered, and as an example of the pedagogy, we opted for theory as a basis for teaching this sharing model of LMS, namely the social constructivism. The latter that allows us to reach a teaching object through the implementation of the courses that are divided into tasks performed by students collaboratively or individually. To achieve a goal we need to reach the model validation step, which is one of the tasks to be performed in our future work; also, we seek a framework that will guide us better to start the development part.

References

1. Ouadoud, M., Chkouri, M.Y., Nejjari, A., EL Kadiri, K.E. (2016) Studying and analyzing the evaluation dimensions of E-learning platforms relying on a software engineering approach. Int. J. Emer. Technol. Learn. (iJET) **11**(1), 11–20. 10p (2016). http://dx.doi.org/10.3991/ijet.v11i01.4924
2. Ouadoud, M., Chkouri, M.Y., Nejjari, A., Kadiri, K.E.E.: Studying and comparing the free e-learning platforms. In: 2016 4th IEEE International Colloquium on Information Science and Technology (CiSt), pp. 581–586 (2016). http://dx.doi.org/10.1109/CIST.2016.7804953
3. Ouadoud, M., Chkouri, M.Y., Nejjari, A., El Kadiri, K.E.: Exploring a recommendation system of free E-learning platforms: functional architecture of the system. Int. J. Emer. Technol. Learn. (iJET) **12**(02), 219–226 (2017). https://doi.org/10.3991/ijet.v12i02.6381
4. Lejeune, A.: IMS learning design: Étude d'un langage de modélisation pédagogique, Revue Distances et Savoirs **2**
5. CEN/ISS WS/LT: Learning Technologies Workshop Survey of Educational Modelling Languages (EMLs).Version 1, September 2002
6. Koper, R.: Modeling units of study from a pedagogical perspective, the pedagogical meta-model behind EML (2001)
7. Burgos, D., Arnaud, M., Neuhauser, P., Koper, R.: IMS learning design. Association EPI, Décembre-2005
8. Martinez-Moyano, I.J.: Exploring the dynamics of collaboration in interorganizational settings. In: Creating a Culture of Collaboration: The International Association of Facilitators handbook, 2006
9. Doise W., Mugny, G.: Le développement social de l'intelligence, vol. 1. Interéditions, Paris (1981)
10. Wake, J.D.: Evaluating the organising of a collaborative telelearning scenario from an instructor perspective œ an activity theoretical approach. Ph.D. dissertation in computer science, Thèse de doctorat, Department of Information Science, University of Bergen, December, 2001
11. El Mhouti, A., Nasseh, A., Erradi, M.: Les TIC au service d'un enseignement-apprentissage socioconstructiviste. Association EPI, janv-2013
12. Henri, F., Rabardel, P., Pastré, P. (eds.): Modèles du sujet pour la conception. Toulouse. Int. J. E-learn. Distance Educ. **22**(1), 101–106 (2005). Octares Éditions, 260p, Aug. 2007
13. Sadiq, M., Talbi, M.: Modélisation des unités d'apprentissage sur des plates-formes de formation à distance. Association EPI, Mar-2010

14. Tonye, E.: Modeling a framework for open and distance learning in sub-Saharan African countries. frantice.net, numéro 2, December 2010
15. Chouchane, K.: Modélisation et réalisation d'une approche pour le m-learning, Magister en Informatique Option Système d'Informatique et de Communication (SIC) 2012, Université Hadj Lakhdar—Batna—Algérienne
16. Brunel, S., Girard, P., Lamago, M.: Des plateformes pour enseigner à distance : vers une modélisation générale de leurs fonctions. In: AIP Primeca 2015, La Plagne, France, 2015
17. née Dahmani Farida, B.: Modélisation basée ontologies pour l'apprentissage interactif - Application à l'évaluation des connaissances de l'apprenant. Ph.D. dissertation in computer science, Computer Science Department, Mouloud Mammeri University of Tizi-Ouzou, Algeria, 28 November, 2010
18. Alfanet project "CopperCore V 3.3," Nov-2008: CopperCore is one of the OUNL's contributions to the Alfanet project
19. Bolton: "RELOAD Project: Editor," United Kingdom: The University of Bolton, The University of Strathclyde and JISC, 2005
20. El-Moudden, F., Aammou, S., Khaldi. M.: A tool to generate a collaborative content compatible with IMS-LD. Int. J. Softw. Web Sci. 11(1), December 2014-February 2015, pp. 01–08
21. Chekour, M., Laafou, M., Janati-Idrissi, R.: L'évolution des théories de l'apprentissage à l'ère du numérique. Association EPI, févr-2015
22. Ouadoud, M., Chkouri, M.Y., Nejjari, A.: LeaderTICE: a platforms recommendation system based on a comparative and evaluative study of free E-learning platforms. Int. J. Online Eng. (iJOE) 14(01), 132–161 (2018). https://doi.org/10.3991/ijoe.v14i01.7865
23. El-Moudden, F., Khaldi, M., Aammou, S.: Designing an IMS-LD model for collaborative learning. IJACSA 1(6), 42–48
24. Ouadoud, M., Nejjari, A., Chkouri, M.Y., Kadiri, K.E.E.: Educational modeling of a learning management system. In: 2017 International Conference on Electrical and Information Technologies (ICEIT), 2017, pp. 1–6. http://dx.doi.org/10.1109/EITech.2017. 8255247
25. Paquette, G.L.: 'Ingénierie Pédagogique: Pour Construire l'Apprentissage en Réseau,' PUQ (2002)
26. Marquet, P.: Lorsque le développement des TIC et l'évolution des théories de l'apprentissage se croisent. Savoirs 9, 105–121 (2005). https://doi.org/10.3917/savo.009.0105
27. Vygotsky, L.S.: Mind in Society: The Development of Higher Psychological Processes, Revised ed. edition. Harvard University Press, Cambridge (1978)
28. Ouadoud, M., Chkouri, M.Y., Nejjari, A.: Learning management system and the underlying learning theories: towards a new modeling of an LMS. Int. J. Inf. Sci. Technol. (iJIST) 2(1), 25–33 (2018)
29. Ouadoud, M., Nejjari, A., Chkouri, M.Y., El-Kadiri, K.E.: Learning management system and the underlying learning theories. In: Innovations in Smart Cities and Applications, 2017, pp. 732–744. https://doi.org/10.1007/978-3-319-74500-8_67
30. Ouadoud, M., Rida, N., Chkouri M.Y.: Designing an IMS-LD model for collaboration space of learning management system. In: 2018 IEEE 5th International Congress on Information Science and Technology (CiSt), 2018, pp. 380–385. https://doi.org/10.1109/CIST.2018. 8596588
31. Ouadoud, M., Chkouri, M.Y.: Generate a meta-model content for disciplinary information space of learning management system compatible with IMS-LD. In: Proceedings of the 3rd International Conference on Smart City Applications, New York, NY, USA, 2018, p. 39:1–39:8. http://dx.doi.org/10.1145/3286606.3286816

32. Ouadoud, M., Chafiq, T., Chkouri, M.Y.: Designing an IMS-LD model for disciplinary information space of learning management system. In: Proceedings of the 3rd International Conference on Smart City Applications, New York, NY, USA, 2018, p. 40:1–40:9. http://dx.doi.org/10.1145/3286606.3286817

Generate a Meta-Model Content for Sharing Space of Learning Management System Compatible with IMS-LD

Mohammed Ouadoud[1(✉)] and Mohamed Yassin Chkouri[2]

[1] LIROSA Lab, UAE, Faculty of Sciences, Tetouan, Morocco
mohammed.ouadoud@gmail.com
[2] SIGL Lab, UAE, National School of Applied Sciences, Tetouan, Morocco

Abstract. The context of this work is that of designing an IMS-LD meta-model of the sharing space of a Learning Management System (LMS). Our approach is to first think about the conditions for creating a real sharing space for LMS between learners, and designing the IT environment that supports this space. In this paper, we try to adapt the IMS-LD model with a sharing meta-model for LMS based on the social constructivism. This adaptation will go through three stages. Firstly, the development of a sharing meta-model of LMS. Secondly, the study of correspondence between the developed meta-model and IMS-LD model. Finally, their transformation to IMS-LD meta-model.

Keywords: LMS · Sharing space · IMS-LD · E-learning platform
Designing an IMS-LD meta-model

1 Introduction

In the 20th century, there was an international movement in favor of e-learning integration in higher education. This movement has been operationalized due to the variety of the educational offer by universities, which most have opted to diversify knowledge dissemination platforms to meet the needs of their target public.

E-learning is promoted through educational platforms: integrated systems offer a wide range of activities in the learning process. Teachers use the platforms to monitor or evaluate the work of students. They use learning management systems (LMS) to create courses, tests, etc. However, the LMS e-learning platforms do not offer personalized services and therefore do not take into account the aspects of personalization such as the level of knowledge, interest, motivation and goals of learners. They access the same resource set in the same way.

In fact, we present an easy sharing meta-model of LMS to create and administer the educational content online. This tool allows the generation and editing structures of websites through database rather than pedagogical models, with a variety of choices that ensures better adaptation to the teaching of the course and learning style. Therefore, it is necessary to find a method to model all LMS types. In order to modeling the sharing space of LMS we have based ourselves upon the IMS-LD specification focusing on learning theory that was judged the most important and relevant to our

M. Ben Ahmed et al. (Eds.): SCA 2018, LNITI, pp. 348–363, 2019.
https://doi.org/10.1007/978-3-030-11196-0_31

modeling, namely the traditional pedagogy. Then, this learning theory which have inspired for a long time the design of computer applications are combined and put into perspective with several emergent pedagogical functionalities to build an original modeling for our sharing space of LMS. Reveals that this proposed modeling that is presented to readers here looks for ways to leverage technology for learning by considering users as being human actors and not human factors [1].

The IMS-LD specification or instructional design engineering uses pedagogical concepts, allowing to model learning units. IMS-LD takes into account a wide variety of teaching models it is there its flexibility. A course plan extract of a general or specific database can be modeled with IMS-LD, through the description of the different roles, activities, environments, methods[1], properties, conditions and notifications. It is used to transform the course plans into formal learning units (UOL) that can be performed with an IMS-LD editor based on an engine such as Coppercore [2]. These executable units can be designed from the beginning using an editor such as Reload [3].

During the last decade, the LMS e-learning platforms have evolved considerably. However, several modeling of LMS platforms has been developed previously [4–8], but they have been abandoned because platforms' life cycle is changing apace. Therefore, we have conducted a comparative and analytical study on free e-learning platforms based on an approach of evaluating the e-learning platforms quality [9–12]. Based on these various research works, which seemed to us incomplete, we proposed a modeling portrait of a designing an IMS-LD sharing meta-model for LMS platform. This latter is anthropocentric and relies on a learning conception that is located at the intersection of the most used learning theory. Indeed, the idea is to orient the design work research towards a great and optimal compatibility between the services offered by e-learning platforms and the needs of all users, particularly learners, for better optimization of online learning.

To concretize our modeling work, we present in the section "Theoretical approach, concepts and related work" the different theories of learning used for the modeling of this space, namely the traditional pedagogy and the social constructivism. We also present the e-learning specification Learning Management Systems—Learning Design (IMS-LD) and related work. Next, in "transformation rules" section, we present the ATL transformation language to perform the transformation between the developed meta-model and IMS-LD model. We also determine in this section, the reasons why, we did not use the ATL language for transformation of models. Finally, we present in the section "Model-driven engineering", the modeling Driven Architecture (MDA). Thus, we try to adapt the proposed meta-model of the sharing space to the equivalent IMS-LD model.

2 LMS and Activity Spaces

2.1 Learning Management System

An LMS (Learning Management System) or e-learning platform is a software including a range of services that assist teachers with the management of their courses. Moreover, as defined by the OVAREP, "the LMS e-learning platform is a computing device that

---

[1] The word "method" used by IMS-LD means the unfolding of the scenario.

groups several tools and ensures the educational lines. Across dedicated platforms to the ODL (open and distance learning), all conduits are preserved and expanded for the learner, tutor, coordinator and administrator within the e-learning platform" [13]. It offers many services allowing the management of content, particularly by creating, importing and exporting learning objects. The set of the available tools in the LMS represent all these services that help in managing the teaching process and the interaction between users such as the access control services, synchronous and asynchronous tools of communication and user administration services.

More precisely these services are linked to the following variety of functionalities:

- The management of pedagogical content (creating, importing and exporting learning objects),
- The creation of individual's personal paths in the training modules,
- The availability of sharing tools,
- The distribution of communication tools,
- The student registration and the management of their files (training tracking and results),
- The distribution of online courses and many other pedagogical resources.

Figure 1 illustrates the general principle of the operation of an e-learning platform LMS by presenting the key features associated with the main actors: learners, teachers, tutors, coordinators, and administrators. The learner can consult and/or download the resources made at his/her disposal by the teacher, he/she can create his/her learning activities while following his/her progress in training. The teacher, who is responsible of one or more modules, can create and manage the educational content he/she wishes to broadcast via the platform. He/she can also build tools for monitoring learners' activities. The tutor accompanies and monitors each learner by providing the tools of communication and collaboration. Concerning the coordinator, he/she ensures the management of the overall system. Finally, the administrator is responsible for the customization of the platform having the rights of the administration deriving from it (system installation, maintenance, access management…).

2.2 LMS and Activity Spaces

The learning management system consists of different activity spaces for activities of teaching and learning [13–15]. Each model represents a space, in these spaces, both teachers and learners can have:

- Disciplinary information space. In disciplinary information space, teachers or tutors can export content from the LMS as IMS/SCORM conformant Content Packages that can be viewed offline, or imported into the LMS. Entire courses or individual course units can be packaged for viewing or redistribution.
- Communication space. Communication is the act of conveying intended meanings from one entity or group to another using mutually understood signs and semiotic rules. Learners can communicate with others through their Inbox using LMS's private mail, through the discussion forums or the chat rooms.

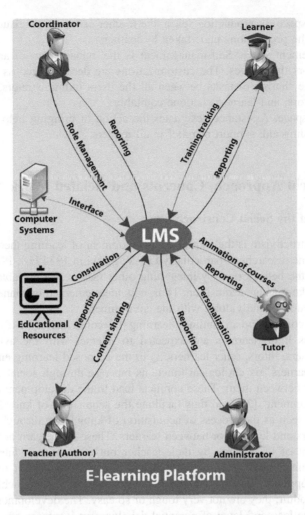

Fig. 1. The general architecture of an LMS

- Collaboration space. Collaboration is the process of two or more people or organizations working together to realize or achieve something successfully [16].
  Collaboration requires leadership, although the form of leadership can be social
  within a decentralized and egalitarian group [17]. Teams that work collaboratively
  can obtain greater resources, recognition and reward when facing competition for
  finite resources.
- Sharing space. Sharing space or exchange space, it allows discussing, to define and
  to follow the implementation of collaborative projects of one or several courses.
- Evaluation space. Evaluation is a process where learners are assessed to determine
  their suitability to take a course. The learners' aptitudes are determined by a variety
  of techniques including interviews, group exercises, presentations, examinations
  and psychometric testing.

- Production space. In production space, the teacher or tutor can create notifications, and assess the productions undertaken by learners.
- Self-management space. Self-management is the management of an LMS by the administrators themselves. The customizations are decentralized as much as possible and the sharing of tasks between all the users (administrators, coordinators, teachers, tutors, and learners) is done equitably.
- Assistance space. Assistance designates the action of bringing help or relief. It is about providing aid, support or relief in all matters.

3   Theoretical Approach, Concepts and Related Work

3.1   LMS and the Social Constructivism

The social constructivism is the fruit of the development of learning theories under the influence of some researchers, particularly Lev Vygotski in 1934 [18, 19], who wanted to depart from the behaviorism by integrating other factors that are able to positively influencing the knowledge acquisition. Thus, new ideas emerged in connection with the possible interaction of individuals with the environment.

The social constructivism outlines learning by construction in a community of learners. In this light, learners are expected to interact with the available human resources (teachers, tutors, other learners…) in the proposed learning environment. In this way, the learners' psychological functions increase through social cognitive conflicts that occur between them. These conflicts lead to the development of the zone of proximal development[2] [20] and thus facilitate the acquisition of knowledge.

Learning is seen as the process of acquisition of knowledge through the exchange between teachers and learners or between learners. These latter learn not only through the transmission of knowledge by their teacher but also through interactions [21]. According to this model, learning is a matter of the development of the zone of proximal development: this zone includes the tasks that learners can achieve under the guidance of an adult; they are not very tough or so easy. The development of this zone is a sign that the learners' level of potential development increases efficiently [22].

The teacher's role is to define precisely this zone in order to design suitable exercises for learners. Furthermore, designing collaborative tasks, which involve discussions and exchange (socio-cognitive conflicts) between learners is so important in this model. Errors are considered as a point of support for the construction of new knowledge.

Based on the social constructivism approach (cf. Fig. 2), the design of the LMS was oriented towards integrating online communication and collaboration tools. In practice, a wide range of platforms, particularly the social constructivist ones, proposes a set of

---

[2] "The distance between actual development level as determined by independent problem solving and the level of potential development as determined through problem solving under adult guidance or in collaboration with more capable peers (Vygotsky, 1978, p. 86)".

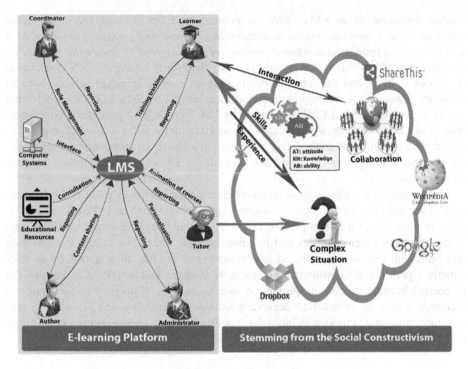

Fig. 2. LMS and underlying social constructivism model

tools, which allow sharing, exchanging and interacting in synchronous and asynchronous mode such as blogs, wikis, forums...

In summary, the ideas of social constructivist authors have highlighted the social nature of learning. Other authors have taken one-step further by emphasizing the distribution of intelligence between individuals and the environment. Furthermore, considering that learning occur in a social context is no longer enough to ensure deep learning. Indeed, working in groups can affect negatively the quality of learning if these following conditions are not taken into consideration: Learning styles, the way groups are formed, interaction modality, and the characteristics of tasks.

3.2 Instructional Management Systems—Learning Design

IMS-LD was published in 2003 by the IMS/GLC. (Instructional Management Systems Global Learning Consortium: Consortium for global learning management systems with training, the original name when IMS was started in 1997 Instructional Management Systems project) [23, 24]. Reminds us of its origins: the source (EML[3]) of the proposed language was assessed by the European Committee for Standardization (CEN) in a comparative study of different SRMS [24, 25], as best suited to satisfy the

---

[3] **EML**: Educational Modelling Language.

criteria definition of an EML. EML is defined by CEN/ ISS as "*an information aggregation and semantic model describing the content and processes involved in a unit of learning from an educational perspective and in order to ensure the reusability and interoperability.*" [24, 26] In this context, the North American IMS consortium undertook a study and provided a specification of such a language, giving birth in February 2003, the Learning Design specification V1.0 (IMS-LD). She adds that proposal, largely inspired EML developed by [24, 27] (OUNL) provides a conceptual framework for modeling a Learning Unit and claims to offer a good compromise between on the one hand to the generic implement a variety of instructional approaches and secondly, the power of expression that allows an accurate description of each learning unit.

This specification allows us to represent and encode learning structures for learners both alone and in groups, compiled by roles, such as "learners" and "Team"(cf. Fig. 3) [24]. We can model a lesson plan in IMS-LD, defining roles, learning activities, services and many other elements and building learning units. The syllabus is modeled and built with resources assembled in a compressed Zip file, then started by an executable ("player"). It coordinates the teachers, students and activities as long as the respective learning process progresses. A user takes a "role" to play and execute the activities related to in order to achieve a satisfactory learning unit. In all, the unit structure, roles and activities to build the learning scenario to be executed in a system compatible with IMS-LD.

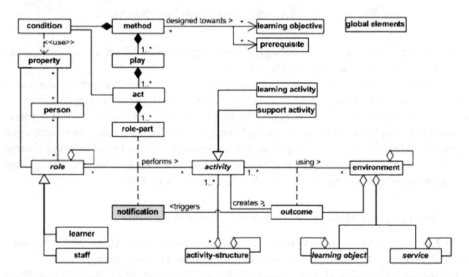

Fig. 3. The conceptual model of Instructional Management Systems—Learning Design [24]

IMS-LD does not impose a particular pedagogical model, but can be used with a large number of scenarios and pedagogic models, demonstrating its flexibility. Therefore, IMS-LD is often called a meta-pedagogic model. Previous eLearning initiatives claim to be pedagogically neutral, IMS-LS is not intended to pedagogical neutrality, but seeks to raise awareness of eLearning on the need for a flexible approach.

IMS-LD has been developed for eLearning and virtual classes, but a course face to face can be done and integrated into a structure created by this specification, as an activity of learning or support activity. If the ultimate goal to create rich learning units, with support to achieve the learning objectives by providing the best possible experience, face-to-face meetings, and any other learning resource are permitted such as video conferencing, collaborative table or any field action research.

IMS-LD uses the theatrical metaphor, which implies the existence of roles, resources and learning scenario itself: one room is divided into one or more acts and conducted by several actors who can take on different roles at different times. Each role is to carry out a number of activities to complete the learning process. In addition, all roles must be synchronized at the end of each act before processing the next act.

3.3 Related Works

Many works emphasized the contribution of learning theories in the design and development of learning systems. The direct application of each of these theories allows particularly providing supporting methods for the design and development of LMS. The learning theory that we presented in this paper for an original conceptual model of sharing has allowed us to design a new space of an LMS whose benefit resides essentially in the richness of the proposed functionalities in a way that fits the needs of all its final users: teachers and learners.

Furthermore, much work has been done in the field of collaborative learning and IMS-LD specification. For example, Dyke [28] explains how models' socio-cognitive interaction are related to the properties of collaborative tools. Ferraris [29] expresses collaborative learning scenarios by teachers animating virtual classrooms to promote the re-use and share teaching practices. He proposes an approach led by the models in accordance with the recommendations of the Model Driven Architecture OMG. He presents a meta-model based on IMS-LD enhanced by the concepts of participation model to capture the richness of the interactions inherent in collaborative activities. Moreover, El-Moudden [26, 30] proposes a designing an IMS-LD model for collaborative learning. His approach is to first think about the conditions for creating a real collective activity between learners, and designing the IT environment that supports these activities.

On the other hand, other research aimed at proposing the modeling of new units, approach, architecture, or adaptive, flexible and interactive e-learning devices. For example, Sadiq proposed the modeling of learning units on eLearning platforms, which relies on the application of the standard IMS-LD in the production of adaptive learning units [4].

Based on these various research works, which seemed to us incomplete, we proposed a modeling portrait of a designing an IMS-LD sharing meta-model for LMS platform. For us, this is not the same case and the same vision as our modelization is more general, it aims on one hand to create an LMS from which teachers can animate virtual groups for the re-use and sharing of teaching practices and on the other hand, the re-use of the content created in other frameworks. Indeed, the idea is to orient the design work research towards a great and optimal compatibility between the services offered by e-learning platforms (LMS) and the needs of all users, particularly learners, for better optimization of online learning.

4 Transformation Rules

4.1 Atlas Language Transformation (ATL)

In their operational ATL, Canals et al. use. [30, 31] State that to deal with the transformation of models; it is difficult and cumbersome to use object languages since we spend so much effort to the development of transformation definitions of Framework for the set work. The use of XSLT as a language if it is more direct and adapted by rest against difficult to maintain [30, 31]. We follow their choice by focusing on the implementation of approaches centered on the MDA (Model Driven Architecture), MDE (Model Driven Engineering) and QVT (Queries View Transformation) tools. Query/ View/ Transformation (QVT) [30, 32] is a standard defined by the OMG. This is a standardized language to express model transformations. QVT is not advanced sufficiently now in its definition for Queries and View aspects. Against transformation by the appearance expressed by MDA approach has resulted in various experiments (e.g. Triskell, ATL...) in both academic and commercial level. To determine the transformation, it is necessary to have tools of transformations. These are based on languages transformations must respect the QVT standard [30, 32] proposed by the OMG [30, 33]. There is an offer of free tools (ATL, MTF, MTL, QVTP, etc.) and commercial (e.g. MIA). We chose ATL (Atlas Transformation Language) from the provision of free tools, to the extent that only ATL has a spirit consistent with OMG/ MDA/ MOF/ QVT [30, 31].

4.2 ATL Description

Atlas Transformation Language (ATL) has been designed to perform transformations within the MDA Framework proposed by the OMG [17, 22, 30]. The ATL language is mainly based on the fact that the models are first-class entities. Indeed, the transformations are considered models of transformation. Since the transformations are considered themselves as models, we can apply their transformations. This possibility of ATL is considered an important point. Indeed, it provides the means to achieve higher order transformations (HOT Higher- Order Processing) [17, 30]. A higher-order transformation is a transformation, including source and target models that are themselves transformations. As ATL is among the languages model transformation respecting the QVT [32] standard proposed by OMG [33], we describe its structure in relation to this standard (QVT).

The study of the abstract syntax of the ATL language is to study two features provided by this language more than rules changes. The first feature, navigation, allows studying the possibility of navigation between meta-models sources and targets. The second feature, Operations, used to describe the ability to define operations on model elements. Finally, the study of the transformation rules is used to describe these types of rules, how they are called and the type of results they return.

- Navigation [22, 30, 33] this feature is offered by ATL language (Object Constraint Language). Navigation is allowed only if the model elements are fully initialized. The elements of the target model cannot be definitively initialized at the end of the execution of the transformation. Therefore, the navigation in the ATL can only be made between elements of the model (or meta-model) source and model (or meta-model) target.
- Operations: [22, 30] this feature ATL is also provided by the OCL (Object Constraint Language). In OCL, operations can be defined on the elements of the model. ATL takes this opportunity to OCL to allow defining operations on elements of the source model and the transformation model [30, 33].

The transformation rules: there are several types of transformation rules based on how they are called and what kind of results they return (cf. Fig. 4).

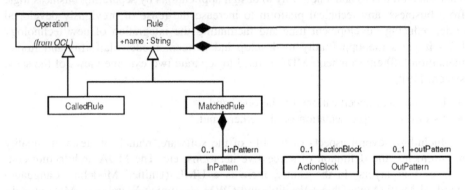

Fig. 4. ATL Transformation rules [30]

CalledRule [22] rule explicitly called by its name and by setting its parameters.

MatchedRule [22] rule executed when a guy (*InPattern*) scheme is recognized in the source model.

The result of a rule may be a set of predefined models (*OutPattern*) or a block of mandatory (*ActionBlock*). If the rule is MatchedRule type and if its result is a set of elements of the target model (*OutPattern*), it was named declarative. If it is CalledRule type whose result is a block of statements, it is then called procedure. Combinations of rules (declarative and imperative) are called hybrid rules [33].

4.3 Synthesis

In this research, we try to adapt the model IMS-LD with a sharing meta-model supporting LMS. This adaptation will go through three stages, first the development of a sharing meta-model; secondly, the study of correspondence between the developed meta-model and IMS-LD model and their transformation to IMS-LD meta-model that reduced the MDA approach in a transformation that is based on the rules implemented in the ATL language. However, we will talk about the IT design of our LMS sharing meta-model without using the transformation that is based on the rules implemented in the ATL language, because we have detected the same problems as in several works [26, 30, 34–36]. In IMS-LD, we do not have the opportunity to build a course, which consists of several chapters. For this, there is at this level a semantic loss.

5   Model-Driven Engineering

5.1   Modeling Driven Architecture (MDA)

In November 2000, the OMG in the field of software engineering consortium of over 1000 companies, has initiated the process MDA [OMG MDA], the concepts-oriented models rather than object-oriented. The Model Driven Architecture MDA [OMG MDA] offers the power of abstraction, refinement and different views of the models. This standard has to add a new way to design applications by separating business logic from business, any technical platform to increase the reuse of previously developed code, reducing development time and facilitating the integration of new technology [37]. It gives the opportunity to develop independent model platforms and implementation [38] environment. MDA is used to separate two extreme views of the same system [19]:

- Its functional specifications on the one hand.
- Its physical implementation on the other hand.

Including several aspects of the life of the software, namely its tests, its quality requirements, the definition of successive iterations, etc. The MDA architecture consists of four layers. In the center, there is a UML (Unified Modeling Language) standard MOF (Meta-Object Facility) and CWM (Common Warehouse Meta-model). The second layer contains the XMI (XML Metadata Interchange) standard for dialogue between the middleware (Java, CORBA,.NET, and Web services). The next layer refers to the services to manage events, transaction security, and directories. The last layer offers specific frameworks in scope (Telecommunications, medicine, electronic commerce, finance, etc.) A designer to create his own application can use UML as it can use other languages. So, according to this architecture independent technical context, MDA proposes to structure the front needs to engage in a transformation of this functional modeling technical modeling while testing each product model [19]. This application model is to be created independently of the target implementation (hardware or software). This allows greater reuse of patterns. MDA is considered an approach with the ambition to offer the widest possible view of the life cycle of the software, not content with only its production. Moreover, this is intended overview described in a unified syntax. One of the assumptions underlying the MDA is that the operationalization of an abstract model is not a trivial problem. One of the benefits of MDA is to solve this problem [39]. MDA proposes to design an application through software chain is divided into four phases with the aim of flexible implementation, integration, maintenance and test:

- The development of a computer model without concern (CIM: Computer Independent Model).
- The manual transformation into a model in a particular technological context (PIM: Platform Independent Model).
- The automatic transformation into a pattern associated with the target implementation of the platform (PSM: Platform Specific Model) model to be refined,
- Its implementation in the target platform.

5.2 Correspondence Between the Terminology of IMS-LD and that of the Sharing Meta-Model

The majority of classes designed in our sharing meta-model for LMS correspond perfectly with the IMS-LD model, which makes possible their transformations to it. The transformation of models is a technique aims to put links between models in order to avoid unnecessary reproductions.

In the Table 1, we have tried to collect all the sharing meta-model classes for LMS and their equivalent at the IMS-LD:

Table 1. Correspondence between the terminology of IMS-LD and that of the sharing meta-model

Sharing meta-model	IMS-LD
Activity	Activity
Sharing document, Dropbox, and project	Learning activity
RSS and notification	Support activity
Task, and subtask of project	Activity structure
Prerequisite	Prerequisite
Project phase	Play
Members, and team	Person
Coordinator, teacher, and tutor	Staff
Learner	Learner
Notification	Notification
Project objective	Learning objective
Course	Learning object

6 Designing a Sharing Meta-Model Based on IMS-LD

In our research, we propose a meta-model for a sharing space designed to achieve the needs of LMS platform, and the needs of teachers and learners. Therefore, we establish the following diagram (cf. Fig. 5) as a first proposal of a sharing meta-model for LMS platform.

Indeed, we will talk about the IT design of our sharing meta-model for LMS. This led us to develop our meta-model in which we will eventually identify the features of

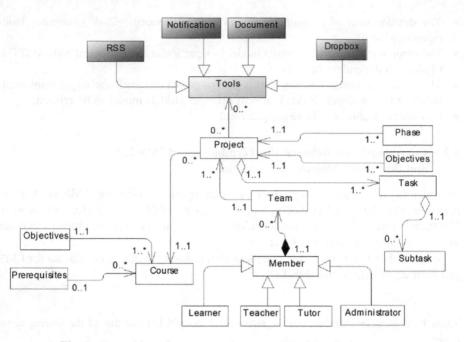

Fig. 5. Proposition of a conceptual meta-model of the sharing space

the constituent entities of our meta-model in which we specify the different classes of our modeling.

Teachers and tutors can create sharing through the following technological advances:

- File sharing. Use file-sharing tool to gather and distribute course documents. Learners can gather their own files in a personal workspace, and hand in assignments from there. Create group workspaces for private file sharing within workgroups.
- Notification. This notification block implements a solution that periodically notifies learners about new content or activities included in an LMS course. This solution will act as a sentinel that detects whether new contents or new activities have been included in the LMS course, and notifies learners, the teachers and/or the tutors about that.
- RSS. The RSS Plugin offers you the possibility to load multiple feeds into a block. The RSS block allows authorized users to configure multiple RSS feeds per instance.
- Assignment DropBox. Extending the File Storage. Teachers, tutors, or learners can create file folders for collecting assignment submissions, collected from all course members, from group members, or from individuals.

7 Conclusion and Perspectives

In our work, we are on the way to the design and modeling a meta-model content for sharing space of Learning Management System compatible with IMS-LD. This design is based on active teaching learner-centered, and as an example of the pedagogy, we opted for theory as a basis for teaching this sharing space of LMS, namely the social constructivism. The latter, that allows us to reach a teaching object through the implementation of the courses that are divided into tasks performed by students collaboratively or individually. To achieve a goal, we need to reach the model validation step, which is one of the tasks to be performed in our future work; also, we seek a framework that will guide us better to start the development part.

References

1. Henri, F.: In: Rabardel, P., Pastré, P. (eds.) (2005) Modèles du sujet pour la conception. Toulouse. Int. J. ELearn. Dist. Educ. **22**(1), pp. 101–106 (2007)
2. Alfanet project: CopperCore V 3.3. Nov-2008. CopperCore is one of the OUNL's contributions to the Alfanet project
3. Bolton: RELOAD Project: Editor. United Kingdom: The University of Bolton, The University of Strathclyde and JISC (2005)
4. Sadiq, M., Talbi, M.: Modélisation des unités d'apprentissage sur des plates-formes de formation à distance. Association EPI, Mar (2010)
5. Tonye, E.: Modeling a framework for open and distance learning in sub-Saharan African countries. frantice.net, numéro 2, Dec (2010)
6. Chouchane, K.: Modélisation et réalisation d'une approche pour le m-learning. Magister en Informatique Option Système d'Informatique et de Communication (SIC), Université Hadj Lakhdar, Batna, Algérienne (2012)
7. Brunel, S., Girard, P., Lamago, M.: Des plateformes pour enseigner à distance : vers une modélisation générale de leurs fonctions. In: AIP Primeca 2015, La Plagne, France (2015)
8. née Dahmani Farida, B.: Modélisation basée ontologies pour l'apprentissage interactif - Application à l'évaluation des connaissances de l'apprenant, Ph.D. dissertation in computer science, Computer Science Department, Mouloud Mammeri University of Tizi-Ouzou, Algeria, 28 Nov (2010)
9. Ouadoud, M., Chkouri, M.Y., Nejjari, A., El Kadiri, KE.: Studying and analyzing the evaluation dimensions of elearning platforms relying on a software engineering approach. Int. J. Emerg. Technol. Learn. (iJET) **11**(1), 11–20 (2016). 10p. http://dx.doi.org/10.3991/ijet.v11i01.4924
10. Ouadoud, M., Chkouri, M.Y., Nejjari, A., Kadiri, K.E.E.: Studying and comparing the free eLearning platforms. In: 2016 4th IEEE International Colloquium on Information Science and Technology (CiSt), pp. 581–586 (2016). http://dx.doi.org/10.1109/CIST.2016.7804953
11. Ouadoud, M., Chkouri, M.Y., Nejjari, A., El Kadiri, K.E.: Exploring a recommendation system of free elearning platforms: functional architecture of the system. Int. J. Emerg. Technol. Learn. (iJET) **12**(02), 219–226 (2017). https://doi.org/10.3991/ijet.v12i02.6381
12. Ouadoud, M., Chkouri, M.Y., Nejjari, A.: LeaderTICE: a platforms recommendation system based on a comparative and evaluative study of free e-learning platforms. Int. J. Online Eng. (iJOE) **14**(01), 132–161 (2018). https://doi.org/10.3991/ijoe.v14i01.7865

13. Ouadoud, M., Nejjari, A., Chkouri, M.Y., Kadiri, K.E.E.: Educational modeling of a learning management system. In: 2017 International Conference on Electrical and Information Technologies (ICEIT), pp. 1–6. (2017). http://dx.doi.org/10.1109/EITech. 2017.8255247

14. Ouadoud, M., Nejjari, A., Chkouri, M.Y., El-Kadiri, K.E.: Learning management system and the underlying learning theories. In: Innovations in Smart Cities and Applications, pp. 732–744. (2017). https://doi.org/10.1007/978-3-319-74500-8_67

15. Ouadoud, M., Chkouri, M.Y., Nejjari, A.: Learning management system and the underlying learning theories: towards a new modeling of an LMS. Int. J. Inf. Sci. Technol. (iJIST) 2(1), 25–33 (2018)

16. WAKE, Jo Dugstad.: Evaluating the Organising of a Collaborative Telelearning Scenario from an Instructor Perspective œ an Activity Theoretical Approach. Ph.D. dissertation in computer science, Dept. of Information Science, University of Bergen, Dec (2001). Thèse de doctorat

17. Combemale, B., Rougemaille, S.: ATL—Atlas Transformation Language. Master 2 Recherche SLCP, module rtm edition (2005)

18. Doise, W., Mugny, G.: Le développement social de l'intelligence (Vol. 1). Paris: Interéditions (1981)

19. Clave, A.: D'UML à MDA en passant par les métas modèles. Technical report, La Lettre d'ADELI n 56, (2004)

20. Vygotsky, L.S.: Mind in Society: The Development of Higher Psychological Processes, Revised ed. edition, Harvard University Press, Cambridge (1978)

21. Spence, M.U.: Graphic Design: Collaborative Processes. A course on collaboration, theory and practice Art 325: Collaborative Processes. Fairbanks Hall, Oregon State University, Corvallis, Oregon (2005)

22. ATLAS group LINA and INRIA Nantes: Atlas transformation language. ATL user manual —version 0.7. Technical report, INRIA University of Nantes, Feb (2006)

23. Lejeune, A.: IMS Learning Design: Étude d'un langage de modélisation pédagogique, Revue Distances et Savoirs, vol. 2

24. Burgos, D., Arnaud, M., Neuhauser, P., Koper, R.: IMS Learning Design. Association EPI, Dec (2005)

25. CEN/ISS WS/LT: Learning Technologies Workshop "Survey of Educational Modelling Languages (EMLs)". Version 1, Sep (2002)

26. El-Moudden, F., Khaldi, M., Aammou, S.: Designing an IMS-LD model for collaborative learning. IJACSA. 1(6), 42–48 (2015)

27. Koper, R.: Modeling Units of Study from a Pedagogical Perspective, the pedagogical meta-model behind EML (2001)

28. Dyke, G., Lund, K.: Implications d'un modèle de coopération pour la conception d'outils collaboratifs. Dec (2006)

29. Ferraris, C., Lejeune, A., Vignollet, L., Jean-Pierre D.: Modélisation de scénarios pédagogiques collaboratifs (2005)

30. El-Moudden, F., Aammou, S., Khaldi, M.: A Tool to Generate a Collaborative Content Compatible with IMS-LD. Int. J Softw. Web Sci. 11(1), 01–08 Dec 2014-Feb 2015

31. Canals, A., Le Camus, C., Feau, M., Jolly, G., Bonnafous, V., Bazavan, P.: Une utilisation opérationnelle d'ATL : L'intégration de la transformation de modèles dans le projet TOPCASED. In: Génie logiciel, pp. 21–26. (2005)

32. OMG/RFP/QVT MOF 2.0 Query/Views/Transformations RFP: OMG, Object Management Group, 28 Oct (2002)

33. Bézivin, J., Dupé, G., Jouault, F., Pitette, G., Eddine Rougui, J.: First experiments with the ATL model transformation language: transforming xslt into xquery. In: OOPSLA 2003 Workshop, Anaheim, California (2003)
34. Ouadoud, M., Chafiq, T., Chkouri, MY.: Designing an IMS-LD model for collaboration space of learning management system. In: 2018 IEEE 5th International Congress on Information Science and Technology (CiSt), pp. 380–385 (2018)
35. Ouadoud, M., Chkouri, MY.: Generate a meta-model content for disciplinary information space of learning management system compatible with IMS-LD. In: Proceedings of the 3rd International Conference on Smart City Applications, New York, NY, USA, p. 39:1–39:8 (2018). http://dx.doi.org/10.1145/3286606.3286816
36. Ouadoud, M., Chafiq, T., Chkouri, MY.: Designing an IMS-LD model for collaboration space of learning management system. In: Proceedings of the 3rd International Conference on Smart City Applications, New York, NY, USA, p. 40:1–40:9 (2018). http://dx.doi.org/10.1145/3286606.3286817
37. Boulet, P., Dekeyser, J.L, Dumoulin, C., Marquet, P.: Mda for soc embeddeb systems design, intensive signal processing experiment. In: SIVOESMDA workshop at UML 2003, San Francisco, Oct (2003)
38. Thi-Lan-anh, D., Olivier, G., Houari, S.: Gestion de modèles: définitions, besoins et revue de littérature. In: Premières Journées sur l'Ingénierie Dirigée par les Modèles, pp. 1–15, Paris, France, 30 Juin- 1 Juillet (2005)
39. Caron, P.A., Hoogstoel, F., Le Pallec, X., Warin, B.: Construire des dispositifs sur la plateforme moodle - application de l'ingénierie bricoles. In: MoodleMoot-2007, Castres, France, 14–15 Juin (2007)

Individualized Follow-up of the Learner Based on the K-Nearest Neighbors (K-NN) Method Embedded in the Retrieval Step of Case Based Reasoning Approach (CBR)

Nihad El Ghouch[1(✉)], El Mokhtar En-Naimi[1], Abdelhamid Zouhair[2], and Mohammed Al Achhab[3]

[1] LIST Laboratory, the Faculty of Sciences and Technologies, UAE, Tangier, Morocco
nihad_elghouch@hotmail.fr, ennaimi@gmail.com
[2] The National School of Applied Sciences, UAE, Al-Hoceima, Morocco
zouhair07@gmail.com
[3] The National School of Applied Sciences, UAE, Tetouan, Morocco
alachhab@gmail.com

Abstract. Learner follow-up in adaptive learning systems requires real-time decision support approaches, using algorithms to predict learner behavior based on the experiences of other learners (learners already classified in groups). We propose an adaptive learning system architecture using the Felder-Silverman learning style model to detect the initial learning profile for each learner in order to provide a learning path based on his profile and the Incremental Dynamic Case Based Reasoning approach based on the exploitation of learning traces in order to follow and to control the behavior of the learner in an automatic and real-time way through the search for similar past experiences using the K-Nearest Neighbors algorithm.

Keywords: Adaptive learning system · Learning style · Learning path
Felder and silverman learning style model (FSLSM) · Incremental dynamic case based reasoning (IDCBR) · Supervised learning machine · K-Nearest neighbors method (K-NN)

1 Adaptive Learning System

The adaptive learning system is an online learning environment that promotes individual learning, providing personalized learning based on the needs and preferences of learners. These systems take into account the profile of the learner (his knowledge, his preferences, his aptitudes …) in the construction of a unique and adapted learning path.

The learning style is considered one of the main individual traits that play a key role in learning. It presents the preferences related to the perception modality, the treatment and organization of information, reasoning, social aspects, etc.

© Springer Nature Switzerland AG 2019
M. Ben Ahmed et al. (Eds.): SCA 2018, LNITI, pp. 364–378, 2019.
https://doi.org/10.1007/978-3-030-11196-0_32

There are several learning styles models that are grouped into three categories [1]:

- Learning style models that focus on preferences for teaching and learning conditions;
- Learning style models that focus on how the learner processes information in terms of privileged means;
- Learning style models that deal with the learner's personality.

There are several adaptive learning systems that are based on learning styles, Table 1 shows some systems:

Table 1. Adaptive learning systems using learning styles

Adaptive learning system	Learning style
CS383 [2]	Felder et Silverman model
TANGOW [3]	Felder et Silverman model
Heritage Alive Learning System [4]	Felder et Silverman model
KiWeaver [5]	Dunn et Dunn model
INSPIRE [6]	Honey et Mumford model

Most adaptive learning systems use methods and techniques that focus primarily on content presentation and navigation [7–9], but they do not make it possible to create dynamically a personalized path and to carry out individualized monitoring of each learner in real time.

Therefore, adaptive learning systems require in addition to the detection of learning style and the use of techniques and methods of adaptation, artificial intelligence approaches to adapt automatically and in real time the learning according to the profile of the learner.

2 Case Based Reasoning

The Case Based Reasoning (CBR) is a paradigm of artificial intelligence that solves a problem by relying on past problem solutions deemed similar to this problem [10]. Solved problems are stored in memory and are a source of exploitable knowledge for future reasoning and reuse.

Solving a problem using the Case Reasoning approach can be achieved through a process that contains a set of steps. These steps are more detailed in [11, 12 and 13].

It should be noted that the traditional cycle of the Case Base Reasoning contains the following steps: Elaboration, Retrieve, Reuse, Revise (See also [14]) and Retain.

Whereas, in the dynamic cycle of Case Base Reasoning (DCBR), the target case is presented by dynamic descriptors that change at time, a new star CBR cycle is proposed by [12, 13] in the Fig. 1: this is the Incremental Dynamic Case Based Reasoning (IDCBR).

This new cycle has led to changes in the order and content of the steps of the traditional CBR cycle (some steps can be re-run several times).

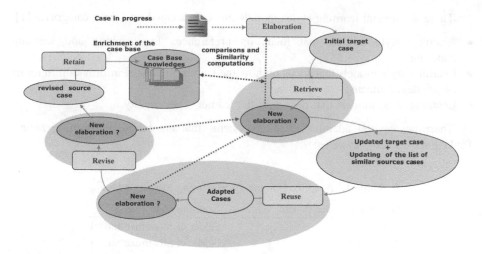

Fig. 1. The IDCBR cycle [12, 13]

The Incremental Dynamic Case Based Reasoning cycle contains two main steps:

- *Initialization step:* this step makes it possible to create the initial target case (the step of elaboration) and a list of similar source cases (the stage of remembering);
- *Dynamic Step:* The detection of the first change of the target case causes the launching of this step which leads first to the creation and activation of a reminder loop (a new development and an update of the list of cases similar sources) with each change of the target case. This loop ends when the system encounters no change. The retrieval loop will be activated when there is a change in the target case level during the adaptation step and the review step. If no, the cycle continues its normal execution.

3   Supervised Machine Learning and Techniques of Supervised Machine Learning

3.1   Supervised Machine Learning

Machine Learning is a subfield of artificial intelligence, it helps to understand the structure of data and integrate it into models that can be understood and used by computer machines [15].

Supervised machine learning involves building a learning game-based model and labels to use it to classify new data [16]. It allows to automatically creating rules from an apprenticeship database containing examples (cases already processed and validated).

3.2 Techniques of Supervised Machine Learning

There are several algorithms and methods used for supervised machine learning, the most used are: supervised learning techniques are [17, 18]:

- K-Nearest Neighbor (KNN);
- Naive Bayes classifier;
- Support Vector Machine (SVM);
- Decision Trees.

The choice of such methods is due to several factors [17–22]: such as the type and number of data to be processed, the execution time...

Table 2 below presents a comparative study of methods and techniques of supervised machine learning.

4 Our Proposed Approach

Our goal is to create an adaptive learning system that allows individualized monitoring of a learner in real time, ie when the learner encounters learning problems, our system must intervene automatically to guide the learner according to his profile and his needs. This intervention uses the learner's traces saved during the learning process, which requires the system to consider the dynamic and incremental change of these traces over time and to provide adequate solutions to the changes.

Our adaptive learning system uses Incremental Dynamic Case Based Reasoning (IDCBR) that integrates the K-Nearest Neighbor (K-NN) algorithm into the Retrieve step, to ensure a personalized learning in real-time based on the learner's profile and the past experiences of other learners by looking for the similarity of learners' behaviors and predicting future behaviors.

4.1 Our Proposed Architecture

Our adaptive learning system must take into consideration the arrival of varied data, whether it is on pedagogical activities or on learners. Therefore, our system must be able to make decisions and deduce new learning situations based on already existing classified data.

Initially, the learner registered in the system, entering personal information, completing the Felder-Silverman Learning Style Test Forms (FSLSM) and Prerequisite Test.

Based on the learning style detected and the prerequisite level of each learner, the system offers him an initial learning path. At this level, the learner begins to learn by following the proposed learning path. During the learning process, the learner may encounter learning obstacles and anomalies, in which case the Incremental Dynamic Case Based Reasoning cycle is triggered by integrating the K-Nearest Neighbor in order to adapt the learning.

Table 2. Comparative study of methods and techniques of supervised machine learning

Method/Technique	Learning	Type of data	Advantages	Inconvenients
K-Nearest Neighbors (K-NN)	• Choice of distance • Choice of number of neighbors • The learning base • Treatment of both regression and classification problems.	• Numeric, categorical	• Deal with problems with a large number of attributes. • Very simple and very used (fast learning) • Adapted to the domains where each class is represented by several prototypes.	• Computer complexity. • Slow prediction • Gourmand method in place memory. • Sensitive to irrelevant and correlated attribute.
Naive Bayes classifier	• Pretreatment desirable • Evaluation of conditional probabilities • Treatment of classification problems.	• Numeric, categorical	• Easy to use • Quick to learn • Very robust	• Model not always simple (because the law is not all-day normal) • A priori knowledge about classes
Support Vector Machine (SVM)	• Construction of non-linear decision rules (boundaries) • Treatment of both regression and classification problems.	• Numeric, categorical	• Adapted to the large number of variables • Possibility of using data structure (strings and trees) as inputs	• Long to learn
Decision Trees	• Integrated pre-treatment • Criteria for evaluating variables	• Numeric, categorical	• Simple and interpretable • Quick to learn • Fast processing of large volumes of data	• Dependence on the starting question

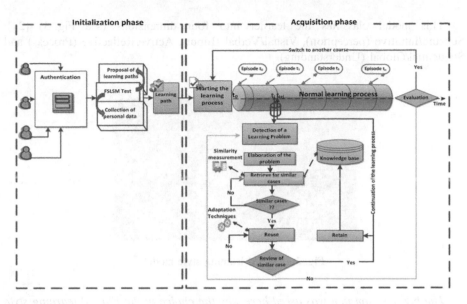

Fig. 2. Proposed architecture

The proposed architecture (Fig. 2) contains two phases:

- **Initialization Phase:** During this phase, the system collects personal information about the learner, detects the learning style of FSLSM, offers an initial learning path based on its style and classifies it into a group of learners.
- **Acquisition Phase:** During this phase, the system must adapt the learning according to the profile of the learner who has difficulty learning. This phase allows dynamic and individual supervision in real time by using the IDCBR approach, which allows through its stages to follow the learner at each moment of his learning process in order to offer him a suitable course.

4.2   Initialization Phase

a. *Collection of personal information*

In our architecture, the learner must be modeled by characteristics describing personal information such as: login, password, last name, first name, age, mail, country, ...

The acquisition of this information can only be done with an identification questionnaire that the learner must fill out at the time of his first contact with the system.

b. *Detection of learning style.*

The system must takes into consideration two types of information: the result of the pre-requisite test and the result of the Felder-Silverman learning style model test [23].

This model classifies the learner into four dimensions (see Fig. 3) [24]: Sensing/Intuitive (perception), Visual/Verbal (Input), Active/Reflective (Process) and Sequential/Global (Understanding).

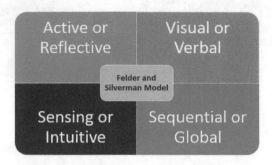

Fig. 3. FSLSM learning style model

The big question that was asked here why the choice of the FSLSM learning style model?

The choice of the Felder-Silverman learning style model is such that this model fulfills most of the criteria required by hypermedia systems [25, 26]:

- The FSLSM model quantifies and calculates learning styles through the Felder-Solomon Learning Style Assessment Index (ILS) in the form of a 44-item questionnaire (11 questions for each). dimensions) whose learner needs and preferences are expressed with values between +11 to -11 for each dimension;
- The FSLSM model provides an accurate evaluation of the learning style (a good degree of validity and reliability);
- The model is suitable for use with a web-based adaptive educational system [27];
- The model is easy to administer.

In our architecture, we are interested in two dimensions: Sensing/Intuitive and Visual/Verbal. The choice of these two dimensions is because our system does not allow the interactions between learners (Active/Reflective) and considers that the guidance and freedom of navigation are implicitly included in our system (Sequential/Global). Therefore, the style of each learner is presented by the Sensing/Intuitive dimension (resource of a learning object) and the Visual/Verbal dimension (format of a learning object).

After detecting the learner's learning style, the learner is asked to answer a pre-requisite test for each course or training in order to know his level of knowledge: Low, Medium, and High.

c. *Proposition of learning path*

The system proposes learning paths (suite of learning objects) based on the correspondence between the FSLSM learning style of each learner and the metadata used to describe the pedagogical objects. These paths must be saved in a learning database.

Some metadata used to describe learning objects include:

- The resource of the educational object: definition, notion, algorithm, example, exercise, quiz.
- The nature of the educational object: text, audio, picture, video.
- Level of difficulty: low, medium, high.

The results obtained from the FSLSM learning style test make it possible to specify the profile of each learner and the corresponding learning objects (Table 3):

Table 3. Different learner profiles taking into account learning objects

Profile	Description	Learning object Resource/Nature
Sensing/Visual	Progress from practice to theory in the form visuel	Example, exercice and quiz in the form of picture, video
Sensing/Verbal	Progress from practice to theory in the verbal.	Example, exercice, quiz in the form of text, audio.
Intuitive/Visual	Progress from theory to practice in the form visuel.	Notion, definition, algorithm in the form of picture, video
Intuitive/Verbal	Progress from theory to practice in the form verbal.	Notion, definition, algorithm in form of text, audio.

According to the previous table, our system classifies the learner according to his profile in 4 groups of profiles (Sensitive/Visual, Sensitive/Verbal, Intuitive/Visual, Intuitive/Verbal) and he offers him an initial learning path to accompany, lead and guide the learner to approach well the Acquisition phase.

4.3 Acquisition Phase

Since the learning process (set of recorded traces) changes dynamically and gradually over time, the system must take into account the arrival of new data in time, which makes it difficult to predict the behavior of the learner. For this reason, we used the approach of Incremental Dynamic Case Based Reasoning that will be based on the observation and analysis of the evolution of traces to predict the behavior of the learner. This analysis is based on decisions corresponding to similar existing learning situations using the supervised learning method K-Plus Nearest Neighbors by looking for learners with similar behaviors using a measure distance.

The choice of the K-Nearest Neighbor method is that we initially determine the learners' classes (the 4 learner profiles mentioned above), we have a learning base that contains all the information about successful learners a course or training (learner profiles, learning traces, final learning paths) using labeled data that describes the learner's behavior. As well as the K-Nearest Neighbors method allows regression by predicting a set of data from a learning base.

a. *Functioning of the Acquisition Phase*

From Fig. 2, we have two learning processes:

- If the learner has not encountered any learning problems, he will have a normal learning process.
- If the learner has had a learning problem, the system must intervene automatically to track the learner by providing learning paths similar to their behaviors (see Fig. 4).

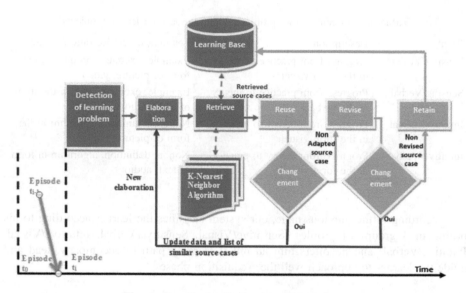

Fig. 4. Functioning de the Acquisition phase

When, the system detects a gap between the intended learning process (the proposed initial learning path) and the actual learning process (the observed learning path) of a learner at each episode t_i, it requires an observation of the learning process and a comparison of the behavior of this learner with the behavior of the other learners carried out during the previous instants (from t_0 to t_{i-1}) in order to find learners with a similar behavior. The latter must be adapted to the behavior of the learner in difficulty.

b. *Elaboration Step*

In our architecture, we use indicators, which we consider relevant, collected from the observation and analysis of learners' traces during the learning process:

- t_i: episode t_i corresponds to a consultation of a learning object;
- **Crs:** Course chosen;
- **VSCrs:** Learning Object;
- **DtVSCrs:** Date of visit of the learning object;
- **NbrVstVSCr:** number of visits of the learning object;
- **DvVSCr:** Duration of visit of the learning object;

- **RVSCrs:** Resource of the learning object (Definition, Notion, Example, Algorithm, Exercise, Quiz);
- **FVSCrs:** Nature of the learning object (Text, Audio, Picture, Video).

The traces must be saved in Logs files at each moment when the learner accesses a learning object by indicating all the necessary information. These tracks have entry points to feed the CBR.

c. *Retrieve Step*

The objective of this phase is to detect the source case(s) most similar to the target case, by looking for the learning path(s) most similar to the target case in the source case base (learning process successfully) by comparing the behavior of the learner in difficulty with other learners. This comparison is based on information that is collected from the observation and analysis of learning traces.

Since, the observed learning path of each learner is composed of a sequence of learning objects numbered i (i = {0 ... n_Obj} with n_Obj: the number of learning objects of each course). Each learning object is consulted at each instant t_i (each instant t_i corresponds to a consultation of a learning object), which can be represented each time by an episode t_i (see Fig. 2).

At each episode ti, the learning traces are recorded by the learners; each trace is defined by a vector Episode$_{(i,j)}$ grouping descriptors already mentioned:

$$\mathbf{Episode}_{(i,j)} = \{\mathbf{t_i}, \mathbf{Appr_j}, \mathbf{Crs}, \mathbf{ObApp_i}, \mathbf{DtVObApp_i}, \mathbf{DvObAppi}, \mathbf{RObAppi},$$
$$\mathbf{FObAppi}, \mathbf{NbrVstObAppi}\}.$$

With:

- **Episode(i,j):** Trace observed at the episode t_i by the learner j;
- **Apprj:** Learner j;

We are interested in grouping all the traces of all learners in the episode t_i (moment t_i) in the form of a matrix, in order to select the learner(s)candidate to be similar at the level of this learning object i.

$$\mathbf{Consultatation\_Episode}_{(i)} = \sum_{j=1}^{n\_App} \mathbf{Episode}_{(i,j)} \qquad (1)$$

With:

- **i:** the number of the learning object viewed (episode t_i);
- **j:** the number of the learner who viewed the learning object i (episode t_i), and successfully completed the course;
- **n_App:** the total number of learners who successfully completed the course;

According to Eq. 1, we will have:

Consultatation_Episode$_{(i)}$

$$= \begin{pmatrix} t_i, \mathrm{Appr}_j, \mathrm{Crs}, \mathrm{ObApp}_i, \mathrm{DtVObApp}_i, \mathrm{DvVObApp}_i, \mathrm{RObApp}_i, \mathrm{FObApp}_i, \mathrm{NbrVstObApp}_i \\ t_i, \mathrm{Appr}_{j+1}, \mathrm{Crs}, \mathrm{ObApp}_i, \mathrm{DtVObApp}_i, \mathrm{DvVObApp}_i, \mathrm{RObApp}_i, \mathrm{FObApp}_i, \mathrm{NbrVstObApp}_i \\ \vdots \\ t_i, \mathrm{Appr}_n, \mathrm{Crs}, \mathrm{ObApp}_i, \mathrm{DtVObApp}_i, \mathrm{DvVObApp}_i, \mathrm{RObApp}_i, \mathrm{FObApp}_i, \mathrm{NbrVstObApp}_i \end{pmatrix}$$

In our system, the goal of the retrieve phase is to find learners who have already completed the training or the course with behaviors similar to the observed behavior of the learner in difficulty in order to assign him the learning path adapted to his behavior. Thus, our goal is to find for each learning object i (episode t_i), the learners (source cases) who are the closest (source cases) of the learner in difficulty (target case), looking for the correspondence between the learner in difficulty and the other learners (successful experiences). This objective will be achieved by using the K-Nearest Neighbors method in order to classify the K closest learners according to indicators that have been deemed relevant.

We must classify the learner in difficulty in each episode t_i in order to detect the class mainly represented by the K nearest learners. To classify learners, by following the next steps:

1. Detect the episode t_i that the learner has had learning problems with;
2. Create the matrices Consultation_Episode$_{(i)}$ for each episode t_i (i from 0 to i-1):

$$\mathbf{Consultatation\_Episode_{(i)}} = \begin{cases} \sum_{j=1}^{n\_App} \mathbf{Episode}_{(0,j)}, \mathbf{i} = 0 \\ \sum_{i=0}^{n\_Obj} \left(\sum_{j=1}^{n\_App} \mathbf{Episode}_{(i,j)} \right), \mathbf{i} > 0 \, et \, i \leq n\_Obj \end{cases}$$

$$(2)$$

3. Determine the number of neighbors "K": in general, the good value of K is $\sqrt{n}$ where n is the population (learning base) [28]. In our case, n is the number of learners who validated the course;
4. Determine the measure of distance to use. In our case, we will use the Euclidean distance d;
5. Apply the K-Nearest Neighbors Algorithm:

K-Nearest Neighbors Algorithm

Input:
Learning data: profile, final learning path, learning traces;
Learning in difficulty: Appx;
nbr_Episode: number of episodes;
nbr_App: number of learners who have successfully completed the course;
k: number of nearest neighbors;
 	
Begin
For i from 0 to nbr_Epi do
  Create the matrix Consultation_Episode$_{(i)}$;
  For j ranging from 0 to nbr_App
    Calculate the distance dist (App$_x$, App$_j$);
    / * *calculate the similarity between the vector Episode$_{(i, Appx)}$ and the different rows of the matrix Consultation_Episode$_{(i)}$;* * /
    D [j] = dist(App$_x$, App$_j$);
    / * *store distances in the table D* * /
  End
  Sort D in ascending order
  For j from 0 to nbr_App
    Neighbor_Index [j];
    / * *recover neighbor index from sorted array D* * /
    T_Neighbors [j];
    / * *retrieve neighbors classes from sorted array D* * /
  End
  p = 0;
  For h from i * k to k * (i + 1) -1 do
    T_Class[h] = T_Neighbors [p]
    / * *class table k first neighbors at each episode* * /
    p = p + 1;
  End
End
For i from 0 to k * nbr_Episode do
  / * *Count the number of occurrences of each class* * /
End
 	
Output:T_Class [App$_x$];
/ * *Get the majority class* * /
End

The result of this algorithm is to recover the majority class of the K-Nearest Neighbors of the learner in difficulty from episode t0 until the episode ti-1.

At each moment the system detects a learning problem in the learning process, it will apply the algorithm of K-NN (retrieval loop [12, 13]).

d. *Reuse Step*

Our system must assign the learner in difficulty to the majority class recovered in the previous step. This learner will inherit the same behavior and learning path of the learner among the K-nearest neighbors of the majority class who has the minimum distance. If the system detects another anomaly in the learning process, it will go back to the previous step (retrieve phase), otherwise it will move on to the next step (revise step).

e. *Revise Step*

The objective of this phase is to evaluate the adapted learning path. In our case, the evaluation is really tested by the observation of the learner's behavior or by the results obtained from an evaluation. If the system detects another anomaly in the learning process, it will go back to the retrieve step, otherwise it will move on to the next step (retain step).

f. *Retain Step.*

When the learner in difficulty has not encountered any new learning problems, this learner has become a new resolved and retained case in the learning base, and is used immediately for the resolution of future problems. The profile of this learner, the learning traces and the learning path will be saved in the learning base as a new experience.

5   Conclusion & Perspectives

Our approach is to design an adaptive learning system that assists the learner in real time during the learning process to provide a learning path that is adapted to their preferences and needs.

Our architecture is divided into two phases, the first phase "Initialization phase" allows to detect the initial profile of the learner through the FSLSM learning style model and to propose a learning path according to his profile. The second phase "Acquisition phase" allows tracking and directing the behavior of the learner in difficulty in real time by looking for similar experiences through the use of the K-Nearest Neighbors method in the step of Retrieve of the Incremental Dynamic Case Base Reasoning approach that is based on the exploitation of learning traces.

Our future work consists in setting up a learning platform based on our proposed architecture, defining the role of the different actors of the platform and developing the different phases in order to test the effectiveness of our approach.

References

1. Curry, L.: An organization of learning style theory and constructs. In: Curry, D.L. (dir.) Learning Style in Continuing Medical Education (pp. 115–123). Council on Medical Education, Canadian Medical Association, Ottawa, Canada (1983)

Individualized Follow-up of the Learner Based on the K-Nearest ...   377

2. Carver, C.A., Howard, R.A., Lanc, W.D.: Enhancing student learning through hypermedia courseware and incorporation of student learning styles. IEEE Trans. Educ. **42**, 33–38 (1999)
3. Paredes, P., Rodriguez, P.: A mixed approach to modelling learning styles in adaptive educational hypermedia. Adv. Technol. Learn. **1**(4), 210–215 (2004)
4. Cha, H.J., Kim, Y.S., Park, S.H., Yoon, T.B., Jung, Y.M. et Lee, J.H.: Learning styles diagnosis based on user interface behaviors for the customization of learning interfaces in an intelligent tutoring system. In: Communication presented at the 8th International Conference on Intelligent Tutoring Systems (ITS 2006). Jhongli, Taiwan (2006)
5. Wolf, C.: iWeaver: Towards an interactive Web-based adaptive learning environment to address individual learning styles. In: Auer, D.M. (dir.) Blended learning: International workshop – Interactive Computer Aided Learning (ICL) 2002. Kassel University Press, Kassel, Germany (2002)
6. Papanikolaou, K.A., Grigoriadou, M., Kornilakis, H., Magoulas, G.D.: Personalizing the interaction in a Web-based educational hypermedia system: the case of INSPIRE. User Model. User-Adap. Inter. **13**(3), 213–267 (2003)
7. Chikh, A.: Une méthodologie de réutilisation en ingénierie du document : Le système SABRA . Thèse de Doctorat d'état en Informatique, INI (2004)
8. Brusilovsky, P.: Methods and techniques of adaptive hypermedia. Adapt. Hypertext User Model. User Adap. Interact. **6**(2–3), 87–129 (1996)
9. Brusilovsky, P.: Adaptive hypermedia. Adapt. Hypertext User Model. User Adap. Interact. **11**, 87–110 (2001). Site http://www.sis.pitt.edu/~peterb/papers.html
10. Amélie cordier.: Gestion des Connaissances pour des Systèmes à Base de Connaissances Hybrides . Mémoire de DEA DISIC, Université de Claude Bernard Lyon I, Laboratoire LIRIS, Juin (2004)
11. Mille A.: Traces bases reasoning (TBR) definition, illustration and echoes with storytelling. Rapport Technique RR-LIRIS-2006-002, LIRIS UMR 5205 CNRS/INSA de Lyon/Université Claude Bernard Lyon1/Université Lumière Lyon 2/Ecole Centrale de Lyon, Jan (2006)
12. Zouhair, A.: Raisonnement à Partir de CasDynamique Multi-Agents: application à un système de tuteur intelligent. Ph.D in computer science, in Cotutelle between the Faculty of Sciences and Technologies of Tangier (Morocco) and the University of Le Havre (France), Oct (2014)
13. El Mokhtar, E.N.-N.A.I.M.I., Zouhair, Abdelhamid: Intelligent dynamic case-based reasoning using multi-agents system in adaptive e-service, e-commerce and e-learning systems. Int. J. Knowl. Learn. (IJKL) **11**(1), 42–57 (2016)
14. Mill. A.: Tutorial CBR: Etat de l'art de raisonnement à partir de cas. Plateforme AFIA'99, Palaiseau (1999)
15. https://www.supinfo.com/articles/single/6041-machine-learning-introduction apprentissage - automatique, published by Metomo Joseph Bertrand Raphael en (2017)
16. Silva, C., et Ribeiro, B.: Inductive Inference for Large Scale Text Classification: Kernel Approaches and Techniques, Vol. 255 (2009)
17. Hastie, T., Tibshirani, R., et Friedman, J.: The Elements of Statistical Learn ing : Data Mining, Inference, and Prediction (2001)
18. Cristianini, N., Shawe-Taylor, J.: An Introduction to Support Vector Machines and Other Kernel-based Learning Methods (2000)
19. Liefooghe, A.: Classification supervisée, pp 30, 3, 2012-2013
20. https://www.researchgate.net/publication/309731330 pp. 12–13, Nov (2016)
21. Hasan, M., Boris, F.: SVM: Machines a Vecteurs de Support ou Separateurs a Vastes Marges ..BD Web, ISTY3.Versailles St Quentin, France. pp 01, 16 janvier (2006)

22. Mathieu-Dupas, E.: Algorithme des k plus proches voisins pondères et application en diagnostic, pp 02,04. Marseille, France, France. (2010)
23. Felder, R.M., Silverman, L.K.: Learning Styles and Teaching Styles in Engineering Education, Presented at the 1987 Annual Meeting of the American Institute of Chemical Engineers, New York, Nov (1987)
24. Sauvé, L., Nadeau, J.-R. et Leclerc, G.: Le profil d'apprentissage des étudiants inscrits dans un certificat de cycle offert à distance et sur campus: une étude comparative. Revue de l'enseignement à distance 8(2), 19–35 (1993). Récupéré du site de la Revue: http://www. jofde.ca/index.php/jde/article/view/218/627
25. Brown, J.S.: New learning environments for the 21st century : Exploring the edge. Change: Magazine High. Learn. 38(5), 18–24 (2006)
26. Popescu, E., Badica, C., Trigano, P.: Description and organization of instructional resources in an adaptive educational system focused on learning styles. In: Advances in Intelligent and Distributed Computing, pp. 177–186. Springer. (2008) ISBN: 978-3-540-74929-5
27. Kuljis, J., Liu, F.: A Comparison of Learning Style Theories on the Suitability for Elearning. In: Hamza, M.H. (ed.) Proceedings of the Iasted Conference on Web Technologies, Applications, and Services, pp. 191–197. ACTA Press (2005)
28. Thirumuruganathan, S.: A Detailed Introduction to K-Nearest Neighbor (KNN) Algorithm (2010)

Integration of an Intelligent System for a University Governance Information System

Majida Laaziri[1]([✉]), Khaoula Benmoussa[1], Samira Khoulji[1], and Kerkeb Mohamed Larbi[2]

[1] Information System Engineering Resarch Group, National School of Applied Sciences, Abdelmalek Essaadi University, Tetouan, Morocco
majida.laaziri@gmail.com
[2] Information System Engineering Resarch Group, Faculty of Sciences, Abdelmalek Essaadi University, Tetouan, Morocco

Abstract. IT plays a positive role in the governance of the university, it is considered as a means of implementing a more effective policy in carrying out the activities and missions of the university through the use of systems university Information System (IS). The establishment of a university governance IS contributes to the improvement of the governance of the university through the modernization of the administrative management, the institutionalization of information, and the provision of the actors of the university. University new procedures improve the quality of information and its management. However, the adoption and deployment of such IS within the university requires support for users when using this new IS, in order to take full advantage of its services and features, and more Choosing a pedagogical method of training is very important in the process of support and accompaniment, to accelerate the adaptation of the user with the new IS. Our contribution in this article and show the usefulness of an implementation an intelligent tutoring system for the effective use of a university governance information system.

Keywords: University governance information system · Change management
SIMARech · SIMACoop · Intelligent system · Moodle

1 Introduction

The governance of public organizations has become a problematic issue today. Recent years have been marked by major changes in the management tools and institutional governance principles of public bodies [1].

The mix of regulations and the need to ensure the effectiveness of public action on the one hand, and the introduction of management tools from the private sphere, on the other hand, have contributed to the emergence of new modes of governance in the public sectors. Especially in universities.

Faced with all these transformations, the university is called upon to strengthen its governance system in order to promote a higher education that is solid enough and allows for a consequent adaptation. The use of new information technologies (IT) is

M. Ben Ahmed et al. (Eds.): SCA 2018, LNITI, pp. 379–390, 2019.
https://doi.org/10.1007/978-3-030-11196-0_33

today considered one of the main mechanisms of governance practices and university management. These tools ensure effective and sustainable governance in the university by improving its sense of responsibility at the economic, ecological, social, cultural, technological and political levels.

IT has become essential to support the growth and sustainability of universities, where the technology infrastructure includes a variety of applications, including information systems [2]. The establishment of information systems (IS) for universities are effective solutions, to ensure good governance of the university, and allows to work independently, transparently, responsibly and adopting a participatory approach which involves all its stakeholders. In addition, sustainable development to the university's missions, improving its performance, its management effectiveness and allowing it to position itself at national and international level. Deploying an IS for strengthening the governance of the Moroccan university is a sustainable governance approach that will allow the university to:

- Ensure a transparent work environment;
- operate with modern management methodologies;
- have reliable and effective information in real time;
- make decisions;
- In addition, share information with internal and external partners.

However, the deployment of an information system is often considered a purely technical project and the organizational and human aspects are sometimes neglected. The arrival of a new computer tool can profoundly modify the trades and the strategy of a university, thus imposing on the users a modification of their practice and their competences. Change management aims to frame this transition so that university users can fully benefit from the functionality of the new tool.

For this reason the support and training of users is very important, because the user will have to acquire sufficient knowledge to use the features of the new IS and adapt to changes in the business that can result. The design of an intelligent tutoring system seems the best solution to meet the needs of users.

2   Digital Strategy of Abdelmalek Essaadi University (AEU) and Modernization of Information Systems

In a more dynamic and competitive governance environment, where the digital transformation of processes and activities is accelerating, the role and importance of Information Technology (IT) for the functioning and governance of universities is a reality undeniable. Like all universities, the Moroccan university is currently called upon to apply the governance approach in its internal management system while respecting both its societal role and on the other hand to benefit from the pervasive technological revolution.

In this national context, Abdelmalek University Essaadi thus understood about the importance of using technology to modernize its governance system.

The establishment of an information system for governance and information sharing is the best solution adopted by AEU after the changes and reforms applied to the higher education sector.

For this purpose the computer infrastructure of the AEU has changed through several actions namely [3]:

- Implementation of a single-entry integration system for all digital services at the university: a digital work environment.
- Establishment of an integration portal dedicated to the winners of the university and companies called "KHIRIJ".
- Development of the information and management system.
- Involvement in the information system "JAMIATI".
- The establishment of a comprehensive integrated university governance information system (SIMAGU) which focuses on the development of higher education and enables it to manage services and business bricks in several sectors of activity (research, university cooperation, professional integration, mobility, training, ...). This system is open to teacher-researchers, responsible for the structures, administrative staff of institutions and universities, and currently includes two brick trades and the others are under construction:
  - Brick business scientific research: Establishment of a Moroccan Information System for the governance of Scientific Research (SIMARech) [4] in order to support researchers and enhance their scientific output.
  - Brick business university cooperation: Establishment of a Moroccan Information System for the governance of University Cooperation (SIMACoop) [5] in order to support governments and universities in the cooperation and partnership plan.

3   University Governance Information System (SIMAGU)

The missions and sectors of a university are the basis of a teaching guided by creativity, innovation and excellence. For the Abdelmaleek Essadi University, the major challenge of its policy is to carry out its fundamental mission, which touches on the sectors of activities of: training, research, student life, international influence and professional integration, consisting of preparing students for their professional and civic life and helping to develop their autonomy, their culture and their critical and ethical sense.

In this context, effective governance of the university's activities has become a mandatory condition for the accomplishment of its missions. For this purpose, the establishment of a governance information system (SIMAGU) is the best solution adopted by the university.

3.1   Simagu

This solution aims at the integration of several building blocks (research, cooperation, professional insertion, training, mobility and management of human resources). The objectives of (SIMAGU):

- Avoid password multiplication: the users concerned by the system can connect with a single password to all the business bricks (IS).
- It makes it possible to modernize the administrative management, to make available to decision-makers dashboards helping to make strategic decisions.
- It offers users an automation of the tasks of the departments concerned by the business bricks in question.
- Improvement of university governance through:
- Access to reliable information more quickly.
- Better structure the prerogatives of the administrative staff.
- Master the management of scientific research: support researchers and enhance their scientific output.
- Master international cooperation: support governments and universities in the cooperation and partnership plan.
- Mastering the management of professional integration: monitoring the integration of graduates and putting university graduates face to face with employers.
- Mastering mobility management: monitoring student mobility.
- Master the management of human resources: follow activities, training.
- Manage bricks trades in an efficient way.

The project initially planned the development of several bricks trades, two of which are already deployed (brick scientific research and brick university cooperation), and the others are under construction.

3.2 Moroccan Information System for the Governance of Scientific Research

Moroccan Information System for the Governance of Scientific Research (SIMARech) was designed to meet the specific needs of Moroccan universities and management bodies, aims to present in a coherent way the institutions, research units within the university, including staff and their scientific activities (publications, patents, events, equipment, etc.) (Fig. 1), as well as the monitoring of financial resources and international activities (partnership agreements, congresses, etc.) [6].

SIMARech provides a description of the existing situation and a needs study for better optimization of human and material resources. It also allows a national evaluation of research to engage in a quantitative assessment process that would provide objective criteria for self-assessment and external peer review. It is configurable and allows for a transcendent role assignment: teacher-researcher, head of a research structure, educational institution and university administrator. Its reporting system makes it possible to print documents, graphs and statistics on various sections specific to research: indexed publications, communications, projects, patents, etc. [4].

3.3 Moroccan Information System for Governance of University Cooperation

Moroccan Information System for Governance of University Cooperation (SIMA-Coop) is a monitoring system designed to support governments and universities in the plan of cooperation, partnership and student exchange and programs. as well as to

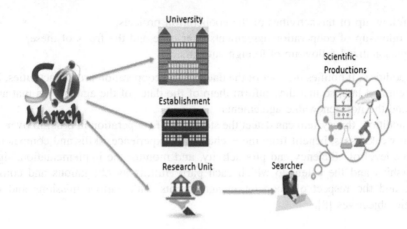

Fig. 1. Generality of SIMarech [4]

improve communication, collaboration and integration between universities and their partners, performance management, strategic planning, as well as to evaluate the performance of research and establish a sound policy for the development of the institutional relationship [5, 7].

SIMACoop offers a range of services for the benefit of the actors of the university and its partners, and mainly, it gives them the possibility to do (Fig. 2):

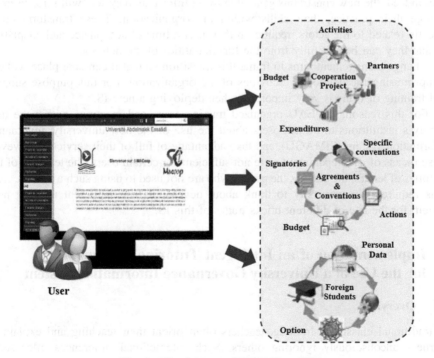

Fig. 2. The features of SIMACoop [5]

- A follow-up of the activities of the cooperation projects,
- A follow-up of cooperation agreements over time and the fruits of these,
- A personalized follow-up of foreign students.

In addition, it notifies the user of the dates of the cooperation project activities, both imminent and late, so that they inform them of the dates of the agreements that are in force and the non-renewable agreements.

In addition, this system can detect the strengths of cooperation in order to overcome its own weaknesses, benefit from the exchange of experience, skills and competences, raise the level of efficiency and productivity, and monitor the implementation. signed partnerships and the extent to which each party fulfills its obligations and commitments, and the respect of cooperation agreements with partner missions and their strategic objectives [8].

3.4 Change Management

In the field of technology, however, the authors have shown that innovation and novelty are not enough to generate organizational change. Indeed, as great as they are, if individuals do not make sense to them, the deployment of new technological tools (IS) will remain purely useless. Thus, for a new technology to generate change, individuals need to make sense of it and then use it to serve their purpose.

The contours of the implementation of a new information system can not therefore be limited to the technical and mechanical description of replacing a tool with a newer one. Indeed, the new constraints generated by an information system will force users to change their practice, their skills within the organization; These transformations, directly related to the actors, require on their part a time of acceptance and adaptation so that they can benefit fully from the functionalities of the new tool.

Change management aims to frame this transition so that it can take place without compromising the day-to-day activities of the organization. For this purpose support and training of users is very important when deploying a new IS.

For this reason, the EAU organized training tours and meetings within the university's institutions to inform users about the use of the new university governance information system (SIMAGU), and take advantage of full of their services. However, these means of accompaniment were not sufficient for certain users (the teachers of the faculties of letters, the elderly, the people who are not used to using such a system, ...). This required the university to think about developing an intelligent tutoring help system for the use of different bricks trades of this IS.

4   Implementation of an Intelligent Tutoring System for the Use of a University Governance Information System

4.1   Overview

In traditional classroom training, teachers often orient their teaching and explain to learners unconsciously ignoring others. Such interactional differences offer some learners better opportunities and incentives to learn, while others can not. According to the author [9] states that Traditional teaching methods are no longer able to meet the needs of today's learners, ICTs offer new opportunities, including the ability to adapt learning to the individual. The introduction and development of ICTs has upset traditional conceptions of learning, confirms the authors [10].

ICT is the showcase of the information society for pedagogy. More generally, this is at the heart of the major major functions of the university: training, professional integration, research, human resources management, management, financial and accounting management [11]. It enriches the training offer and is closer to the needs of learners. ICTs allow new teaching and learning modalities, both in face-to-face and distance learning. They make it possible to increase the range of action of the universities by a wider diffusion of the knowledge transmitted towards the learners, that they are in initial or continuous formation. In addition, they improve the quality of teaching in different ways, they facilitate the acquisition and appropriation of knowledge through better access to educational resources, an enrichment of these resources, more stimulating teaching relationships and greater involvement of learners in the learning process [12]. Being increasingly collaborative network technologies, they also make it possible to acquire skills related to the new organizational model (team work, decision autonomy, coordination skills, modularity, etc.) and thus prepare students for the job market. job. The quality of the training is assessed through the acquisition, on the one hand, of the disciplinary knowledge and, on the other hand, the skills required by the new organizational model [2].

The application of these new information and communication technologies to the field of training has led to the creation of this new learning mode called e-learning [13]. It is based on access to online, interactive and sometimes personalized training delivered via a network or other electronic media. This access helps to develop learners' skills, while making the learning process independent of time and place [14, 15]. The intelligent tutoring system (ITS) is part of the range of e-learning applications, it is a coaching system or tutoring uses techniques derived from artificial intelligence based on the theory of learning and cognition.

The importance of intelligent tutoring systems (ITS) has increased rapidly in recent decades. Intelligent tutoring systems aim to support and improve learning processes in certain problem areas, taking into account the knowledge and individuality of the student as in one-to-one tutoring [16, 17]. According to the author [18], ITS are computer-based teaching systems that have content in the form of a knowledge base (which specifies what should be taught), teaching strategies (which specify how to teach this content) as well as knowledge about the level of the learner in the content, in order to dynamically adapt their teaching. These systems mimic the human tutor and offer the benefits of individual tutoring. Such systems provide a more natural learning process and allow the learner to learn on their own, thus providing them with high quality instruction.

In general, the intelligent tutoring system consists of four main modules [19] (Fig. 3) : the expert module, the student module, the tutoring module and the interface environment.

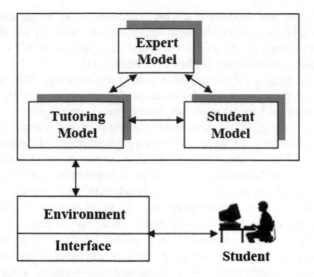

Fig. 3. The different components of an ITS

- The expert module: It is the module that brings together all the knowledge necessary for the teaching process, we speak of knowledge related to the expertise of the field. In general, an expert module must also have know-how, that is, expertise on how to solve domain problems.
- The learner module: This module makes it possible to identify, for a learner, his or her current level of understanding of the field of knowledge. According to the authors [20], the implementation of the learner model is essential for the adaptation of the learning system to the needs of learners. Student behavior is compared to an expert domain.
- The pedagogical module or tutor: this module implements pedagogical strategies to teach the knowledge of a given field. Any educational strategy must be based on formal pedagogical and psychological principles. The needs of the learner must be identified and taken into account in the teaching strategy. Overall, this module serves three main functions:
  - it controls the presentation of the content to the learner,
  - he must be able to answer the questions of the learner,
  - It must be able to determine when learners need help and what kind of help they need.
- The interface module: it allows an effective interaction with the students. It refers to the way of presenting the content.

The assessment of knowledge acquisition is a primary task in the intelligent tutoring system to adapt learning materials and activities to the level of learner capacity [21, 22]. based on a systematic process of data collection, analysis and interpretation. Several techniques can be used by the system, such as tests exercises, frequently asked questions, etc.

4.2   The Proposed Intelligent System

Our contribution is to design an adaptive intelligent tutoring system which will be added in the global information system of university governance (SIMAGU), to support all actors of the university (teachers and administrative staff) during the use of different business bricks of this new IS. This ITS will satisfy the following characteristics: profitability: to produce more significant results from a quality point of view;

- reusability: reuse already existing educational resources;
- flexibility: easy to implement on several different platforms;
- adaptability: adapt to several learner profiles;
- Interactivity: managing the learners' reactions during the training.

There are four types of actors who can access the two information systems and connect with a single password (teacher-researchers, responsible of structures, administrative staff of institutions and universities), their roles differ from one another. Business Brick (SI) to Other, which means that each role will be associated with a profile in the ITS, representing all the services that the learner can access [23, 24]. This determination of the profiles plays an important role in the design of this intelligent tutoring system in order to select the training in which the learner will participate.

The ITS has a resource base that groups all the usage trainings associated with each role in each business core (IS), which can be tracked by the learners enrolled in this training. In addition, each of these learners is subjected to a set of controls during the registration to be able to know their preliminary achievements and to present them among the attributes of their profile whose objective is to provide to each learner a list of the formations adapted to their profile.

First, in order to follow a defined training, the system must evaluate the preliminary knowledge of the learner of certain concepts defined by the expert, and after having followed this training, the learner must have acquired certain concepts, educational purpose of this training. Then this different information will be recorded and used when searching for the optimal route. Therefore, the learner profile will be the initial state before the training follow-up and the educational objective will be the target end state of the profile after the training follow-up.

In order to adapt the training to the profile of the learner, the system will present a descriptive profile of the profile containing the information describing his achievements, his intellectual abilities and his motivations. His intellectual abilities will allow him to classify him according to three classes beginner, intermediate or advanced, and his motivations will direct his training towards the application or the theory.

Our tutoring system will be divided into different modules based on the classic modules of an intelligent tutoring system listed above (Fig. 3). Its operation in general can be summed up in three essential steps: to model the pedagogical resources as well as the educational objective in compatible formats, the results obtained will be called later by the system adapter. Then the space learners will receive this information from the learner, selects his profile in the database and sends it back to the adapter as well as the purpose of this training. In the space of the expert, which reserves the integration of the new resources in the database, saves the information of the latter in a form by the

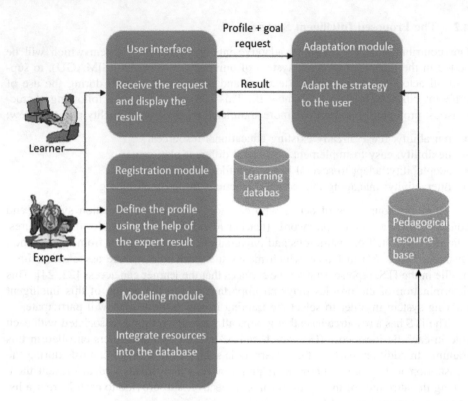

Fig. 4. Initial architecture of ITS

modeller. The latter facilitates the reuse of these resources. Figure 4 shows an initial architecture of the ITS.

4.3   Choice of the Technical Tool

There are many educational development solutions for online training that enable easy communication of information and interoperability of applications.

The choice of a platform to develop our intelligent tutoring system is motivated by problems of maximum adaptability to the needs and criteria of this system. Indeed, the Moodle platform perfectly meets these needs, and enjoys a good reputation thanks to its ease of use, its free and especially the possibility of developing and integrating new tools through its open source license.

Moodle is a platform for setting up online courses and websites. It benefits from an active development and designed to favor a socio-constructivist training framework. Moodle is made freely available as free software, according to the GPL license. Which means that it has a copyright, and developers have a number of freedoms. They have the right to copy, use and modify Moodle as long as they agree [25, 26].

Moodle works on all computers that can run PHP and can implement a database (especially MySQL). The term "Moodle" was originally an acronym for "Modular

Object-Oriented Dynamic Learning Environment". This terminology is especially useful to educational programmers and theorists, also "moodle" is a verb that describes how to lazily wander through something, do things when that programmer agrees best, a nice way to act that often leads to reflection and creativity. This term therefore applies to the way Moodle was developed, as well as the way students and teachers approach learning and teaching in an online course [13].

The Moodle platform will help us develop a teaching system that will be fully automated and adaptable to the needs of each teacher and administrative staff of the university, which will give this system the main features of the intelligent tutoring computer system.

5 Conclusion

The development of digital and its social and economic integration reinforces the strategic position of information systems in the university environment, the information systems facilitate the daily tasks of the users, and they guarantee a sustainable development to the missions of the university, in improving its performance, efficiency and enabling it to position itself at national and international level [27].

However, the introduction of changes within the university is not always easy, it causes problems at the level of resistance to change. This requires support for users after the deployment of a new IS within the university.

To this end, in this article we have emphasized the usefulness of the establishment of a system of intelligent tutoring for the use of a university governance information system, which is currently instrumented by the use of ICT and concepts of artificial intelligence.

References

1. Eddine, A.S.: Governance and Management of University in Morocco. pp. 45–62 (2018)
2. Tikam, M.V.: Impact of ICT on Education. Int. J. Inf. Commun. Technol. Hum. Dev. 5(4), 1–9 (2013)
3. UAE: Services Numériques Transversaux de l'Université Abdelmalek Essaâdi." [Online]. Available: https://www.ensat.ac.ma/guide/08servicesnumeriquestransversaux.htm
4. Benmoussa, K., Khoulji, S., Laaziri, M., Larbi, K.M.: Web information system for the governance of. 8(4), 3287–3293 (2018)
5. Laaziri, M., Benmoussa, K., Khoulji, S., Larbi, K.M.: Information system for the governance of university cooperation. 8(5), 3355–3359 (2018)
6. Benmoussa, K., Laaziri, M., Khoulji, S., Mohamed Larbi, K.: SIMARECH 3: a new application for the governance of scientific research. Trans. Mach. Learn. Artif. Intell. 5(4) (2017)
7. SIMACoop.: [Online]. Available: http://simacoop.uae.ac.ma/
8. Laaziri, M., Benmoussa, K., Khoulji, S., Mohamed Larbi, K.: SIMACoop: a framework application for the governance of university cooperation. Trans. Mach. Learn. Artif. Intell. 5 (4) (2017)

9. Ebrahimi, R., Science, I.: The Effect of Information and Communications Technology (ICT) on Teaching Library and Information Science. no. Apr, 2006–2009 (2009)
10. Garner, R., Gillingham, M.: Students' knowledge of text structure. J. Read. Behav. **19**(2), 247–259 (1987)
11. Chevalier, Y.: Système d'information et démocratie à l'université. pp. 55–66 (2016)
12. Pelgrum, W.J., Law, N.: Les TIC et l'education dans le monde: tendances, enjeux et perspectives (2004)
13. AZOUGH, S.: E-Learning Adaptatif: Gestion Intelligente Des Ressources Pédagogiques Et Adaptation De La Formation Au Profil De L'Apprenant (2014)
14. E-learning : Présentation, aspects, enjeux et avenir. [Online]. Available: https://www.procomptable.com/papier_recherche/mmbf.htm
15. L'évolution des environnements technologiques du e-Learning. [Online]. Available: http://blog.wikimemoires.com/2013/01/levolution-des-environnements-technologiques-du-e-learning/
16. De Lièvre, B., Depover, C., Dillenbourg, P.: Quelle place accorder au tuteur système et au tuteur humain dans un processus d'industrialisation ? Distances et savoirs **3**(2), 157–181 (2005)
17. Bessagnet, M., Canut, M., Bessagnet, M., Canut, M.: Sur la conception de tuteurs intelligents … To cite this version : HAL Id : edutice-00001256. (2005()
18. Murray, T.: Authoring Intelligent Tutoring Systems : An Analysis of the State of the Art. Int. J. Artif. Intell. Educ. (1999)
19. Burns, H.L., Capps, C.G.: Foundations of intelligent tutoring systems : an introduction. In: Foundations of Intelligent Tutoring Systems. (1988)
20. McCalla, G.I., Greer, J.E.: Granularity-Based Reasoning and Belief Revision in Student Models. Student Model. Key to Individ. Knowledge-Based Instr. (1994)
21. Ramírez-Noriega, A., Juárez-Ramírez, R., Martínez-Ramírez, Y.: Evaluation module based on bayesian networks to intelligent tutoring systems. Int. J. Inf. Manage. **37**(1), 1488–1498 (2017)
22. Ferreira, H.N.M., Brant-ribeiro, T., Ara, R.D., Dorc, F.A., Cattelan, R.G.: An Automatic and Dynamic Knowledge Assessment Module for Adaptive Educational Systems. (2017)
23. Continuous, W., Conference, A.: Integration of e-learning modules for a particular information system within higher education. pp. 535–548 (2006)
24. Benmoussa, K., Laaziri, M., Khoulji, S., Kerkeb, M.: Intelligent system for the use of the scientific research information system. Int. J. Adv. Comput. Sci. Appl. **9**(6), 132–138 (2018)
25. MoodleDocs.: [Online]. Available: https://docs.moodle.org/34/en/Main_page
26. Benyounes, B.: Évaluation de la fonctionnalité et l'utilisabilité de la plateforme d'apprentissage Moodle : une approche du génie logiciel. (2009)
27. Nurcan, S., Rolland, C.: 50 ans de Système d ' Information : de l' automatisation des activités individuelles à l ' amélioration des processus et la création de valeur ajoutée 2 . Evolution diachronique des Systèmes d ' Information. pp. 1–27

LMS 3.0: A Collaborative Learning Management System Based on Web 3.0 Concepts

Abderrahim El Mhouti[1,3]([✉]), Mohamed Erradi[2,3],
and Azeddine Nasseh[2,3]

[1] Faculty of Sciences and Technologies of al-Hoceima, Ctre Ait Youssef Ou Ali,
Morocco
abderrahim.elmhouti@gmail.com
[2] ENS of Tetouan, Martil, Morocco
[3] SIIPU Lab, Abdelmalek Essaadi University, Tetouan, Morocco

Abstract. Over the last years, collaborative learning has been introduced as a solution to the lack of sufficient support to the large number of students in e-learning environments, particularly LMS (Learning Management System). Nevertheless, current collaborative learning puts a heavy emphasis on group discussion, but provides inadequate support for individual learners to promote teamwork during the whole learning process. The aim of this chapter is to propose an intelligent LMS based on Web 3.0 concepts and able to improve student's collaboration in higher education context. The intelligent LMS called "LMS 3.0" is based on emergent technologies used in the Web 3.0 era. It integrates a multi-agents system responsible to the tracking of the levels of collaboration and productivity of each student and group, and therefore aids tutors to make optimal decisions to encourage collaborative learning. On the other hand, in order to optimize the large-scale resource management, the proposed LMS 3.0 is based on a cloud computing architecture.

Keywords: E-learning · E-learning 3.0 · Collaborative learning
Web 3.0 · LMS 3.0 · Multi-Agents system · Cloud computing

1 Introduction

Due to the easy access to higher educational resources via the Internet, the number of students using e-learning environments has been increasing at an enormous rate. With these new orientations, the number of teachers supporting students in the e-learning process has become insufficient, which has led to teaching problems within these e-learning environments. In this context, collaborative learning has been adopted as a solution to this problem [1].

Indeed, e-learning featured by collaboration is becoming more and more important in higher education both for learners and teachers [2]. Collaborative learning is a typical example of non-traditional learning strategies which are easily supported by new technologies, and that can be more effective than traditional approaches [3].

M. Ben Ahmed et al. (Eds.): SCA 2018, LNITI, pp. 391–407, 2019.
https://doi.org/10.1007/978-3-030-11196-0_34

However, supporting collaborative learning in e-learning environments is not always evident. Many e-learning systems do not provide new opportunities for development of collaborative learning practices. This applies to Learning Management Systems (LMS), which is the traditional approach to ensure the e-learning process. With the new orientations of collaborative learning, LMSs remain focused on course management purposes and has a limited impact on pedagogy and collaboration between students and teachers [2]. LMS, as traditional approach to e-learning organized as courses, puts emphasis on some aspects of group discussion, but provides inadequate support for individual learners by sustaining them to collaborate during the whole learning process [1]. While social software including blogs, wikis, social networking sites, and social bookmarking sites etc. are adopted by many students and educators to meet their emerging needs in educations, LMS offers restricted tools to track and obtain information about the level of collaboration and productivity of students and groups [4]. These limitations make LMS inefficient to support collaborative learning.

This chapter presents a new model of intelligent Learning Management System using Web 3.0 technologies. The aim is to support collaborative learning practices in this type of e-learning systems, widely used in higher education. The proposed intelligent LMS, called "LMS 3.0", provides teachers information about the level of collaboration and productivity of each student and group. Therefore, teachers exploit data provided to promote collaborative learning, avoid students' isolation and motivate them to use collaboration tools.

Thus, the LMS 3.0 is based on Web 3.0 concepts such as the Multi-Agents System (MAS) technologies and cloud computing services. On the one hand, in order to collect and manage data about students' and groups' activities and assist teachers to exploit these data to promote students' collaboration, the LMS 3.0 integrates a MAS ensuring the tracking of students' and groups' collaboration and productivity levels.

On the other hand, in order to optimize the large-scale software and hardware resources management, the proposed system uses cloud computing technology. Thus, the proposed intelligent LMS is deployed on a cloud computing architecture.

The reminder of this chapter is organized as follows: Sect. 2 gives an overview of the main concepts on which this work is based, including: e-learning, e-learning 3.0, collaborative learning, Web 3.0, MAS and cloud computing technology. To solve the open issues, Sect. 3 introduces the LMS 3.0 framework. This section starts with a global description of the proposed framework and therefore presents its features, its modeling and its conceptual architecture. Therefore, Sect. 4 presents a prototype system developed. Conclusions and future work can be founded in the last section of this chapter.

2 Related Works

2.1 E-Learning in Higher Education

E-learning concept: overview. E-learning refers to the use of electronic media (computers, networks, web sites …) and information technologies in education to deliver information for education where students and instructors are separated by distance, time, or both. The aim is to enhance the student's learning experience and performance [5].

Reference [6] define e-learning as an Internet-based learning process that uses computer technologies and Internet to design, implement, manage, support and extend learning, which will not replace traditional education techniques, but will greatly enhance the efficiency of education.

In higher education, e-learning is widely enclosure of all kinds of educational technology in learning and teaching. The term e-learning is widely synonymous with online education, technology-enhanced learning, computer-based training, computer-assisted instruction, web-based training, and virtual education. These other alternative names dwell on a specific component, aspect or delivery method [7]. However, the evolution of the Web to Web 2.0 and the influence of new practices on the Web have resulted in a new array of services, which can be collectively termed "e-learning 2.0".

E-learning systems. An e-learning system is a software application, system or platform for flexible learning. Its aim is the realization of learning process theory: organization of contents and resources, delivery of educational courses and training programs, tracking, documentation and administration.

In higher education, the key features of an e-learning system includes learning management, courses management, assessment, reporting, content management, content creation and content delivery. Also, these tools provide facilities for managing the learning experience, communicating the intended learning experience and facilitating tutors' and learners' involvement in that experience. There are various types of e-learning systems used in higher education. The most used systems are LMS (Learning Management Systems), VLE (Virtual Learning Environment), LCMS (Learning Content Management System), etc.

E-learning 3.0. Because the Web has always been considered as a major tool for the development of e-learning platforms, e-learning has naturally accompanied the evolution of Web standards. As new technologies are developed, the e-learning platforms try to integrate them, and this trend actually explains why e-learning has changed over the years.

Indeed, the evolution of e-learning is closely linked to the evolution of Web technologies. In recent years, the evolution of web technologies has led to significant changes in e-learning practices. The nature of the Internet evolved from a static environment (Web 1.0) to a highly dynamic, interactive, collaborative and intelligent media (Web 3.0), enabling end users to run software applications, collaborate, share information and to create new services online.

As the evolution of Web technologies, the evolution of e-learning has given birth to three generations: e-learning 1.0, e-learning 2.0 and e-learning 3.0. Thus, just as the Web has changed from "read only" to "read-write" and therefore to "read-write-collaboration", the concept and methods of e-learning have shifted from simply translating educational material in online support, to entirely new educational approaches that focus on active participation, interaction and student collaboration (Miranda et al. 2014).

Web 3.0 involves transforming the Internet from a platform of global interactivity and information sharing to an open, intelligent and effective tool for information management. As Web 2.0 tools and principles have reshaped e-learning systems, the same is true for Web 3.0 tools. Practitioners and researchers have explored the possibilities that this reality may bring, especially under the influence of powerful Web 3.0 concepts, such as distributed computing, artificial intelligence, cloud computing, etc.

Thus, e-learning 3.0 transposes the boundaries and boundaries of traditional education systems in a much more complete and revolutionary way, as learning becomes a personalized activity that individuals can manage themselves [8]. The key tool to achieve this is collaborative filtering, as it will facilitate not only obtaining the required information, but also the development of massive multi-user platforms and features, leading to an explosion of collaborative learning [9].

2.2   Collaborative Learning in LMS

Collaborative learning for higher education is a learning strategy embodies the application of new technologies and where several students interact with each other in order to achieve their common goals. This strategy uses learning environments supported by the computer network to carry out the collaborative learning, in the form of teamwork, between teachers, tutors and students, based on their discussion, cooperation and communication [10], using various interaction tools offered by e-learning systems such as chat, forum, wikis, social networking, etc.

Indeed, collaborative learning refers to a second phase of e-learning based on Web 2.0 and emerging trends in e-learning. Collaborative learning is inspired by the popularity of Web 2.0, which places increased emphasis on social learning and use of social software tools [11].

In higher education, LMSs are designed to provide e-learning course management service to teachers, educators and administrators. They provide also some course-based collaboration and interaction services to teachers and students. By focusing on the second aspect (collaboration and interaction), LMS puts emphasis on some aspects of group discussion, but provides inadequate support for individual learners and the relationship between actors in a course is temporal and unequal (Fig. 1).

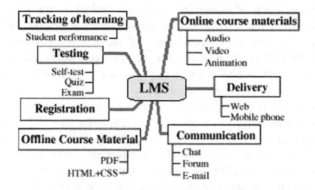

Fig. 1. Collaboration and interaction in LMS [12]

In LMS, since collaboration and interaction is oriented at all members of a course and mostly based on text and photos, the form and effect of collaboration is relatively simple and limited. The duration of communication is the same as the corresponding course which is usually some months. Consequently, everything in the space of a course will be of no use after the end of the course, and there is no opportunity for students/teachers to accumulate personalized social and knowledge network [13].

2.3   Web 3.0 Concepts

It is widely accepted that the Web has evolved steadily over the years. The first Web tools were simple, but as information technology and Internet speeds evolved, new tools emerged, creating a user-centric interactive space where information was shared by all [8].

From its beginnings in the 1990s, the concept of the Web has evolved to its transformation into Web 3.0 [14, 15]. From a Web that aims to connect people to the Internet (Web 1.0), the web concept has grown into a way for users to actively participate through collaboration, control and sharing of information (Web 2.0 and 3.0). Thus, the Web 1.0 is founded to provide information, the Web 2.0 is developed to ensure the overload of the information whereas the Web 3.0 is created to ensure the control of the information [16].

Web 3.0, also known as the Semantic Web or Web of Data, is the transformed version of Web 2.0 with technologies and features such as intelligent collaborative filtering, cloud computing, big data, interconnected data, openness interoperability and intelligent mobility. While Web 2.0 is about social networking and mass collaboration between the creator and the user, Web 3.0 refers to intelligent applications that use natural language processing, machine learning and reasoning [17].

Web 3.0 is a very important emergent concept of the modern Internet that appeared in the year 2008. This is the name of the new version of the Internet which is articulated around the Semantic Web and the intelligence artificial. Web 3.0 is considered the third generation of the Internet, an evolved version of the static version of the Web (Web 1.0) to the Smart Web through Web 2.0, where the user does not just store and disseminate data but is interested in their understanding by reasoning about their meanings by machines and software agents. On the other hand, this version of the Web is, to date, considered as "the Internet of Things", i.e. it is the objects serving users that communicate with servers via sensors over the Internet. There is thus a relationship between the physical universe and the digital universe.

The aim of Web 3.0 is to make the Web and Artificial Intelligence (AI) coexist using a set of technologies, tools, standards, protocols, standards, etc. These technologies enable intelligent machines and agents to conduct online reasoning and processing in a cooperative and automatic manner on Web content. In fact, the amount of data generated by the Internet by users and web agents is very important and can reach levels that could not be managed by current tools and technologies. The goal of the Web 3.0 technologies is to lighten the management of the large mass of data generated by the use of the Internet which, without the integration of these tools, can lead to blockages of the treatments and to the saturation of the global network.

Web 3.0 is an evolution of Web 2.0 and its properties. This means that aspects of interactivity and cooperation are still part of Web 3.0. It also means that users are not completely replaced by machines, but they have, in addition to roles played in Web 2.0, additional roles related in particular to the supervision, control and configuration of applications and intelligent agents deployed in work environments. These users are withdrawn in certain situations to let the machines perform the first treatments before their validation. With these new directions, machines play important roles as a means of helping to perform treatments or pretreatments on the Internet when human intervention is not necessary.

In terms of the technologies used, Web 3.0 uses technologies such as RDF (Resource Description Framework), SPARQL (Query Language for RDF), OWL (Ontology Web Language) and SKOS (Simple Knowledge Organization System). These technologies help structure information so that programs such as Web spiders and Web crawlers can search, discover, gather and analyze information from the Web [17].

2.4   MAS for E-Learning Purposes

Overview. A MAS is a society organized, constituted by semi-autonomous agents, which interacts with others, aiming to resolve collaboratively some problems, or to achieve some individuals or collectives goals. The agents may be homogeneous or heterogeneous and have common goals or not, but still maintain a degree of communication between them.

The main component of a MAS is the "agent". An agent, which is a specific programming entity, is defined as "one who acts" or "one who acts in place of" [18]. An agent could be described by the following features: autonomous, reactive, communicative, learning, cooperative, or mobile. Its structure includes a set of modules (Fig. 2).

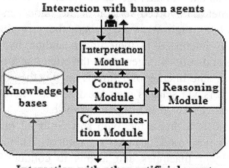

Fig. 2. Architecture of an artificial agent

The communication module allows the agent to communicate with other artificial agents in the system. The control module is based on a description of the agent's behaviors toward the messages that can be received from the other agents. The reasoning module uses the agent's knowledge and a set of reasoning rules allowing it to accomplish its role. Finally, an optional module called interpretation module associated to agents having an interaction with human actors (learner, teacher and tutor) because its main function is the interpretation of the human agents' actions [19].

MAS in e-learning. With the popularization of Internet, the use of MAS technology in the design and modeling process of teaching environments has evolved spectacularly in parallel. Experience has shown that a large quantity of projects has emerged in the world where information technology are almost been embedded [20].

Indeed, the major potential of MASs relates their ability to support personalized and collaborative learning. In e-learning, one observes a move from a personal learning environment into a space where students are taking more control on their learning, either as monolithic applications to help students to manage their resources and collaboration, or as a collection of online tools. This online learning environment has a lot of potential to support personalized and collaborative learning, because there is no need for a central authority to orchestrate collaborations and learning activities [21]. Recently, various collaborative e-learning systems based on agent model have been developed. Interesting results have been achieved by pedagogical agents regarding student motivation and companion human agents (teachers, tutors, etc.) acting sometimes as mediator of the learning process [22].

2.5   Cloud Computing for E-Learning

The concept of cloud computing. As a new kind of advanced technology, cloud computing is a promising technology that refers to the use of applications and services that run on a distributed network using virtualized resources and accessed by common Internet protocols and networking standards [23]. Cloud computing has been recently emerging as a compelling paradigm of the present century, for managing and delivering services over the Internet. It takes the technology, services, and applications that are similar to those on the Internet and turns them into a self-service utility.

The five essential characteristics of cloud computing paradigm are [24]:

- On-demand self-service: users can request and manage their own computing resources on demand;
- Broad network access: cloud services are offered over the Internet or private networks and accessed through standard mechanisms;
- Resource pooling: computing resources are pooled together to serve multiple consumers according to theirs demands;
- Scalability and elasticity: cloud services and computing platforms offered can be rapidly and elastically provisioned and can be scaled across various concerns;
- Measured service: cloud computing resources usage can be measured, controlled, and reported providing transparency for both the provider and consumer of service.

In terms of services, there are many services provided by cloud computing. The three basic service models (or layers) are Software as a Service (SaaS), Platform as a Service (PaaS) and Infrastructure as a Service (IaaS) (Fig. 3).

In a SaaS model, software applications are offered as services on the Internet rather than as software packages to be purchased by individual customers. In PaaS, an

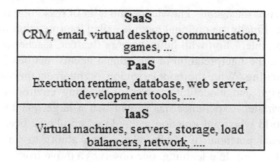

Fig. 3. Cloud computing layers

operating system, hardware and network are provided to support the entire application development lifecycle. The IaaS model provides just the hardware resources (such as storage), computing power (such as CPU and memory) and network as services [25].

From the deployment models perspective, there are four ways used to deploy a cloud computing environment delivering cloud services [23]:

- Public cloud: it is provisioned for public and open use alternatively for a large industry group. It is owned by an organization selling cloud services;
- Private cloud: it is provisioned for the exclusive use by a single organization. The cloud may be managed by that organization or a third party;
- Hybrid cloud: it combines multiple clouds (private, community or public) where those clouds retain their unique identities, but are bound together as a unit;
- Community cloud: it is one where the cloud has been organized to serve a common function or purpose.

Cloud computing for e-learning: convenience and benefits. Today, cloud computing has become an alternative and attractive model for delivering IT services with which most universities would like to be incorporated with for various reasons.

Firstly, the use of computers in higher education institutions is increasingly intensive. Thus, these institutions are seeking to provide free or low-cost alternatives to expensive and exclusive tools. In this sense, cloud computing has promoted the growth of e-learning systems with its pay as you go model. It offers to universities a new type of business model where the services that are delivered become computer resources.

Secondly, cloud computing has a great potential to increase the efficiency and availability of e-learning systems. Due to the scalability and cost reduction, cloud services allow implementing easier, faster and less expensive e-learning solutions.

Also, cloud computing supports the creation of a new generation of e-learning systems able to run on a number of hardware devices, while the data is stored in the cloud. It enables also the implementation of data mining techniques when large databases are being used so that meaning can be extracted from data [26].

Moreover, adopting cloud-based e-learning allows managing educational and technical tasks better. By adopting cloud computing, universities become responsible for the content creation, management and delivery, while the cloud services providers are responsible for the construction as well as management of e-learning system [27].

Finally, Cloud computing responds to the aim for which e-learning environments were deployed in higher education: collaborative learning. Thus, cloud solutions create virtualized resources that can be made available to users for collaboration purposes. This model enhances productivity and provides a learning environment where students could share educational resources and actively collaborate.

3 LMS 3.0: Towards a Learning Management System Based on Web 3.0 Concepts

In order to promote collaborative learning practices in LMS, this chapter proposes an intelligent LMS designed according to a multi-agents approach and based on cloud computing services. This section describes features and agents integrated in the proposed platform and provides its conceptual architecture.

3.1 Description

LMS 3.0 is an intelligent collaborative learning platform designed as a MAS supported LMS. In order to overcome the limitations of LMS in Web 3.0 era, so that students may collaborate on how best to gain knowledge, LMS 3.0 is not a traditional LMS that only makes available simple tools to manage courses and communicate, but it is an intelligent Learning Management System using Multi-Agents System and cloud services potentials to enhance collaborative learning practices.

Thus, LMS 3.0 is designed on Web 3.0 technology in order to ensure two types of roles supporting collaborative learning:

- it incorporates some artificial intelligence aspects by integrating a MAS. Multiple agents interact to collect and exchange information so that students may collaborate on how best to gain knowledge.
- it is deployed in a cloud computing infrastructure, which make possible that cloud services can support collaborative learning by allowing storage and management of all data and learning resources, especially, data collected by the MAS in order to provide a database of learners' activities traces.

In the LMS 3.0, teachers and tutors can interact with students and groups, track their level of collaboration and productivity, know collaboration tools used by students, etc. As result, teachers and tutors can analyze and exploit collected information, guide, encourage and motivate students to learn collaboratively.

3.2 LMS 3.0 Features

To define the LMS 3.0 features, we have identified the roles and needs of actors involved in the proposed system, especially, tutors, teachers, students and groups.

For student' roles, a student can learn, collaborate and interact with his group members. For needs, student must obtain information on its progress in learning, on the fulfillment of its tasks, on its performance and on the performance of its whole group. This means that the LMS 3.0 must be able to identify three types of students' status. Table 1 shows different students' status identified.

Table 1. Students' status

Student' status	Description
Active student	Student who performs his learning activities and who communicates in a regular manner
Passive student	Student who connects but do not communicates and do not realizes activities
Collaborative student	Active student who seeks assistance from others and who responds to them

As students' status, LMS 3.0 must be able to identify the groups' status. Table 2 shows different groups' status identified.

Table 2. Groups' status

Group' status	Description
Active group	If more than half of the students are active, the group is an active group
Passive group	If more than half of the students are passive, the group is a passive group
Collaborative group	If more than half of the group is collaborative, the group is collaborative group

On the other hand, to support collaborative learning, tutors and teachers must intervene to estimate the students' progress, regulate the learning activity and collaboration tools used as well as the frequency of use of these tools. Thus, the LMS 3.0 must be able to:

- determine (in addition to the students'/groups' status) absent/present students over a period in a group. Therefore, the tutor can then try to contact the absent students.
- identify if a student/group performs an activity or not in order to enable the tutor to measure the productivity of students and groups.
- determine software collaboration tools used by each student or group. This allows to tutors to know whether students/group benefit at best of tools available.
- identify the number of times where a collaboration tool has been used by each student or group. This allows to tutors to know the most appreciated tools for students.

3.3   MAS Modeling and Functioning

LMS 3.0 is designed according to a multi-agents approach. This means that each of the system' features, mentioned above, must be carried out using one or more artificial agents. Thus, the MAS modeling must include a set of artificial agents. Each agent is responsible for the management of roles and needs of one or more actors or groups in the system.

Firstly, to ensure student' tasks and collect traces about students' activities, the proposed system must be able to monitor learning activities performed by students as well as collaboration tools manipulated. These roles will be played by an agent called "Student_Agent". This agent will ensure three functions that concern the student:

- assists student by offering an interface which makes the learning task easier. It contains a student's profile.
- tracks all student's activities during a learning session. It collects the traces of a student and stores them in a database.
- collects traces about collaboration tools consulted by a student and saves data obtained in a database. It gives also statistics on the use of these tools.

Secondly, to ensure tutor' tasks (assists students, adapts learning scenario to their needs, encourages collaborative work, helps and motivates struggling students), the system defines a "Tutor_Agent" which assists tutor to realize his tasks by giving councils to learners and following-up their learning processes.

Thirdly, to ensure teacher' tasks (design and production of courses contents, knowledge transfer, assessment), LMS 3.0 defines a "Teacher_Agent" which offers an interface to teacher in order to assist him. In addition, the systems must integrate a "Pedagogical_Agent" that consists to present to students pedagogical contents adapted to theirs profiles. Also, an "Assessment_Agent" must be designed to allow assessment of students' knowledge level using a set of tests tools and exercises.

Fourthly, the proposed system must integrate an "Activities_Agent" which ensure the tracking of all actions realized by the student during a learning session. This agent collects students' traces and stores them in the database.

Fifthly, a "Tools_Agent" must be designed in order to collect traces about collaboration tools consulted by student (chat, forum, wiki …) and saves the collected data in the cloud-based database. This agent gives also statistics on the use of these tools (duration of use, traffic of connections to tools used …).

Finally, LMS 3.0 must manage the cloud-based database that stores all collected information. It must also manage and coordinate interactions between different agents. Thus, each of these roles will be played by an agent:

- *Database_Agent*: manages data stored in the database.
- *Coordinator_Agent*: manages and coordinates interactions between agents deployed and the cloud-based database.

Figure 4 shows the Multi-Agents System modeled.

The life cycle of the MAS starts when a faculty member connects to the LMS 3.0. The three personal agents are the artificial substitute for each user in the platform. Thus, in each connection to the platform, any user passes through his personal agent. All other agents are agents that manage the system features.

If the user is a student, Student _Agent allows managing the pedagogical act. This agent is in permanent relation with the Assessment_Agent responsible for evaluation, and the Activities_Agent that collect all students' traces and sends them to the Coordinator_Agent to save them in the cloud-based database. For this, Coordinator_Agent,

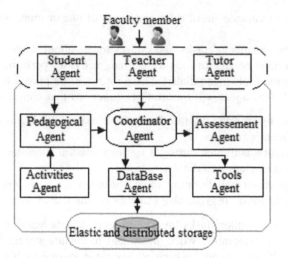

Fig. 4. Agents modeled in LMS 3.0

which manages the interactions between agents, calls Database_Agent to save learners' traces retrieved.

The Database_Agent manages data collected which contains the list of students/groups (absent or present) as well as their connection time, their state (passive, active) and their level of collaboration and productivity. It also contains information about collaboration tools used by each user. The cloud-based database contains other information concerning the platform users (profile, account, etc.). Users employ Database_Agent whenever they need access to the database to know the members of their group, integrate a new group or form a new group.

Finally, Tools_Agent is responsible for collecting traces about the use of collaboration tools. Indeed, it is important for users to know if these tools are sufficiently used. Tools_Agent brings them the answer. Whenever a user uses a tool, this agent sends this information to the Coordinator_Agent to include it in the database.

3.4 LMS 3.0 Architecture

The LMS 3.0 architecture includes the two main components introduced in this work to enhance and promote collaborative learning practices: a learning management system and a multi-agents system. On the other hand, the system architecture is based on a cloud computing infrastructure.

The combination between these three main aspects (LMS, MAS, cloud services) pushes us towards an architecture based on several layers. Thus, the LMS 3.0 architecture is divided into three main layers:

- *Infrastructure layer*: it is as a dynamic and scalable physical host pool composed of information infrastructure including (Internet/Intranet, system software, information management system and some known software/hardware);

- *Services Layer*: containing software, infrastructure an platform services (SaaS, IaaS and PaaS). It offers an unified interface for students developers;
- *Application Layer*: includes interactive courses and sharing teaching resources. It provides content production, content management, content delivery, collaborative learning, assessment and management features.

Figure 5 describes the Collaborative Learning Management System architecture.

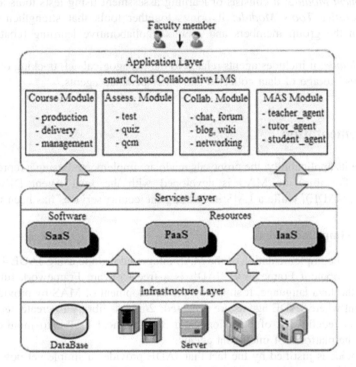

Fig. 5. LMS 3.0 architecture

The Infrastructure Layer is placed in the bottom layer of cloud services and represents the information infrastructure and all resources used. It works as a dynamic and scalable physical host point and contains information management system, Internet, system software and some common software and hardware. Learning resources and data collected (by MAS deployed) about learners' activities are stored in this layer. This layer provides also the basic computing power such as physical and CPU memory.

The Services Layer is the virtual machines and services implemented within the platform. It delivers three levels of cloud services: SaaS, Paas and IaaS. Faculty members use software via the Internet, not to need a one-time purchase for software and hardware, and not to need to maintain and upgrade.

The Application Layer is the layer through which we integrate the e-learning practices in the cloud computing model. This layer provides content production and delivery technology, collaborative learning, assessment and management features.

Faculty members can simply use their Internet browsers to upload course content, create new courses, and communicate with students. This is all done through a secure LMS, which gives educators the ability to store information on the cloud.

Application Layer includes four main modules:

- *Course Module*: it provides training and interactive courses. It consists of content production and delivery, resources sharing and management components;
- *Assessment Module*: it consists of learning assessment using tests tools and quiz;
- *Collaborative Tools Module*: it brings together tools that strengthen the links between the group members and promote collaborative learning (chat, forums, wiki);
- *MAS Module*: it includes agents related to the pedagogical act, tracking of learners' activities, storage of data collected and other personal agents.

4   Implementation

To validate its applicability, the proposals made are implemented using a representative prototype. For this, the MAS is developed with the Java Agent DEvelopment Framework (JADE), while a LMS integrating the course services has been used.

4.1   MAS Implementation

In LMS 3.0, the implementation of the MAS integrated is done using JADE 4.4.0 (Java Agent DEvelopment Framework). JADE is a free software Framework fully implemented in the Java language. It simplifies the development of MAS by providing: 1/an environment where JADE agents are executed; 2/a class library to create agents using heritage and redefinition of behaviors; and 3/a graphical toolkit to monitoring and managing the platform of intelligent agents.

This choice is justified by the fact that JADE provides a simple yet powerful task execution and composition model and allows developers to quickly develop distributed applications using multi-agents principles. It supports coordination between several agents FIPA and provides standard implementation of the communication language FIPA-ACL, which facilitates the communication between agents.

In JADE, creating an agent is done through the programming of class that inherits from the class jade.core.Agent. This class must have the setup() method which is called to initialize the agent. At the creation of the agent, an identifier is assigned to it using the jade.core.AID class. This identifier object is of the form: < nickname > @ <platform-name > (Fig. 6).

To manage the system database, the JDBC (Java DataBase Connectivity) API has been used to ensure the connection to database independently of the DBMS (Database Management System) used. DataBase_Agent communicates with the system database using SQL queries. On the other hand, MySQL was chosen as DBMS.

When the tutor starts the execution of his agent, an interface appears and asks him to choose the nature of the query that he wishes to formulate (queries relating to a 'group', to a "student", to the status of realization of an "activity"). After the execution

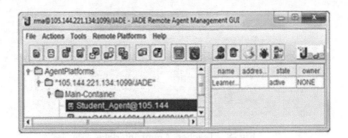

Fig. 6. JADE framework interface

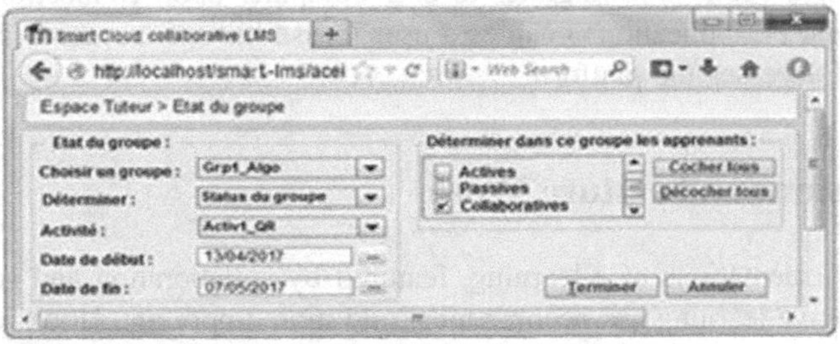

Fig. 7. Tutor_Agent interface

of the request, the tutor can see the result appear. Figure 7 shows the interface between the tutor and his agent to run queries concerning a "group".

It is the same when the student starts executing his agent. The interface of the Student_Agent allows the learner to input her preferences and parameters and manage her profile. Student can also obtain his level of productivity in a learning session.

4.2 LMS Implementation

The application layer is implemented using a LMS because the work realized in this chapter focuses on the extension of LMS features to support collaborative learning and not a work that aims to establish a new platform. This choice is motivated by the fact that existing LMSs, although they have certain shortcomings, they have very interesting and advanced features. Therefore, it seems more realistic and effective to integrate the proposed MAS in an existing LMS than trying to develop a new LMS. Thus, following a comparative study of the main existing LMS, and view the ease of use and pedagogical flexibility that it brings, the choice fell on the Moodle LMS.

Moodle is a LMS developed in PHP language and works under Linux, Windows and Mac, according to the combination Apache (HTTP server) and MySQL (database server). It incorporates tools very useful in collaboration between learners. Figure 8 shows the interface of the Teacher_Agent used in LMS 3.0.

The LMS 3.0 is a web application with a database server. To access the system and achieve its pedagogical activities, the user needs an Internet browser. Every user

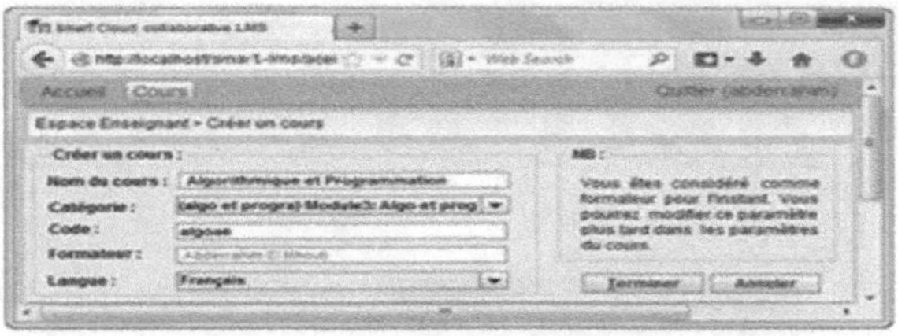

Fig. 8. Teacher_Agent interface

(students, tutors and teachers) has an agent encoded as a Java application located on the server. This Java application migrates to user computer upon connection (when running). This type of applications enables a user to use his agent regardless of the post from which it accesses the platform.

5   Conclusions and Future Work

With the new orientations of e-learning, featured by collaboration and social interaction, collaborative learning has become more and more important. However, according to the huge growth of the number of students and resources needed, existing LMS fall short to support collaborative learning and respond on demand.

In this chapter, by taking advantage of Web 3.0 technologies, MAS and cloud computing, we have proposed an intelligent LMS called "LMS 3.0" supporting collaborative learning in higher education context. The MAS integrated provides tutors and teachers with information about the realization of students' and groups' activities as well as theirs level of collaboration and productivity. As result, tutors exploit data obtained to motivate students to collaborate, to maintain the group and make it dynamic, and to well conduct the collaborative e-learning sessions. To validate its applicability, the proposals made are implemented using a representative prototype.

As part of continuity of this work, we have planned to experiment the collaborative LMS in real conditions of e-learning in higher education with a group of students and teachers.

References

1. Liao, J., Wang, M., Ran, M., Yang, S.J.H.: Collaborative cloud: a new model for e-learning. Innovations Educ. Teach. Int. **51**(3), 338–351 (2014)
2. Du, Z., Fu, X., Zhao, C., Liu, Q., Liu, T.: Interactive and collaborative e-learning platform with integrated social software and learning management system. In: Lu, W., Cai, G., Liu, W., Xing, W. (eds.) ITSE 2012. LNEE, vol. 212, pp. 11–18. Springer (2013)
3. Bouroumi, A., Fajr, R.: Collaborative and cooperative e-learning in higher education in Morocco: a case study. Int. J. Emerg. Technol. Learn. **9**(1), 66–72 (2014)
4. Kurcz, J., Chang, T.W., Graf, S.: Improving communication and project management through an adaptive collaborative learning system. In: 15th IEEE International Conference on Advanced Learning Technologies, Hualien, Taiwan (2015)

5. Keller, C., Hrastinski, S., Carlsson. S.A.: Students' acceptanc of e-learning environments: a comparative study in Sweden and Lithuania. Int. Bus. 395–406 (2007)
6. Jain, A., Chawla, S.: E-learning in the cloud. Int. J. Latest Res. Sci. Technol. 2(1), 478–481 (2013)
7. Sneha, J.M., Nagaraja, G.S.: Virtual learning environments: a survey. Int. J. Comput. Trends Technol. 4(6), 1705–1709 (2013)
8. Miranda, P., Isaias, P., Costa, C.J.: E-learning and web generations: Towards web 3.0 and E-learning 3.0. In: Proceedings of the 4th International Conference on Education, Research and Innovation IPEDR, vol. 81, pp. 92–103. IACSIT Press, Singapore (2014)
9. Wheeler, S.: e-Learning 3.0 - Learning with e's. Steve Wheeler Blogspot. http://www.steve-wheeler.co.uk/2009/05/connectivism-dead.html. (2009)
10. Wang, J., Sun, Y.H., Fan, Z.P., Liu, Y.: A collaborative e-learning system based on multi-agent. In: 1st International Workshop on Internet and Network Economics, China (2005)
11. Rosen, A.: E-Learning 2.0: Proven practices and emerging technologies to achieve results. AMACOM, New York (2009)
12. Cavus, N.: The evaluation of learning management systems using an artificial intelligence fuzzy logic algorithm. Adv. Eng. Softw. 41(2), 248–254 (2010)
13. Al-Ajlan, A., Zedan, H.: Why moodle. In: 12th IEEE international workshop on future trends of distributed computing system, pp. 58–64. IEEE Press, Washington (2008)
14. Berners-Lee, T., Hendler, J., Lassila, O.: The semantic web. Sci. Am. 284(5), 34–43 (2001)
15. Berners-Lee, T.: Past, Present and future. IEEE Comput. 29(10), 69–77 (1995)
16. Rego, H., Moreira, T., García-Peñalvo, F.J.: AHKME eLearning Information System: A 3.0 Approach. Int. J. Knowl. Soc. Res. (IJKSR). 2(2), 73–81 (2011)
17. Hussain, F.: E-Learning 3.0 = E-Learning 2.0 + Web 3.0?. IOSR J. Res. Method Educ. (IOSR-JRME). 3(3), 39–47 (2013)
18. Franklin, S., Graesser, A.: Is it an agent, or just a program?: a taxonomy for autonomous agents. In: 3rd International Workshop on Agent Theories, Architectures, and Languages. Springer-Verlag, Budapest, Hungary (1996)
19. Lafifi, Y., Bensebaa, T.: Supporting learner's activities in a collaborative learning system. Int. J. Instr. Technol. Distance Learn. 4(3), 3–12 (2007)
20. Bennane, A.: Tutoring and multi-agent systems: modeling from experiences. Inf. Educ. 9(2), 171–184 (2010)
21. Bentivoglio, C.A., Bonura, D., Cannella, V., Carletti, S., Pipitone, A., Pirrone, R., Rossi, P. G., Russo, G.: Intelligent agents supporting user interactions within self-regulated learning processes. J. e-Learning Knowl. Soc. 6(2), 27–36 (2010)
22. Webber, C., Bergia, L., Pesty, S., Balacheff, N.: The Baghera project: A multi-agent architecture for human learning. In: Workshop Multi-Agent Architectures for Distributed Learning Environments, pp. 12–17. San Antonio, TX, USA (2001)
23. Bhure, G.C., Bansod, S.M.: E-learning using cloud computing. Int. J. Inf. Comput. Technol. 4(1), 41–46 (2014)
24. Wang, L., Laszewski, G.: Scientific cloud computing: early definition and experience. In: 10th International Conference on High Performance Computing and Communications. China (2008)
25. Oladimeji, I.W., Folashade, I.M.: Design of cloud-based e-learning system for virtual classroom. Int. J. Sci. Appl. Inf. Technol. 5(1), 1–6 (2016)
26. Fernández, A., Peralta, D., Herrera, F., Benítez, J.M.: An overview of E-learning in cloud computing. In: Workshop on learning technology for education in cloud (LTEC'12), pp 35–46. Springer, Berlin (2012)
27. Sharma, M.K., Rana, S.: G-cloud (e-Governance in cloud). In: 5th National Conference. INDIACom (2011)

Peer Assessment Improvement Using Fuzzy Logic

Mohamed El Alaoui[1(✉)], Khalid El Yassini[2], and Hussain Ben-Azza[1]

[1] Moulay Ismail University, ENSAM Meknes, Morocco
mohamedelalaoui208@gmail.com, hbenazza@yahoo.com
[2] Faculty of Science Meknes, Moulay Ismail University, Meknes, Morocco
khalid.elyassini@gmail.com

Abstract. Peer assessment, consists of a prearrangement between learners to consider and specify the level, value, or quality of a product or performance or other equal-status learners. The practice imposes itself when trying to evaluate a large number of students, teachers are practically obliged to use peer assessment, especially in Massive Open Online Courses (MOOCs). However, the novice students, unlike their teachers, are not formed to assess others contributions. Therefore, their evaluations are unreliable and may be biased. Here we try to improve the peer assessment outcome, using fuzzy logic to model opinions, those opinions are weighed according to their validity, then aggregated in order to achieve consensus, hence reliable evaluation.

Keywords: Peer assessment · Validity · Reliability · Group decision making
Massive open online course · Fuzzy logic · Weighting opinions

1 Introduction

Peer assessment is automatically linked to new learning methods, especially in MOOCs. However, George Jardine, professor at the University of Glasgow from 1774 to 1826, investigated peer assessment in writing [1].

Actually, the learning needs are changing rapidly, imposing new ways of learning, to which the classical educative system must adapt [2, 3]. MOOCs might be an appropriate solution to the emerging learning necessities such as overcrowded classes, college affordability to under developed counties and so forth [4].

Since MOOCs are not just a distribution of educative content online [5], the process requires a pedagogical evolving scheme, including exercises and quizzes, raising the question of how can thousand students be assessed by a handful staff or teachers? While Multiple Choice Questions (MCQ) can easily be implemented to evaluate quantitative assessments, they cannot cover all learning aspects. Writing, analyzing are among other aspects that cannot be evaluated through any form of MCQ. Hence, the use of qualitative criteria is irrefutable.

Facing the large number of users in MOOCs, the evaluation of thousands learners can either function by a very sophisticated software or via a low cost and widely used solution, which is peer review by other participants [6]. In order to encourage

© Springer Nature Switzerland AG 2019
M. Ben Ahmed et al. (Eds.): SCA 2018, LNITI, pp. 408–418, 2019.
https://doi.org/10.1007/978-3-030-11196-0_35

participants grade each other's, a reward is proposed, as supplementary points in bonus. However, certain students can assess at random, just to get the bonus points.

Noting that not all divergent assessments are due to dishonesties. The personal background of each assessor, influence mainly his/her opinion [7]. This is particularly true in online courses that can involve students from different cultures.

In the opposite side, some authors claimed that MOOC learners aren't supposed to be assessors [8]. While others maintained, that facing the lack of developed software that can evaluate quantitative tasks. The large number of learners cannot be assessed except by peer reviewing [9]. Furthermore, this process may enhance the learners engagement and comprehension [10]. Mulder et al. [11] studied the benefits effects of peer reviews on the learning process result. They also linked the peer review content to students' perceptions and assessment grades. Meek et al. [12] investigated peer reviewing adequacy to MOOCs.

However, peers assessments aren't trustworthy [13], since the judges are novices evaluating other novices, that may share the same errors and misconceptions. Ashton and Davies [14] showed that a predefined scoring rubrics can result on more reliable assessments. Several authors insisted on the persistent validity and reliability issues [15–20] James et al. [21] studied different aggregation operators aiming to release biased marks. In the same sense, we propose a method to permit elevating consistent opinions. Since the new referees have only a vague conception about the studied subjects. We review peer-reviewing methods in Sect. 2. The necessary background from fuzzy logic is presented in Sect. 3. In Sect. 4, we explain the proposed methodology and how it is meant to improve validity and reliability. Section 5, contains an illustrative example. While Sect. 6 concludes this work.

2 Peer Evaluation

We have been familiarized during our respective learning processes to assessments. However, this task is not as simple as it may appear, it can demonstrate the teachers' abilities as well as students' level [22]. Imposing a high quality assessment outcome [23]. The astronomos number of MOOCs learners reduces the feasible assessment methodologies. Various learning assessment methodologies were proposed for MOOCs [24, 25]. Since quantitative results are assessed through MCQ [26], qualitative ones such as writing and analyzing in this context can only be assessed via a developed software or peers [27]. Stating that MOOCs are supposed to be free for people that do not want a certificate and the exorbitant cost to develop efficient software, the peer-reviewing alternative imposes itself in actual context. An overview of assessment techniques in large groups is presented in [28].

Peer reviewing have numerous applications [29–32] it can be used to improve performances [33], and even to evaluate teachers themselves [34, 35]. Noting that teachers evaluations themselves are to be reexamined [36], how can we trust learners assessments! Assuming that learners assimilate well the studied subject and the assessment procedure, collusions can bias the peer evaluation outcome. Song et al. [37] studied two types of collusions, in the first, which they call small circles, a small number of students who are reviewing each other's work, are attempted to increase the peer reviews grades. While in the second, called pervasive collusion, the evaluators give high marks to all, without knowing each other necessarily. The authors [37, 38] have proposed an algorithm to estimate the inflation caused by colluders.

Peer evaluation benefits also are undeniable [29, 39], as well as its drawbacks [40–42]. Since the actual context, which can rapidly change [43], imposes peer reviewing, several authors tempted to enhance the current methods [6, 13, 14, 16, 27, 44, 45]. First let us define self-assessment and peer assessment.

Self-assessment consists of judging its own work according to given criteria. While in peer assessment the participants are asked to review their peers' work, the peer assessment methods can contain only peer assessment as it may include also self-assessment. Peer assessment can be defined also as a process during which students consider the quality of a peer's work or performance, judge the extent to which it reflects targeted goals or criteria, and make suggestions for revision [46]. A comprehensive review of pros and cons of self and peer assessment was developed in [47]. A larger comparison involving self, peer and teachers assessment was proposed in [48].

Jackson and Marks [44] tried to improve post-graduated masters student by assessed reflections and grade withholding, they showed in their experience that a slight majority was sustaining grade withholding believing it help them analyze carefully the delivered feedback. They also showed that feedback quality correlated with grades improvements.

Staubitz et al. [27] maintained on the usefulness of peer assessment, and the difficulties encountered such as grading accuracy and rogue reviewers. They requested encouraging the practice by permitting self-assessed bonus point.

Gamage et al. [45] claimed that non-blind peer assessment can afford a high quality review, since permitting extended and consistent debates and avoiding harsh reviews that can be sent behind the vail of anonymity.

Numerous authors [6, 14, 40] insisted on the role of helping assessors by predefined notation scales, which can improve reliability and validity of the expressed reviews.

While Jones et al. [49] investigated peer assessment in the absence of judgement criteria in advance mathematics. Orsmond et al. [50] studied the importance of predefined criteria in peer assessment process for biology.

Suen [13] claimed that there is a necessity to combine several peer assessment methods in order to achieve accurate results.

Li [51] examined the role of anonymity and training on students' performances and perceptions in the process of peer assessment. While Sridharan et al. [52] evaluated the peer assessment effectiveness form the students' perspective. Recent studies on the anonymity impact were presented in [53, 54].

Cho et al. [40] questioned instructor's evaluation reliability. According to the authors, teachers are attempted to accelerate the evaluation process, especially when evaluating a large number of students. They can also modify evaluation criteria due to the same reasons. While students, are classically given a much smaller set of contributions to evaluate.

Overviews of peer assessment methods and future development are proposed in [55–57].

3 Fuzzy Logic Tools

A legitimate question can be formulated as follows: How to model assessors' opinions? Especially when their knowledge is vague and imprecise. Fuzzy logic [58] aims to model human uncertainties, particularly in decision making situations [59]. Capuano et al. [60] uses fuzzy sets to model assessors opinions in a decision making situation aiming to overcome unreliable opinions. Wu [34] uses fuzzy C-means to analyze English teachers' data in order to promote their evaluation ability and accuracy. Similarly Lubis et al. [61] used the same algorithm to measure learners satisfaction aiming to predict course completion of future students. Ospina-Delgado and Zorio-Grima [62] showed thru a fuzzy approach that the low intensiveness level in MOOCs is due to the absence of prestigious faculties.

Several methods have chosen fuzzy numbers to model opinions [63–67]. A fuzzy number (FN) $\widetilde{A}$ [68] is a fuzzy set defined by its membership function $\mu_{\widetilde{A}}: \mathbb{R} \to [0, 1]$. We restrict ourselves to trapezoidal fuzzy numbers (TrFN) given by 4-tuples $\widetilde{A}(a^1, a^2, a^3, a^4)$ where $a^1 \leqslant a^2 \leqslant a^3 \leqslant a^4$ and represented by:
  where

$$\mu_{\widetilde{A}}(x) = \begin{cases} \frac{x-a^1}{a^2-a^1} = \mu_{L_{\widetilde{A}}}(x) & \text{if } a^1 \leqslant x \leqslant a^2 \\ 1 & \text{if } a^2 \leqslant x \leqslant a^3 \\ \frac{x-a^4}{a^3-a^4} = \mu_{R_{\widetilde{A}}}(x) & \text{if } a^3 \leqslant x \leqslant a^4 \\ 0 & \text{else.} \end{cases}$$

Considering two TrFNs $\widetilde{A}(a^1, a^2, a^3, a^4)$ and $\widetilde{B}(b^1, b^2, b^3, b^4)$ the fuzzy inverse (Eq. 1), fuzzy addition (Eq. 2), fuzzy multiplication (Eq. 3), and fuzzy division (Eq. 4) are defined as follows:

$$1/\widetilde{B} = \left(1/b^4, 1/b^3, 1/b^2, 1/b^1\right) \tag{1}$$

$$\widetilde{A} \oplus \widetilde{B} = \left(a^1 + b^1, a^2 + b^2, a^3 + b^3, a^4 + b^4\right) \tag{2}$$

$$\widetilde{A} \otimes \widetilde{B} = \left(a^1 * b^1, a^2 * b^2, a^3 * b^3, a^4 * b^4\right) \tag{3}$$

$$\widetilde{A}ø\widetilde{B} = \left(a^1/b^4, a^2/b^3, a^3/b^2, a^4/b^1\right) \tag{4}$$

In the sequel we focus on enhancing validity and reliability of peer assessments.

4 Proposed Approach

Validity [40], can be measured as the correlation between students' assessments and teacher's evaluations.

We assume that the teacher's evaluation are fair. Remembering, that the teacher's objective, especially while assessing a large group of students, is to reduce his workload. He will be asked to assess a small number of students, to which, all other assessors' estimations will be compared. Hence, as a first step, we compute the similarity between $N : 1 \leq i \leq N$ student's judgments and the teacher's evaluation, according to all $K : 1 \leq j \leq K$ criteria. In addition, even if a student's judgement is valid according to a certain criterion, it is not necessary the case in accordance to the hole criteria list. Hence, each student evaluation validity, will be measured in accordance to all criteria.

Step 1: set the similarities between the teachers' fuzzy evaluation and the ith students' fuzzy judgements e_{ij} according to all criteria of evaluation, using the similarity proposed in [64].

Asking students to asses all other $N - 1$ participants, is practically infeasible even in class, the question is not conceivable in MOOCs context. Hence, each student will be evaluated by n peers, such that $n \ll N - 1$.

Step 2: each student $i : 1 \leq i \leq n$ expresses his opinion by a trapezoidal fuzzy number $\widetilde{R}_{ij}\left(r_{ij}^1, r_{ij}^2, r_{ij}^3, r_{ij}^4\right)$ in accordance to all criteria of evaluation.

Reliability [40] can be driven from notes spread, intuitively if an evaluated work is assessed as very good and very bad in the same time, we can conclude automatically that the reviews are inconsistent with each other's. A modified version of Lee's algorithm [65] was proposed in [69] to solve the following problem (Eq. 5), aiming to attribute adequate weights to opinions $\widetilde{R}_{ij}\left(r_{ij}^1, r_{ij}^2, r_{ij}^3, r_{ij}^4\right)$ in order to achieve consensus $\widetilde{R}_j\left(r_j^1, r_j^2, r_j^3, r_j^4\right)$.

$$\min_{M \times I\!R^4} \sum_{\substack{i=1 \\ j=j_0}}^{n} w_i^m * \left(c - e_{ij} * S\left(\widetilde{R}_{ij}, \widetilde{R}_j\right)\right), \tag{5}$$

where, $M = \left\{ \begin{array}{l} W = (w_1, w_2, \ldots, w_n), w_i \geq 0, \\ \sum_{i=1}^{n} w_i = 1, \end{array} \right\}$

m is a positive integer $m > 1$, n number of evaluators, $\widetilde{R}_{ij}$ the ith evaluation in accordance to the jth criterion, $\widetilde{R}_j$ the consensus in accordance to the jth criterion, $S\left(\widetilde{R}_{ij}, \widetilde{R}_j\right)$ the similarity between the ith decision and the consensus [65], c is a real number $c > 1$.

For all $K : 1 \leq j \leq K$ criteria of evaluation,

Step 3: fix the initial weights $W^{(0)}\left(w_1^{(0)}, \ldots, w_n^{(0)}\right)$ verifying $0 \leq w_i^{(0)} \leq 1$ and $\sum_{i=1}^{n} w_i = 1$. The iterations will be marked by $l = 0, 1, \ldots$

Step 4: compute

$$\widetilde{R}_j^{(l+1)} = \frac{\sum_{i=1}^{n} e_{ij} * w_i^{(l)m} \otimes \widetilde{R}_{ij}}{\sum_{i=1}^{n} e_{ij} * w_i^{(l)m}}$$

Step 5: compute

$$w_i^{(l+1)} = \frac{\left(1/\left(c - e_{ij} * S\left(\widetilde{R}_{ij}, \widetilde{R}_j^{(l+1)}\right)\right)\right)^{1/(m-1)}}{\sum_{i=1}^n \left(1/\left(c - e_{ij} * S\left(\widetilde{R}_{ij}, \widetilde{R}_j^{(l+1)}\right)\right)\right)^{1/(m-1)}}$$

Step 6: if $\left\| W^{(l+1)} - W^{(l)} \right\| \le \varepsilon$ stop, else set $l = l+1$ and go to step 4.

5   Illustrative Example

We adapt the example provided in [70] concerning peer evaluation. In this example, the student #A is evaluated by three participants (#B, #C and #D) and a teacher #T according to 7 criteria of evaluation (Table 2). The linguistic variables used in the evaluation process are presented in (Table 1).

Table 1. Linguistic variable for each assessment

Linguistic grade	Abbreviation	Membership function
Poor	P	(0, 0, 1.5, 3.5)
Below average	BA	(0.5, 2.5, 3.5, 5.5)
Average	A	(2.5, 4.5, 5.5, 7.5)
Above average	AA	(4.5, 6.5, 7.5, 9.5)
Excellent	E	(6.5, 8.5, 10, 10)

Table 2. An example of peer assessment evaluation

Evaluation criteria	Evaluators			
	#B	#C	#D	#T
C_1 : Participated in group meetings	AA	E	E	E
C_2 : Communicated constructively to discussion	E	E	E	AA
C_3 : Generally was cooperative in group activities	AA	AA	AA	AA
C_4 : Contributed to good problem-solving skills	AA	AA	A	A
C_5 : Contributed useful ideas	E	E	A	E
C_6 : Demonstrated good interest to task given	AA	AA	E	E
C_7 : Prepared drafts of report in good quality	E	E	AA	AA

Step 1: compute the similarities between each student opinion and the teacher's evaluation, using the similarity proposed in [64]. (Table 3).

Step 2: the students (#B, #C and #D) evaluates another participant #F (Table 4).

Step 3-6: using (Eq. 5) compute the overall results for student #F in accordance to all criteria (Table 4).

Comparing the criteria C_2, C_6 and C_7, we can observe that the participant #F was evaluated as Excellent by two other participants and above the average by the last one.

Table 3. An example of peer assessment evaluation measuring validity

Evaluation criteria	Validity measures		
	#B	#C	#D
C_1	0.5818	1	1
C_2	0.5818	0.5818	0.5818
C_3	1	1	1
C_4	0.5625	0.5625	1
C_5	1	1	0.3800
C_6	0.5818	0.5818	1
C_7	0.5818	0.5818	1

Table 4. An example of peer assessment evaluation 2

Evaluation criteria	Evaluators			Results
	#B	#C	#D	
C_1	A	AA	E	(5.2521, 7.2521, 8.4767, 9.5781)
C_2	E	AA	E	(5.8567, 7.8567, 9.1959, 9.8392)
C_3	AA	E	AA	(5.0871, 7.0871, 8.2339, 9.6468)
C_4	AA	AA	A	(2.9708, 4.9708, 5.9708, 7.9708)
C_5	E	AA	A	(5.3791, 7.3791, 8.6166, 9.6664)
C_6	AA	E	E	(6.2669, 8.2669, 9.7086, 9.9417)
C_7	E	E	AA	(4.9922, 6.9922, 8.1152, 9.6230)

However, the final results were clearly different from one criteria to another, according to the validity of each opinion.

Observing criterion C_7 according to which the slowest score was obtained between the three. The lowest evaluation was delivered by the most valid evaluator, explaining the final result.

The difference between C_2 and C_6 is debatable, it can be seen as in C_6 the assessment Excellent is more trustworthy, as it can be considered that none of the evaluation in C_2 are more important than the other. In such situation a normalization can bias the results, which explain our choice to use the validities without normalization.

6   Conclusion

In this study we proposed a methodology to improve peer reviews validity and reliability. To permit considering human uncertainty, all opinions were modeled by trapezoidal fuzzy numbers. Primary, a reference contribution is evaluated by the teacher and the students. Supposing the teacher's evaluation is fair, the students' judgement validity is deducted from their similarities to the teacher's one in accordance to all criteria. Then reliability is achieved by minimizing the sum of weighted incoherencies between the expressed opinions and the consensus for each criterion.

We assumed in this work that assessors are honest. Hence we restricted ourselves to validity and reliability. In future works, we can consider other aspects such as collusions.

Other aspects have to be investigated such as the compatibility between the linguistic variable representing opinions and the metrics used. Different similarities and parameters are to be considered to target the best fitting parameters.

Since, type 2 fuzzy sets permits a better representation of uncertainty, generalization to type 2 fuzzy sets are to be done.

References

1. Lewis Gaillet, L.: A Foreshadowing of modern theories and practices of collaborative Learning: The work of scottish rhetorician george Jardine. In: Presented at the 43rd Annual Meeting of the Conference on College Composition and Communication, Cincinnati OH, Mar 19 (1992)
2. García-Peñalvo, F.J., Fidalgo-Blanco, Á., Sein-Echaluce, M.L.: An adaptive hybrid MOOC model: disrupting the MOOC concept in higher education. Telemat. Inform. **35**, 1018–1030 (2018)
3. Giovannella, C., Martens, A., Zualkernan, I.: Grand challenge problem 1: people centered smart "cities" through smart city learning. In: Grand Challenge Problems in Technology-Enhanced Learning II: MOOCs and Beyond. pp. 7–12. Springer, Cham (2016)
4. Haber, J.: MOOCs. The MIT Press, Cambridge, Massachusetts (2014)
5. Alario-Hoyos, C., Pérez-Sanagustín, M., Delgado-Kloos, C.G.H.A.P., Muñoz-Organero, M., Rodríguez-de-las-Heras, A.: Analysing the impact of Built-In and external social tools in a MOOC on educational technologies. In: Scaling up Learning for Sustained Impact. pp. 5–18. Springer, Berlin, Heidelberg (2013)
6. Formanek, M., Wenger, M.C., Buxner, S.R., Impey, C.D., Sonam, T.: Insights about large-scale online peer assessment from an analysis of an astronomy MOOC. Comput. Educ. **113**, 243–262 (2017)
7. Ho, D., McAllister, S.: Are health professional competency assessments transferable across cultures? a preliminary validity study. Assess. Eval. High. Educ. **43**, 1069–1083 (2018)
8. Wilson, M.J., Diao, M.M., Huang, L.: 'I'm not here to learn how to mark someone else's stuff': an investigation of an online peer-to-peer review workshop tool. Assess. Eval. High. Educ. **40**, 15–32 (2015)
9. Usher, M., Barak, M.: Peer assessment in a project-based engineering course: comparing between on-campus and online learning environments. Assess. Eval. High. Educ. **43**, 745–759 (2018)
10. Bordel, B., Alcarria, R., Martín, D., Sánchez-de-Rivera, D.: Improving MOOC student learning through enhanced peer-to-peer tasks. In: Digital Education: Out to the World and Back to the Campus. pp. 140–149. Springer, Cham (2017)
11. Mulder, R., Baik, C., Naylor, R., Pearce, J.: How does student peer review influence perceptions, engagement and academic outcomes? A case study. Assess. Eval. High. Educ. **39**, 657–677 (2014)

12. Meek, S.E.M., Blakemore, L., Marks, L.: Is peer review an appropriate form of assessment in a MOOC? Student participation and performance in formative peer review. Assess. Eval. High. Educ. **42**, 1000–1013 (2017)
13. Suen, H.K.: Peer assessment for massive open online courses (MOOCs). Int. Rev. Res. Open Distrib. Learn. **15**, 312–327 (2014)
14. Ashton, S., Davies, R.S.: Using scaffolded rubrics to improve peer assessment in a MOOC writing course. Distance Educ. **36**, 312–334 (2015)
15. Love, K.G.: Comparison of peer assessment methods: reliability, validity, friendship bias, and user reaction. J. Appl. Psychol. **66**, 451–457 (1981)
16. Cho, K., Schunn, C.D., Wilson, R.W.: Validity and reliability of scaffolded peer assessment of writing from instructor and student perspectives. J. Educ. Psychol. **98**, 891–901 (2006)
17. Speyer, R., Pilz, W., Van Der Kruis, J., Brunings, J.W.: Reliability and validity of student peer assessment in medical education: a systematic review. Med. Teach. **33**, e572–e585 (2011)
18. Schunn, C., Godley, A., DeMartino, S.: The reliability and validity of peer review of writing in high school AP english classes. J. Adolesc. Adult Lit. **60**, 13–23 (2016)
19. Salehi, M., Masoule, Z.S.: An investigation of the reliability and validity of peer, self-, and teacher assessment. South. Afr. Linguist. Appl. Lang. Stud. **35**, 1–15 (2017)
20. Yoon, H.B., Park, W.B., Myung, S.-J., Moon, S.H., Park, J.-B.: Validity and reliability assessment of a peer evaluation method in team-based learning classes. Korean J. Med. Educ. **30**, 23–29 (2018)
21. James, S., Pan, L., Wilkin, T., Yin, L.: Online peer marking with aggregation functions. In: 2017 IEEE International Conference on Fuzzy Systems (FUZZ-IEEE). pp. 1–6 (2017)
22. Kearney, E.M.: Assessing learning. In: On Becoming a Teacher. pp. 85–89. Sense Publishers, Rotterdam (2013)
23. Sale, D.: Assessing learning. In: The Challenge of Reframing Engineering Education. pp. 59–80. Springer, Singapore (2014)
24. Ettarres, Y.: Evaluation of online assignments and quizzes using Bayesian networks. In: Innovations in Smart Learning. pp. 39–44. Springer, Singapore (2017)
25. Govindarajan, K., Boulanger, D., Seanosky, J., Bell, J., Pinnell, C., Kumar, V.S., Kinshuk.: Assessing learners' progress in a smart learning environment using bio-inspired clustering mechanism. In: Innovations in Smart Learning. pp. 49–58. Springer, Singapore (2017)
26. Zhu, M., Sari, A., Lee, M.M.: A systematic review of research methods and topics of the empirical MOOC literature (2014–2016). Internet High. Educ. **37**, 31–39 (2018)
27. Staubitz, T., Petrick, D., Bauer, M., Renz, J., Meinel, C.: Improving the peer assessment experience on MOOC platforms. In: Proceedings of the Third (2016) ACM Conference on Learning @ Scale. pp. 389–398. ACM, New York, NY, USA (2016)
28. Rust, C.: A briefing on assessment of large groups. In: LTSN Generic Centre: Assessment Series (2001)
29. Bali, M.: A new scholar's perspective on open peer review. Teach. High. Educ. **20**, 857–863 (2015)
30. Soh, K.C.: Peer review: has it a future? Eur. J. High. Educ. **3**, 129–139 (2013)
31. Millard, W.B.: The wisdom of crowds, the madness of crowds: rethinking peer review in the web era. Ann. Emerg. Med. **57**, A13–A20 (2011)
32. Clase, K.L., Gundlach, E., Pelaez, N.J.: Calibrated peer review for computer-assisted learning of biological research competencies. Biochem. Mol. Biol. Educ. Bimon. Publ. Int. Union Biochem. Mol. Biol. **38**, 290–295 (2010)
33. Purcell, M.E., Hawtin, M.: Piloting external peer review as a model for performance improvement in third-sector organizations. Nonprofit Manag. Leadersh. **20**, 357–374

34. Wu, J.: Empirical analysis of evaluation of english teachers' educational ability under MOOC environment. In: 2018 International Conference on Intelligent Transportation, Big Data Smart City (ICITBS). pp. 303–306 (2018)
35. Yin, Z.: Educational ability evaluation of japanese language teacher under MOOC environment. In: 2018 International Conference on Intelligent Transportation, Big Data Smart City (ICITBS). pp. 299–302 (2018)
36. Koç, E.S.: An evaluation of the effectiveness of committees of teachers according to the teachers' views, ankara province sample. Procedia - Soc. Behav. Sci. **174**, 3–9 (2015)
37. Song, Y., Hu, Z., Gehringer, E.F.: Collusion in educational peer assessment: How much do we need to worry about it?. In: 2017 IEEE Frontiers in Education Conference (FIE). pp. 1–8 (2017)
38. Gielen, S., Dochy, F., Onghena, P., Struyven, K., Smeets, S.: Goals of peer assessment and their associated quality concepts. Stud. High. Educ. **36**, 719–735 (2011)
39. Luo, H., Robinson, A.C., Park, J.-Y.: Peer grading in a MOOC: reliability, validity, and perceived effects. J. Asynchronous Learn. Netw. **18**, 1–14 (2014)
40. Derrick, G.: The Evaluators' Eye: Impact Assessment and Academic Peer Review. Palgrave Macmillan (2018)
41. Roberts, T.S. (ed.): Self, Peer and Group Assessment in E-learning. Information Science Publishing, Hershey, PA (2006)
42. Zheng, Q., Chen, L., Burgos, D.: Emergence and development of MOOCs. In: The Development of MOOCs in China. pp. 11–24. Springer, Singapore (2018)
43. Waks, L.J.: The Evolution and Evaluation of Massive Open Online Courses: MOOCs in Motion. Palgrave Macmillan US (2016)
44. Jackson, M., Marks, L.: Improving the effectiveness of feedback by use of assessed reflections and withholding of grades. Assess. Eval. High. Educ. **41**, 532–547 (2016)
45. Gamage, D., Whiting, M., Rajapakshe, T., Thilakarathne, H., Perera, I., Fernando, S.: Improving Assessment on MOOCs Through Peer Identification and Aligned Incentives. pp. 315–318 (2017). ArXiv170306169 Cs
46. Lui, A., Andrade, H.: Student Peer Assessment. In: Gunstone, R. (ed.) Encyclopedia of science education, pp. 1003–1005. Springer, Netherlands, Dordrecht (2015)
47. Adachi, C., Tai, J.H.-M., Dawson, P.: Academics' perceptions of the benefits and challenges of self and peer assessment in higher education. Assess. Eval. High. Educ. **43**, 294–306 (2018)
48. Alias, M., Masek, A., Salleh, H.H.M.: Self, peer and teacher assessments in problem based learning: are they in agreements? Procedia - Soc. Behav. Sci. **204**, 309–317 (2015)
49. Jones, I., Alcock, L.: Peer assessment without assessment criteria. Stud. High. Educ. **39**, 1774–1787 (2014)
50. Orsmond, P., Merry, S., Reiling, K.: The importance of marking criteria in the use of peer assessment. Assess. Eval. High. Educ. **21**, 239–250 (1996)
51. Li, L.: The role of anonymity in peer assessment. Assess. Eval. High. Educ. **42**, 645–656 (2017)
52. Sridharan, B., Muttakin, M.B., Mihret, D.G.: Students' perceptions of peer assessment effectiveness: an explorative study. Account. Educ. **27**, 259–285 (2018)
53. Pitt, E., Winstone, N.: The impact of anonymous marking on students' perceptions of fairness, feedback and relationships with lecturers. Assess. Eval. High. Educ. **43**, 1183–1193 (2018)
54. Rotsaert, T., Panadero, E., Schellens, T.: Anonymity as an instructional scaffold in peer assessment: its effects on peer feedback quality and evolution in students' perceptions about peer assessment skills. Eur. J. Psychol. Educ. **33**, 75–99 (2018)

55. Wahid, U., Chatti, M.A., Schroeder, U.: A systematic analysis of peer assessment in the MOOC era and future perspectives. In: Presented at the eLmL 2016, The Eighth International Conference on Mobile, Hybrid, and On-line Learning Apr 24 (2016)
56. Gielen, S., Dochy, F., Onghena, P.: An inventory of peer assessment diversity. Assess. Eval. High. Educ. **36**, 137–155 (2011)
57. Xiong, Y., Suen, H.K.: Assessment approaches in massive open online courses: possibilities, challenges and future directions. Int. Rev. Educ. **64**, 241–263 (2018)
58. Zadeh, L.A.: Fuzzy sets. Inf. Control **8**, 338–353 (1965)
59. Bellman, R.E., Zadeh, L.A.: Decision-making in a fuzzy environment. Manag. Sci. **17**, B141–B164 (1970)
60. Capuano, N., Loia, V., Orciuoli, F.: A fuzzy group decision making model for ordinal peer assessment. IEEE Trans. Learn. Technol. **10**, 247–259 (2017)
61. Lubis, F.F., Rosmansyah, Y., Supangkat, S.H.: Experience in learners review to determine attribute relation for course completion. In: 2016 International Conference on ICT For Smart Society (ICISS). pp. 32–36 (2016)
62. Ospina-Delgado, J., Zorio-Grima, A.: Innovation at universities: a fuzzy-set approach for MOOC-intensiveness. J. Bus. Res. **69**, 1325–1328 (2016)
63. El Alaoui, M.: SMART grid evaluation using fuzzy numbers and TOPSIS. IOP Conf. Ser. Mater. Sci. Eng. **353**, 012019 (2018)
64. El Alaoui, M., Ben-Azza, H., Zahi, A.: New multi-criteria decision-making based on fuzzy similarity, distance and ranking. In: Proceedings of the Third International Afro-European Conference for Industrial Advancement—AECIA 2016. pp. 138–148. Springer, Cham (2016)
65. Lee, H.-S.: Optimal consensus of fuzzy opinions under group decision making environment. Fuzzy Sets Syst. **132**, 303–315 (2002)
66. Chen, C.-T.: Extensions of the TOPSIS for group decision-making under fuzzy environment. Fuzzy Sets Syst. **114**, 1–9 (2000)
67. Chen, C.-T., Lin, C.-T., Huang, S.-F.: A fuzzy approach for supplier evaluation and selection in supply chain management. Int. J. Prod. Econ. **102**, 289–301 (2006)
68. Skalna, I., Rębiasz, B., Gaweł, B., Basiura, B., Duda, J., Opiła, J., Pełech-Pilichowsk, T.: Advances in Fuzzy Decision Making—Theory and Practice. Springer International Publishing (2015)
69. El Alaoui, M., Ben-Azza, H., El Yassini, K.: Optimal weighting method for fuzzy opinions. In: Presented at the International Conference on Industrial Engineering and Operations Management, Paris, France July 26 (2018)
70. Chai, K.C., Tay, K.M., Lim, C.P.: A new fuzzy peer assessment methodology for cooperative learning of students. Appl. Soft Comput. **32**, 468–480 (2015)

Serious Games Adaptation According to the Learner's Motivational State

Othman Bakkali Yedri[✉], Abdelali Slimani, Lotfi El Aachak, and Mohamed Bouhorma

Computer Science, Systems and Telecommunication Laboratory (LIST), Tangier, Morocco

{othmanbakkali, slimani.abdelali, lotfi1002, bouhorma}
@gmail.com

Abstract. Multi-agent systems (MAS) have demonstrated their relevance many times in the design of computer simulations or video games where a number of autonomous entities evolve in a complex and dynamic environment. Serious Games (SG) represent a new discipline at the frontier of simulation and gaming. We believe that a category of SG, intended to foster the learner motivation in a gamified environment, they introduce new and challenging problems for the community. Although several studies focused on providing an enjoyable game experience and ensure the effective learning outcomes, the learner motivation differs from the leaner to another and requires an adaptation of difficulty level to player competence. The current research studies the adaptation through serious game and proposes an adaptive model based on SMA architecture to solve the motivation issues, and then it illustrates the proposal for the professional job interview training and discusses the obtaining results.

Keywords: Serious games · Adaptation · Player motivation · Multi-Agents system · Fuzzy logic · Classification algorithms · Clustering · E-learning Flow theory

1 Introduction

Serious game has become one of the most popular teaching tools used by tutors to ensure both good knowledge and skills transfer to the learners [1]. In order to train employees for complex situations or hardly realizable in reality, serious games have been widely exploited in different sectors, such as communication, sensitization and so on. It is gradually gaining the education sector and now integrates the world of education. It then presents itself as a potential training resource which can be integrated into a global training system and places the learner in an immersive learning context [2–7].

The primary issue highlighted from learning on serious game is related to the learner motivation. It differs from the leaner to another and requires an adaptation of difficulty level to player competence. Indeed, the adaptation of game content and strategies according to the learner performance can beneficial [8], and provide a high effectiveness of these tools for education field [9, 10]. Adaptation of level difficulties on

© Springer Nature Switzerland AG 2019
M. Ben Ahmed et al. (Eds.): SCA 2018, LNITI, pp. 419–436, 2019.
https://doi.org/10.1007/978-3-030-11196-0_36

the play experience aims to promote the player motivation and enhance learning; leaner participates on the right level adequate to their competencies, and ensures the training on the best progress [11–14].

In the purpose to build a game as close as possible to the professional job interview situation, we began by understanding the professional situation of the job interview profession from an analysis of the activity. It relies on the field of professional didactics and assumes to analyze the activity, then the tasks and all the technical, social and organizational conditions in which is inserted.

The current training involves the prediction of the player' emotional state though serious game experience and ensure the optimal motivation. It analysis the recruitment interview and proposes a serious game to learn their activities and provides adaptation through the play experience. In this perspective, we propose a conceptualization of their architecture based on the MAS.

The current research introduces concepts of the multi-agents system (MAS), and then studies the adaptation model of the flow theory to promote the player motivation in serious game. The paper proposes an adaptation model based on multi-agents system to predicate the emotional state of player and ensure the optimal motivation, it proposes in addition an illustration of the proposal for the professional job interview training and discusses the obtaining results.

2   Theoretical Background

2.1   Multi-Agents SYSTEMS: Concepts and Reference Theories

Artificial intelligence is a very large field. But their goals are really concrete. Achieving systems that can be adapt to a dynamic environment, such as robotics for space exploration, is one of them. The realization of expert systems is also part of it, particularly in medicine or industry.

However, Artificial intelligence has several current including one that allows apprehending, manipulating and designing forms of intelligence: multi-agent systems. This current of artificial intelligence also aims to allow the distribution of intelligence, allowing a more open conception of so-called intelligent systems.

Generally, the system should be adequate when conceiving it. Although, dynamics and openness triggered unexpected events, that's why the system must be able to adapt to several changes [15].

In our research, we found a number of definitions concerning agents. According to S. Russell et P. Norvig:

«An agent is any entity that can be considered as perceiving its environment through sensors and that acts on this environment through effectors.» [16].

According to M. Wooldridge,

«An agent is a computer system that is situated in some environment, and that is capable of autonomous action in this environment in order to meet its design objectives.» [17].

J. Ferber has a very detailed definition taking into account all the components of an agent:

«An agent is a physical or virtual entity (a) that is able to act in its environment, (b) can communicate directly with other agents, (c) is driven by a set of trends (in the form of individual objectives or a function of satisfaction or survival, which it seeks to optimize), (d) who has own resources, (e) who is able to perceive (but in a limited way) his environment, (f) who has only a partial representation of this environment (and possibly none), g) who possesses skills and offers of services, (h) who may possibly reproduce, (i) whose behavior tends to meet his objectives, taking into account the resources and the skills available to them, and according to their perception, their representations and the communications they receive.» [18]

Depending on the type of agent used, we will speak of cognitive systems or reactive systems. Thus, in the MAS, the programming of the agents is concentrated largely on systems whose formal specification was a prerequisite for all programming and modeling approaches, with a level of abstraction allowing understanding the system in the unique form of agents system. The notions of assistance to users and guidance by agents have been taken into account only through the prism of this programming.

For J. Ferber, a multi-agent system is the realization of "electronic or computer models composed of artificial entities that communicate with each other and act in an environment" [18]. It states that agents capable of communicating grouped together in a community form allow many interactions. From these interactions appear structures organized in this community of agents. These structures define and then influence the behaviors of agents and at the same time the behaviors of the mufti-agents system. Moreover, J. Ferber himself gives us his definition of a multi-agent system:

«A multi-agent system (or MAS) is a system composed of the following elements: (1) An environment E, that is to say a space usually with a metric; (2) A set of objects O. These objects are located, that is to say that, for any object, it is possible, at a given moment, to associate a position E. These objects are passive, it is that is, they can be perceived, created, destroyed and modified by the agents; (3) A set of agents, which are particular objects (A < O), which represent the active entities of the system; (4) A set of relations R which unite objects (and thus agents) between them; (5) A set of Op operations allowing the agents of A to perceive, produce, consume, transform and manipulate objects of O; (6) Operators to represent the application of these operations and the reaction of the world to this attempt to change, which will be called the laws of the universe.» [18]

However, due to the unexpected and even unknown context changes, it is necessary to reconsider the planning and guidance of the agents in order to adapt them to the systems. Learner behavior analysis identifies the main factors limiting its evolution and allows decision-makers to develop new strategies to ensure a good transfer of information. In this study, we focused on the emotional state of the player based on two essential factors: 'Challenges', 'Abilities' in order to keep the player in an optimal emotional state.

2.2 Adaptation Through Serious Games

The adaptive difficulty through serious game aims to promote the player motivation and enhance learning by providing the right level of challenge in real time. It ensure the learner to complete their training on the best progress [19]. Several previous researches focused in general game designing, motivation and assessment design for more effectiveness and optimization in order to, achieve player's satisfaction. Serious games highlights two main approaches to provide an adapted content: (1) the dynamic adaptation (online) focuses on the content adjustment related to learner's performance in time of play see [20], (2) the offline adaptation propose an adjustment of game content based on a sufficient information collected at the begin of game [21]. The assessment of the learner's competency will be more accurate when he or she is actually performing the task.

An adaptation technique refers to the set of concepts, structures and computerized means used to implement an adaptation process. We can identify three generic levels by following the MVC decomposition (Fig. 1):

Fig. 1. Adaptation layers

Adapting game presentation focuses on a dynamic personalization of the user screen, son, and feedback management [22]. The presentation component can provide an adaptation adjusting the frequency of screens and the graphic levels. In this purpose, the adaptation engine must implements the required mechanics to keep the player attention, this contribute, in the context of a use for educational purposes for example, to have beneficial consequences on their pedagogical motivations [23].

The game control defines the specific business and game rules adequate to respond the players' actions. Adapting the control component provides new rules and modification of existing rules. It can for example modify parameters of game difficulty like speed and challenge [23].

The adaptation of content provides a dynamic information related to the play context, pedagogical goals and the player skills [23]. It can provide an automatic generation of dialogues and narrative texts or of sound ambience.

2.3   Learner's Motivation

Motivation is often presented in a central construct of learning theories. Motivation is defined as the tensor of internal and external forces, directed or not directed by a goal, that influence a person cognitively, affectively or behaviorally [24]. It is a process that influences triggering, direction, intensity, persistence and frequency of behavior or attitudes [24].

In cognitive psychology, the motivation for success depends on the individual's desire and expectations, personal sense of self-efficacy, and strengthening or support in the social environment. It is defined as a continuum that goes from one extreme, the non-motivation, to another, the intrinsic motivation (which comes from the individual himself), going through the extrinsic motivation (which comes from stimuli external to the individual).

One of the main benefits of using serious gaming in class is the overall positive impact on student motivation. If we think that the novelty aspect of the introduction of the game in class comes into play, the few studies on the use of games over the long term show that the object "Game serious" [25]. Indeed, an adapted game gives regular returns to the student on his actions, thus maintaining his motivation [26].

Flow theory [27] is based on a symbiotic experience between challenges and the skills that need to be implemented to address them. However, flow experience arises when skills are not overtaken or underused, when the challenge is optimal. When the individual plunges into the flow, the involvement in the activity is such that he forgets time, fatigue and everything around him except the activity itself. In this state, the individual operates to the maximum of his abilities and for the flow experience. The activity is performed for itself (as defined in the context of intrinsic motivation), even if the goal is not yet attaint (Fig. 2).

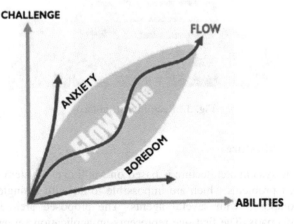

Fig. 2. Flow state [28]

3   Serious Game Adaptation

The main objective of this current paper is to adapt a serious game based on several parameters related to the learner's motivational state during the game and the adaptation will be done by using a variety of machine learning algorithms, e.g. "Fuzzy logic, Clustering, decision tree". In this section, we will describe the proposed serious game and the establishment of the hall system.

3.1   The Proposed Serious Game

The purpose of our game Recruitment protocol game is to prepare the player as good as possible for his job interview or job promotion, providing him with an interview and various real situations. The learner should manage to face each situation by selecting right answers in the right moment. The player must choose among several answers the one that seems to him the most adapted, thus becoming aware of the rules of politeness, adaptability, and motivation and so on, in order to be implemented during a job interview (Fig. 3).

Fig. 3. Recruitment protocol

3.2   System Architecture

As mentioned the system architecture is based on Multi agent system (MAS). In order to solve complex problems which are impossible to do with a single agent, we proposed MAS composed with several agents. The proposed architecture system is composed by two parts. The first one represents an application server where the proposed serious game is deployed. The second part is established by using JADE framework [29]. The proposed multi-agents system Fig. 4 is composed of three main agents. The first one is the predictor agent, its role is to predict the learner competencies during the game sequence, and this agent is established by using artificial Fuzzy logic. The second is to predict and save data in database. The third one is the decision maker, it takes a decision to adapt the game according to both learning difficulty and competencies of the player.

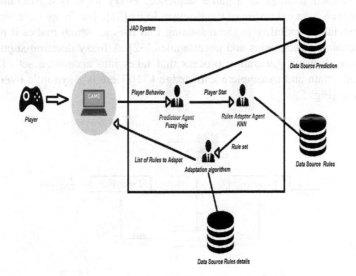

Fig. 4. Proposed adaptive multi agent system

We provided to use machine learning algorithms in order to produce a model capable of predicting a learner's emotional state.

Machine learning can be summarized by the definition of a mathematical model which finds its parameters thanks to the statistical properties of a data set [30].

Collected data saved into the database, as shown in the Table 1, these informations will be operated also by the tools of learning analytics and educational data mining, to have a global view on the progression of all the learners.

Table 1. Quiz database attributes

Quiz
✓ Age
✓ Sexe
✓ Session
✓ Question
✓ Response
✓ Difficulty
✓ Result
✓ Average of time response
✓ Number of wrong answers
✓ Game abort
✓ Objectives

The fuzzy logic agent

The fuzzy logic agent is an agent of MAS that predict learner's engagement according to their feelings in a game sequence. Fuzzy logic is a machine learning algorithm which is an extension of the Boolean logic [31]. The fuzzy logic thus confers a very appreciable flexibility to the reasoning that uses it, which makes it possible to take into account inaccuracies and uncertainties [32]. A fuzzy decision-support model is a real-time decision generation process that takes into account a set of different, evolving, uncertain and incomplete knowledge [32]. Here is a synoptic overview of a fuzzy system (Fig. 5):

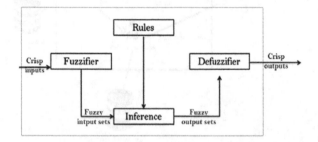

Fig. 5. Synoptic overview of a fuzzy system

In our emotional state case of study, we need to redefine membership functions for each fuzzy subset of each of our three variables as presented in Table 2:

Table 2. Language variables

Input	Competence	Competent, non-competent
	Difficulty	Low, Medium, High
Output	Emotional state	Anxiety, Bored, Flow

Let V be a variable (quality of the questions, difficulty, competence, etc.), X the range of values of the variable and TV a finite or infinite set of subsets blurred [32]. A linguistic variable corresponds to the triplet (V, X, TV) [32]. Linguistic variables are presented below in Figs. 6, 7 and 8.

In fuzzy logic, fuzzy reasoning, also called approximate reasoning, is based on fuzzy rules that are expressed in natural language using the linguistic variables that we have defined earlier [32].

The result of applying a fuzzy rule therefore depends on three factors:

– The definition of fuzzy involvement chosen;
– The definition of the membership function of the fuzzy set of the proposition located at the conclusion of the fuzzy rule;
– The degree of validity of the propositions situated in premise.

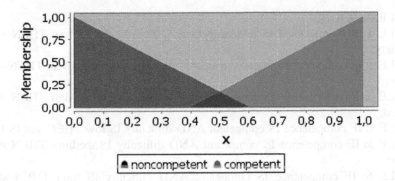

Fig. 6. Language variable 'Competence'

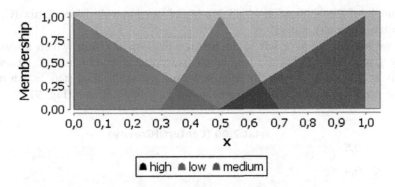

Fig. 7. Language variable 'Difficulty'

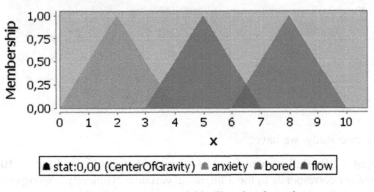

Fig. 8. Language variable 'Emotional state'

As we have defined the fuzzy AND, OR, and NOT operators, the premise of a fuzzy rule may very well be formed by a conjunction of fuzzy propositions. The set of rules of a fuzzy system is called the decision matrix. Here is one of our caseof study:

RULEBLOCK No1
RULE 1: IF competence IS noncompetent AND difficulty IS low THEN stat IS anxiety;
RULE 2: IF competence IS noncompetent AND difficulty IS medium THEN stat IS flow;
RULE 3: IF competence IS noncompetent AND difficulty IS high THEN stat IS bored;
RULE 4: IF competence IS competent AND difficulty IS low THEN stat IS bored;
RULE 5: IF competence IS competent AND difficulty IS medium THEN stat IS flow;
RULE 6: IF competence IS competent AND difficulty IS high THEN stat IS anxiety;
END_RULEBLOCK

Example of the application of all 3 rules of our input decision matrix (Competence = 0.6, Difficulty = 0.33)

As we see, all we have left to do is make the final decision, which tip we will actually give, knowing that the learner competencies is rated 6 out of 10 and the difficulty is 3,33 on 10. Defuzzification is considered as the final step, which moves from the fuzzy set of conclusions aggregation to a single decision (Fig. 9).

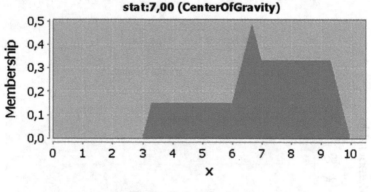

Fig. 9. Defuzzification

In our case study, we have:

- the input is 'the skill is rated 6 out of 10 and the difficulty 3.33 out of 10';
- the fuzzifier corresponds to the 3 linguistic variables 'Difficulty', 'Competence' and 'Emotional State';
- defuzzifying is the part where the defuzzification method (COG method) comes into play;
- the output corresponds to the final decision: the emotional state is 7 '.

The predictor based on KNN

In machine learning there are different types of classifiers that have been developed, always with the aim of achieving a maximum degree of accuracy and efficiency, each with its advantages and disadvantages. However, they share common characteristics. The construction of the majority of them involves two main stages. First, the definition of the function that associates a document with a value between 0 and 1 that represents its degree of membership of the category. This function is called CSV for Categorization Status Value and takes different forms depending on the type of classifier. The second step, but not least, is to determine a threshold that will be used in the final decision, whether or not a document will be accepted or rejected from the category. If the CSV function returns a value greater than the threshold for a document, then it is decided to associate it with the category. Several methods for determining the threshold are possible and the choice of one of them can significantly influence the performance of a classifier.

In artificial intelligence, the k nearest neighbors method is a supervised learning method. In this framework, we have a learning database consisting of N input-output pairs. To estimate the output associated with a new input x, the method of the k nearest neighbors consists in taking into account (identically) the k learning samples whose input is closest to the new input x, according to a Distance to be defined [33].

For example, in a classification problem, one will retain the class most represented among the k outputs associated with the k inputs closest to the new input x.

The algorithm (as described in [34, 35]) can be summarized [36] as:

1. A positive integer k is specified, along with a new sample [36]
2. We select the k entries in our database which are closest to the new sample [36]
3. We find the most common classification of these entries [36]
4. This is the classification we give to the new sample [36] (Table 3).

Table 3. Knowledge database attributes

Attributes	Values
Difficulty	{1, 2, 3,4,5}
Competence	{1,2,3,4,5,6,7,8,9,10}
Flowstat	{flow, bored, anxiety}
gRules	{1, 2, 3}

The attributes "Inputs" that feed the classificatory algorithm are: the result of the predictor agent "calculated through the fuzzy logic, according to the learner's competence and the game difficulty. By cons the classes "Outputs" are: collected data in the knowledge database.

One of the characteristics of instance-based learning is that there is no construction of an explicit description of the function to learn. The advantage is that we do not estimate the function for the entire space once, but we estimate it rather locally and differently for each new instance. The disadvantage is that although the training cost of the classifier is low because it only memorizes training examples, the cost of classifying new instances can be high, since that is when all calculation is done. However, good indexing of examples helps a lot to overcome this problem.

Decision adaptor

The third agent in the proposed system is called the decision maker; its role is to take decisions, in order, to adapt the serious game according to several parameters like the result of the second agent "calculated through the KNN", the Competence, the difficulty and learner's emotional state all these parameters will build dynamically a decision (Table 4).

Table 4. KNN rules

Rule level	Description
1	The game is very difficult
2	The game is very easy
3	Nothing to do

The proposed algorithm is:

Afterwards the KNN agent will modify the chosen parameter of "Recruitment protocol" game, according to this automated modification, the game will be adapted.

Interaction between agents:

As we have seen previously, the models interact and exchange data that can be extracted from the simulation or deduced from reasoning internal to the model.

Step 1: Observe Using the interface, the system analyzes the learner's interactions. Necessary information is: the learner's actions, the elements observable by the learner and their movements.

Step 2: Detect and identify the levels (FuzzyLogicAgent). The system analyzes the learner's actions (learner model) and compares them to the actions to be performed. This confrontation makes it possible to detect a possible the average of learner's motivational state.

Step 3: The output of step 2 is considered as an input of RulesAdaptorAgent with competence and difficulty. The rules adaptor (RulesAdaptorAgent) uses a learner model (difficulty, session, level, errors, etc.) and the domain model (knowledge on organizational structures), a mechanism simulating pedagogical reasoning advocates the audience adapted to the situation.

Step 4: Provide a game adaption (AdaptorAgent) according to the new set of game rules generated by KNN agent.

Step 5: Representing adapted game in the virtual environment.

In the next section, we will discuss and interpret the obtained results, in order to evaluate in a large scale the impact of the adaptation of the learning process through serious games.

4   Result and Discussion

In this part, we will put into experiment the MAS proposed model. The main idea is to adapt the game sequences according to learner emotional state. Obtained data (50 players) have been analyzed by using some machine learning algorithms as shown above.

First of all, we provide to present the distribution of request of an individual in a game sequence. There answers will be clustered by using Data ming algorithm. An algorithm often used in educational data mining to cluster learners' performances, the criteria used to cluster the learners are "id_session, Competence, right answers, wrong answers, and difficulty" [37, 38], Fig. 10.

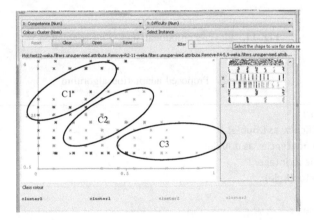

Fig. 10. Player's emotional state before adaptation

In Fig. 10, we have three clusters as defined:

C1: Anxiety (31%)
C2: Flow (36%)
C4: Flow (9%)
C3: Boredom (24%)

According to the given results 45% of the learners are satisfied but the others have the ability to quit the game according to their negative feelings. According to the obtained result, we have decided to adapt the game in order to ensure a balance between competence and question difficulty by using multi agent system as presented above. The combination of these elements and the implementation of intelligence in the game results a feeling of well-being. This feeling creates an order in our state of consciousness and strengthens the structure of self. Self-development and good learning transfer occurs only when the experience is practiced as positive by the learner.

As present in the Fig. 11 there is a harmony between competence and question difficulty. The combination of these elements results a feeling of well-being that the

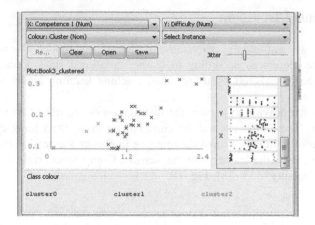

Fig. 11. Player's emotional state after adaptation

Table 5. Proposed adaptation algorithm

```
Start
Difficulty as Double;
Competencies as double;
Rule as integer;
Switch(Rule)
{
Case 1:
// get game difficulty
// get player competencies
// get corresponding difficulty according to the player competencies
Break;
Case 2:
// get game difficulty
// get player competencies
// get corresponding difficulty according to the player competencies
Break;
Case 3:
// get questions with the same level of difficulty
Break;
Default:
Break;
}
End
```

mere fact of being able to feel it justifies a great expense of energy. This feeling creates an order in our state of consciousness and strengthens the structure of self. Self-development occurs only when the interaction is experienced as positive by the person (Table 5).

As presented in Table 6, we have taken these parameters into consideration:

Table 6. Before and after game adaptation

Heading level	Average of the time spent per session	Average of the right answers	Average of the wrong answers	Average of abort
Before adaptation	7,3	5,1	4,6	3,5
After adaptation	4,5	8,3	2,1	1,1

- Average time spent per session
- Average of the wrong
- Average of right answers
- Average of the game aborts mad by the learners

During the different phases of the experiment, we can see that the link between the individual interest before and after the first game adaptation, then between the first and the second experience has slightly progressed. On the other hand, we also note a remarkable progression concerning the links between the difficulty and learner's progress.

Our results highlight that while showing links between their structures (around 4 dimensions: cognitive absorption, alteration of the perception of time, lack of concern about the self and well-being), the flow felt in a situation of ordinary education is different from that felt in the context of using a serious game.

In light of this study, we can conclude that the serious game adaptation improves the optimal learning experience of learners. As we might expect, the alteration of the perception of time is related to the situational interest provoked by the introduction of serious play during the training. We formulate the hypothesis that this contributes to increasing their working time, to reinforcing their concentration in the work, while at the same time providing them with good -being what can be a predictor of a reengagement in the activity, in order to perceive again all the pleasant elements of this optimal experience.

5 Conclusion and Future Works

A serious game provides an additional mean to increase learning interest, coaching and evaluation of user performance, It offers a rich tapestry by using a number of tools and techniques that make them more and more benefits. This research proposed the prediction of the learners' emotional state through serious game experience and provided an implementation of the adaptation process to enhance learning and ensure their optimal motivation. It analyzed the proposal serious game to learn the recruitment interview and provided adaptation through the play experience.

The implementation of such MAS allowed the improvement of the learning process by attracts the learner interest, and motivate him to continue learning as the game progression. With the obtained result the proposed solution will be more interesting with a large scale of the data saved on knowledge base. Adaptation process, used in this paper, is based on the learner emotional state that can adapt a variety of game parameters in order to improve learners' immersion without interaction of the instructors. The proposed solution can be used also to evaluate the learner engagement in a more deterministic manner.

References

1. Belahbib, A., Elaachak, L., Bouhorma, M., Yedri, O.B., Abdelali, S., Fatiha, E.: Serious games adaptation according to the learner's performances. Int. J. Electr. Comput. Eng. IJECE 7(1), 451–459 (2017)
2. 3DiTeams | Medical | Team Training | DoD Patient Safety Program | TeamSTEPPS. [Online]. Available: http://www.virtualheroes.com/portfolio/Medical/3DiTeams. Accessed: 10 Jun 2017
3. Robert, S., Graham, S.R.: A simulation learning approach to training first responders for radiological emergencies. In: Proceedings of the 2007 Summer Computer Simulation Conference, SCSC 2007, San Diego, California, USA (2007)
4. Greitzer, F.L., Kuchar, O.A, Huston, K.: Cognitive science implications for enhancing training effectiveness in a serious gaming context. J. Educ. Resour. Comput. 7(3), p. 2–es, Nov (2007)
5. Prensky, M.: Digital game-based learning. McGraw-Hill, New York (2001)
6. De Gloria, A., Bellotti, F., Berta, R.: Serious Games for education and training. Int. J. Serious Games 1(1), Feb (2014)
7. Zielke, M.A., et al.: Serious games for immersive cultural training: Creating a living world. IEEE Comput. Graph. Appl. 29(2), 49–60 (2009)
8. Roberts, D.L., Isbell, C.L.: A survey and qualitative analysis of recent advances in drama management. Int. Trans. Syst. Sci. Appl. Spec. Issue Agent Based Syst. Hum. Learn. 4(2), 61–75 (2008)
9. Vermeulen, H., Gain, J.: Experimenal Methdology For Evaluating Learning In Serious Games
10. Mortara, M., Catalano, C.E., Fiucci, G., Derntl, M.: Evaluating the effectiveness of serious games for cultural awareness: the icura user study. In: De Gloria, A. (ed.) Games and Learning Alliance, vol. 8605, pp. 276–289. Springer International Publishing, Cham (2014)
11. Alvarez, J., Djaouti, D.: An introduction to Serious game Definitions and concepts. Serious Games Simul. Risks Manag, p. 11 (2011)
12. Loh, C.S., Anantachai, A., Byun, J., Lenox, J.: Assessing what players learned in serious games: in situ data collection, information trails, and quantitative analysis. In: 10th International Conference on Computer Games: AI, Animation, Mobile, Educational & Serious Games (CGAMES 2007), pp. 25–28 (2007)

13. Bellotti, F., Kapralos, B., Lee, K., Moreno-Ger, P., Berta, R.: Assessment in and of serious games: An overview. Adv. Hum. Comput. Interact. **2013**, 1–11 (2013)
14. Dehem S., et al.: Assessment of upper limb motor impairments in children with cerebral palsy using a rehabilitation robot and serious game exercise. In: The 5th IEEE Conference on Serious Games and Applications for Health (2017)
15. Bernon, C., Camps, V., Gleizes, M.-P., Picard, G.: Engineering adaptive multi-agent systems: The adelfe methodology. Agent-Oriented Methodol. pp. 172–202 (2005)
16. Russell, S., Norvig, P.: Intelligence artificielle: Avec plus de 500 exercices. Pearson Education France (2010)
17. Wooldridge, M.: Intelligent agents. (1999)
18. Ferber, J.: Les systèmes multi-agents: vers une intelligence collective. InterEditions (1995)
19. De Gloria, A., Bellotti, F., Berta, R.: Serious Games for education and training. Int. J. Serious Games. **1**(1), Feb (2014)
20. van Oostendorp, H., van der Spek, E.D., Linssen, J.M.: Adapting the Complexity Level of a Serious Game to the Proficiency of Players. EAI Endorsed Trans. Serious Games. **14**(2), e5:1–e5:8 May (2014)
21. Lopes, R., Bidarra, R.: Adaptivity challenges in games and simulations: a survey. IEEE Trans. Comput. Intell. AI Games **3**(2), 85–99 (2011)
22. Dias, R., Martinho, C.: Adapting content presentation and control to player personality in videogames. In: Proceedings of the 8th International Conference on Advances in Computer Entertainment Technology. pp. 18:1–18:8, New York, NY, USA, (2011)
23. Hocine, N., Gouaïch, A., Di Loreto, I., Abrouk, L.: Techniques d'adaptation dans les jeux ludiques et sérieux. Rev. Sci. Technol. Inf.-Sér. RIA Rev. Intell. Artif. **25**(2), 253–280 (2011)
24. Karsenti, T.P., Thibert, G.: The Interaction between Teaching Practices and the Change in Motivation of Elementary-School Children. (1998)
25. Malone, T.W.: Toward a theory of intrinsically motivating Instruction*. Cogn. Sci. **5**(4), 333–369 (1981)
26. Whitton, N.: Game Engagement Theory and Adult Learning. Simul. Gaming **42**(5), 596–609 (2011)
27. Csikszentmihalyi, M., LeFevre, J.: Optimal experience in work and leisure. J. Pers. Soc. Psychol. **56**(5), 815–822 (1989)
28. FlowFig1.png (298 × 246).: [Online]. Available: http://edutechwiki.unige.ch/fmediawiki/images/4/46/FlowFig1.png. Accessed 10 Apr 2017
29. Jade Site | Java Agent DEvelopment Framework
30. Kotsiantis, S.B., Zaharakis, I.D., Pintelas, P.E.: Machine learning: a review of classification and combining techniques. Artif. Intell. Rev. **26**(3), 159–190 (2006)
31. Zadeh, L.A.: Fuzzy sets. Inf. Control **8**(3), 338–353 (1965)
32. Dernoncourt, F., Sander, E.: Fuzzy logic: between human reasoning and artificial intelligence. Masters Thesis ENS Ulm (2011)
33. Vipani, R., Hore, S., Basak, S., Dutta, S.: Detection of epilepsy using Hilbert transform and KNN based classifier. pp. 271–275 (2017)
34. [Online]. Available: http://www.lkozma.net/knn2.pdf. Accessed 31 Oct 2017
35. Sutton, O.: Introduction to k nearest neighbour classification and condensed nearest neighbour data reduction. Univ. Lect. Univ. Leic. (2012)
36. Feature Extraction of Diabetic Retinopathy Images.: [Online]. Available: http://research.ijcaonline.org/ncet/number1/ncet1410.pdf. Accessed 24 Dec 2017

37. Bakkali Yedri, O., Elaachak, L., Belahbib, A., Zili, H., Bouhorma, M.: Motivation analysis process as service applied on serious games. Int. J. Inf. Sci. Technol. 2(1) (2018)
38. Bakkali Yedri, O., El Aachak, L., Belahbib, A., Zili, H., Bouhorma, M.: Learners' motivation analysis in serious games. In: Ben Ahmed, M., Boudhir, A.A. (eds.) Innovations in Smart Cities and Applications. Vol. 37, pp. 710–723. Springer International Publishing, Cham (2018)

Service Oriented Computing and Smart University

Ouidad Akhrif[✉], Younès EL Bouzekri EL Idrissi, and Nabil Hmina

Ibn Tofail University, System Engineering Laboratory, ENSAK,
Kenitra, Morocco
akhrif.ouidad@univ-ibntofail.ac.ma

Abstract. Digitization is a catalyst for transforming a traditional university that is primarily based on human intelligence and that implements good human-human learning practices, towards an intelligence university that uses artificial intelligence and which integrates intelligent learning platforms. Admittedly, the traditional university boasts of human intelligence that has served many learning techniques for many years. However, the content and the organization of higher education should change with the evolution of society and communications technologies, this need has emerged a smart system that conform to the referential smart city system, in terms of infrastructure, interactions, reasoning and visualization. The importance of service oriented computing (SOC) is reflected in the flexible integration of contextual data collection, anticipation, prediction and recommendation services, into a university system to make it intelligent and proactive, through the reuse, the composition and the context awareness of the services.

Keywords: Smart city · Smart university · Smart service · Service oriented computing

1 Introduction

In the current context of "information overload" and with the development of web and communications technologies, the world becomes submerged by information and data and individuals are increasingly affected by these changes. There is, therefore, a need for designing an instrumented, interconnected and intelligent city, called "Smart City".

To improve the quality of life of its citizens, anticipate their needs in education, transport, and health, solve energy and environmental problems, and ensure sustainability in terms of services, the smart city (SC) adopts a strategy that is mainly based on the integration of Information and Communication Technologies (ICT). The aim is to promote its ability to learn, respond and recommend relevant services for these citizens.

The main goal of Smart City (SC) is to provide efficient services for its citizens in order to satisfy their needs. This presupposes not only a finer management of these different services but also more coordination between them, more communication, more integration and more involvement of the citizens who will furthermore contribute to enriching the services, their applications, and their functionalities. This requires a vision to design, discover and deploy services.

© Springer Nature Switzerland AG 2019
M. Ben Ahmed et al. (Eds.): SCA 2018, LNITI, pp. 437–449, 2019.
https://doi.org/10.1007/978-3-030-11196-0_37

"Service-Oriented Computing (SOC) is the computing paradigm that utilizes services as fundamental elements for developing applications/solutions" [1] which are fast, inexpensive, interoperable, scalable and massively distributed. The main benefit of Service Oriented Computing (SOC) is the flexible integration of contextual data collection, anticipation, prediction and recommendation services into a city system to make it smart and proactive.

To integrate services components in a smart city system (SCS), several approaches were developed. Our research will focus on the integration of services within the smart city (SC) and more specifically in a Smart University. Along with this, a set of basic components of a Smart University (SU) will be described.

In our case, we are interested in studying universities according to the vision of Service Oriented Computing (SOC) in order to identify the opportunities and the constraints adapted to this level. To present our vision, this paper is organized into five sections: The second section presents smart service in the smart city (SC). The third section deals with the SOC-based environment. The fourth section talks about the shift from digital university to smart university and the last one is concerned with SOC-based university opportunities.

2 Smart Service in the Smart City

The growing prevalence of technology-driven concepts, such as the "Internet of things" or "Industry 4.0", holds substantial potentials for service sectors worldwide [2]. One of the challenges of this kind of environments is to allow mobile users to seamlessly consume and often combine functionalities offered by software and hardware resources anywhere and at any time [3].

Moreover, the idea of conglomeration of services as a goal that requires the generation of transversal platforms to manage the multiple services involved in the smart cities ecosystem has been evolving [4].

2.1 Smart City Layers

Smart cities are urban environments that exploit heterogeneous data received from sensors to deliver intelligent services for the benefit of users.

Sensors: They are entities (Citizen, RFID, GPS, IR, camera, laser scanners, etc.) which are interconnected under an "Internet of Things" (IoT) infrastructure; for example, environmental sensors can collect data from the environment, such as noise and air quality [5].

Data: They are two different types of the collected data: the first type is automatically observed by the city sensors, while the second is declared by the citizen, who acts as mobile sensors in their daily interactions with the city. These collected data (heterogeneous, structured, and unstructured) can be represented in a structured form through a process of processing, analysis, and processing (Hadoop allows noise filtering, data classification and information extraction).

Smart Services: Smart Computing Technologies (SCT) solutions provided by the Smart City to allow smart interactions between citizen and his environment and satisfy user requirements and needs. These include the following:

Smart City Energy Service: It's a service that effectively controls the electrical use of the entire city.

The smart municipal energy: service can adjust the unnecessary electricity consumption by using the threshold value of electricity use across the smart city service platform and the smart home service platform.

Smart City Water Service: It's a service to effectively manage the water use of the entire city.

Our study will focus on the layer of smart services in order to highlight its importance and relevance in the city.

2.2 Citizen-Centric Smart Service

The smart service represents the application layer provided by a smart city, where the citizen interacts as:

Consumer: exploits functionalities or services provided by the Smart City to improve the quality of his life;

Partner: contributes to producing smartness in the city;

Designing smart services is an asset that allows citizens to discover their environment and have a participative and collaborative life. Its role is the creation of an environment where any citizen can get any service anywhere and anytime through any device.

2.3 Smart City-Centric Smart Service

The smart city is a collection of Smart Computing Technologies (SCT) applied to the critical infrastructure components and services, which include city administration, education, healthcare, public safety, real estate, transportation, and utilities to make it more intelligent, interconnected, and efficient [6].

The combination of these services brings along Smart City applications for active and autonomous adaptation by using the benefits of contextual information.

2.4 Smart Service Properties

In a smart city context, a smart service is characterized by a number of properties that are related to the environment, mobility and invocation, like:

User-centric: based on the specific context and the preferences of the user;

Ubiquitous: accessible everywhere and from any devices;

Highly integrated: based on the integration of services and data from several and different applications or on the social cooperation of multiple users [7];

Adaptive: a smart service can flexibly sense, understand, and adapt to the user needs. To achieve a successful personalized service, two fundamental requirements are needed. The first is the ability to understand the behavior of the users and the second is the ability to adapt efficiently, to the user's changing behavior over time [8];

Context awareness: to provide services to occupants, context data is also needed [9];

Open: To realize open services, developer site is disclosed to the public. Everyone can bring their innovation to utilize Internet of things (IoT) devices and develops new applications, by opening the device API to the public, developers can bring their innovation and develop new applications to provide service to occupants [9].

The implementation of a Smart Service (SS) requires a service-oriented environment that offers opportunities for flexibility, portability, independence, and adaptation. This environment is called Service Oriented Computing (SOC) and is elaborated in the next section.

3 Service Oriented Computing-Based Environment

3.1 Definition

The Service Oriented Computing (SOC) can be defined as the paradigm that uses services as fundamental elements for developing fast, inexpensive, interoperable, scalable, and massively distributed applications [1]. It is a service that is a stand-alone, loosely coupled feature unit designed to address the interoperability issue between heterogeneous applications and architectures.

The basics characteristics of the service are:

Independent technology: Service should be invoked through standardized technologies. This implies that invocation mechanisms (protocols, descriptions and discovery mechanisms) should conform to widely accepted standards;

Loose coupling: The concept of service behaves as a "black box": nothing of its internal content is visible to customers, only its access interface.

Location Transparency: Service should have their definitions and location information stored in a repository such as UDDI and should be accessible by a variety of clients who can locate and invoke services regardless of their location.

Services can offer high-level abstractions for organizing large-scale applications in open environments so they help implement and configure software applications in a way that improves productivity and application quality. By focusing on the intrinsically linked themes of service foundations, service composition, service management and monitoring in addition to service design and development, researchers can establish the necessary links between these different activities (Table 1):

Table 1. Assignment of activities to actors [1]

Actor/activity	Binding	Discovered	Selection	Invocation	Publication	Composition	Management
Provider					✓		
Customer	✓	✓	✓	✓			
Aggregator						✓	
Operator							✓

The concept of service defines two main actors: the provider that provides the implementation of the services and the client that can be an application, a process or a customer. Through the description of the service, the customer can designate the supplier to invoke his service.

3.2   Service Oriented Computing and Smart City

The Smart City is an emerging generation of systems that bring together innovative technologies and heterogeneous infrastructures. In order to provide intelligent services for citizens, the need to integrate smart technologies such as cloud computing, learning analytics, big data, internet of things (IoT), wearable technology and etc., entails a complexity related to the adoption and the integration of these technologies into a so-called intelligent system.

Therefore, the introduction of service orientation solves this complexity and integrates all these technologies into a unified system through:

- The Service Oriented Computing paradigm allows the software-as-a-service concept to expand to include the delivery of complex business processes and transactions as a service while permitting applications to be constructed on the fly and services to be reused everywhere and by anybody [1].
- Allow mobile users to seamlessly consume and often combine functionalities offered by software and hardware resources anywhere and at any time [3].
- Services provide a uniform and ubiquitous information distributor for a wide range of computing devices (such as handheld computers, PDAs, cellular telephones) and software platforms; they constitute the next major step in distributed computing [1].

The application of Service Oriented Computing (SOC) as a paradigm enables designing Smart City Services (SCS) as software solutions independents to heterogeneous infrastructures, technologies, and architectures. This ability allows the creation of an ecosystem of services that communicate through a standard interfaces, composed and orchestrated. In fact, it is an orientation that allows implementing services in order to build a Smart City.

3.3   Service Oriented Computing Implementations in the Smart City

Smart city platforms are designed to provide and agglutinate heterogeneous services to support a variety of application domains [4] as (Table 2):

The integration of the concept of Service Oriented Computing (SOC) and its application to Smart City's business model, can offers a smart services ecosystem able to enhance life citizen and anticipate his needs. The use of the services does not concern only the web service but they can be applications, agents or just functions.

According to this context, all entities of the smart city must base on the service orientation to take advantages related to reusability, interoperability and adaptability, this is a reason to adopts and develops smart services for a university which has a digital system meeting functional and communication needs.

Table 2. Services in the smart city

Service	Application domain	Service type
Web-based robotics programming [10]	Education	Web service
Services designed to collect context data [11]	Traffic congestions	Application for smartphones, tabletPCs, or computers
Mobile Agent Service Model for Smart Ambulance [12]	Telemedecine	Mobile agent
Parking occupancy detection [12]	The traffic flows	Application for mobile phone
Context-aware services [5]	Governance	Application for mobile phone
Services for development of smart services as a smart space [13]	E-Tourism	Based on the multi-agent architecture
A smart city service platform performs centralized management of the smart home [14]	Smart home	Collaborative services that store context information

4 From Digital University to Smart University

4.1 Digital University: Constraints

Today, with the advent of the internet and the increasing use of digital tools, the digitization of higher education institutions has progressively imposed itself.

Faced with these new challenges and being aware of the need to give a place to digital, universities are obliged to remain permanently looking for performance and modernity. The aim is to satisfy a new generation of students who are very connected and that privilege digital tools to learn, get informed and stay in touch with the outside world.

Certainly, digitalization has transformed the university a lot through the new communication channels, the permanent interconnection, the digital media, and the dematerialization of education through the emergence of new concepts, namely e-learning. However, the digital experience has shown its limits. Among which we can mention the dehumanization of the formation. Take the example of e-learning, the trainee is alone in front of his screen. There is no interaction, no exchanges with a trainer and no sharing experiences with other trainees. The other disadvantage lies in the fact that, as there is no physical trainer, motivation is not necessarily present.

Starting from this point, we will have to introduce a completely intelligent system that meets the requirements of the current university: a modern university that adapts, anticipates, recommends and proposes.

4.2   Smart University: The Emergence of a Smart System

The primary focus of smart universities is in the education area, but they also drive the change in other aspects such as management, safety, and environmental protection.

The availability of newer and newer technology reflects on how the relevant processes should be performed in the current fast-changing digital era. This leads to the adoption of a variety of smart solutions in university environments to enhance the quality of life and to improve the performances of both teachers and students [15], and to provide self-learning, self-motivated and personalized services which learners can attend courses at their own pace and are able to access the personalized learning content according to their personal difference [16].

Starting from the context of the trainee (information on his profile, his academic or professional career, his ambitions …), the system can recommend courses that will interest him and therefore it will enable him to be motivated through suggestions and proposals. It is about intelligent learning, which is an opportunity that allows offering a training course for the professions of tomorrow, to produce smart student for the society, and which evolves with the needs and the changes of the society.

To reach this goal, modern sophisticated smart devices, smart systems, and smart technologies create unique and unprecedented opportunities for academic and training organizations in terms of new approaches to education, learning, and teaching strategies, services to on-campus and remote/online students, set-ups of modern classrooms and labs [17].

To approach education and how we teach various types of students differently [15], we must develop smart universities based on Smart education, Smart pedagogy, Smart classroom, Smart learning, and Smart campus. The table below describes, in a summary way, each concept and its opportunities (Table 3).

The opportunities offered by a smart university require a whole smart system that implement and demonstrate significant maturity at various "smartness" levels such as (1) adaptation, (2) sensing (awareness), (3) inferring (logical reasoning), (4) self-learning, (5) anticipation, and (6) self-organization and re-structuring [18] (Fig. 1).

The importance of Service Oriented Computing can be seen in the integration of layers of services that serve to: contextual data collection, anticipation, prediction and recommendation, and makes a university system intelligent and proactive. This contribution will be dealt detailed in the following section.

5   SOC-Based University: Opportunities

5.1   Service in Smart University

We cannot talk about smartness without addressing the notion of the service-oriented; its capacity allows a smooth movement from a traditional university to a smart university. The modern university, taking into account the student in these decisions based on his profile, focus, and ambitions, is able to offer a training curriculum for future professions to produce students for society as they evolve with the needs and changes of society.

Table 3. Smart university-based approach

Approach	Smart education	Smart campus	Smart classroom	Smart learning	Smart pedagogy
Opportunities	• Adaptive learning programs; • Collaborative technologies and digital learning resources; • Personalized services and seamless learning experience; [18]	• Smart building management with automated security control and surveillance; • Protective and preventative health care; • Social networking and communications for work collaboration; • Green and ICT sustainability with intelligent sensor management systems;	• Learning Disabilities; • Speech or Language Impairments; • Visual Impairments; • Hearing Impairments; • Reading comprehension; • Writing comprehension; [15]	• Provides the necessary learning guidance, hints, supportive tools or learning suggestions in the right place, at the right time and in the right form; [19]	• Learning-by-doing (including active use of virtual labs); • Collaborative learning; • e-books; • Learning analytics; • Adaptive teaching; • Student-generated learning content; • Serious games- and gamification-based learning; • Flipped classroom; • Project-based learning; • Bring-Your-Own-Device; • Smart robots (robotics) based learning; [15]

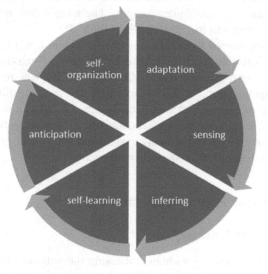

Fig. 1. Smartness level in the smart system

Indeed, the characteristics of the service, namely loose coupling, flexibility, independence, portability, allow reaching these objectives and encouraging the developers to integrate them in a university system according to different aspects:

Supporting service: aimed at student's possible problems and difficulties in the autonomous learning process. To treat students as the center, to satisfy students' needs as the guidance, it consists of five parts:

Information consulting services;
Training of learning strategies;
Real-time supervision service;
Psychological consultation;
Affective social interaction [10].

Learning service: provides an additional means for reaching educational objectives, and academic credit is appropriate for service activities when learning objectives associated with the service are identified and evaluated. Faculty who use service-learning discover that it brings new life to the classroom, enhances performance on traditional measures of learning, increases student interest in the subject, teaches new problem-solving skills, and makes teaching more enjoyable.

Ubiquitous learning service (U-Conferencing, and for U-Communication) [20];
Implementing course using services [10];
Tools and infrastructure services.

Partner service: services are more relevant to the quality of education resources, extending learning beyond classrooms; student-centered learning and national competitiveness [21], it can also provide information service of integrated convenient life [22]. Moreover, The academic field has a lot to integrate the services to take advantage of the ample opportunities offered by these services, indeed the integration of services solve problems related to collaborative teaching and partnerships.

Smart university adopts the Service Oriented Computing (SOC) environment to take advantage of the wide range of opportunities offered by the services; indeed the integration of services enhances the education efficiency, effectiveness, and productivity. Which facilitator of higher-order thinking skills, learner-centered support, self-directed learning, tailored learning and supporting decision [21]. On the other hand, it contributes to reaching the level of intelligence of a system, as we will show in the following section.

5.2   The Contribution of Service Oriented Computing to Achieve a Smart University

Service integration represents the layer key in the production of personalized and equitable services for all the stakeholder of a university that serve accessibility, self-study, collaboration and partnerships. All features of the service allow him to contribute to increase the smartness level of a university system, as we will develop in the following table (Table 4):

As primary requirement to design a smart system, the service orientation is indispensable in the implementation of intelligent systems and especially in the university systems. Indeed, the role of services has shown much importance to achieve a smart system. However, it can be observed that the use of services as anticipatory mechanisms is limited, and given the importance of their role, our use case will focus on a study that will address this axis in terms of anticipatory service layers.

Table 4. Service Oriented Computing contribution in a smart university

Article	Adaptation	Sensing (awareness)	Inferring	Self-learning	Anticipation	Self-organization
[20]	✓	✓				✓
[23]			✓	✓		
[10]		✓	✓	✓		✓
[24]	✓	✓				
[25]	✓	✓	✓			
[22]	✓	✓	✓			
[21]	✓	✓	✓	✓		

5.3 Study Case: Smart Certification System

Many services based on the Service Oriented Computing can be offered by the smart university to promote the use of smart technologies and the integration of Information and Communication Technologies (ICT), in order to successfully satisfy student's needs related to learning, accessibility, and capabilities. In our use case we will present a conceptual view to highlight the integration of services in the development of Smart Certification System (SCS) as an anticipatory service.

Its main goal is to predict the student's future career, based on his profile, skills, and ambitions, browse certifications categories, and offer the most appropriate certifications that will validate his core knowledge, and will enhance his technical credibility.

By interpreting the student's intention, the Smart Certification System (SCS) serves to anticipate his needs, therefore, our smart services system must be designed on a Service oriented Computing paradigm in order to emerge a system that use **Supporting service, Learning service** and **Partner service** in order to design recommendation services based on student's intention. The instantiation of Smart Certification System (SCS) is shown in Fig. 2.

The Smart Certification System invoke the inputs services to takes information that is useful in the processing step in order to design recommendation services based on student's intention.

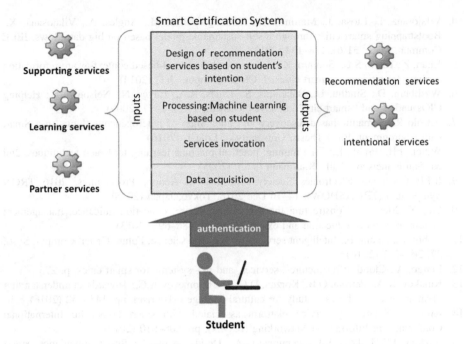

Fig. 2. Smart certification system

6 Conclusion

In this paper we present and discuss the role of service to reach smartness in a Smart University (SU), this is a Service oriented environment that allows to add layers of service to a system in order to make it smart and proactive. Our study focused on the integration of services within a Smart University (SU). Along with this, a set of basic components of a Smart University (SU) has been described. Admitting that the integration of anticipation mechanisms as a smart service is limited, we proposed a Smart Certification System (SCS) based on the Service Orientation Computing (SOC) that invoke Supporting service, Learning service and Partner service, related to a student in order to design recommendation services based on student's intention.

References

1. Papazoglou, M.P.: Service-oriented computing: Concepts, characteristics and directions. In: Proceedings of the Fourth International Conference on Web Information Systems Engineering, WISE, pp. 3–12 (2003)
2. Bullinger, H.-J., Neuhuttler, J., Nagele, R., Woyke, I.: Collaborative Development of Business Models in Smart Service Ecosystems. (2017), pp. 1–9
3. Urbieta, A., González-Beltrán, A., Ben Mokhtar, S., Anwar Hossain, M., Capra, L.: Adaptive and context-aware service composition for IoT-based smart cities. Future Gener. Comput. Syst., **76**, 262–274 (2017)

4. Vilajosana, I., Llosa, J., Martinez, B., Domingo-Prieto, M., Angles, A., Vilajosana, X.: Bootstrapping smart cities through a self-sustainable model based on big data flows. IEEE Commun. Mag., **51**(6), 128–134 (2013)
5. Khan, Z., Kiani, S.L., Soomro, K.: A framework for cloud-based context-aware information services for citizens in smart cities. J. Cloud Comput., **3**(1) (2014)
6. Washburn, D., Sindhu, U., Balaouras, S., Dines, R.A., Hayes, N., Nelson, L.E.: Helping CIOs understand "smart city" initiatives. Growth, **17**(2), 1–17 (2009)
7. Petrolo R.: Semantic-based discovery and integration of heterogeneous things in a Smart City environment. Ph.D Thesis, University of Lille 1, (2016)
8. Witten, I.H., Frank, E.: Data mining: practical machine learning tools and techniques, 2nd ed. San Francisco, Calif.: Kaufmann [u.a.], (2005)
9. IEEE Consumer Electronics Society: The IoT in Action: Proceedings, 2016 TRON Symposium (TRONSHOW): 14–16 Dec 2016, Tokyo, Japan (2016)
10. Liu, X., Mei, Y.: Constructing the supporting service, education guidance, management system of network education and examination, pp. 688–692 (2013)
11. Dobre, C., Xhafa, F.: Intelligent services for big data science. Future Gener. Comput. Syst., **37**, 267–281 (2014)
12. Longo, A.: Cloud infrastructures, services, and IoT systems for smart cities. p. 275
13. Kulakov, K.A., Petrina, O.B., Korzun, D.G., Varfolomeyev A.G.: Towards an understanding of smart service: The case study for cultural heritage e-Tourism, pp. 145–152 (2016)
14. Kim, E.: Smart city service platform associated with smart home. In: International Conference on Information Networking (ICOIN), pp. 608–610 (2017)
15. Bakken, J.P, Uskov, V.L., Penumatsa, A., Doddapaneni, A.: Smart universities, smart classrooms and students with disabilities. In: Uskov, V.L., Howlett, R.J., Jain, L.C. (eds.) Smart Education and e-Learning 2016, vol. 59, pp. 15–27. Springer International Publishing, Cham (2016)
16. Kim, T., Cho, J.Y., Lee, B.G.: Evolution to smart learning in public education: a case study of korean public education. In: Ley T., Ruohonen, M., Laanpere, M., Tatnall, A. (eds.) Open and Social Technologies for Networked Learning, vol. 395, pp. 170–178. Springer Berlin Heidelberg, Berlin, Heidelberg (2013)
17. Uskov, V.L., Bakken, J.P., Pandey, A., Singh, U., Yalamanchili, M., Penumatsa, A.: Smart university taxonomy: features, components, systems. In: Uskov, V.L., Howlett, R.J., Jain, L. C, (eds.) Smart Education and e-Learning 2016, vol. 59, pp. 3–14. Springer International Publishing, Cham (2016)
18. Heinemann, C., Uskov, V.L.: Smart University: Literature Review and Creative Analysis. In: Uskov, V.L., Bakken, J.P., Howlett, R.J., Jain, L. C (eds.) Smart Universities, vol. 70, pp. 11–46. Springer International Publishing, Cham (2018)
19. Gros, B.: The design of smart educational environments. Smart Learn. Environ., **3**(1), (2016)
20. Dludla, A.G., Bembe, M.J., Byambaakhuu, B., Abdulai, M.-S., Rho, J.J.: System architecture for ubiquitous live video streaming in university network environment, pp. 1–5 (2013)
21. Tantatsanawong, P., Kawtrakul, A., Lertwipatrakul, W.: Enabling future education with smart services, pp. 550–556 (2011)
22. Shi, A., Wang, X., Chen, Y., Yu, J.: Construction of University Campus Public Information System of Service Design, pp. 427–430 (2017)
23. Wang, Y., Chen, Y., Tong, X., Lee, Y., Yang, J.: Robot as a service in information science & electronic engineering education, pp. 223–228 (2017)

24. Adamko, A., Balazs, B., Krisztian, E., Attila, F., Kristof, H.N., Norbert, K.-F.: Smart campus service link: adaptation and interaction planes for campus and university environments, pp. 000271–000276 (2017)
25. Msaed, S., Pernelle, P., Ben Amar, C., Carron, T.: Service oriented approach for modeling and generic integration of complex resources in MOOCs, pp. 1–8 (2016)

Towards a Mobile Serious Game for Learning Object Oriented Programming Paradigms

Elaachak Lotfi$^{(\boxtimes)}$, Bakkali Yedri Othman,
and Bouhorma Mohammed

Laboratory of Informatics Systems and Telecommunications (LIST),
Abdelmalek Essaadi University, Tétouan, Morocco
lotfil002@gmail.com

Abstract. As it is known the majority of beginners in software development encounter several difficulties to learn and understand programming paradigms including object oriented programming. This lack of comprehension is due to the complexity and abstraction of some concepts like: polymorphism, inheritance, etc. However, the use of serious games in such situations can handle this kind of issues and make experience of learning more enjoyable and beneficial for the learners. In this perspective of research we aim in this paper to present a mobile serious game for teaching object programming concepts in a fun and easy way, the proposed game will be dedicated for learners from different backgrounds.

Keywords: Serious games · OOP · Mobile · Assessment

1 Introduction

The huge success that know the field of mobile applications due to the big number of users that use such applications around the word which keeps increasing, has encouraged mobile developers to develop several video games dedicated for mobile, including mobile serious games usually used by young students [1]. Since mobile phones can be used by them in anytime anywhere that can offer new opportunities for teaching and learning the programming language paradigms [2].

Thanks to their potential, mobile serious games can be used to teach both procedural and object oriented programming languages [3–5] e.g. "C, C++, JAVA, Python, etc.". They can play a major role to simplify notions of both abstract and complex programming concepts for the beginners, who do not feel that the learning environment is motivating and are not interested in computer programming [6], the learning process via this king of video games can offer to them an opportunity to learn new skills in easy and fun way and allow them to save time and improve their programming abilities during a game sequence.

As it is known the effectiveness of any serious game is related to its ability to transmit knowledge to the learners. Assessment is among the techniques that can be introduced into serious games to evaluate learners' knowledge about many basics and concepts; in addition, assessment can be decisive to prove this effectiveness by analyzing the obtained learning outcomes.

© Springer Nature Switzerland AG 2019
M. Ben Ahmed et al. (Eds.): SCA 2018, LNITI, pp. 450–462, 2019.
https://doi.org/10.1007/978-3-030-11196-0_38

The remainder of this paper is organized as it follows. Sect. 2 presents a literature review regarding the other realization concerning teaching programming skills through serious games and assessment approaches used in this kind of video games. Sect. 3 describes the design and development of OOP serious Game and discusses assessment mechanism used in the OOP serious game to evaluate learners' knowledge in the field. Sect. 4 focuses and the obtained results, analysis of the learner outcomes and also the discussion about the impact of such approach in the learning process. A general conclusion with the future research works will conclude this paper in Sect. 5.

2 Literature Review

During these last few years, many programming serious games have been created; despite that learning a programming language requires logical thinking to understand both abstract and complex concepts, serious games have handled this problem by giving the learners a variety of easy scenarios that explain complex notion. This section will discuss some of these realizations.

2.1 Object Oriented Programming Concepts

Object oriented programming is a programming paradigm that has been designed around objects, reader than procedures. The object oriented approach brought a new path, giving more importance to the objects [7]. This programming paradigm is based on several concepts e.g. "object, class, inheritance, polymorphism".

Object: object is basic and main entity of OOP, it's an instance of class, for developing any program, and there is necessity of an object in OOP [8]. It acts as intermediate between the program and the methods which are added with it. Objects contain private attributes as we do not want any hacker or unauthorized person to gain access to this data [9].

Class: Class is the definitions for the data format and procedures for a given type or class of object; may also contain data and class methods. Classes form the basic development unit of any software [10]. It uses attributes and functions to operate the given data [11]. Class is the key element in software monitoring and implementing the concepts of OOP [12].

Inheritance: Inheritance allows classes to be arranged in a hierarchy that represents relationships. Inheritance gives any user the provision to utilize the same code in his program just by changing the definition of its variables [13]. This technique allows easy reuse of the same methods and attributes definitions, in addition to potentially mirroring real world relationships in an intuitive way.

Polymorphism: polymorphism is a mechanism triggered when calling code can be agnostic as to whether an object belongs to a super class or one of its descendants. Polymorphism gives the way to represent variables as a group or a common way to represent the variables [14]. We could define only one singular code for many different types of variables to perform the action of execution quickly. By allowing common methods to work for various objects, avoid the complexity and bind objects.

2.2 Serious Games for Teaching Programming Languages and Paradigms

As mentioned above, many programming serious games have been created to teach beginners the basics and notions of programming languages and paradigms. T. jordine et al. [3] have developed a mobile device based serious games for teaching and learning Java programming language, the game has been experimented in two universities in Germany and UK.

IPlayCode is another serious game introduced by J. zhang team from the university of West YorkShire UK [4], iPlayCode aims at simplifying the learning of programming by including entertainment, they highlighted that learning programming languages is abstract and complex at university-level.

A. Yassine et al. [5] from University Sidi Mohamed Ben Abdellah Morocco, have developed also a new game based learning "Perbo" to teach scientific courses in general and especially programming languages, the proposed game handles the problem related to learn one of the most difficult concepts in C language programming, the pointers.

According to literature, there is a variety of serious games that have been developed to resolve specific problems in relation of learning specific concepts of programming paradigms, and they prove their effectiveness in consonance with the obtained results. By cons, there is a lack concerning serious games that teach object oriented programming concepts independently from the programming languages like: java, C#, C++, etc.

2.3 Assessment in Serious Games

Many studies have questioned the effectiveness of serious games [15, 16]. According to those reviews, among the decisive factors to prove the effectiveness of a serious game there is the verification of knowledge acquisition through game sequences, this acquisition process is also called assessment. Assessment describes the process of using data to demonstrate that the learning goals and pedagogical objectives are actually being met [17]. We can consider assessment as either summative whereby it is conducted at the end of a learning process and tests the overall achievements, and formative whereby it is implemented and present throughout the entire learning process and continuously monitors progress and failures [18]. Michael and Chen in their study [19] describe three primary types of assessment: completion assessment, in-process assessment, and teacher assessment. The first and second correspond to summative and formative assessments, respectively. By cons the teacher assessment focuses on the instructor's observations and judgments of the student and his actions.

There are other methods and techniques used to asses learning in serious games, like summative assessment, which is accomplished with the use of tow kind of tests pre and post, the aim of those testes is to measure changes in educational outcomes after modifications to the learning process such as testing the effect of a new teaching method [20].

The method that is very often used in post-assessment consists in testing a players' knowledge about what they learned by way of a test, questionnaire or teacher evaluation. It is the simplest method to implement. But it relies on the opinions of the learner and does not depend on all the information that can be collected regarding what happened during the game sequence [21].

Assessment of learning requires a systematic approach to determine a learner's achievements and detect learning issues. Standardized assessment methods often take less time, their results are readily interpretable and they are easier to administer [22]. Furthermore, standardized tests lack the flexibility necessary to adjust or modify materials for certain groups e.g. "high- or low-performing groups", and therefore may lead to loss of sensitivity for certain groups [23].

Many studies have explored how in-game assessment can provide more beneficial and reliable assessment and emerging interests reflect the need of alternative tool that overcame the limitations of standardized approach [24, 25]. It refers to analyzing how a learners play in order to assess their cognitive development. Serious games have the advantage in this type of assessment; they can easily track every learner's move and decision [19].

In general, serious games can contain in-game tests of effectiveness; they can allow a learner during his progression to accumulate points and experience to face new topics in the next levels and stages. Finally, in-game assessment employs alternative, less intrusive and less obvious forms of assessment which could become a game element itself [26].

3   OOP Serious Game

OOP Serious Game is a video game that teaches the concepts of object programming language, it is inspired by the zoo environment, including several animals, drag and drop GamePlay and both puzzle and strategy game genres are used in this game. The principal of the game consist to create and class animals and understand different forms of methods e.g. "voices, actions, behaviors, etc.", all this while respecting the OOP paradigms.

3.1   Game Design of OOP SG

Game scenario: The scenario of OOP SG "Fig. 1" is composed of four immersions during each immersion the learner should do some tasks in order to understand one concept from OOP concepts, if the learner do well the immersion he will win some bonus as rewards if not he will be punished.

GamePlay: Drag & Drop is a GamePlay used in this game, it is a technique adapted to serious games. The interest of the drag & drop lies in the fact that it is suitable for beginner learners. Using a mouse or just the finger on a touch screen, the player can drag and drop game elements to do actions. For example, the player must drag pieces of the puzzle and place them correctly in order, to create a full picture.

Main Characters: As mentioned before the game was inspired from the zoo environment which a variety of animals play the role of the main characters "Fig. 2" in each Immersion that belong to the game scenario.

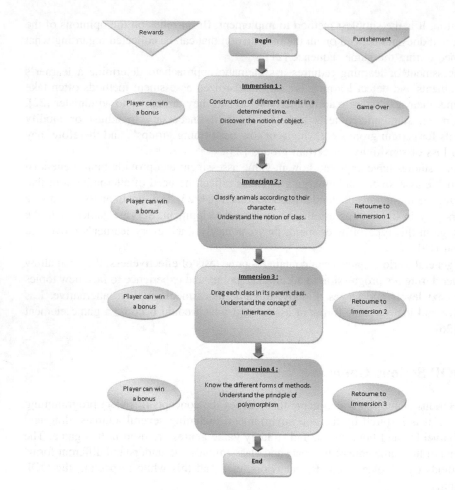

Fig. 1. Flowchart of Scenario of OOP SG

3.2 Level Design

The short definition of level design is a game development discipline that involves the creation of video game levels, maps, missions or stages. This is done using some sort of level editor "Software" or by hand drawing "draw level prototypes". The proposed serious game was composed of four levels, according to immersions already established on flowchart of scenario described above.

Level 1 "Fig. 3", the main objective of this level is to discover the notion of an object through the construction of three animals where each animal is considered as an object.

Level 2 "Fig. 4", the main objective of this level is to understand the notion of class, by classifying animals into identical groups.

Fig. 2. Main characters of OOP SG

Fig. 3. Hand drawing of level 1

Level 3 "Fig. 5", the main objective of this level is to understand the notion of inheritance, by dragging and dropping animals according to their parent class e.g. "fishes, reptiles, mammals, etc.".

Level 4 "Fig. 6", the main objective of this level is to understand the notion of polymorphism, that is, an animal can manifest itself in several forms.

Fig. 4. Hand drawing of level 2

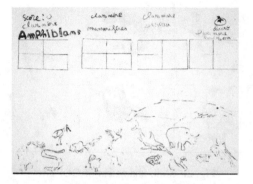

Fig. 5. Hand drawing of level 3

Fig. 6. Hand drawing of level 4

3.3 In-Game Assessment of OOP SG

The assessment strategy adopted to evaluate learners' knowledge during the game progression is In-game assessment. In each level of the game the learner will be evaluated concerning one of the four OOP concepts.

For example, in Level 1 and 2 the learner should understand both object and class notions, so the assessment mechanism consists in these two levels, evaluating the ability of the learners to construct objects based on different elements placed randomly on the screen then classify them according to their corresponding classes by using drag and drop GamePlay.

By cons, in level 3 and 4, inheritance and polymorphism are the concepts that the learner will be evaluated for, therefore, the mechanism of assessment adopted in these two levels are respectively dragging the correct object in a set of corresponding spices, and placing the right method that corresponds to one action performed by a specific animal to the right place. Both scoring and timing mechanisms from the Gamification techniques are used to control the learners' performances during all game's levels.

4   Results and Discussion

The OOP SG was developed by using Unity Game engine [27] more precisely the C# programming language. It is a 2D serious game that teaches and evaluate knowledge about the object oriented programming concepts for the new software developers, the game style adopted during the development phase is a combination of two styles strategy and puzzle and the GamePlay of the hall game is based on Drag & Drop mechanism, in game assessment is the method adopted to evaluate learners' knowledge about the field in each level of the proposed serious game. The OOP SG has several scenes, including Graphical user interface as main menu "Fig. 7" and the four levels in Fig. 8.

Fig. 7. Graphical User Interface of POO SG

Level one the learner should drag and drop elements to create objects displayed on the screen "object concept", level two he should classify objects according to their classes "class concept", level three he will move animals and place them to the corresponding parent classes depending to their species "inheritance" and the last one he should move methods to perform animals' actions "polymorphism concept".

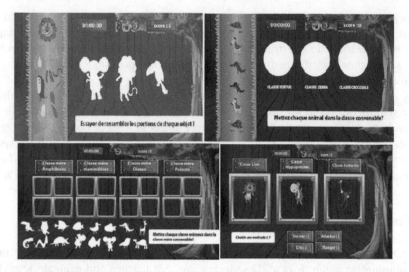

Fig. 8. Game Levels of POO SG

A sample of 250 players have played OOP SG until the writing of this paper, from different ranges of age and gender, according to their feedback the majority of them are satisfied, the flow chart below describes some statistics about the use of the proposed game, the Fig. 9 represents the rate of use of games by gender, we note that 143 male used the game against 107 female used it.

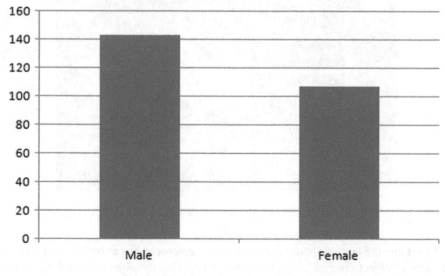

Fig. 9. Flowchart players' number/gender

In Both Table 1 and Fig. 10 there is a graph that presents the number of learners according to their age, the majority of players have between 8 and 30 years old, we have diversified the range of persons that have tested the proposed serious game.

Table 1. Number of learners according to their age

Age	Number of learners
8	10
10	23
15	17
18	24
20	29
22	103
25	33
30	7
31	2
35	2

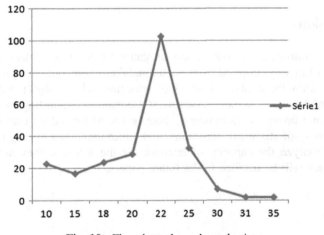

Fig. 10. Flowchart players' number/age

The graph in Fig. 11 presents the progression of the learners according to the four game levels; in level one 196 learners have done this level, then quit the game or continue to the next level, 157 of them have done the second level, by cons, 107 learners have done the third level and 83 have succeeded the game in totality.

We notice that even young learners have succeed certain levels, more precisely levels one and two, according to the results 83 of learners have passed all levels successfully so they have mastered all the object oriented programming basics. Unfortunately is not sufficient to tack such decision therefore to get a clear view about learning progression according to the obtained results we must use advanced

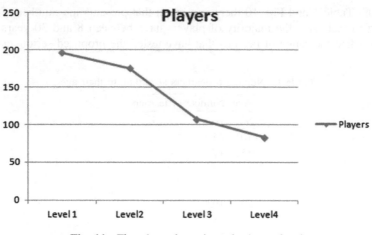

Fig. 11. Flowchart players' number/game level

techniques and algorithms inspired from learning analytics, educational data mining and data visualization.

5 Conclusion

This paper has introduced a mobile serious game for teaching object oriented programming in a fun way, inspired from the animals' environment. The proposed game evaluates the knowledge about four OOP concepts "class, object, inherence and polymorphism" by using in-game assessment mechanism, according to the obtained results we cannot prove its effectiveness, there is a need for a deep learning analytics that uses a variety of statistical, algorithms and educational data mining techniques to measure and analyze the capacity of learning for the learners, they are among the perspectives that will be detailed in the future papers.

References

1. Khalaf, S.: Apps solidify leadership six years into the mobile revolution, Apr-2014. [Online]. Available: http://blog.flurry.com/bid/109749/Apps-Solidify-Leadership-Six-Years-into-the-Mobile-Revolution
2. Frohberg, D., Göth, C., Schwabe, G.: Mobile learning projects—a critical analysis of the state of the art: Mobile learning projects. J. Comput. Assist. Learn. **25**(4), 307–331 (2009). https://doi.org/10.1111/j.1365-2729.2009.00315.x
3. Jordine, T., Liang, Y., Ihler, E.: A mobile device based serious gaming approach for teaching and learning java programming. Int. J. Interact. Mobile Technol. **9**(1) (2015)
4. Zhang, J., Lu, J.: Using mobile serious games for learning programming. In: The proceedings of The Fourth International Conference on Advanced Communications and Computation (INFOCOMP 2014), Paris, France, 2014

5. Yassine, A., Chenouni, D., Berrada, M., Tahiri, A.: A Serious Game for Learning C Programming Language Concepts Using Solo Taxonomy. **12**(03) (2017). Accessed 24 May 2018
6. Mitamura, T., Suzuki, Y., Oohori, T.: Serious games for learning programming languages. In: IEEE International Conference on Systems, Man, and Cybernetics (SMC), pp. 1812–1817. IEEE (2012)
7. Laimek, R., Pawgasame, W.: In: Asian Conference on Internal ballistics simulation based on object oriented programming, Defence Technology (ACDT) (2015)
8. Reyna, A.M., et al.: Object-oriented programming as an alternative to industrial control. In: 9th International Conference on Electrical Engineering, Computing Science and Automatic Control (CCE) (2012)
9. D'Andrea, R.J., Gowda, R.G.: Object-oriented programming: concepts and languages. In: Proceedings of the IEEE 1990 National Aerospace and Electronics Conference (NAECON) (1990)
10. Weisfeld, M.: The importance of object-oriented programming in the era of mobile development. (2013) Available from: http://www.informit.com/articles/article.aspx?p=2036576
11. Seban, R.R.: An overview of object-oriented design and C++. In: Proceedings of 1994 IEEE Aerospace Applications Conference (1994)
12. Butler, S., et al.: Mining java class naming conventions. In: 27th IEEE International Conference on Software Maintenance (ICSM) (2011)
13. Vedpal, N., Chauhan, H., Kumar, A.: Hierarchical test case prioritization technique for object oriented software. In: International Conference on Contemporary Computing and Informatics (IC3I) (2014)
14. Milojkovic, N., et al.: Polymorphism in the spotlight: studying its prevalence in java and smalltalk. In: IEEE 23rd International Conference on Program Comprehension (ICPC) (2015)
15. Gosen, J., Washbush, J.: A review of scholarship on assessing experiential learning effectiveness. Simul. Gaming **35**(2), 270–293 (2004)
16. Kulik, A.A.: School mathematics and science programs benefit from instructional technology. United States National Science Foundation (NSF), National Center for Science and Engineering Statistics (NCSES), InfroBrief NSF-03-301, November 2002, http://www.nsf.gov/statistics/infbrief/nsf03301/
17. Chin, R., Dukes, R., Gamson, W.: Assessment in simulation and gaming: a review of the last 40 years. Simul. Gaming **40**(4), 553–568 (2009)
18. Boston, C.: The concept of formative assessment. Pract. Assess. Res. Eval. **8**(9) (2002)
19. Michael, D., Chen, S.: Proof of learning: assessment in serious games, October 2005, http://www.gamasutra.com/view/feature/2433/proof_of_learning_assessment_in_.php
20. Dugard, P., Todman, J.: Analysis of pre-test-post-test control group designs in educational research. Educ. Psychol. **15**(2), 181–198 (1995)
21. Shute, V., Ventura, M., Bauer, M., Zapata-Rivera, D.: Melding the power of serious games and embedded assessment to monitor and foster learn ing: flow and grow. In: Ritterfeld, U., Cody, M., Vorderer, P. (eds.) Serious Games: Mechanisms and Effects, pp. 295–321. Routledge, Taylor and Francis, Mahwah, NJ, USA (2009)
22. Clarke, M.M., Madaus, G.F., Horn, C.L., Ramos, M.A.: Retrospective on educational testing and assessment in the 20th century. J. Curriculum Stud. **32**(2), 159–181 (2000)
23. Test, F.: What's wrong with standardized tests? (May 2012). http://www.fairtest.org/facts/whatwron.htm
24. Short, E.J., Noeder, M., Gorovoy, S., Manos, M.J., Lewis, B.: The importance of play in both the assessment and treatment of young children. In: Russ, S., Niec, L. (eds.) An

Evidence-Based Approach to Play in Intervention and Prevention: Integrating Developmental and Clinical Science, Guilford, London, UK (2011)

25. Kaugars, A.S., Russ, S.W.: Assessing preschool children's pretend play: preliminary validation of the affect in play scale-preschool version. Early Educ. Devel. **20**(5), 733–755 (2009)

26. Bente, G., Breuer, J.: Making the implicit explicit: embedded measurement in serious games. In: Ritterfield, U., Cody, M.J., Vorderer, P. (eds.) Serious Games: Mechanisms and Effects, pp. 322–343. Routledge, New York, NY, USA (2009)

27. Unity Game Engine: available in : https://unity3d.com/

Computer Vision in Smart Cities

3D Modeling of Flood Areas

Souhaib Douass[(⊠)] and M'hamed Ait Kbir

LIST, CED STI, UAE, Tangier, Morocco
souhaib.douass@gmail.com, m.aitkbir@fstt.ac.ma

Abstract. For excellent governance of cities and to better manage floods, 3D simulation of flood zones, allows city managers to consider upstream a plan to prevent any risk that may come from floods To avoid the limitation of cartographic representations in two dimensions, we will build from geographic coordinates obtained from Google MAP API [1] a three-dimensional models of a given zone with a web-compatible visualization. Our approach is based on three steps: Data collection, 3D processing and modeling, the visualization of 3D flooding, and finally analysis. Two simulations, for the same region, with a rainfall that attains two water levels 1 and 3 m.

Keywords: 3D geographic information system · Flood areas · 3d modeling
Simulation of flooding · Visualization 3D of flooding first section

1 Introduction

With all the development in the infrastructures of the cities, several of them are always threatened by the risk of increase of level of water during a flood and which cause damage in the roads, the buildings, and the electrical and telecom installations.

we worked recently in a 3D visualization and analysis in GIS geographic information requirements in the Smart City. We discussed GIS application in Smart Cities and possibilities.

given by the GIS analysis methods applied on 3D visualization and we took Tangier Geographic Information as an example [2].

In our research in flooding management, we focus on simulation and visualization of 3D flood areas, for better flood risk mitigation and risk reduction.

Existing works based on the same concept: 3d flood-risk models of government infrastructure [3], This research explores LiDAR data and the application of 3D modeling in order to provide an analysis of the risk of floods on government buildings and utilities. LiDAR data provides a cheaper, faster and denser multidimensional coverage of features for 3D mapping. 3d GIS for flood modeling in river valleys [4], The objective of this study is implementation of system architecture for collecting and analyzing data as well as visualizing results for hydrodynamic modeling of flood flows in river valleys using remote sensing methods, tree-dimensional geometry of spatial objects and GPU multithread processing. Three-dimensional (3d) modeling for flood communication [5], The aim of this study was to examine the use of a 3D model for the purpose of communicating predicted flood levels in residential areas.

© Springer Nature Switzerland AG 2019
M. Ben Ahmed et al. (Eds.): SCA 2018, LNITI, pp. 465–471, 2019.
https://doi.org/10.1007/978-3-030-11196-0_39

To study flooding, different solutions have been developed, as satellite, aerial photogrammetry and LiDAR images processing.

Existing visualization solutions do not provide a better perspective of endangered areas; the model of the buildings represented by polygons and the 3d model of the terrain do not exist. however the visualization solution represented by 3D simulation of flooding, the analysis can be better done and allows the city managers to have good answers in relation to the various flood scenarios.

In this paper, we worked on a 3D simulation of the floods, from the geographical data extracted from Google MAP API [1], passing by a 3D modeling of terrain and finally the 3d visualization of the flooding. The analysis was done in relation to the two scenarios for the same region, with a rainfall that attains two water levels 1 and 3 m.

2   Data Collection and Processing

Google MAP API helps to get everything we need to build location based applications, it is divided into three major parts : Maps, navigation and Places. In this paper, we worked with :

- The Drawing Tools [6] on the Map provides a graphical interface for users to draw polygons, rectangles, polylines, circles, and markers on the map.

In our works, we used rectangles markers, to get the geographic coordinates of a given area, each position is defined by longitude and altitude (Fig. 1).

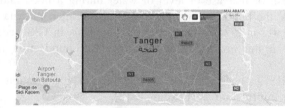

Fig. 1. Determining the geographical area based on the Google map drawing tools

After we have divided this rectangular area into a matrix of two dimensions, each matrix element contains a position in longitude and latitude, then through the Google Maps API Elevation Service [7] we obtained, the height in meters of each position.

- The Elevation service provides elevation data for locations on the surface of the earth in meter unity.
- The distance between two points is the length of the shortest path between them. This shortest path is called a geodesic. On a sphere all geodesics are segments of a great circle. To compute this distance, we use functions of Google Geometry Library [8].

3   Implementation, Simulation and Result

3.1   Implementation

In this paper we took as region an area on the Tangier region with the following geographical coordinates:

The latitudes are between the values 35.73 and 35.77; the longitudes between the values −5.84 and −5.76.

The following figure shows the region selected by Google Map Drawing Tools:

To calculate the dimensions of this region we used the Google Map Geometry Library and especially the distance calculation function between two geographical points.

Spherical namespace [9] contains spherical geometry utilities allowing you to compute angles, distances and areas from latitudes and longitudes. The real dimensions of this region are 6980 m over 4187 m.

We converted the real dimensions to a scale dimension according to this formula:

$$unitScal = \frac{realDimension}{modelDimension}$$

To standardize the 3D terrain model, we set the model width to widthStep, and for the model height; we divided the real height on a scale unit

$$heightModel = \frac{heightReal}{unitScal}$$

To obtain the elevations, we have devised the region in a matrix of segmentWidth * segmentHeight, and by the Google Map Elevation service we obtained the segmentWidth * segmentHeight elevations. And in order to normalize theses elevations we used this formula:

$$ElvModel = \frac{ElvReal - ElvMin}{ElvMax - ElvMin} * unitScal$$

ElvModel   This is the height of a position in a 3D terrain model
ElvReal   This is the real elevation in meter compared to the level of the sea
ElvMin   This is the real minimum elevation of this region in meters compared to the level of the sea
ElvMax   This is the real maximum elevation of this region in meters compared to the level of the sea

The result provided a 3D model based on the vertices of a geometric plane, and we assigned all elevations to the third dimension of our 3D terrain model.

3.2   Simulation and Results

We took as widthStep for this simulation 1000, and we have devised the region in a matrix of 100 * 59.

This is the most important part of this work, the floods were modeled in 3D visualizations and its effect can be discussed in more comprehensive terms.

To achieve this goal we have developed a web application based on threejs [10] library that we can render the 3D scene of the region under consideration.

For the 3D flood visualization, we did two simulations, for the same region selected in Fig. 2, In the first simulation we assume a rainfall that gives a water level of 1 m, and in the second case, we let's simulate a rainfall that gives a water level of 3 m.

The following Figs. 3 and 4 illustrate the two simulations:

The red bar corresponds to the district AL AMAL defined by the geographical coordinates, latitude: 35.75 and longitude: −5.79.

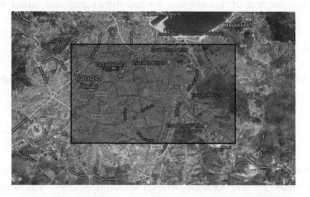

Fig. 2. Zone area by Google map drawing tools

Fig. 3. 3D Simulation of flooding for a level of water 1 m

The red bar corresponds to the district IDRISSIA defined by the geographical coordinates, latitude: 35.76 and longitude: −5.78.

Fig. 4. 3D Simulation of flooding for a level of water 3 m

In the case where the water level reaches 1 m, we find that the region of the district AL AMAL will be the most vulnerable zone to the floods.

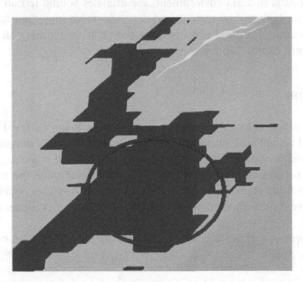

In the case where the water level reaches 3 m, the flood can extend to the area of the district IDRISSIA

Water simulation in a 3D environment, accumulates behind terrain for flooding to occur, flows from high to low regions and depends on an important factor, the water level. 3D visualization of flood plays an important role in avoiding all types of risks, especially in areas exposed to rising water levels.

4 Conclusion

This work is interested to an application for 3D visualizing of flood situations that provides a good insight with respect to the 2D visualization. This provides a better tool for analyzing and preparing for emergency measures. It also presents a real situation that can easily be understood. Cities managers can now through the use of 3D flood models, a better flood analysis which is close to reality. It is also possible to simulate flood for different pluviometer water levels in order to produce different flood scenarios.

For the perspective of this work, we plan to work on a simulation of floods related to watercourses riparian areas.

References

1. https://developers.google.com/maps/
2. Douass, S., Kbir, M.A.: 3D GIS for Smart cities, ACM digital library ISBN: 978-1-4503-5211-6, (2017)
3. Addaa, P., Mioc, D., Anton, F., Mcgillivray, E., Morton, A., Fraser, D.: 3d flood-risk models of government infrastructure, isprs-archives, April (2008)
4. Tymkow, P., Karpina, M., Borkowski, A.: GIS system for 3d visualization of hydrodynamic modeling of flood flows in river valleys, isprs-archives-xli-b8-175, (2016)
5. Bogetti, S.: Three-dimensional (3d) modeling for flood communication, thisis, university of gävle, Sweden, June (2012)
6. https://developers.google.com/maps/documentation/javascript/drawinglayer

7. https://developers.google.com/maps/documentation/javascript/elevation?hl=en
8. https://developers.google.com/maps/documentation/javascript/geometry?hl=en
9. https://developers.google.com/maps/documentation/javascript/reference#spherical
10. https://threejs.org/(03/2018)

A Review of Digital Watermarking Applications for Medical Image Exchange Security

A. Hassani Allaf$^{(\boxtimes)}$ and M. Ait Kbir

LIST Laboratory, Faculty of Sciences and Technologies, Tangier, Morocco
a.hassani@uae.ac.ma, m.aitkbir@fstt.ac.ma

Abstract. In this paper, we present the usage of watermarking techniques in medical imaging, and especially for security goals, the watermarking is considered a great solution to protect the personal data of patients during the medical images and telemedicine data exchange. This paper is devised in two part. The first one is reserved for an overview on image watermarking with a presentation of the most important requirements of watermarking (robustness, imperceptibility and capacity). We offer also the general scheme of watermarking with the two essential phases and different types of attacks. Furthermore, we present a classification of watermarking techniques based on various parameters such as: insertion domain, human perception and detection methods, in the end of the section we display some metrics and benchmarks for analysis the performance of the watermarking technique. The second part is reserved for the usage of watermarking techniques in medical imaging especially for integrity verification, authentication and data hiding, we also discuss a literature review on watermarking techniques for medical image. In addition we present the concept of telemedicine and telehealth fields and the importance of watermarking in the modern health care.

Keywords: Watermarking · Medical imaging · Medical image watermarking (MIW) · E-health · Telemedicine

1 Introduction

Security of medical information today is a necessity, derived from legislative rules, gives rights to the patient and duties to the health professionals. The modern health care systems today are based on sharing the medical information in unsecured networks like the internet to be accessible by the doctors or the specialists. Medical image is one of this information, it's considered as a main core in the telemedicine field and used for clinical analysis and medical diagnostics. Any modification of the medical image content of the will affect in the specialist diagnosis. For these reasons, it is necessary to provide the security conditions of interchanges in order to guarantee the integrity and authenticity of the medical images during the transmission. The watermarking techniques it's a great solution to protect the medical image, especially with the advancement of watermarking techniques.

© Springer Nature Switzerland AG 2019
M. Ben Ahmed et al. (Eds.): SCA 2018, LNITI, pp. 472–480, 2019.
https://doi.org/10.1007/978-3-030-11196-0_40

2   Overview on Watermarking

2.1   What is Watermarking?

The digital watermarking concept describes methods and technologies that embed some information (a watermark) into the host object such as an image, video, audio or any other digital data, without altering its visual quality. Digital watermarking can be used for: copyright protection, database indexing, authentication, broadcast monitoring, medical imaging and many other applications.

2.2   Watermarking, Steganography and Cryptography

The watermarking concept is related with two fields: cryptography and steganography, and the three concept classified under data security system field (Fig. 1).

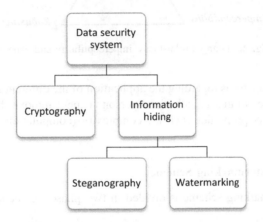

Fig. 1. Data security system field

Cryptography is a method to sending an encrypted message that only authorized person can decode, and when the message is decrypted it is not protected anymore and this is the main different between cryptography and watermarking.

Steganography used for hide the existence of a message within another object (image, video, audio) known as a data [1] In order to be undetectable, while the goal of watermarking is to embed a message in a way that it cannot be removed.

2.3   Watermarking Requirements

The most vital properties for digital watermarking are outlined as beneath. The importance of every property may change according to the field of application.

Robustness. One of the most important properties, it's the capacity of a watermarking method to resist against different attacks like removal attack, affine transforms attacks such as translation, rotation and scale or compression etc.

Imperceptibility. Imperceptibility or invisibility is the perceptual similarity between the original signal and the watermarked signal.

Capacity. Watermarking capacity is an evaluation of how much information can be hidden within a digital data. The capacity relies on the size and the type of the host signal.

J.Fridrich [2] presents the contradictory relation between the imperceptible, robustness and the capacity by a triangle (Fig. 2).

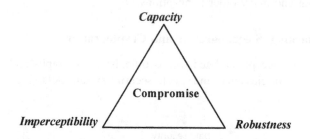

Fig. 2. Triangle robustness, imperceptibility and capacity

Security. This factor is regarding the application of the different kinds of keys, such as public or private, so that unauthorized persons cannot remove the watermark.

There are others properties such as: complexity (computation time), false alarm, etc. [3].

2.4   General Watermarking Scheme

The digital watermarking scheme is divided in two phases: embedding and detection (Fig. 3).

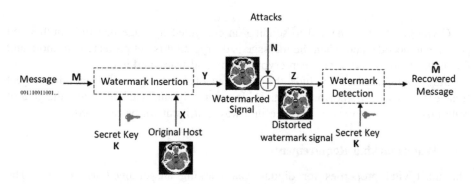

Fig. 3. General cycle life of watermarking

Embedding. In this step, the watermark $W = f(M, K)$ embedded in the signal host (X) with the help of an insertion algorithm for producing a watermarked signal (Y).

Detection (Extraction). The detection is made to extract the watermark from the distorted watermark signal (Z) and to decode the message (M) if is exist.

2.5 Attack

The concept of attack means any modification in the resulted signal, the attacks can be classified into four main groups [4]:

- Removal Attacks: attacks with intent to remove watermarks from the host signal.
- Geometrical attack: that attempt to damage the embedded watermark without removing it, like rotation, cropping and scaling.
- Cryptographic attacks: aim of cracking the security methods in watermarking schemes.
- Protocol attacks: aim attacking the entire scheme of the watermarking.

2.6 Classification of Watermarking Techniques

The watermarking techniques can be classified in many categories (Fig. 4) and upon various parameters like domain of insertion, human perception or method of detection [5–7].

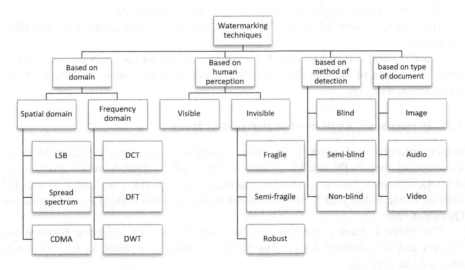

Fig. 4. Classification of watermarking techniques

Classification based on domain of insertion. Most of watermarking techniques can be distinguished into two approaches, those in the spatial domain (insertion in the pixel level) and those in the frequency domain (transformed domain) like Discrete Cosine Transform (DCT), Discrete Fourier Transform (DFT) and Discrete Wavelet Transform.

A basic comparison between watermarking methods spatial domain and frequency domain are presented in Table 1.

Table 1. Comparison between spatial and frequency domain

	Spatial domain	Frequency domain
Complexity	low	high
Capacity	high	low
Robustness	low	high
Applications	Authentication, integrity verification, etc.	Copy control, etc.

Classification based on human perception. In this classification there are two types of watermarks, the visible one like logos or text that clearly identifies the owner of the image or video. The second one is the invisible watermarks which cannot be perceived by the human sensory system. An invisible watermark can be either fragile, semi-fragile or robust against various attacks (Sect. 2.5).

Classification based on method of detection. The embedded data can be detected with three method blind, semi-blind and no-blind, the first method detect the mark without the use of the original signal, the second method use just some information, and the third method require the original signal to detect the mark.

The non-blind method is most robust to attacks in comparison with the blind method [6].

Classification based on type of document. Watermarking methods can be classified by the type of the host signal this can be image, audio, video, text, script or any other digital data.

2.7 Performance Analysis of Watermarked Image

Imperceptibility evaluation. Watermarked image perceptibility is evaluated by using some quality metrics [8] like Mean Squared Error (MSE), Peak-Signal-to-Noise Ratio (PSNR), Structural Similarity (SSIM), Euclidean distance (ED), and other metrics [9] such as Image Fidelity (IF), Normalized Mean Square Error (NMSE) and Correlation Quality (CQ).

The PSNR is most commonly used to measure the ratio of noise between the original and watermarked image. The PSNR formulation is given in Eq. (1) and is measured in decibels.

Where MAX_I is the maximum possible pixel value of the original image I (for an image represented with 8 bits the $MAX_I = 255$).

MSE is the Mean Square Error between original and watermarked image is measured by the Eq. (2).

Where $I(i, j)$ represent the original image, $I_w(i,j)$ is the watermarked image and $M \times N$ is the dimensions of the image.

To have acceptable perceptual value, the PSNR should be greater than 30 dB (Table 2).

Table 2. Formulation of some quality metrics

Metrics	Formulation
Peak-Signal-to-Noise Ratio	$PSNR(I, I_w) = 10 \times log_{10} \frac{(MAX_i^2)}{MSE}$ (1)
Mean square error	$MSE = \frac{1}{M \times N} \sum_{i=0}^{N-1} \sum_{j=0}^{M-1} (I(i,j) - I_w(i,j))^2$ (2)
Image fidelity	$MSE = \frac{1}{M \times N} \sum_{i=0}^{N-1} \sum_{j=0}^{M-1} (I(i,j) - I_w(i,j))^2$ (3)
Correlation quality	$CQ = \frac{\sum_{i,j} I(i,j) \times I_{(i,j)w}}{\sum_{m,n} I(i,j)}$ (4)
Euclidean distance	$ED(I, I_w) = \sum_{i=1}^{M} \sum_{j=1}^{N} (I(i,j) - I_w(i,j))^2$ (5)

Robustness evaluation. There are many benchmarks (tools) for testing the robustness of image watermarking algorithms like: StirMark benchmark realized by Kutter el al [9], CheckMark benchmark suggested by Voloshynovskiy et al. [10], OptiMark benchmark developed by Nikolaidis et al. [11]. These benchmarks of evaluation used a bank of attacks, in order to try to destroy the embedded watermark.

3 Medical Image Watermarking (MIW)

3.1 Basic Concepts in Telehealth, Telemedicine and Medical Imaging

Telehealth involves the use of information and communications technologies to provide and support the health care services. Telehealth applications include telemedicine, telediagnosis, teleconferences, telesurgery and others medical applications.

Telemedcine defined as the provision of medical services at a distance [12] where the patients and the specialists are disjointed by physical distance, services like Teleradiology Telepharmacy, Telecardiology and many other services in emergency situations. The first transformation of a medical information in history, it was between Alaska and Australia by 1930 [13].

Medical imaging refers to technologies and process that used for the visualization of some parts of body, tissues, or organs in order to use in clinical diagnosis, treatment or disease monitoring. Imaging has become today an essential component in many fields of medicine like X-Ray image, Computed tomography image and Positron emission tomography.

3.2 Applications of Watermarking in Medical Imaging

The watermarking techniques can be used in medical images mainly for:

- **Integrity Verification.** The purpose of this application is to find out whether any modification has been done upon the medical image or not, such as translation, rotation and scale or compression. In this application, fragile or semi-fragile watermarking algorithms should be applied, which are not robust against content modification.

- **Authentication.** Is an important application in the medical field for assuring the authenticity of patient information during the transformation of medical image.
- **Data hiding.** Are used to hide a large amount of data secretly into the medical image like the electronic patient report (EPR).

3.3   Medical Image Watermarking Requirements

In addition to the basic requirements of a typical watermarking system, as previously explained, some other specific features are needed for the medical watermarking system. These are explained below:

Intactness of ROI. Medical image have two regions, interest region and non-interest region. The watermarking schemes should not affect the ROI adversely. Distorted ROI will lead to wrong diagnosis.

Authentication. Identification of the image source and verification that the image belongs to the correct patient.

Integrity Control. The ability to verify that the image has not been modified without authorization.

3.4   Watermarking Techniques for Medical Images

Medical image it's usually formed of two regions, region of interest (ROI) and region of no interest (RONI).

Abhinav Shukla [14] suggested a scheme with three processes : embedding, extraction and authentication process (Fig. 5) to embedding the watermark the no interest region of medical image to avoid any distortions in the image content (Fig. 6).

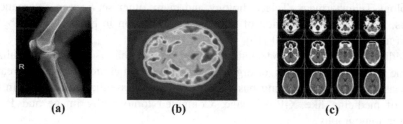

| (a) | (b) | (c) |

Fig. 5. a X-ray of the knee, **b** PET scan of the human brain, **c** CT of the human brain

Memon et al. proposed a blind hybrid watermarking method [15] which embeds two watermarks, the fragile mark it's inserted in the region of the interest (ROI) for tamper detection and authentication, the robust mark is used for security and confidentiality of patient information and it's inserted into the region of the non interest (RONI).

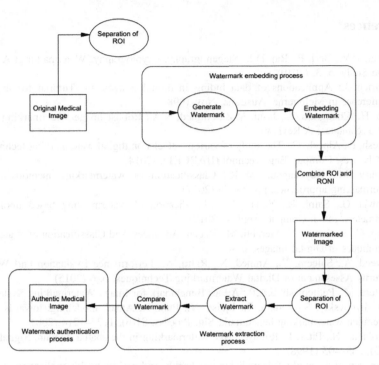

Fig. 6. Processes scheme of insertion a watermark in ROI region [14]

Gunjal et al. [16] inserted a hospital logo into RONI region of medical image, with used the Discrete Wavelet Transform method and they got a good results with a correlation factor equals to 1 and PSNR up to 48.53 dBs.

Al-Qershi et al. proposed a watermarking scheme [17] combines two methods: Difference Expansion (DE) and Discrete Wavelet Transform (DWT), for authentication and data hiding usages. In this scheme Qershi et al. inserted tamper detection and recovery in RONI region by the usage of DWT technique and the patient's data were inserted into the ROI region with DE technique, with a hiding capacity up to 0.46–0.50 bpp.

Hajjaji et al. [18] proposed a multi-layer method inspired by works of Vassaux et al. [19, 20] and CDMA technique. Hajjaji inserted the signature of the institution, patient data and the secret key in different types of medical imaging. And he gets a good result for radiography image.

4 Conclusion

The fast advancement of the health care systems and the needs of sharing the medical data between doctors, specialists and hospitals. All these challenges make the protection of patient's data on the cap of priorities of any future e-health system. And now with the advancement of watermarking field, the sharing of the medical information like medical images can become more secure.

References

1. Desai, H.V., Beri, P., Raj, D.J.: Steganography, Cryptography, Watermarking: A Comparative Study. p. 3, (2010)
2. Fridrich, J.: Applications of data hiding in digital images. In: Tutorial for the ISSPA, Conference in Melbourne, Australia, Nov-1998
3. Tao, H., Chongmin, L., Jasni, M.Z., Ahmed, N.A.: Robust Image Watermarking Theories and Techniques: A Review
4. Mitesh, P., Alpesh, C.: The study of various attacks on digital watermarking technique. Int. J. Adv. Res. Comput. Eng. Technol (IJARCET), (2014)
5. Boreiry, M., Keyvanpour, M.-R.: Classification of watermarking methods based on watermarking approaches. pp. 73–76 (2017)
6. Chawla, G., Saini, R., Yadav, R.: Classification of watermarking based upon various parameters. Int. J. Comput. Appl. 4 (2012)
7. Song, C., Sudirman, S., Merabti, M.: Recent Advances And Classification of Watermarking Techniques in Digital Images, p. 6
8. Naveed, A., Saleem, Y., Ahmed, N., Rafiq, A.: Performance Evaluation and Watermark Security Assessment of Digital Watermarking Techniques, p. 6 (2015)
9. Kuttera, M., Petitcolasb, F.A.P.: A Fair Benchmark for Image Watermarking Systems p. 14
10. Voloshynovskiy, S., Pereira, S., Iquise, V., Pun, T.: Attack modelling: towards a second generation watermarking benchmark. Sig. Process. 81(6), 1177–1214 (2001)
11. Nikolaidis, N., Pitas, I.: Robust image watermarking in the spatial domain. Signal Process. 66(3), 385–403 (1998)
12. Weinstein, R.S., et al.: Telemedicine, telehealth, and mobile health applications that work: Opportunities and Barriers, Am. J. Med. 183–187, (2014)
13. Karen, M.Z.: Telemedicine: History, Applications, and Impact on Librarianship. US National Library of Medicine National Institutes of Health (1996)
14. Abhinav, S., Chandan, S.: Medical image authentication through watermarking. Int. J. Adv. Res. Comput. Sci. Technol. (2014)
15. Memon, N.A., Asmatullah, C., Mushtaq, A., Zulfiqar Ali, K.: Hybrid watermarking of medical images for ROI authentication and recovery. Int. J. Comput. Math., 2057–2071, (2011)
16. Gunjal, B.L., Mali, S.N.: ROI based embedded watermarking of medical images for secured communication in telemedicine, 6(8), 6 (2012)
17. Al-Qershi, O.M., Khoo, B.E.: Authentication and data hiding using a hybrid ROI-based watermarking scheme for DICOM images. J. Digit. Imaging 24(1), 114–125 (2011)
18. Hajjaji, M.A., Ridha, H., Abdellatif, M., El-Bey, B.: 'Tatouage des Images Médicales en Vue d'Intégrité et de Confidentialité des Données', p. 7
19. Boris, V.: 'Technique multicouches pour le tatouage d'images et adaptation aux flux vidéo MPEG-2 et MPEG-4', Grenoble INPG, (2003)
20. Vassaux, B., Bas, P., Chassery, J.M.:'Tatouage d'images par étalement de spectre : apport de la technique CDMA en mode multicouche' 8 (2000)

A Chaotic Cryptosystem for Color Images Using Pixel-Level and Bit-Level Pseudo-Random Permutations

Said Hraoui[1]($\boxtimes$), Faiq Gmira[2], Fouad Mohammed Abbou[3],
A. Oualidi Jarrar[1], and Abdellatif Jarjar[4]

[1] Department of Mathematics, Faculty of Sciences Dhar El Mahraz, Sidi Mohamed Ben Abdellah University, LSO, P.O. Box 1796, Atlas-Fez, Morocco
Said.hraoui@usmba.ac.ma, ajarrarl@yahoo.fr
[2] Laboratory of Modeling Applied to Economy and Management, Faculty of Economics, Hassan II University, Casablanca, MAEGE, Ain Sbaa, Casablanca, Morocco
faiqgmira@hotmail.com
[3] School of Science & Engineering, Al Akhawayn University, Ifrane, Morocco
F.Abbou@aui.ma
[4] Secondary School Moulay Rachid, Taza, Morocco
abdoujjar@gmail.com

Abstract. This paper, we present a novel color image encryption algorithm based on pseudorandom permutations intra and inter pixels. Firstly, confusion by modification of pixel values in the image is applied by a permutation via a bijective function in the ring $\mathbb{Z}/8\mathbb{Z}$. Secondly, permutation is applied to all pixels in the image by the same function but this time in the ring $\mathbb{Z}/n\mathbb{Z}$ to dispel the redundancy on the entire image. Finally, a second chaotic confusion is applied to the different pixels in the image. The results show that the algorithm effectively reduces the correlations between plain and ciphered image and therefore can encrypt the color image more effectively.

Keywords: Image encryption · Bit-level permutation · Chaos
Confusion · And diffusion

1 Introduction

With the rapid development of the Internet and the advent of smart phones that use a large amount of private information such as digital images, information security issues have become more serious and challenging.

Indeed, due to the size of the data and redundancy of digital image, traditional cryptographic algorithms, such as the Data Encryption Standard (DES) and Advanced Encryption Standard (AES) are not suitable for image encryption [1, 2]. In this context, in order to meet these security requirements, new encryption algorithms of image using different techniques have been proposed, including those based on chaos theory [3–5].

In the literature, a variety of encryption algorithm based on permuting inter pixel in the image, to change pixel value have been proposed [6, 7]. Zhu et al. have proposed a

© Springer Nature Switzerland AG 2019
M. Ben Ahmed et al. (Eds.): SCA 2018, LNITI, pp. 481–491, 2019.
https://doi.org/10.1007/978-3-030-11196-0_41

bit-level scheme for image encryption based on boot Arnold cat map and logistic map [8]. Xiang et al. proposed an encryption scheme of selective image that is used to encrypt the four high bits of each pixel and leaves the lower four bits unchanged [9].

Diaconu et al. [13] Have proposed a new confusion strategy based on the inter-intra pixel permutation. This architecture considerably reduces the redundancy of the image encryption scheme based on the Fridrich structure [14]. Indeed, the authors conclude that the distribution of the bits of each bit plane is more uniform. This is due not only to the reduction of the correlation between the adjacent upper bit planes but also to the modification of the pixel values.

Despite the advantages of permutation, it has a number of inherent limitations. The ciphers based only on the permutations, discloses certain essential characteristics of the plaintext, as the frequency distribution of the symbols in the plaintext. Also, when the size of the plaintext is limited, i.e. the number of possible arrangements for the raw text elements is less than the key space, the number of effective key may be reduced and as a result, the mapping of permutation may be disclosed.

Taking into account what is mentioned above, in this paper, a new image encryption scheme based on a chaotic system is proposed. Based on the substitution-permutation architecture. The schema uses the enhanced logistic map which has a very good uniform distribution. The proposed algorithm consists of a double permutation. The first one, intra-pixels allowing the change of the values of each pixel. The second, changes the position of each pixel. Finally, in order to improve the statistical performances, a layer of confusion is realized via the enhanced logistic map.

The rest of the paper is organized as follows: in the second section, the proposed encryption/decryption algorithm is presented. The analysis of the performance of the proposed image encryption system is discussed in the third section. Finally, the fourth section concludes the paper.

2   Cryptosystem Block Diagramag

The crypto-systems schematized in Fig. 1 is a cipher with an architecture based on three layers. The first is the layer of confusion (substitution) of the crypto-system; which allows making the relationship between the key and the very complex cipher is performed first by a bijective pseudorandom permutation in the ring $\mathbb{Z}/8\mathbb{Z}$. While the second layer of permutation, whose role is to dispel the redundancy of the plaintext by spreading over the cryptogram, is performed by permutations of pixels in each RGB channel respectively by a bijective function in the ring $\mathbb{Z}/n\mathbb{Z}$, where n is the size of the image. Finally, a third layer creating a confusion effect is achieved by using the operator 'exclusive or' between the image obtained and an enhanced logistic map.

The proposed crypto-system is a symmetric encryption algorithm with the architecture illustrated in Fig. 1. The principle of the technique is based on permutations intra-inter pixels then confusion by a chaotic map. The steps in the encryption method are as follows. First, a confusion allowing the modification of the values of each pixel via a permutation by a bijective function. Secondly, a pseudo-random permutation of pixels by bijective functions, for dividing the cryptographic characteristics of the substitution layer, is applied. Finally, confusion by the operator or exclusive is applied.

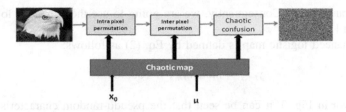

Fig. 1. The architecture of image encryption and decryption

2.1 The Logistic Map

The logistic map is a second-degree polynomial [1, 10] widely used in different arrears, expressed by the following recurrence relation:

$$u_{t+1} = \theta u_t(1 - u_t) \tag{1}$$

where $\theta \in [0, 4]$ is a control parameter of the logistic map, the variable $u_t \in [0, 1]$ with t is the iterations number used to generate the iterative values. When varying the parameter of the logistic map θ, we see that for a good choice of the initial condition u_0, the chaotic nature occurs only when $\theta \in [3.57, 4]$ as shown in Fig. 2.

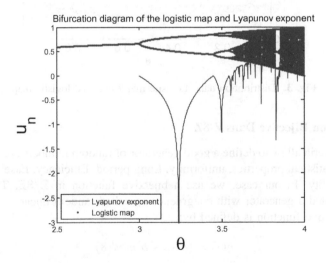

Fig. 2. Bifurcation diagram of the logistic map and its Lyapunov exponent

The logistic map is used due it presents several advantages such as simple structure, easy to implement in digital systems due its discrete nature, low implementation resources, low memory consumption and high-speed data generation [12]. Nevertheless, the logistic map has some disadvantages when it is used in cryptography such as chaotic ranges discontinues, distribution not uniform, small space key and periodicity in chaotic ranges [12]. Consequently, we are going to use the enhanced logistic map proposed in [11].

The enhanced logistic map is defined by Eq. (2) as follows:

$$y_{t+1} = mod(u_{t+1} \times 100,000, 1) \tag{2}$$

As shown in Fig. 3, it can be seen that the pseudo-random characteristics of the enhanced logistic map presents a good uniform distribution than the original logistic map.

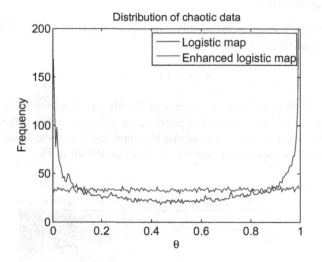

Fig. 3. Distribution data: Logistic map/Enhanced logistic map

2.2   Fonction Bijective Dans Z/8Z

The main criteria allow to define a good generator of random numbers are respectively, their good statistical properties, uniformity, Long period, Efficiency, Ease of setting-up and separability. In our case, we use a bijective function in $\mathbb{Z}/8\mathbb{Z}$. This bijective function created a generator with congruence so called random linear.

The bijective function is defined by:

$$g(x) = (a * x + b)mod(8) \tag{3}$$

where a and b are integer.

An example of mode of permutation is described by the Fig. 4 below:

Is the bijective function of $\mathbb{Z}/8\mathbb{Z}$ defines by:

$$g(x) = 5 * x + 1 \, mod(8) \tag{4}$$

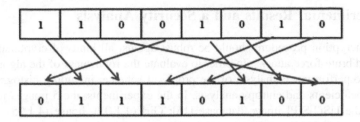

Fig. 4. Example of permutation

The permutation of the bits of different pixel, generated by the function of the Eq. (4) is illustrated as follows:

2.3 The Scheme of Image Encryption and Decryption

The encryption algorithm proposed in this paper is based on permutation–confusion architecture. The initial value x and the control parameter p of logistic map are used as secret key. The first confusion is based on bit-level permutation described in Sect. 2. While the inter pixel, permutation used the same bijective function but this time in the ring $\mathbb{Z}/n\mathbb{Z}$. Finaly, a confusion process is realized by using the bitwise XOR operator between the permuted image and the logistic map. The algorithm can be described in the scheme illustrated in Fig. 5:

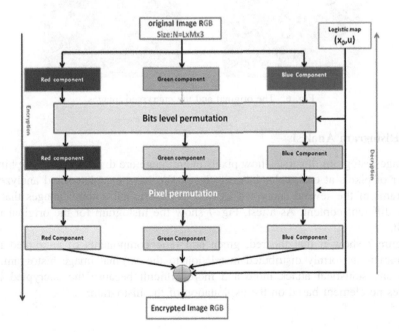

Fig. 5. Encryption/decryption scheme

3   Experimental Results and a Security Analysis

A good encryption procedure should be robust against all kinds of cryptanalytic, statistical and brute-force attacks. In order to evaluate the robustness of the algorithm, we discuss the performance analysis of the proposed scheme, including histograms, correlation coefficients and entropy analysis. In the experiments, the 5 images for testing are from the USC-SIPI image database [15]: Girl (4.1.01), house (4.1.05), Mandrill (4.2.03), Lena (4.2 04) and Peppers (4.2.07).

3.1   Visual Testing

The objective of visual testing is to have confirmation of the absence of similarities between the original image and the encrypted one. The visual test was performed on the Mandrill test image of 512 × 512 pixels, 24 bits. Figure 6 shows the test image, as well as the encrypted image. By comparing them, it can be said that there is no visual information referring to the original image.

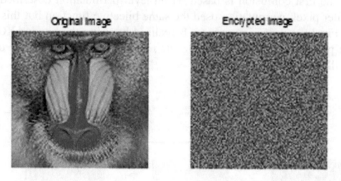

Fig. 6. The original and the encrypted images

3.2   Histogram Analysis

An image-histogram illustrates how pixels in an image are distributed by graphing the number of pixels at each color intensity level. We have calculated and analyzed the histograms of the several encrypted as well as its original colored images that have widely different content. As a test, Fig. 7 show the histogram for the original image Madrill.

Figure 7 showed that the red, green and blue components of encrypted image histogram is uniformly distributed in relation to the original image histogram. This makes any statistical attack more and more difficult because the encrypted image provides no element based on the exploitation of the histogram.

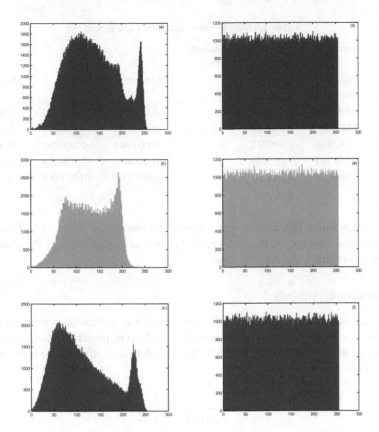

Fig. 7. Frames **a–c** respectively, show the histograms of red, green and blue channels of the plain image. Frames **d–f** respectively, show the histograms of red, green and blue channels of the encrypted image

3.3 Information Entropy Analysis

Image information entropy can measure the distribution of image gray values. The more uniform the gray value distribution is, the bigger the information entropy is. The less information of the original image can be obtained from the gray value distribution of the cipher-image by the attacker, the higher security the encryption algorithm has. Image information entropy is defined as:

$$H = -\sum_{i=1}^{256} p_i \log(p_i) \tag{5}$$

where p_i is the probability of the gray value. The ideal value of the cipher information entropy is 8. The information entropy of the cipher-image for the 5 images for testing generated by the proposed algorithm is given in Table 1.

Table 1. Entropy and correlation coefficient

Image	Entropy	Correlation coefficient for encrypted image			
	Original image	Cipher image	Horizontal	Vertical	Diagonal
Girl	6.898139	7.998946	0.007740	−0.005254	−0.006280
House	7.068625	7.998960	0.001188	0.008772	−0.000049
Mandrill	7.762436	7.999778	−0.001744	−0.000786	0.002803
Lena	7.750197	7.999769	0.004489	−0.006314	0.006204
Peppers	7.669825	7.999747	−0.006168	−0.001688	−0.001078

The entropy value obtained is very close to the theoretical value of 8. This means that information leakage in the encryption process is negligible and the encryption system is secure upon the entropy attack.

3.4 Correlation Analysis

Another statistical attack technique is the analysis of the correlation coefficient [1, 16], which allows to evaluate the correlation between adjacent pixels and hence assess the robustness of the algorithm. We calculate the correlation coefficient for a sequence of adjacent pixels by the formula below:

$$r_{xy} = \frac{\text{cov}(x, y)}{\sqrt{D(x)} \sqrt{D(y)}} \tag{6}$$

where x, y. e two vectors formed respectively by the values of the image's selected sequence pixels and the values of their adjacent pixels. The terms $\text{cov}(x, y)$, $E(x)$, $D(x)$ are calculated by the following formulas:

$$E(x) = \frac{1}{N} \sum_{i=1}^{N} x_i \tag{7}$$

$$D(x) = \frac{1}{N} \sum_{i=1}^{N} [x_i - E(x)]^2 \tag{8}$$

$$\text{cov}(x, y) = \frac{1}{N} \sum_{i=1}^{N} [x_i - E(x)][y_i - E(y)] \tag{9}$$

Where N is the number of adjacent pixels selected in the image to calculate the correlation coefficient. x_i and y_i are, respectively, the elements of x and y.

Furthermore, in order to resist statistical attack, an efficient cryptosystem must have low correlation of any two adjacent pixels in the ciphered image. From Table 1, it can be seen that our encryption algorithm satisfies zero correlation for all test images.

Therefore, it's difficult to recover the original image from the encrypted image without knowing the encryption key.

The representative results of correlations and the correlation coefficients are well illustrated in Fig. 8.

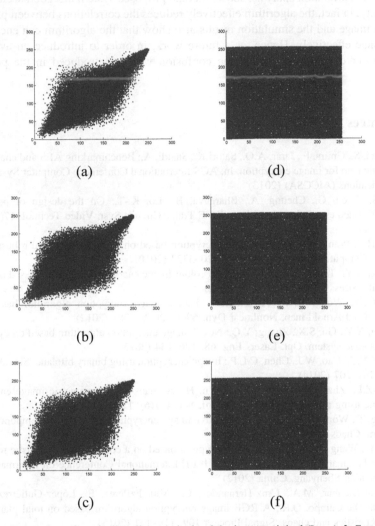

(a)   (d)

(b)   (e)

(c)   (f)

Fig. 8. **a** Correlation of vertically adjacent pixels of the image original (Lena), **b** Correlation of horizontally adjacent pixels of the original image, **c** Correlation of diagonally adjacent pixels of encrypted Lena, (d) Correlation of pixels

4 Conclusions

In this paper, a novel color image encryption based on intra-inter pixel level permutation is presented. The security analysis, which includes histogram, information entropy and correlation analysis, shows that the proposed system has good security and complexity. In fact, the algorithm effectively reduces the correlations between plain and ciphered image and the simulation results also show that the algorithm can encrypt the color image effectively. Hence, as a future work, in order to introduce an avalanche effect to our cryptosystem, a chaotic confusion will be introduced in the proposed scheme.

References

1. Hraoui, S., Gmira, F., Jarar, A.O., Satori,K., Saaidi, A.: Benchmarking AES and chaos based logistic map for image encryption. In: ACS International Conference Computer Systems and Applications (AICCSA) (2013)
2. Li, S., Chen, G., Cheung, A., Bhargava, B., Lo, K.-T.: On the design of perceptual MPEGVideo encryption algorithms. IEEE Trans. Circuits Syst. Video Technol. 17(2), 214–223 (2007)
3. Liu, H.J., Wang, X.Y.: Color image encryption based on one-time keys and robust chaotic maps. Comput. Math Appl. 59(10), 3320–3327 (2010)
4. Wang, X.Y., Teng, L., Qin, X.: A novel colour image encryption algorithm based on chaos. Signal Process 92(4), 1101–1108 (2012)
5. Ye, G.D., Wong, K.W.: An efficient chaotic image encryption algorithm based on a generalized Arnold map. Nonlinear Dyn. 69(4), 2079–2087 (2012)
6. Wang, X.Y., Gu, S.X., Zhang, Y.Q.: Novel image encryption algorithm based on cycle shift and chaotic system. Opt. Lasers Eng. 68, 126–134 (2015)
7. Zhou, Y.C., Cao, W.J., Chen, C.L.P.: Image encryption using binary bitplane. Signal Process 100, 197–207 (2014)
8. Zhu, Z.L., Zhang, W., Wong, K.W., Yu, H.: A chaos-based symmetric image encryption scheme using a bit-level permutation. Inf. Sci. 181(6), 1171–1186 (2011)
9. Xiang, T., Wong, K.W., Liao, X.: Selective image encryption using a spatiotemporal chaotic system. Chaos 17(3), 023115 (2007)
10. Dai, Y., Wang, X.: Medical image encryption based on a composition of logistic maps and chebyshev maps. In: Proceeding of the IEEE International Conference on Information and Automation Shenyang, China (2012)
11. Murillo-Escobar, M.A., Cruz-Hernández, C., AbundizPérez, F., López-Gutiérrez, R.M., Acosta Del Campo, O.R.: A RGB image encryption algorithm based on total plain image characteristics and chaos. Signal Process 109, 119–131 (2015)
12. Murillo-Escobar, M.A., Cruz-Hernández C., Cardoza-Avendaño, L., Méndez-Ramírez, R.: A novel pseudorandom number generator based on pseudorandomly enhanced logistic map. Nonlinear Dyn. 87(1):407–425 (2017)
13. Diaconu, A.V., Ionescu, V., , Lopez-Guede, J.M.: Security analysis of a new Bit-Level permutation image encryption algorithm. In: International Conference on EUropean Transnational Education, pp. 595–606. Springer International Publishing, (2016)

14. Fridrich, J.: Symmetric ciphers based on twodimensional chaotic maps. Int. J. Bifurcat. Chaos **8**, 1259–1284 (1998)
15. University of Southern California, Signal and Image Processing Institute, The USC-SIPI Image Database, Available at: ⟨http://sipi.usc.edu/database/⟩
16. Wang, X., Wang, Q.: A novel image encryption algorithm based on dynamic S-boxes constructed by chaos. Nonlinear Dyn. **75**, 567–576 (2014)

Text Line and Word Extraction of Arabic Handwritten Documents

Asmae Lamsaf[1(✉)], Mounir Aitkerroum[1], Siham Boulaknadel[2], and Youssef Fakhri[1]

[1] LaRIT, Faculty of Sciences-Ibn Tofail University, Kenitra, Morocco
{asmaelamsaf,maitkerroum}@gmail.com, fakhri@uit.ac.ma
[2] IRCAM, Madinat al Irfane, Rabat-Instituts, Rabat, Morocco
siham_06@yahoo.fr

Abstract. The documents of Arabic handwritten contain text lines and words. Words are often a succession of sub-words (characters, connected components) separated by spaces, in Arabic handwritten its spaces are divided into two types: the first type represents the spaces that separate two connected components of the same word (within-word), the second type are spaces that separate two connected components from two consecutive words(between-words). We detect the second type for word extracting. Word extraction based on the classification of spaces detected and extracts between-words spaces to segment the text into words. In this paper, we present a method for segmenting Arabic handwritten text into lines and words, to make our method of word extraction more optimal, we compute the threshold of spaces for each line, the threshold is not fixed in the document, each line is associated its classification threshold spaces. Before segmenting the text into words, it is necessary to segment it into text lines in order to apply our method to each line. To extract the lines, the preprocessing is applied to the text images in order to apply the proposed method for the line segmentation step. Our system is applied on the benchmarking datasets of the Arabic handwriting database for text recognition (AHDB) and the experimental results are very promising as we achieved a success word extraction rate of 87.9%.

Keywords: Word extraction · Arabic handwriting · Recognition of arabic handwriting · Lines segmentation · Handwriting analysis

1 Introduction

Handwriting considered the easiest way of communication between humans and machines. However, the understanding of handwriting by a machine is more difficult than reading its symbolic representation. It is necessary to transform handwriting into its symbolic representation. This transformation is performed by a system of automatic recognition of handwriting. The recognition of Arabic handwriting includes the recognition of characters, words and texts. In text recognition, it is necessary to segment the text into lines and then the text lines into words or characters in order to recognize these words or characters.

© Springer Nature Switzerland AG 2019
M. Ben Ahmed et al. (Eds.): SCA 2018, LNITI, pp. 492–503, 2019.
https://doi.org/10.1007/978-3-030-11196-0_42

The first step proposed in the literature for recognizing Arabic handwritten words is the segmentation of words into characters in order to recognize these characters. But the separation of characters is a delicate and expensive operation, as the writings are varied, and the letters often connected to each other. We don't need segmentation into characters for text recognition; it is enough to recognize the words without segmenting them into characters. Word extraction remains the major problem in increasing the performance of the Arabic handwriting recognition system, as it is an important step in making recognition systems more efficient and accurate.

Words are often a succession of sub-words (characters, connected components) separated by spaces, in Arabic handwritten its spaces are divided into two types: the within-word space is the space between sub-words of the same word and the space between two consecutive words called between-word space. The spaces in Arabic handwritten texts do not respect any rule because each person has his own style of writing so each writer has his own way of writing between-word spaces. Word extraction consists of detecting between-word spaces. In the case of Arabic handwriting where there is no sufficient between-word space, separation becomes a difficult task. Most of the techniques proposed for the word extraction in the literature consider the measurement of a threshold to classify the spaces between the words (between-word space) and between the connected components of the same word (within-word space). Three types of thresholds used: a manually determined threshold for predefined images [1] and an adaptive threshold for each image (handwritten document) [2], the latter represents the most tolerant and robust method, and a threshold determined by grouping spaces between connected components into three distinct groups according to the size of the connected components [3].

In this paper, we presented in the second section the previous works proposed in the literature for word and text line extraction. In the third section we present the proposed method explaining the steps to follow for creating a word extraction system from a text image. The first step is the preprocessing which consists of preparing the input images of text at the next step of lines segmentation; we proposed a new method for extracting text lines, and finally the step of extracting words. Before concluding the paper, we discuss the result obtained.

2   Previous Work

2.1   Lines Segmentation

Text line segmentation aims to isolate text line and characters within lines in the text image. The methods used in the literature, Jayant Kumar et al. [9] have proposed a graph-based method for extracting lines of handwritten text. It is very robust to non-uniform asymmetry and to variations in font size. Zhixin Shi et al. [10] proposed a method based on a generalized adaptive local connectivity map (ALCM) using a directional filter. It is effective for fluctuating, touching or crossing lines of text. Ouwayed et al. [11] applied morphological analysis to the final letters of Arabic words. The proposed system has been evaluated on overlapping documents and is very effective. Muna Khayyat et al. [12] proposed a method based on morphological dilation with a dynamic adaptive mask. It is evaluated on the basis of handwritten Arabic data, which contains multi-inclined and touching lines.

Dinges et al. [13] proposed a new method based on a local group to extract inclined and curved handwritten text lines in images of Arabic documents. It is based on the detection of connected components to classify them as Arabic word pieces (PAW) or diacritic using support vector machine (SVM). The results of the method show successful line segmentation of documents of different authors and styles.

2.2   Word Extraction

The segmentation of Arabic handwritten line into words is an important step to make the recognition systems more efficient and accurate. Word Extraction is an operation based on the assumption that writing is naturally divided into words and the between-word space is larger than the within-word space, where the value of the space between these sub-words does not respect any rule. The word extraction consists of detecting these between-word spaces. However, in the case of handwriting where there is no sufficient between-word space, separation becomes a difficult task.

The work for the recognition of Arabic writing began more than two decades ago. Almuallim and Yamaguchi [4] proposed a structural recognition technique for Arabic handwritten words that were segmented into text lines. Blows were classified and combined with characters according to their characteristics. However, their system has shown a failure in most cases because of the incorrect segmentation of words.

Most of the techniques proposed for word extraction in the literature consider a special measure of the gap between the connected successive components and define a threshold for classifying the spaces between the words (between-words space) and between the connected components of the same word (within-word space). The most works in the literature of the word extraction of Arabic handwriting, in AlKhateeb et al. [1], these thresholds are used by manually analyzing more than 200 images containing more than 250 words, derived from all IFN/ENIT databases. Methods that do not use prior knowledge and do not adapt to the properties of the document image would be more robust [5]. Al-Dmour and Fraij [2] introduced a method in which segmentation thresholds are calculated from each handwritten document. This makes the method adaptive and robust when applied to the Arabic handwritten nature, the results tested on the Arabic Handwritten Text Recognition (AHDB) reference databases, with a correct extraction rate 84.8%. Al-Dmour and Zitar [3] proposed a method that is based on grouping the spaces between connected components into three distinct groups according to the size of the connected components to determine an optimal threshold.

In this work, the objective of our system is to segment the text image in words in order to extract words, before to segment into words it is necessary to segment it into text lines. The proposed method for segmenting into lines is based on the analysis of horizontal projections. For the word extraction, we propose a method that classifies the spaces between the connected components of the same word (within-word) and the spaces between the connected components of two consecutive words (between-words), using a threshold calculated for each line, each line has its own classification threshold of spaces.

3 Method Proposed

In this work, we propose a method of lines segmentation and word extraction of Arabic handwritten text image. The input data of the system are text images written in a different way by several people. After the preprocessing on the text images which makes it possible to clean the images to the following steps, we segment the text image into text line images which aims to isolate the lines of the text image, which allow us to segment these lines into word images that consists of extracting the connected components in the text line, in order to determine the distances between them for computing the threshold for each line and classifies these distances into two types: between-word distances (spaces between two connected components of two words consecutive), and within-word distances (space between two connected components of the same word) (Fig. 1).

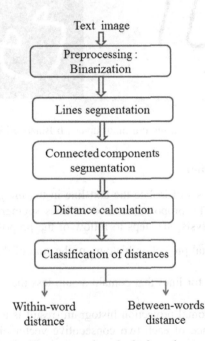

Fig. 1. The proposed method of word extraction

3.1 Preprocessing

The input images are poorly prepared, and includes noise that can cause problems in the next phase (the segmentation step), to simplify the processing of the segmentation step, we prepare these images before segmenting into text line, for converting the image in cleaned version, the operation used for this step is the binarization which consists of separating the background of the text if the original image is in grayscale or in color in order to produce a two-tone image: white in the background, and black for the text or the opposite. In a grayscale image, each pixel is associated with a luminosity ranging from 0 to 255. We used the global approach (global thresholding) which aims to compute a single threshold for the entire image. Pixels with a gray level darker than the threshold are set to black and the others to white.

We use the method of Otsu, which separates the pixels of the image into two classes. A pre-calculation of the gray level histogram is performed, then the separation into two classes is done from the moments of the mean and the standard deviation [6] (Fig. 2).

Fig. 2. Binarization: **a** Input image, **b** Binarized image

3.2   Text Line Extraction

Text line extraction consists of isolate the text line in the image and isolate the characters inside the lines. The proposed method of lines segmentation is based on the horizontal projection analysis, the steps to follow of the proposed method are:

(1) Compute the horizontal projection histogram: the sum of the pixels of all the lines of the image,
(2) Affect the value of 0 for lines that contain a sum less than or equal to a threshold $l = 12$,
(3) Recalculate the horizontal projection histogram (result is a vector),
(4) Compute the difference of each two consecutive vector element of the previous step (result is a difference vector),
(5) Compute the local maximum and local minimum of difference vector,
(6) Affect the neighborhood lines of local maximum and minimum lines the value of 1,
(7) Separate the lines:
   a. Determine the sum of the rows (result is the sum vector),
   b. Look for the lines with the sum greater than the column number of the text image (the lines of separations between the lines of text).

Algorithm 1: Text line extraction
Input: Text image

- Compute the horizontal projection histogram
- Affect the value of 0 for lines that contains a sum less than or equal to a threshold l
- Compute the horizontal projection histogram
- Compute the difference of vector result
- Compute the local maximums and local minimums of difference vector
- Affect the neighborhood lines of local maximum and minimum lines the value of 1
- Determine the sum of the rows
- Extract the lines of separation of lines (the lines which the sum > column number of image

Output: the text image segmented into lines

The result of the lines segmentation algorithm in the Fig. 3:

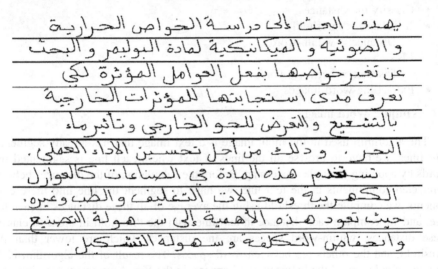

Fig. 3. Result of the line extraction algorithm

3.3   Words Extraction

In this work, we present a new method for extracting words in text image, our approach is based on the calculation of threshold of distances between connected components for each text line, and classification of these distances using this threshold. After segmenting text into lines, we have to browse all the lines where each line has its own classification threshold. The average of all distances between connected components in each line saved as the threshold for classification, we added a **b** value to the mean to increase the performance of the threshold calculation after affecting multiple values to be and choose the best among them.

$$\sum_{i=1}^{N} \frac{d_i}{N} + b \tag{1}$$

With: **S** the threshold of the text line, d_i the distance number i in the current line, **N** the total number of spaces between the connected components of the text line, **b** the value added to increase the performance of the system.

The algorithm used for this method writes as follows:

Algorithm 2: Word extraction
Input: Text image segmented into lines

- Apply for each text line
- Segment the text line into connected components (CCs)
- Compute the distances between CCs
- Compute the average of distances = threshold
- Classify the distances
- **If** distance < threshold
    Distance = within-word
  Else
    Distance = between-words
- Extract the word images

  Output: Word images

The algorithm used in the input image is a text image that has already segmented into lines, we have to browse all the lines of text where each line is segmented into words by applying multiple processing, Before the calculation of the distances between connected components in the text line, we have to segment the line into these components, we compute the threshold using the average of all these distances in each text line, and we compare each distance in text line with the threshold in order to classify these distances, within-word spaces are the distances which are lower than this threshold, and the others are between-word spaces. The result of the algorithm is the number of columns of the between-word spaces, in the Fig. 4 we present the image of the result.

Segmentation into connected components. The input image of this step is a text image that has already segmented into text lines. For computing the distances between connected components; we have to browse all the lines of text where each line is segmented into these connected components. The algorithm applied to extract these components written as follows:

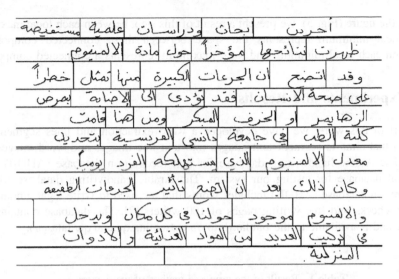

Fig. 4. Result of the word extraction algorithm

Algorithm 3: Connected components (CCs) segmentation Input: Text line image

- Binarize the image of text line
- Compute the vertical projection histogram
- Binarize the result vector
- Compute the difference of vector
- Compute local maximum et local manimum
- **For** i = 1 to number of local maximum
    Segment the CC number i (between ith local maximum and ith local minimum)
- **End for**

Output: Text line segmented into CC

Fig. 5. Connected components segmentation

In the figure (Fig. 5), we present a result of this step, between each local maximum and local minimum represents the space between these connected components, between each local minimum and local maximum represents a connected component.

4   Experiments and Results

As part of this work, we have developed a word extraction and lines segmentation system that aims to segment text images into text line images and word images. We used The Arabic handwriting database for text recognition database (AHDB) international database for testing our system. The Arabic handwriting database for text recognition database (AHDB) includes words that are used to write legal amounts on Arabic checks and free writing pages of 100 writers [7]. The database contains 105 forms, and is available to the public (http://handwriting.qu.edu.qa/dataset/).

Table 1. Result of the proposed method of word extraction

Image no.	Number of words	Misplaced words	Segmentation rate (%)
1	70	7	90
2	90	7	92.2
3	58	5	91.3
4	92	9	90.2
5	29	5	82.7
6	88	11	87.5
7	38	4	89.4
8	92	10	89.1
9	59	8	86.4
10	33	5	84.8
11	76	6	92.1
12	96	11	88.5
13	62	10	83.8
14	56	3	94.6
15	50	8	84
16	80	11	86
17	56	10	82
18	85	9	89.4
19	50	3	94
20	67	6	91
21	55	9	83.6
22	43	7	83.7
23	101	10	90
24	46	6	86.9
25	51	7	86
Average	**1683**	**187**	**87.9**

The programming language used is MATLAB 2014a, the test image number is 25 text images written by several writers.

We applied the proposed method of text line segmentation for each text image to extract text line, for each line we compute the thresholds of distances between connected components to extract the words, so the number of thresholds computed for a text equal to the number of the text lines, we have update the threshold for each line.

To evaluate the performance of our system, we applied the algorithm on 25 images of Arabic texts written by several writers, we computed the success rate of result obtained, and the most errors are overlapping letters or the badly written words. The results of our system are very promising as we achieved a success lines segmentation rate of 99% and word extraction rate of 87.9%. Table 1 presents the result of the proposed method of word extraction.

We compare the result of our approach to those for a previously published method. Table 2 compares the success rates of published systems of lines segmentation in the literature. Our result of word extraction is the best result proposed in the literature, since the threshold calculation method for each line becomes the determined threshold more accurate than the thresholds calculate for each document. Table 3 compares the success rates of published systems of word extraction in the literature.

Table 2. Lines segmentation rates

Method	Rate(%)
Ouwayed and Belaıd [11]	96.88
Kumar et al. [9]	96
Khayya et al. [12]	96.3
Proposed method	**99**

Table 3. Word extraction rates

Method	Rate (%)
AlKhateeb et al. [1]	85
AlDmour and Faij [2]	84.3
AlDmour and Zitar [3]	86.3
Proposed method	**87.9**

5   Conclusion

In this paper, we presented the proposed system of text line and word extraction from an Arabic handwritten text. The proposed method of text line segmentation is based on the horizontal projection analysis, and the method of word extraction based on the classification of distances between connected components according to the threshold of these distances, the threshold computed for each line in the text, the system classify the distances into between-words or within-word spaces, the result of the system is the identification of separation between words in each line using the classification by the calculated threshold. The system is applied on the AHDB database and achieves a good result with a segmentation rate of 87.9%.

References

1. AlKhateeb, J.H., Jiang, J.J., Ren, J., Ipson, S.: Interactive knowledge discovery for baseline estimation and word segmentation. Recent advances in technologies (2009)
2. Al-Dmour, A., Fraij, F.: Segmenting Arabic handwritten documents into text lines and words. Int. J. Adv. Comput. Technol. (IJACT) **6**(3), 2014 (2014)
3. Al-Dmour, A., Abu Zitar, R.: Word extraction from Arabic handwritten documents based on statistical measures. Int. Rev. Comput. Software **11**(5), 2016 (2016)
4. Al-Muallim, H., Yamaguchi, S.: A method of recognition of Arabic cursive handwriting. Pattern Anal. Mach. Intell. **9**(1987), 715–722 (1987)
5. Papavassiliou, V., Stafylakis, T., Katsouros, V., Carayannis, G.: Handwritten document image segmentation into text lines and words. Pattern Recogn. **43**(1), 369–377 (2010)
6. Otsu, N.: A threshold selection method from gray-level histograms. IEEE Trans. Syst. Man Cybern. **9**, 62–66 (1979)
7. Al-ma'adeed, S., Elliman, D. . Higgins, C.A., Campus, J.: A data base for Arabic handwritten text recognition research. In: Proceedings of the Eighth International Workshop on Frontiers in Handwriting Recognition (IWFHR'02) (2002)
8. Aouadi, N., Echi, AK.: Word extraction and recognition in Arabic handwritten text. Int. J. Comput. Inf. Sci. **12**(1) (2016)
9. Kumar, J., Abd-Almageed, W., Kang, L., Doermann, D.S.: Handwritten Arabic text line segmentation using affinity propagation. In: Proceeding(s) of DAS 10 Proceedings of the 9th IAPR International Workshop on Document Analysis Systems, pp. 135–142 (2010)
10. Shi, Z., Setlur, S., Govindaraju, V., Setlur, S., Govindaraju V.: A steerable directional local profile technique for extraction of handwritten Arabic text lines. In: ICDAR, pp. 176–180 (2009)
11. Ouwayed, N., Belaïd, A.: Separation of overlapping and touching lines within handwritten Arabic documents. In: Proceeding(s) of the 13th International Conference on Computer Analysis of Images and Patterns, CAIP. 9, pp. 123–138 (2009)
12. Khayyat, M., Lam, L., Suen, C.Y., Yin, F., Liu, C-L.: Arabic handwritten text line extraction by applying an adaptive mask to morphological dilation. In: Proceeding(s) of 10th IAPR International Workshop on Document Analysis Systems, pp. 100–104 (2012)
13. Dinges, L., Al-Hamadi, A., Elzobi, M.: A locale group based line segmentation approach for non uniform skewed and curved Arabic handwritings. In: 12th International Conference on Document Analysis and Recognition (ICDAR), IEEE (2013)
14. Yousif, I., Shaout, A.: Off-Line handwriting Arabic text recognition: a survey. Int. J. Adv. Res. Comput. Sci. Software Eng. **4**(9) (2014)
15. Ouwayed, N., Belaïd, A.: A general approach for multi-oriented text line extraction of handwritten document. Int. J. Doc. Anal. Recogn., Springer Verlag (2011)
16. Abdullah, S., AL-Nassiri, A., Salam, R.A.: Off-Line Arabic handwritten word segmentation using rotational invariant segments features (2008)
17. Elnagar, A., Bentrcia, R.: A recognition-based approach to segmenting Arabic handwritten text. J. Intell. Learn. Syst. Appl. 93–103 (2015)
18. Lawgali, A.: A survey on arabic character recognition. Int. J. Signal Process. Image Process. Pattern Recogn. **8**(2) 401–426 (2015)

19. Lorigo, L., Govindaraju,V.: Off-line Arabic handwriting recognition: a survey. IEEE Trans. Pattern Anal. Mach. Intell. **28**(05) 712–724 (2006)
20. Parvez, M.T., Mahmoud, S.A.: Offline Arabic handwritten text recognition: a survey. ACM Comput. Surv. **45**(2) (2013)
21. Boulid, Y., El Youssfi, E.M., A. SOUHAR. Reconnaissance de l'écriture manuscrite arabe en mode hors ligne (2016)
22. El Abed, H., Märgner, V.: The IFN/ENIT-database—a tool to develop Arabic handwriting recognition systems. In: 9th International Symposium on Signal Processing and Its Application (2007)
23. http://handwriting.qu.edu.qa/dataset/
24. Menasri, F.,: Contributions à la reconnaissance de l'écriture arabe manuscrite, Thèse Université Paris Descartes (2008)
25. Ouchtati, S., Redjimi, M. ., Bedda, M.: Recognition of the Arabic handwritten words of the algerian departments. Int. J. Comput. Theory Eng. **6**(2) (2014)
26. Abuzaraida, M.A., Zeki, A.M., Zeki, A.M.: Online recognition of Arabic handwritten words system based on Alignments matching Algorithm. In: Proceedings of the International conference on computing, Mathematics and statistics, Springer Nature Singapore (2017)
27. Khémiri, A., KacemEchi, A., Belaid, A., Elloumi, M.: A system for off-line Arabic handwritten word recognition based on bayesian approach. In: 15th International Conference on Frontiers in Handwriting Recognition (2016)
28. Ebrahinpour, R., Amini, M., Sharifizadehi, F.: Farsi handwritten recognition using combining neural networks based on stacked generalization. Int. J. Electr. Eng. Inf. **3**(2) 146–160 (2011)
29. Nouar, F., Aissaoui, M.E., Seridi, H.: Approche globale pour la reconnaissance de mots arabes manuscrits par combinaison parallèle de classifieurs. In: Proceedings des Journées des Jeunes Chercheurs en Informatique (JCI) (2008)
30. Alkhoury, I.: Arabic handwritten word recognition based on Bernoulli mixture HMM, Master Thesis, University of Valencia (2010)
31. Mohamed, K.: Reconnaissance de formes appliquée à l'écriture Arabe manuscrite par des multiclassifieurs, thesis (2010)
32. Boukerma, H.: Combinaison de classifieurs flous pour la reconnaissance de l'écriture arabe manuscrite, Master Thesis, (2010)

Intelligent Systems

A New Hybrid Framework Based on Improved Genetic Algorithm and Simulated Annealing Algorithm for Optimization of Network IDS Based on BP Neural Network

Zouhair Chiba[✉], Noreddine Abghour, Khalid Moussaid,
Amina El omri, and Mohamed Rida

LIMSAD Labs, Faculty of Sciences, Hassan II University of Casablanca,
20100 Casablanca, Morocco
chiba.zouhair@gmail.com, {noreddine.abghour,
khalid.moussaid,amina.elomri,mohamed.rida}@univh2c.ma

Abstract. Nowadays, network security is a world hot topic in computer security and defense. Intrusions, attacks or anomalies in network infrastructures lead mostly in great financial losses, massive sensitive data leaks, thereby decreasing efficiency and the quality of productivity of an organization. Network Intrusion Detection System (NIDS) is an effective countermeasure and high-profile method to detect the unauthorized use of computer network and to provide the security for information. Thus, the presence of NIDS in an organization plays a vital part in attack mitigation, and it has become an integral part of a secure organization. In this chapter, we propose to optimize a very popular soft computing tool widely used for intrusion detection namely, Back Propagation Neural Network (BPNN) using a novel hybrid framework (IGASAA) based on Improved Genetic Algorithm (IGA) and Simulated Annealing Algorithm (SAA). Genetic Algorithm (GA) is improved through optimization strategies, namely Parallel Processing and Fitness Value Hashing, which reduce execution time, convergence time and save processing power. Experimental results on KDD CUP'99 dataset show that our optimized ANIDS (Anomaly NIDS) based BPNN, called "ANIDS BPNN-IGASAA" outperforms the original ANIDS BPNN, ANIDS BPNN optimized by using only GA and several traditional and new techniques in terms of detection rate, false positive rate and it is very much appropriate for network anomaly detection.

Keywords: Network intrusion detection system · Back propagation neural network · Genetic algorithm · Simulated annealing algorithm · Learning rate Momentum term · Parallel processing · Fitness value hashing

1 Introduction

Over the past decade, the use of the Internet and its applications has been increasing at a fast pace. The evolving trends of Cloud Computing, Mobile Technologies, Big-data & Data Mining, Internet of Things (IoT) have blurred the perimeter separating corporate networks from the wider world. In fact, on one hand, Internet has brought huge potential

© Springer Nature Switzerland AG 2019
M. Ben Ahmed et al. (Eds.): SCA 2018, LNITI, pp. 507–521, 2019.
https://doi.org/10.1007/978-3-030-11196-0_43

for business and has a profound impact on people's lives and ways of working at the same time, but on the other hand, it poses lots of security risks and threats. A variety of viruses, malwares, security vulnerabilities, attacks have caused the loss of users, enterprises, governments, even international security. In the coming years, cyber attacks will almost certainly intensify. McAfee; the American global computer security software company, in its 2018 Threats Predictions Report previews [1] that "Attackers will target less traditional, more profitable ransomware targets, including high net-worth individuals, connected devices, and businesses. This pivot from the traditional will see ransomware technologies applied beyond the objective of extorting individuals, to cyber sabotage and disruption of organizations. The drive among adversaries for greater damage, disruption, and the threat of greater financial impact will not only spawn new variations of cybercrime "business models," but also begin to seriously drive the expansion of the cyber insurance market". In addition, SonicWall, originally a private company headquartered in San Jose, California, and Dell subsidiary from 2012 to 2016, states in its 2018 SonicWall Annual Threat Report [2] that " Cybercriminals are turning to highly effective weapons like ransomware, infostealers, IoT malware, mobile threats and SSL/TLS-encrypted malware to target all organizations around the world. Now is the time to add new cyber defenses to your security arsenal to stay proactive against both known and unknown threats". These reports indicate that today, it has become more substantial to defend IT infrastructure from security issues. Security has now become necessary due to intensive use of information technologies in day to day life with greatly confidential commercial and personal data being transferred over the network.

One of the main approaches to ensure information security and cyber security has been the development and deployment of network intrusion detection systems (NIDS). NIDS has emerged as one of the most common parts for every network security infrastructures and an important tool to detect the unauthorized use of computer network and to provide the security for information [3]. It is the process of identifying various events occurring in a network and analyzing them for the possible presence of intrusion [4]. Thus, it plays a crucial role in maintaining a safe and secure network.

Generally, there are two main techniques to conduct intrusion detection: signature-based detection (misuse detection) and anomaly-based detection. Both anomaly and misuse based systems have their own advantages and limitations. Misuse systems fail to detect new attacks and derivative of known attacks; they can only detect the attacks which are predefined with their signatures in misuse/signature database. Whereas anomaly based systems are able to detect novel attacks, which have abnormal patterns without prior knowledge, since the classification model established by this approach has the generalization ability to extract intrusion pattern and knowledge during the training phase [3]. Nevertheless, number of false alarms (false negatives and false positives), i.e., the possibility misclassification of an attack and legitimate/normal pattern is high. In signature-based IDS there is very less chance of wrong classification except the incapability of detecting new attacks. Hence currently, although misused detection techniques are most commonly used in practice, but there are good features of detecting novel attacks in anomaly-based systems, which can give better results when there is reduction in the false alarms. Therefore, a lot of research is going on in building anomaly-based systems to minimize false positives and false negatives using soft computing and other techniques [5].

Hence, our goal is to build an effective and efficient NIDS based on anomaly approach using a very popular soft computing tool widely used for intrusion detection namely Back propagation Neural Network (BPNN), optimized by means of a novel hybrid framework (IGASAA) that combines Improved Genetic Algorithm (IGA) and Simulated Annealing Algorithm (SAA). The role of developed framework is to find the optimal or near optimal values of two relevant parameters of BPNN, namely Learning rate and Momentum term in order to improve performance of BPNN classifier. GA is improved through optimization strategies, namely Parallel Processing and Fitness Value Hashing, which reduce execution time and convergence time, and lower computational cost. SAA was incorporated to IGA with the purpose to optimize its heuristic search. In fact, SAA reduces slowly mutation rate and crossover rate during IGA process that pushes it to focus its search on areas of search space where fitness values are higher. The aim of our proposed IDS is to detect network attacks (known attacks, and unseen attacks), while ensuring higher detection rate, lower false positive rate, higher accuracy and higher precision with an affordable computational cost and low execution time.

The rest of this paper is organized as follows: Sect. 2 introduces previous NIDS approaches. Next, Sect. 3 explains the background of this chapter such BPNN, SAA, GA and some optimization strategies of GA like Parallel processing and Fitness value hashing. Section 4 presents the proposed system in detail, describes its work, and provides the framework of our model. Experimental results and analysis are given in Sect. 5. Finally, Sect. 6 ends with the conclusions.

2 Literature Review

Wang et al. [6] have proposed a novel IDS based on Advanced Naive Bayesian Classification (NBC-A), which combines Naïve Bayesian Classification (NBC) and ReliefF Algorithm. The ReliefF Algorithm has been utilized by authors to give every attribute of network behavior in KDD'99 dataset a weight that reflects the relationship between attributes and final class for better classification results. The proposed IDS was divided into two processes: in the training process, the train set includes the known network behavior data and the marked classes, goes through the pre-processing which is consists of discretization and feature selection. Finally, Relief algorithm is used to weight the features to get NBC-A. In the test process, after application of discretization on the test set which contains unknown network behavior data, NBC-A is used to get behavior classification results. Experimental results shows that the True Positive rate of NBC-A (98.40%) is greatly higher than NBC, and the False Positive rate of NBC-A (8.2%) is lower than the NBC, which means that NBC-A has better performance than NBC in intrusion detection performance. Nevertheless, the False Positive rate obtained by NBC-A remains relatively high, thus, the proposed model should be enhanced.

Sangve and Thool [7] have implemented an Anomaly-based Network Intrusion Detection System (ANIDS) based on hierarchical clustering algorithm, multi-start metaheuristic algorithm and genetic algorithm. The hierarchical clustering technique was used to divide training dataset to reduce time and processing complexity. After applying the clustering algorithm, the metaheuristic method with genetic algorithm play an substantial role to select multiple initial start points, to create number of detectors, to calculate the radius limit of hypersphere detector and to remove redundant detectors to give final output i.e., anomaly or normal. The experimental results show that hierarchical clustering yields minimum false positive rate and detector generation time as compared to k-mean clustering. However, using only false positive rate and detectors generation time measurements is not sufficient to evaluate the performance of proposed IDS, whereas other evaluation criteria should be used to assess the ability of this IDS to detect intrusions, namely True Positive Rate.

Aminanto et al. [3] have proposed an anomaly detection system to detect network intrusions based on Ant Clustering Algorithm (ACA) and Fuzzy Inference System (FIS). The system works in two phases. In the first phase that is the training phase, ACA is implemented in order to cluster training dataset into different clusters, namely normal clusters and abnormal clusters to construct a labeled dataset. Afterwards, this dataset is presented to the Fuzzy Inference System (FIS) in the second phase, named classification phase. In the classification phase, FIS goes into action; it combines two distance based methods to detect anomalies in new monitored data. Experimental results show that proposed scheme is very effective to detect both known and unknown attacks (TPR = 92.11%). However, it provides high FPR (10.3%).

Ma et al. [8] have proposed a novel approach called KDSVM, which utilized the K-mean clustering technique and advantage of feature learning with deep neural network (DNN) model and strong classifier of support vector machines (SVM), to detect network intrusions. KSVM algorithm consists of three phases. In the first phase, the training dataset is separated and clustered into k subsets based on the cluster centers of k-means algorithm in a bid to find more knowledge and patterns from similar clusters. In the second phase, the sub train datasets are trained by kth DNNs, the number k is the value of clusters, this take DNNs that have learned various characteristic of each cluster centers. In the third phase, testing dataset is distanced with Huffman function by the same cluster centers generated in the first phase, and then the sub testing datasets obtained are applied to detect intrusion attack type by completely trained per DNN which top layer used SVM classifier. Lastly, the outputs of every DNN are aggregated for the results of intrusion detection classifiers. The experimental results show that the KDSVM outperforms SVM, BPNN, DBN-SVM and Bayes tree models in terms of detection accuracy and it has the lower error rate than these four methods. However, limitations of the KDSVM include the DNN parameters of weights and threshold of the every layer, and the SVM parameters that necessitate to be optimized by heuristic algorithms.

In order to overcome the slowness of training speed and local optimality issues of the back propagation neural network applied into the network intrusion detection system (NIDS), Chang et al. [9] have proposed a new algorithm named Simulated Annealing Back Propagation (SABP), incorporating Back Propagation Neural Network (BPNN) with Simulated Annealing Algorithm (SAA). Compared to the climbing algorithm, SAA introduces a random factor in the search process. In particular, it has a certain probability to jump out of the local minimum, thus it is able to find out the globally optimal solution. Further, the random factors could be used to accelerate the training process. In the process of training of BPNN, the proposed algorithm determines by a probability function of SAA whether to generate a new network and complete the iterative training or just to complete the iterative training. If a new network is determined to be produced, the bias and weight parameters of the neurons in the network are adjusted, and the other parameters remain the same as in the original network. To limit the way the network parameters changes when generating a new network a "backtracking" method is also integrated in that hybrid algorithm, where "backtracking" means that the algorithm will record the optimal network within a certain number of training times and choose the better one for follow-up training after comparing the optimal network to the current network. Experimental results show that SABP outperforms BPNN in terms of the training speed and the error rate. In fact, the error rate of SABP is 5.58%, and the training time average is 14.31 s. In contrast, the error rate of BPNN is 10.33%, and the training time average is 60.24 s. However, the error rate remains relatively high and other important performance measurement was not used as to say the detection rate.

3 Related Background

This section provides the necessary background to understand the problem in hand. First subsection shed the light on BPNN. Next subsection presents SAA. Then, the third subsection introduces and describes the operation of a standard GA, followed by a presentation in the last subsection of some optimization strategies of GA, namely Parallel Processing and Fitness Value Hashing.

3.1 Back Propagation Neural Network (BPNN)

Back Propagation Neural Network (BPNN) is a special type of neural network. It is also named error back propagation neural network, and it is a multilayer feed forward neural network, which uses Multi-layer Perception as network architecture and Back Propagation Learning Algorithm as training or learning algorithm. Back Propagation network learns by example. You provide the algorithm examples of what you want the network to do and it changes the network's weights so that, when training is achieved, it will give you the required output for a particular input [10]. BPNN with a strong self-learning ability, be able to do adaptive calculations, it is a large scale nonlinear adaptive systems. For every kinds of neural networks, BPNN is relatively mature.

Role of Learning Rate and Momentum Term in Learning Phase of BPNN
The learning process of BPNN can be divided into two parts: *the forward propagation process* and *the error back propagation process*. Durant this process, Back Propagation Learning Algorithm uses two prominent and crucial parameters namely Learning rate and Momentum term to adjust the weights of connections between neurons in neural network.

- **Learning rate**: The learning rate is a relatively small constant that indicates the relative modification in weights. If the learning rate is too low, the network will learn very slowly. However, if the learning rate is too high, the network may oscillate around minimum point, overshooting the lowest point with each weight adjustment, but never actually reaching it. Habitually, the learning rate is very small, located in the interval [0; 1].

- **Momentum term**: The introduction of the momentum term is used to accelerate the learning process by "encouraging" the weight changes to continue in the same direction with larger steps. Furthermore, the momentum term prevents the learning process from settling in a local minimum by "over stepping" the small "hill". Typically, the momentum term has a value between 0 and 1 [11].

3.2   Simulated Annealing Algorithm

Simulated Annealing is a meta-heuristic and a popular search algorithm that has proven to be effective in solving many difficult problems, including NP-hardy combinatorial problems by an analogy to statistical mechanics. The idea of SA comes from a paper published by Metropolis et al. [12] in 1953. The algorithm in this paper simulated the cooling of material in a heat bath. This is a process known as annealing in statistical mechanics. It is a physical process which is often performed in order to relax the system to a state with minimum free energy. In this process, a solid in a heat bath is heated up by increasing the temperature of the bath until the solid melts into liquid, and then the temperature is lowered slowly. In the liquid phase, all particles of the solid arrange themselves randomly. In the ground state, the particles are arranged in a highly structured lattice and the energy of the system is minimal. The ground state of the solid is obtained only if the maximum temperature is sufficiently high and the cooling is performed sufficiently slow. Otherwise, the solid will be frozen into a metastable state rather than into the ground state.

3.3   Genetic Algorithm

John Holland in 1970 has introduced familiar problem solving algorithms called Genetic Algorithms (GAs) which are based on the principles of biological development, natural selection and genetic recombination. GAs are computational intelligence techniques and search procedures often used for optimization problems. A potential solution to the problem is encoded in chromosome (each chromosome represents an individual) like data structure. Each parameter in a chromosome is called as gene. Genes are selected according to our problem definition. These are encoded on bits, character or numbers. The set of generated chromosome is called a population [13]. An evaluation function is used to calculate the goodness of each chromosome according to the desired solution; this function is known as "Fitness Function". From one generation to the next, GA evolves the group of chromosomes to a new population of quality individuals through selection, crossover (recombination to produce new chromosomes), and mutation operators until a global optimum solution is found at the end of GA process (convergence).

3.4   Optimization Strategies for Genetic Algorithm

With the fitness function, typically being the most processing demanding component of genetic algorithm (GA), it makes sense to focus on improvement of the fitness function to see the best return in performance. In this section, we will explore two optimization strategies that are used in this work to improve performance of GA by optimizing the fitness function, namely **Parallel Processing** and **Fitness Value Hashing**.

Parallel Processing

One of the easiest approaches to achieve a performance enhancement of GA is by optimizing the fitness function. The fitness function is typically the most computationally expensive component; and it is often going to be the bottleneck of GA. This makes it an ideal candidate for multi-core optimization. By using multiple cores, it's possible to compute the fitness of numerous individuals simultaneously, which makes a tremendous difference when there are often hundreds of individuals to evaluate per population. Java 8 provides some very useful libraries that make supporting parallel processing in our GA much easier. Using *Java's IntStream*, we can implement parallel processing in our fitness function without worrying about the fine details of parallel processing (such as the number of cores we need to support); it will instead create an optimal number of threads depending on the number of cores available in our multi-core system. Hence, by using parallel processing, fitness function will be able to run across multiple cores of the computer, consequently, it is possible to considerably reduce the amount of time the GA spends evaluating individuals and, so reduce the overall time of execution of GA, and accelerate convergence process [14].

Fitness Value Hashing

Fitness Value Hashing is another strategy that can reduce the amount of time spent computing fitness values by storing previously calculated fitness values in a hash table [14]. During running of GA, solutions found previously will occasionally be revisited due to the random mutations and recombinations of individuals. This occasional revisiting of solutions becomes more common as GA converges and begins to find solutions in an increasingly smaller area of the search space. Each time a solution is revisited its fitness value needs to be recalculated, wasting processing power on recurrent, duplicate computations. Luckily, this can be easily fixed by storing fitness values in a hash table after they have been computed. When a previously visited solution is revisited, its fitness value can be retrieved from the hash table, avoiding the need to recalculate it.

4　The Proposed System

This section presents the proposed system in detail, describes its work, explains the role of Simulated Annealing Algorithm in this system and provides the framework of our model.

4.1　The Approach of Our Novel Proposed System

In order to enhance the performance of our previous and best obtained ANIDS [10] based on BPNN classifier presented in Table 1, we have implemented and integrated to that ANIDS a hybrid framework of optimization based on Improved Genetic Algorithm (IGA) and Simulated Annealing Algorithm (SAA). GA is improved through optimization strategies/techniques, namely Parallel Processing and Fitness Value Hashing, which decrease execution time, convergence time and save processing power (Sect. 3.4). The purpose of developed framework is searching the optimal values of Learning rate (LR) and Momentum term (MT) parameters, which influence the performance of BPNN classifier. In fact, LR and MT are two important parameters in the learning phase of BPNN, which impact the convergence and performance of that classifier. Since their values are between 0 and 1, we thought to use IGA & SAA in order to find the ideal/optimal values of those parameters in the interval [0; 1], which is a large space of real numbers. As shown in Fig. 2, illustrated our approach, IGA process begins with a randomly generated population of individuals (potential solutions) represented by their chromosomes; each chromosome takes the form of a pair of values (Learning rate, Momentum). Then, this population evolves through several generations by means of genetic operations such elitism, selection, recombination (crossover) and mutation until stopping or optimization criteria of IGA is met. At each generation, each chromosome (chromosome = pair (Learning rate, Momentum)) of current generation is evaluated by passing it as parameter to ANIDS obtained in our precedent work [10], after using Min-Max normalization to convert the two substrings representing Learning rate and Momentum term into values between 0 and 1. This IDS firstly goes through the learning phase, then passes to the test/evaluation phase and returns the values of performance metrics calculated at the end of evaluation phase. Among those performance metrics, we have selected the pertinent of them to serve as "Fitness Function" for evaluation of goodness of chromosomes. From one generation to the next, IGA converges towards the global optimum through genetic operations cited previously. Finally, the best individual (chromosome) is picked out as the final result once the optimization criterion is met. In our work, termination condition adopted for IGA is the maximum number of generations (100 generations). Hence, the best chromosome resulted correspond to the optimal or near-optimal values of the pair (Learning rate, Momentum), which ensuring high detection rate and low false alarm rate.

Table 1. Parameters values and performance of our best previous ANIDS [10]

		Value
Parameters	Number of attributes	12 (feature selection algorithm: a modified Kolmogorov–Smirnov correlation based filter)
	Normalization	Min-Max
	Architecture (I–H–O)	12–10–1
	Method of calculating number of nodes in hidden layer	H = 0.75 * Input + Output
	Activation function	Sigmoid
Performance	Accuracy	98.66%
	Precision	99.62%
	FPR	1.13%
	TPR (DR)	98.59%
	F-score	0.99
	AUC	98.73%

For successful use of IGA, the two following key elements must be well defined:

- **Chromosome encoding/representation**: In our study as indicated by Fig. 1, we have chosen the binary representation for chromosomes. At the beginning of IGA process; when the initial population of individuals was created, 50 binary strings of length 40 bits representing the chromosomes are generated randomly. Afterwards, each string is divided into two substrings of 20 bits. The first binary substring is then converted in decimal value, and this value is normalized using the Min Max normalization to get a value between 0 and 1, which will serve as the Learning rate of our IDS. Similarly, the foregoing approach is applied to the second binary substring to obtain Momentum term of our IDS, with a value within the interval [0; 1].

1	0	0	0	1	0	1	0	0	0	1	1	1	0	1	1

Learning rate (20 bits)   Momentum (20 bits)

Fig. 1. Structure of GA or IGA chromosome

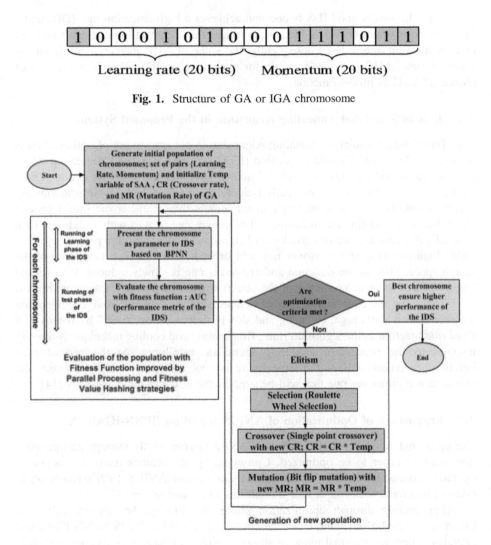

Fig. 2. Workflow of proposed system

- **Fitness Function or Evaluation Function**: Among the values returned by the IDS at the end of the evaluation phase, we have chosen the AUC metric as a score (fitness function) of individuals to assess their adaptability to the optimization problem. The AUC metric [10] is by definition the ability to avoid misclassifications of network packets, and from our point of view, it represents a good trade-off between the DR (Detection Rate) metric and the FPR (False Positive Rate) metric. In effect, this is due to the fact that AUC is the arithmetic mean of the DR and TNR (1-FPR) as shown by Eq. 1 of the AUC:

$$AUC = (DR + TNR)/2 = (DR + (1 - FPR))/2 \tag{1}$$

As it is known, a good IDS is one that achieves a high detection rate (DR) and a low false positive rate (FPR). In fact, as the value of the DR measure augments and that of FPR measure diminishes, consequently, the value of AUC increases. So, from our point of view, AUC is the best metric for evaluating an IDS. That is the reason of choice of AUC as fitness function.

4.2 Role of Simulated Annealing Algorithm in the Proposed System

The aim of using Simulated Annealing Algorithm in our framework of optimization is to optimize Improved Genetic Algorithm (IGA) process. Simulated Annealing Algorithm is a hill climbing algorithm which initially accepts worse solutions at a high rate; then as the algorithm runs, it gradually reduces the rate in which worse solutions are accepted. One of the easiest methods to implement this characteristic into a genetic algorithm is by updating the mutation and crossover rate to start with a high rate then gradually decrease the rate of mutation and crossover as the algorithm progresses. This initial high mutation and crossover rate will drive IGA to search a large area of the search space. Then as the mutation and crossover rate is slowly reduced, IGA should begin to focus its search on areas of the search space where fitness values are higher.

To vary the mutation and crossover rate/probability, we have used a temperature variable, which starts high, or "hot", and slowly decreases, or "cools" by means of a *Cool rate* function as the algorithm runs. This heating and cooling technique is directly inspired by the process of annealing found in metallurgy. At the end of each iteration/generation of IGA, the temperature is cooled slightly, which decreases the mutation and crossover rate that will be used in the next generation of IGA [14].

4.3 Framework of Optimization of ANIDS Based on BPNN-IGASAA

Our optimized ANIDS (ANIDS BPNN-IGASAA) passes firstly through an optimization stage in order to be optimized. Consequently, it becomes ready to operate in operation/normal mode. The framework of our system ANIDS BPNN-IGASAA in optimization mode consists of four modules as illustrated in Fig. 3.

After passing through optimization phase and finding the optimal values of Learning rate and Momentum term, the optimized ANIDS (ANIDS BPNN-IGASAA) operates in operation/normal mode as shown in Fig. 4 to classify connection instances extracted from KDD'99 cup test dataset.

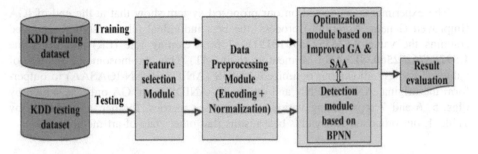

Fig. 3. Framework of optimized ANIDS based on BPNN-IGASAA in optimization stage

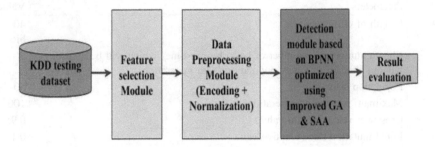

Fig. 4. Framework of optimized ANIDS based on BPNN-IGASAA in operation/normal mode

5   Experimental Results and Analysis

For experimental set up, we have used a computer with a Core-i7 2700 K CPU, and 16 GB of DDR3, and utilized two independent subsets extracted from two KDD cup 99 datasets, namely *kddcup.data_10_percent.gz 10% dataset* (training dataset) and *corrected.gz KDD dataset* (testing dataset). Table 2 shows the details of our datasets and Table 3 presents the parameters of the proposed module of optimization based on IGASAA for our previous ANIDS.

Table 2. Distribution and size of training and testing datasets

Dataset	Category					
	Normal	DoS	Probe	R2L	U2R	Total
Training dataset	15,000	60,000	2490	165	30	77,685
Testing dataset	20,000	59,245	585	170	260	80,260

The experiments conducted on our proposed system show that at the end of IGA (Improved Genetic Algorithm) process, the best individual (chromosome) generated contains the value **9.536752259018191E−7** for Learning rate (LR) and the value **1.592637627256038E−4** for Momentum term (MT). This chromosome or this pair of values (LR, MT) allows our optimized ANIDS (ANIDS BPNN-IGASAA) to outperform the original ANIDS BPNN and optimized ANIDS using GA only, as shown by Figs. 5, 6 and 7 representing various performance metrics. Further, as indicated by Table 4, our proposed IDS yields best results that other state-of-art methods.

Table 3. Parameters of our framework of optimization based on improved genetic algorithm and simulated annealing algorithm for our previous ANIDS

	Parameters	Value
GA	Length of chromosomes	40 bits
	Elitism number: the number of best chromosomes which will be copied without changes to a new population (next generation)	5
	Population size	50
	Maximum number of generations	100
	Crossover rate (constant value)	0.95
	Initial mutation rate (adaptive value)	0.1
SAA	Initial temperature	1.0
	Cooling rate	0.001

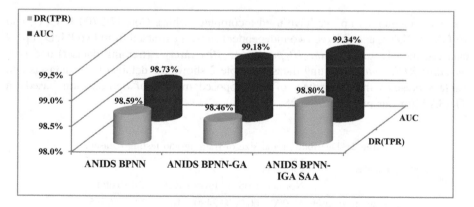

Fig. 5. Detection rate (DR) and AUC (Ability to avoid false classification)

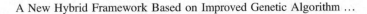

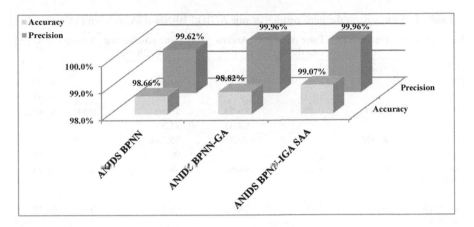

Fig. 6. Accuracy and precision

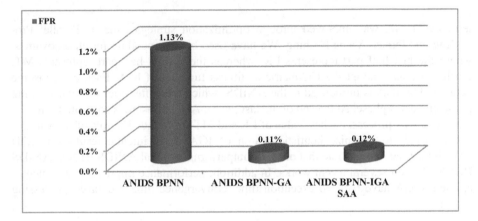

Fig. 7. False positive rate (FPR)

Further, using of IGA in our framework IGASAA, obtained by application of optimization strategies to the fitness function of a standard GA, namely *Parallel Processing* and *Fitness Value Hashing* have brought several benefits. These advantages are: 85% reduction of execution time compared to a standard GA, acceleration of the IGA convergence process and save of processing power.

6 Conclusions

In order to enhance the performance of our previous ANIDS based BPNN [10], we have developed for it a framework of optimization based on Improved GA (IGA) and Simulated Annealing Algorithm (SAA) with the purpose of searching the optimal values of critical parameters for BPNN, namely Learning rate (LR) and Momentum

Table 4. Comparison the performance of our ANIDS BPNN-IGASAA with other works

Research work	Precision (%)	False positive rate (FPR) (%)	Accuracy (%)	True positive rate (TPR)/DR (%)	F-score	AUC (%)
[15]	97.2	1.7		97.3	0.972	97.8
[5]		5.56	99.39	94.05		94.25
[3]		10.03		92.11		90.91
[16]	95.23	0.56	96.53	76.50	0.848	87.97
[8]			92.03	91.35		
[17]	99.456		95.299	95.764		
[18]		25		90		82.50
Our ANIDS BPNN [10]	99.62	1.13	98.66	98.59	0.99	98.73
Our ANIDS BPNN-GA	99.96	0.11	98.82	98.46	0.99	98.73
ANIDS BPNN-IGASAA	99.96	0.12	99.07	98.80	0.99	99.34

term (MT). GA was improved through optimization strategies that is Parallel Processing and Fitness Value Hashing. We have used binary encoding for chromosomes; where the first half part represents LR, whereas the second half part represents MT. Further, we have adopted AUC metric as fitness function of IGA. Each chromosome generated by IGA is introduced to the ANIDS, which thereafter goes through learning phase and a test phase. At the end of the last phase, AUC measure is computed. Finally, at the end IGA process, the best value of LR and MT are found. The role of SAA in this framework is to optimize heuristic search of IGA. Experimental results on KDD CUP'99 dataset demonstrate that our IDS outperforms original ANIDS BPNN, ANIDS BPNN-GA and several recent works. In addition, performance improvement strategies applied to GA have reduced execution time, convergence time and saved processing power.

References

1. McAfee Company: McAfee Labs 2018 Threats Predictions Report. https://securingtomorrow.mcafee.com/mcafee-labs/2018-threats-predictions
2. SonicWall Enterprise: 2018 SonicWall Annual Threat Report. https://cdn.sonicwall.com/sonicwall.com/media/pdfs/resources/2018-snwl-cyber-threat-report.pdf
3. Aminanto, M.E., Kim, H., Kim, K.M., Kim, K.: Another fuzzy anomaly detection system based on ant clustering algorithm. IEICE Trans. Fundam. Electron. Commun. Comput. Sci. **100**(1), 176–183 (2017)
4. Ashok Kumar, D., Venugopalan, S.R.: A novel algorithm for network anomaly detection using adaptive machine learning. In: Saeed, K., Chaki, N., Pati, B., Bakshi, S., Mohapatra, D. (eds.) Progress in Advanced Computing and Intelligent Engineering. Advances in Intelligent Systems and Computing, vol. 564, pp. 59–69. Springer, Singapore (2018)

5. Lokeswari, N., Chakradhar Rao, B.: Artificial neural network classifier for intrusion detection system in computer network. In: Satapathy, S., Raju, K., Mandal, J., Bhateja, V. (eds.) Proceedings of the Second International Conference on Computer and Communication Technologies, vol. 381, pp. 581–591. Springer, New Delhi (2016)
6. Wang, Y. et al.: A novel intrusion detection system based on advanced naive bayesian classification. In: Long, K., Leung, V., Zhang, H., Feng, Z., Li, Y., Zhang, Z. (eds.) 5G for Future Wireless Networks. 5GWN 2017. Lecture Notes of the Institute for Computer Sciences, Social Informatics and Telecommunications Engineering, vol. 211, pp. 581–588. Springer, Cham (2018)
7. Sangve, S.M. Thool, R.C.: ANIDS: Anomaly network intrusion detection system using hierarchical clustering technique. In: Satapathy, S., Bhateja, V., Joshi, A. (eds.) Proceedings of the International Conference on Data Engineering and Communication Technology. Advances in Intelligent Systems and Computing, vol. 468, pp. 121–129. Springer, Singapore (2017)
8. Ma, T., Yu, Y., Wang, F., Zhang, Q., Chen, X.: A hybrid methodologies for intrusion detection based deep neural network with support vector machine and clustering technique. In: Yen, N., Hung, J. (eds.) Frontier Computing. FC 2016. Lecture Notes in Electrical Engineering, vol. 422, pp. 123–134. Springer, Singapore (2018)
9. Chang, C., Sun, X., Chen, D., Wang, C.: Application of back propagation neural network with simulated annealing algorithm in network intrusion detection systems. In: Sun, S., Chen, N., Tian, T. (eds.) Signal and Information Processing, Networking and Computers. ICSINC 2017. LNEE, vol. 473, pp. 172–180. Springer, Singapore (2018)
10. Chiba, Z., Abghour, N., Moussaid, K., El Omri, A., Rida, M.: A novel architecture combined with optimal parameters for back propagation neural networks applied to anomaly network intrusion detection. Comput. Secur. 75, 36–58 (2018)
11. Multi-Layer Perceptron. http://www.cse.unsw.edu.au/~cs9417ml/MLP2
12. Metropolis, N., Rosenbluth, A.W., Rosenbluth, M.N., Teller, A.H., Teller, E.: Equation of state calculations by fast computing machines. J. Chem. Phys. 21(6), 1087–1092 (1953)
13. Chaudhary, V.R., Bichkar, R.S.: Detection of intrusions in KDDCup dataset using GA by enumeration technique. Int. J. Innov. Res. Comput. Commun. Eng. 3(3), 2365–2369 (2015)
14. Jacobson, L., Kanbe, B.: Genetic Algorithms in Java Basics, pp. 143–144. Apress, New York (2015). https://doi.org/10.1007/978-1-4842-0328-6
15. Aslahi-Shahri, B.M., Rahmani, R., Chizari, M., Maralani, A., Eslami, M., Golkar, M.J., Ebrahimi, A.: A hybrid method consisting of GA and SVM for intrusion detection system. Neural Comput. Appl. 27(6), 1669–1676 (2016)
16. Hamamoto, A.H., Carvalho, L.F., Sampaio, L.D.H., Abrão, T., Proença Jr., M.L.: Network anomaly detection system using genetic algorithm and fuzzy logic. Expert Syst. Appl. 92, 390–402 (2018). https://doi.org/10.1016/j.eswa.2017.09.013
17. Sharma, R., Chaurasia, S.: An enhanced approach to fuzzy C-means clustering for anomaly detection. In: Somani, A., Srivastava, S., Mundra, A., Rawat, S. (eds.) Proceedings of First International Conference on Smart System, Innovations and Computing. Smart Innovation, Systems and Technologies, vol. 79, pp. 623–636. Springer, Singapore (2018)
18. Borah, S., Panigrahi, R., Chakraborty, A.: An enhanced intrusion detection system based on clustering. In: Saeed, K., Chaki, N., Pati, B., Bakshi, S., Mohapatra, D. (eds.) Progress in Advanced Computing and Intelligent Engineering. Advances in Intelligent Systems and Computing, vol. 564, pp. 37–45. Springer, Singapore (2018)

A Semantic Method to Extract the User Interest Center

Ibtissam El Achkar[✉], Amine Labriji, and Labriji El Houssine

Laboratory of Technological Information and Modeling, Faculty of Sciences Ben
M'sick, University Hassan II, Casablanca, Morocco
Ibtissam.elachkar-etu@etu.univh2c.ma,
Labriji73@hotmail.com, labriji@yahoo.fr

Abstract. With the evolution of information systems and the large mass of
heterogeneous data on the web, problems such as cognitive overload and disorientation (Conklin and Begeman in GIBIS a hypertext tool for team design
deliberation, pp. 247–251, [1]) are starting to appear, and to overcome them
several techniques have been invented to customize the information system, in
order to improve search engine performance as well as recommendation
engines. (Zeng et al. in Temporal User Profile Based Recommender System.
Artificial Intelligence and Soft Computing. ICAISC 2018, [2], Su et al. in Music
Recommendation Based on Information of User Profiles, Music Genres and
User Ratings. Intelligent Information and Database Systems. ACIIDS 2018, [3])
have shown that the best techniques to provide relevant results to the specific
needs of the user are based on the use of user profiles and more specifically its
interests. Generally the selection of interesting documents to a user is done on
the basis of his area of interest, inferred from the information about the user or
his user profile. Thus the calculation of the interest center is one of the essential
elements for a relevant research. In this article we will introduce a new method
of extracting the user's interests from his knowledge, based on the structure of
an ontology to deduce the user's interest's center, taking into account the
semantic links between the graph's concepts.

Keywords: User profile · Interest center · Semantic web · Recommender
systems · Adaptive information systems · Ontology · Similarity measurement
Conceptual graph

1 Introduction

The rapid development of new information and communication technologies as well as
the rise of the web, confronted us with a very large mass of heterogeneous information.
The masses of accessible information have been steadily increasing, and the volumes of
documents storing them are rising very rapidly, because of this increase in the volume
of information, we arrive at a paradoxical situation, never has there been so much
information available, but finding in this accumulation what we are looking for,
becomes more and more difficult, which leads us to both major problems: Cognitive
overload and disorientation [1]. The problem of cognitive overload is related to the
difficulty that can have the user to select the information corresponding to his specific

© Springer Nature Switzerland AG 2019
M. Ben Ahmed et al. (Eds.): SCA 2018, LNITI, pp. 522–534, 2019.
https://doi.org/10.1007/978-3-030-11196-0_44

needs facing a large amount of data. For example, a programmer who submits the query "java" to a search engine and which is waiting for results related to the programming language java, will have a difficulty to choose the results that interests him in a mass of results where is intermingle those related to programming and those related to the Java region in Indonesia and those related to the Java Parisian dance. The problem of disorientation is related to the fact that the user does not know which path to follow when navigating via user interface, for example when browsing a website. So the problem is no longer the availability of information but the ability to select information that responds to the specific needs of the users. So, the conception and implementation of effective tools, allowing the user to have access only to the information he deems relevant, becomes an absolute necessity.

Studies by Bradford and Marshall [4] have shown that the majority of users have no idea about how the information systems work; as a result, they do not express their needs. On the other hand, users usually use only a few words (5 or 6 at most) to formulate their queries, which gives unfinished specifications on their information needs. Thus, the analysis of the user interests reveals a particular importance. Indeed, it is by knowing perfectly his interests that it will be possible to offer him a relevant information for his research.

In this paper, we present our method of extracting the user's interest center, using the graph structure of an ontology as a reference in order to use the semantic links between the concepts. This article is organized as follows: First of all, we present the user profile (its modeling, its representations and its uses), then we present the ontologies and their importance, and the ontology that we well use on our approach, after that we introduce our methodology of work for extracting the user interests centers, with its various axes, in particular the representation of the user's knowledge as a conceptual graph, the similarity functions used between their concepts as well as the calculation method, used and finally we give a conclusion with our perspectives.

2   Related Work

2.1   User Profile Definition and Modeling

2.1.1   User Profile

The user profile represents a collection of personal data associated with a specific user that describes a set of attributes. These data can be extracted from different sources such as: the user's browsing activities [5], his social networks [6–8] etc. These attributes can include geographic location, academic and professional experience, goals (short-term and long-term), behavior, interests, preferences, etc. [9]. A profile can be relative to a single person, or to a group of people with commonalities, such as members of a working group.

Generally a user profile can be built by two methods: either by the user himself what is called explicit profile, or automatically from his interactions data with the system, in this case it is called implicit profile. This last sort is the most common, since the manual entry of parameters (preferences, interests...) by the user can be a tiring task for him and can take a lot of time to express his needs.

2.1.2 User Modeling

User profile data is represented according to the needs and objectives of each information system. In general, they are stored in the knowledge base of the system as a
property-value pair. Once the data is structured, they are used by the data mining
algorithms to build the user profile. Several techniques of data mining can be used [10]
classified the techniques used according to the models of user profiles to build: interests
modeling, intentions modeling and behavior modeling (Fig. 1).

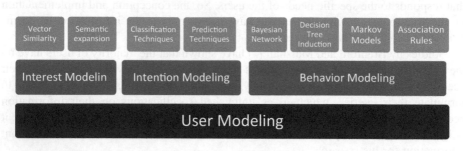

Fig. 1. User modeling

- **Interest modeling.**

   This type of modeling is characterized by the definition of the user interests by a
function fct (i) which gives the degree of interest or disinterest of a user for an item i by
analyzing his previous behavior [11].

- **Intention modeling.**

   An intention here is the purpose for which the user uses the information system.
The modeling of user's intentions is to build a model that will identify the purpose of
each user of the information system. For example, customers of an e-commerce website
can be divided into two groups: those who really aim to buy and those who don't aim
to buy. Intention modeling is largely based on classification techniques with predefined
categories [12].

- **Behavior modeling.**

   Consists of analyzing user behaviors via browsing histories or transactions they
perform on Web servers [12], in order to determine recurring navigation courses, to
validate marketing strategies or to check the relevance of marketing campaigns. Among
the methods of this type of analysis we cite: The Markov chains to predict future URLs
that will be visited by the user [13].

2.2 User Interests

The interest user expresses his area of expertise or its area of exploration. It can be
defined by a set of keywords (concepts), or a set of logical expressions (queries). In
many approaches the importance of each concept is defined by a weighing of the

interests keywords. The ontology of the domain completes the definition of the interest by explaining the semantics of certain terms. For example, we can explicitly define Apple 'a computer ' in the profile and not 'a fruit'. The interest can be seen as a virtual screening which reduces the amount of information to be taken into account. Therefore any request issued by the user will be enriched with the keywords or predicates defining the user interests to provide him the most appropriate results to his needs.

2.2.1 User Interests Representation
In the literature, there are three mains techniques to represent the user interest:

– The Set Representation.

This approach consists of representing the user's profile by packets of weighted terms, also known by vector representation by analogy to the vector model of Salton [14] on which it is based. These packages of terms, reflecting the user interests can be grouped differently according to the approach taken to exploit the user's profile. We distinguish four major approaches that represent the user profile based on this kind of representation: Representation by keyword list [15], like the case of web portals: MyYahoo, InfoQuest, representation by vector of weighted terms for each interest [16], representation by a set of vectors of terms: weighted (or not) independent [17], representation by definition of a relation of order between the interest of the profile, in this case we talk about preferences [18]. The weighting of terms is usually based on the tf. idf schema commonly used in information retrieval [19].

– The Representation by semantic networks.

In order to solve the problem of polysemy of terms inherent to the set representation, a first solution consists in representing the profile by a network of weighted nodes in which each node represents a concept reflecting the user interest. This type of representation offers the double advantage of structuring and associative representation (relationships between terms) allowing considering all aspects representative of the profile.

The interests are often represented by node-pair relationships in which each node contains a term derived from implicit data used to build the profile. The arcs linking the nodes are created on the basis of co-occurrences between these terms. However, the separate representation of each word by nodes in the semantic network is not precise enough to decline the different meanings of the user interests. Another possibility is to exploit external sources such as ontologies to establish links between nodes, hence the appearance of conceptual representation.

– The conceptual representation.

The representation of the profile in this approach highlights one or more semantic relations between the profile information. According to a general context [20], this representation offers an interesting alternative to the semantic networks approach. Indeed, current works tend to represent the profile in the form of ontology of personal concepts based on the knowledge contained in ontologies rather than building user profiles only from the implicit data collected from its interactions [21]. The conceptual representation is similar to the semantic network representation in the sense that represents the user's interests through a network of conceptual nodes. However, in the conceptual approach, the nodes correspond to abstract domains representing the user interests, contrary to the semantic approach where the interests are represented by specific words, or a set of relevant words.

From association of the user interest to the concepts of the domain ontology, we obtain a profile represented in the form of a hierarchy of concepts. The implicit data from user interactions is classified into these concepts and the user interest in such concepts is recorded. (It's the type of representation that inspires our work). The semantic representation approaches exploit ontology of reference to represent interests of the user according to weighted concept of ontology vectors used. We mention the hierarchy of concepts "Yahoo" or the ODP as the most evidence sources used in this type of approach.

2.3 The Uses of the User Profile

The user profile is widely used in many fields and applications because of its great importance, such as the improvement of processes and the processing of information, as the case of information adaptive systems where we integrate the user profile or more specifically theirs interests in order to solve two major problems as we have mentioned in the introduction of this article: cognitive overload and disorientation. In information adaptive systems there are two main types of systems:

2.3.1 Personalized Information Search System

Information Retrieval (IR) refers to the set of methods, procedures and techniques for the acquisition, organization, storage, retrieval and selection of information (data, texts, images, videos). Unlike the recommendation systems, the user expresses his need for information by a query. In a conventional IR system, the user's query (weighted wordlist) is matched with the list of indexed documents (weighted wordlist) in order to return relevant documents corresponding to the user's query (search engines): Google search, AOL, Yahoo!, etc.). In a hypertext environment like the Web, documents (Web pages) are not indexed only on their content (keyword list), but also via additional metrics based on the structure of links between pages, Google's PageRank algorithm [22] is one of the best known in this context. The queries expressed by users in conventional IR systems are generally short and may contain ambiguities [23]. So the personalized IR systems come to improve the cognitive overload problems of conventional IR systems by integrating the user profile in these mechanisms. According to [24] this integration can be done in three main ways in the information retrieval process:

- Customized selection of information: consists on integrating the user profile parameters into the similarity function between the user's request and each document, as the case of [25].
- Reformulation or expansion of the request: consists in introducing in the structure of the request the terms coming from the user profile [26].
- Reordering the results: consists of using the user's profile terms to reorder the results from a conventional IR system, [1].

These systems are classified into three scopes: individualized systems, community-based systems and aggregate level systems [27].

2.3.2 Recommendation Systems

The purpose of recommendation systems is to present the pieces of information that may be likely to interest the user. The most common case in this kind of system is the recommendation based on filtering by contents using the user interests in e-commerce sites that sell a very large variety of products, while each customer potential is very often interested in a very limited number of products. A content recommendation system will compare the profile of resources (e.g. product) with the user's profile (interests) to select the resources corresponding to their specific needs. The filter used to select the relevant resources for the user is a similarity function between the user's profile and each of the resources.

3 Our Approach: Extraction of User's Interests Centers Based on the Structure of an Ontology

After defining the user profile and touch its importance in the different phases of information adaptive systems, defining the user interests centers is an essential research area in this field. Indeed, the deduction of the interests' centers will allow the information retrieval system to target the specific needs of the user by using it in the different phases. Unfortunately, the existing methods of calculating the interests' centers do not take into account the semantic links between the concepts of the user graph, so we have represented the user profile in conceptual graph in order to use the power of the mathematical methods, which allowed us to deduce the user interest's centers using the semantic relationships between concepts. In this section we present our method of extraction the interests centers based on a semantically structured user profile inspired from the structure of an ontology. First, we present the ontologies and especially the ODP ontology that we will use in our approach, then we present our method of extracting the interests' centers.

3.1 Ontology

Ontology refers to a set of hierarchically related semantic resources. It allows to describe the knowledge of a specific domain and to present the relations between the concepts as well as to give the missing rules and axioms to the semantic networks. The fundamental purpose of ontology is to semantically process information. There are several ontologies designed to list the content of web pages for an easier navigation by users. We cite for example online portals such as Yahoo,[1] Lycos,[2] and ODP[3] for Open Directory Project which is an open concept hierarchies in RDF format widely adopted

---

[1] www.yahoo.com.

[2] www.lycos.com.

[3] www.dmoz-odp.org.

by many systems such as OBIWAN (Ontology Based Information Web Agent Navigation) [28], Personae [29], and since the ODP is the largest and most complete directory of the web published and edited by experts of the domain, we use it as a source of semantic knowledge in the process of access to information.

ODP data is available in two RDF files: (**Structure.rdf** and **Content.rdf**), the first one contains the tree structure of the ODP ontology, and the second lists the resources or web pages associated with each concepts. The concepts are organized hierarchically where high level concepts represent general concepts and low level concepts represent specific concepts as shown in Fig. 2.

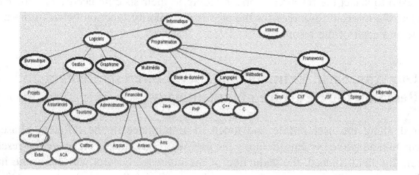

Fig. 2. Extract from the architecture of the ODP ontology

Each concept of the ODP is represented by a title and a description describing in general the content of the associated web pages, and each web page is also associated with a title and a description describing its content. The concepts of ontology are connected by relations of different types such as "is a", "symbolic" and "related". The links "is a" allows us to pass hierarchically from generic concepts to more specific concepts. "Symbolic" links support the multi-classification of pages in several concepts and allow the user to navigate between concepts related semantically without resorting to general concepts. "Related" links are labeled "see also" permit to point to the concepts dealing with the same theme without having web pages in common.

3.2   The User Profile as a Graph Concept

Our Purpose is to represent the user profile as a set of semantically related concepts comprising concepts of several specific levels by representing the user profile using the same structure of the "Structure.rdf" file to exploit the architecture of this ontology. Thus we define the file "StructProfil.rdf" constituting the user profile graph containing all the knowledge concerning the user, in order to consider these knowledges as concepts.

So, the structure of this graph G = (V, E, W) is constituted by a hierarchical component formed by the links of type "is-a" and a non-hierarchical component formed by links of different types predefined in the ODP ontology, where:

- V is a set of weighted nodes, representing the concepts used by the user,
- E is a set of arcs between the nodes of the graph V,
- W (Ci) is the weight of the concept Ci; it represents the number of use of this concept by the user.

To extract the interests centers of a user we will first calculate the importance of each concept of his graph, after that we will extract his interests centers by using the maximum mathematical function.

3.2.1   Calculation of Concepts Importance

Two concepts are neighboring or adjacent if they are connected by one arc, one is the father or the son of the other. The neighborhood of a concept c, denoted V (c), is the set of his neighboring vertices. So we use this definition to define the importance of a concept by the following formula:

$$Ip(c) = \sum_{a \in V(c)} poids(a)sim(a, c) \tag{1}$$

where poids(a): is the weight of the concept a, and sim(a, c) is the similarity between the concepts a and c.

3.2.2   Extraction the User Interests Center

The calculation of the concepts importance's by (1) will allow us to extract the most relevant interests for the user or very specifically his interests' centers by the following formula:

$$IP(C_0) = Max\{Ip(c), c \in P\} \tag{2}$$

where P is the set of concepts that constitutes the user profile graph, and $IP(C_0)$ is a finite number that represents the concept C_0.

3.2.3   Methodology of Work

The following schema summarizes our work methodology (Fig. 3).

3.3   Evaluation

We want to apply our method of extraction user's interests centers based on the structure of the ontology of reference which one mentioned previously.

So we assume the following set of words used by a user in a given information system, our goal is to extract his interest center from this cloud of word (Fig. 4).

We began by assigning weights to these words using the tf.idf function commonly used in information retrieval [30], and then, structured them as a graph of concepts using the same tree structure of ODP ontology previously studied in order to respect the semantic links between words, which led us to the following graph (Fig. 5):

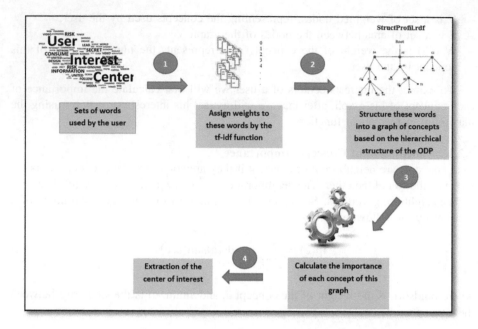

Fig. 3. Process of extracting the user's interest center

Fig. 4. Set of words used by the user

After that, we calculated the importance of each graph's concept by the formula (1). In this step, we have chosen to calculate this measure by two different similarity function: Labriji measurement, and WU and palmer measurement, in order to verify if we will obtain the same result.

The Similarity Functions Chosen

The similarity function is a function characterized by values between 0 and 1, which allows probabilistic interpretations of similarity. The most intuitive similarity measure

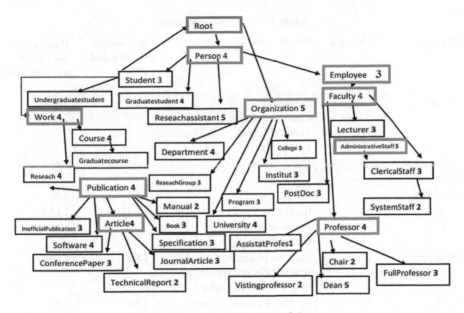

Fig. 5. The conceptual graph of the user

of objects in ontology is their distances. These measures use the hierarchical structure of the ontology to determine the semantic similarity between the concepts. Among the most used measurement classified under this category we have:

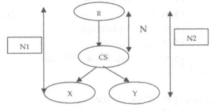

– The measurement of Wu and Palmer [31]

$$Sim_{WP}(X, Y) = \frac{2N}{N_1 + N_2 + 2N} \qquad (3)$$

– The measurement of Labriji [32]:

$$Sim_{Lab}(X, Y) = \frac{2N}{N_1 + N_2 + 2N + N.N_1.N_2} \qquad (4)$$

Our choice for those functions is based on their results in the improvement of similarities between the concepts located on the same hierarchy of an ontology.

Table 1. Concepts importance

Concept	Weight	Neighbourhood	Ip(concept) By Wu and Palmer measurement	Ip(concept) By Labriji measurement
Article	4	Publication, ConferencePaper, TchnicalReport, JournalArticle	10.82	10
Administrative Staff	5	ClericalStaff	2.72	2.66
Employee	3	Person, Faculty	5.84	5.86
Faculty	4	Employee, PostDoc, Professor, Lecture, ClericalStaff	13.45	13.54
Organization	5	Root, Depatement, ResearchGroup, University, Program, Instut, College	13.2	13.32
Work	4	Student, Research, Course	8.38	9.2
Person	4	Root, GraduateStudent, RessearchAssistant, Employee, Student	10.01	10
Professor	4	AssistaProfs, Faculty, Chair, Visiting Professor, FullProfessor, Dean	14.98	14.97
Publication	4	Research, unofficielPublication, Software, Article, Specification, Book, Manual	**20.31**	**20.21**

Table 1 shows the concepts' importance:

By applying the formula (2), we deduce that the most important interest in this graph is the concept "Publication".

4 Conclusion

Since researchers began to focus on the evolution of information systems, approaches to extracting user interests are multiplying and changing day after day, such us using categorized web browsing history to estimate the user's latent interests for web advertisement recommendation [33], or the analysis of users' browsing behaviors to evaluate the degree of their interests [34], There are even approaches that go as far as using user's data on social networks to extract its interests [35].

Certainly most of these approaches improve very remarkably the function of the information systems, but on the other hand they do not take into account the semantic aspect during the extraction of these interests, hence the strength of our approach. We propose in this article a semantic computation of the interests using the structure of a well-known and rich ontology. Indeed, our method can be used in all phases of an information retrieval system including the enrichment phase of the query, the search phase or even the indexing phase.

We plan to define a threshold to choose the most relevant interests centers for the user by an experimentation, and to use this approach in our semantic search meta-search engine.

References

1. Conklin, J., Begeman, M.L.: GIBIS A Hypertext Tool for Team Design Deliberation, pp. 247–251 (1987)
2. Zeng, W., Du, Y., Zhang, D., Ye, Z., Dou, Z.: Temporal user profile based recommender system. In: Rutkowski, L., Scherer, R., Korytkowski, M., Pedrycz, W., Tadeusiewicz, R., Zurada, J. (eds.) Artificial Intelligence and Soft Computing. ICAISC 2018. Lecture Notes in Computer Science, vol 10842. Springer (2018)
3. Su., J.H., Chin, C.Y., Yang, H.C., Tseng, V.S., Hsieh. S.Y.: Music recommendation based on information of user profiles, music genres and user ratings. In: Nguyen, N., Hoang, D., Hong, TP., Pham, H., Trawiński, B. (eds.) Intelligent Information and Database Systems. ACIIDS 2018. Lecture Notes in Computer Science, vol 10751. Springer (2018)
4. Bradford, C., Marshall, I.W.: A bandwidth friendly search engine. Proc. IEEE Int. Conf. Multimed. Comput. Syst. **2**, 720–724 (1999)
5. Gasparetti, F.: Modeling user interests from web browsing activities. Data Min. Knowl. Discov. **31**(2), 502547 (2017)
6. Al-Qurishi, M., Alhuzami, S., AlRubaian, M. et al. User profiling for big social media data using standing ovation model. Multime. Tools Appl. **77**(9), 11179–11201 (2018). Springer
7. Zhang, L., Fu, S., Jiang, S., Bao, R., Zeng Y. (2018) A fusion model of multi-data sources for user profiling in social media. In: Natural Language Processing and Chinese Computing. NLPCC 2018, vol. 11109. Springer
8. Frikha, M., Mhiri, M., Gargouri, F.: Using social interaction between friends in knowledge-based personalized recommendation. In: 2017 IEEE/ACS 14th International Conference on Computer Systems and Applications (AICCSA), IEEE, Tunisia (2018)
9. Hassan, O., Habegger, B., Brunie, L., Bennani, N., Damiani, E.: A discussion of privacy challenges in user profiling with big data techniques: the EEXCESS use case. In: 2013 IEEE International Congress on Big Data, pp. 25–30 (2013)
10. Gao, M., Liu, K., Wu Z.: Personalisation in web computing and informatics: theories, techniques, applications, and future research. Inf. Syst. Front. 607–629 (2010)
11. Frias-Martinez, E., Magoulas, G., Chen, S., Macredie, R.: Automated user modeling for personalized digital. Int. J. Inf. Manag. 234–248 (2006)
12. Sarukkai, Link prediction and path analysis using Markov chains. Comput. Netw. **33**, 377–386, (2000)
13. Jung, S.Y., Hong, J.H., Kim, T.S.: A statistical model for user preference. IEEE Trans. Knowl. Data Eng. 834–843 (2005)
14. Schubert, P., Koch, M.: The power of personalization: customer collaboration and virtual communities. In: Proceedings of the Eighth Americas Conference on Information Systems (AMCIS), pp. 1953–1965 (2002)

15. Freitag, D., Joachims, T., Mitchell, T., Armstrong, R.: WebWatcher: a learning apprentice for the World Wide Web. In: Proceedings of the 1995 AAAI Spring Symposium on Information Gathering from Heterogeneous, Distributed Environments, March (1995)
16. Tebri, H., Boughanem, M., Chrisment, C., Tmar, M.: Incremental profile learning based on a reinforcement method. In: SAC'2005-20th ACM Symposium on Applied Computing, Santa Fe, New Mexico, USA, pp. 1096–1101, mars (2005)
17. Pazzani, M., Muramatsu, J., Billsus, D.: Syskill Webert: identifying interesting web sites. In: Proceedings of the Thirteenth National Conference on Artificial Intelligence (1996)
18. Kieling, W., Endres, M., Preisinger, T.: The BNL ++ Algorithm for Evaluating Pareto Preference Queries (2006)
19. Salton, G., Yang, C.S.: On the specification of terms values in automatic indexing. J. Doc. **29**, 351–372 (1973)
20. Huhns, M.N., Stevens, L.M.: Personal ontologies. IEEE Internet Comput. **3**, 85–87 (1999)
21. Chaffee, J., GAUCH, S.: Personal ontologies for web navigation. In: Proceedings of the Ninth International Conference on Information and Knowledge Management, CIKM, 2000
22. Brin, S., Page, L.: The anatomy of a large-scale hypertextual Web search engine. Comput. Netw. ISDN Syst. **30**, 107–117 (1998)
23. Ruthven, I., Lalmas, M.: A survey on the use of relevance feedback for information access systems. J. Knowl. Eng. Rev. **18**, 95–145 (2003)
24. Zemirli, N., Boughanem, M., Tamine-Lechani, L.: Exploiting multi-evidence from multiple user's interests to personalizing information retrieval. In: IEEE 2nd International Conference on Digital Information Management (ICDIM), France (2008)
25. Speretta, M., Gauch, S.: Personalized search based on user search histories. Web Intell. 622–628 (2005)
26. Daoud, M., Tamine, L., Boughanem, M.: Towards a graph based user profile modeling for a session-based. Knowl. Inf. Syst. **21**(3), 365–398 (2009)
27. Rami Ghorab, M., Zhou, D., OConnor, A., Wade, V.: Personalised information retrieval: survey and classification. User Model. User-Adapt. Interact. **23**(4), 381443 (2013)
28. Pretschner, A., Gauch, S.: Ontology based personalized search. In: Proceedings 11th International Conference on Tools with Artificial Intelligence, Chicago (2002)
29. Tanudjaja F., Mui, L.: Persona: a contextualized and personalized web search. In: Proceedings of the 35th Hawaii International Conference on System Sciences (2002)
30. Salton, G., Yang, S.C.: The specification of terms values in automatic indexing. J. Doc. **29**(4), 351–372 (1973)
31. Wu, Z., Palmer, M.: Verb semantics and lexical selection. In: Proceedings of the 32 nd Annual Meeting of the Associations for Computational Linguistics, pp. 133–138 (1994)
32. Labriji, A., Abdelbaki, I., Reddahi, N., Namir, A., Boudou, A.: Conceptual similarity measure. J. Theor. Appl. Inf. Technol. **83**(2), 291–298 (2016)
33. Siriaraya, P., Yamaguchi, Y., Morishita, M.: Using categorized web browsing history to estimate the user's latent interests for web advertisement recommendation. In: 2017 IEEE International Conference on Big Data (Big Data), Boston, MA, USA (2018)
34. Lv, J.: User interest degree evaluation models. In: 2017 13th International Conference on Natural Computation, Fuzzy Systems and Knowledge Discovery (ICNC-FSKD), Guilin, China (2018)
35. Ko, H.G., Ko I.Y., Kim T., Lee D., Hyun S.J.: Identifying user interests from online social networks by using semantic clusters generated from linked data. In: Sheng Q.Z., Kjeldskov J. (eds.) Current Trends in Web Engineering. ICWE 2013. Lecture Notes in Computer Science, vol 8295. Springer, Cham (2013)

Assessment of Physicians' Knowledge on Ionizing Radiation Exposure During Pediatric Computed Tomography: Case of Hassan II Hospital Center of Agadir

Mohamed El Fahssi[1]([⊠]), Slimane Semghouli[1,2], Mustapha Massaq[3], Oum Keltoum Hakam[4], and Abdelmajid Choukri[4]

[1] Higher Institute of Nursing Professions and Health Techniques, Agadir, Morocco
elfahssi.mohamed@gmail.com
[2] Laboratory Information Systems & Vision (LabSiV), Department of Physics, Faculty of Sciences Ibn Zohr, Agadir, Morocco
[3] Laboratory of Physics of Condensed Matter and Nanomaterials for Renewable Energy (LPMC-Nano), Department of Physics, Faculty of Sciences Ibn Zohr, Agadir, Morocco
[4] Nuclear Physics and Techniques Research Unit, Department of Physics, University of Ibn Tofail, Kenitra, Morocco

Abstract. The main objective of this study is to explore the physician's knowledge on patient radiation exposure during pediatric Computed Tomography (CT). A questionnaire from various questions was developed during this work. The questions have focusing on prescribers' practices and their knowledge of patient radiation protection. It was distributed and completed by 44 doctors of various medical specialties in the regional hospital center of Agadir. Forty-four questionnaires survey were analyzed. Only 2% of our prescribers have correctly estimated the delivered dose during a standard thoracic radiological examination for a 1 year old child. While 84% of our physicians have underrated the dose delivered during a thoracic CT scan for a 5-year-old child. One for five prescribers assigns a dose of irradiation higher than 0.01 mSv during a thoracic Magnetic Resonance Imaging (MRI) exam for child. Finally, 34% of physicians choose CT scan (not indicated) as the first choice for abdomino-pelvic mass and epileptic syndromes.

Keywords: Medical X rays exposure · CT scan · Physician knowledge Radioprotection

1 Introduction

Medical imaging has become an indispensable tool for modern medicine, and global needs are increasing. Medical applications of ionizing radiation Participates in an improvement of medical practices, and bring a real benefit in terms of health [1].

© Springer Nature Switzerland AG 2019
M. Ben Ahmed et al. (Eds.): SCA 2018, LNITI, pp. 535–541, 2019.
https://doi.org/10.1007/978-3-030-11196-0_45

However, they constitute, with natural radiation, the main sources of exposure of people. Therefore, control of dose increases remains a priority objective for radiation protection of patients. Indeed, the world today has a very remarkable increase in imaging activity and the average equivalent dose is constantly increasing. Thus, the risk of reaching the upper limit of low doses, 100 mSv in adults and 50 mSv in children could mean a real epidemiological risk of cancer [1].

Approximately 74.6 million diagnostic procedures were performed in France in 2007, an increase of 26% compared to 2001 [2]. While the average effective dose received following medical exposure per year and per capita (1.3 mSv) increased by 57% compared to 2007 (0.83 mSv) [2]. In 2015, the Nuclear Safety Authority (ASN) declared that even though Computed Tomography account for only 10% of imaging, they account for 58% in 2007 and 71% in 2012 of the dose delivered to the population. For this reason the prescriber of the scanner must have a perfect knowledge of the principles of radiation protection which are defined by the International Commission of Radiological Protection [2].

However many studies indicate that primary care providers are unaware of the hazards associated with the use of radiation. Physicians who are responsible for requesting radiological examinations tend to underestimate the actual doses involved, have poor knowledge about the possible risks to the health of populations and do not discuss the potential risks of CT scans with their patients [3].

Therefore, it is important that physicians who request imaging be well trained to decide whether diagnostic imaging is appropriate, but also to have a clear understanding of the associated risks [4].

The need of data on medical radiation exposure [5] and on doctors' knowledge and their related practices in Morocco [6] pushed us to undertake this study. Given the diversity in experience, the aims of this study is to estimate the perception of the risks to patients from radiation exposure by the physicians, their background training, and their practices witch requiring diagnostic imaging and informing patients of risks.

2 Materials and Methods

The data used in this study were collected from various departments of the Regional Hospital Center of Agadir. A questionnaire was distributed to 44 physicians of various medical specialties. Various questions related to their practice of prescriptions and their knowledge on radiation protection of patients. The questionnaire is divided into four sections.

The first section deals with the demographic data of the physicians participating in the study. While the second section talks about doctors' knowledge of radiation doses delivered in conventional imaging in adults and children. The doctors' knowledge of radiation doses delivered by CT imaging in adults and children is explored in the third part. The last part focuses on the basic and continuing training of practitioners in radiation protection of the patient.

3   Results

The data of surveys were analyzed using Statistical Package for the Microsoft Office Excel 2007 and the results will be presented as follow (Table 1).

Table 1. Distribution of the study group per specialties'

Specialties'	N	%
Internal doctors	20	46
Medical specialists	16	36
General practitioners	08	18

Among the 44 participants in this study, 46% are internal, 36% are specialists and 18% are general practitioners. These doctors have more or less short experiments in practice of medical imaging. 45% of them had an experience of less than 5 years. Yet a good number of professionals have accumulated experience between 5 to 10 years (17%) and the rest (37%) have experience between 10 and 15 years (Fig. 1).

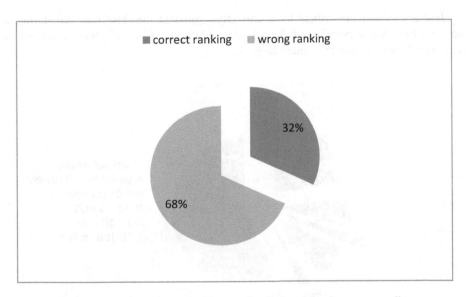

Fig. 1. Evaluation of various classes of radiation doses by our prescribers

68% of our physicians were unable to categorize the dose received following exposure to the following radiation sources in increasing order: one year of natural irradiation, standard chest x-ray, chest CT, pulmonary MRI, standard abdominal X-ray (Fig. 2).

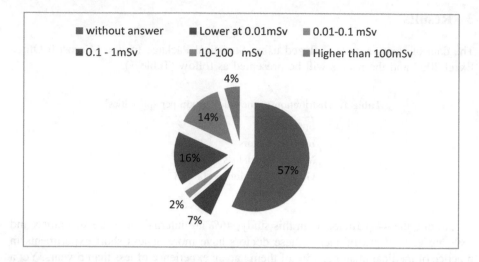

Fig. 2. Assessment of the average dose received during a standard chest radiograph for a 1 year old child by our prescribers

Just 2% of our prescribers have correctly estimated the dose received during a standard lung X-ray examination for a 1 year old child, while 34% overestimated this dose and 57% have not responded (Fig. 3).

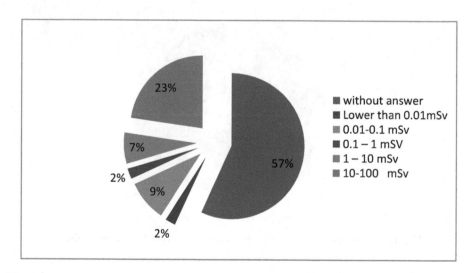

Fig. 3. Assessment of the average dose received during thoracic CT scan for a 5 year old child by our prescribers

The dose received during a chest CT scan for a child of 5 years old was correctly estimated by 7% of our doctors, while 14% underestimated it and 80% of them do not have an answer (Fig. 4).

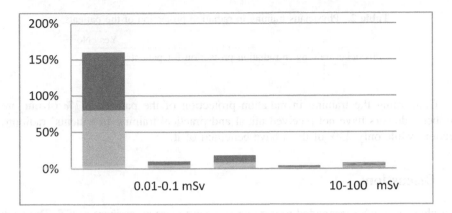

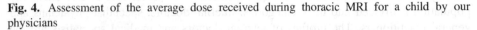

Fig. 4. Assessment of the average dose received during thoracic MRI for a child by our physicians

The dose received during an MRI examination of the chest for the child was estimated to be less than 0.01 mSv by 80% of our prescribers, while 20% of them answered that it is greater than 0.01 mSv (Fig. 5).

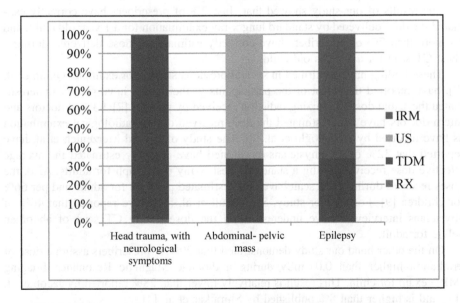

Fig. 5. The first-choice examination chosen for each symptom by our physicians

98% of prescribers indicate to their patients a CT scan for head trauma with neurological signs. while 34% of physicians prescribe a CT scan (not indicated) for abdominal and pelvic masses. Finally, all prescribers do not prescribe CT in case of epilepsy (Table 2).

Table 2. Physicians training in radiation protection of the patients

	Yes	No
Physician training in radiation protection for patients	9	35

Concerning the training in radiation protection of the patients, 77% of our pre-scribers' doctors have not received initial and practical training in patients' radiopro-tection, while only 23% of them have benefited of it.

4 Discussions

All physicians who responded to our questionnaire were prescribers of CT. The study population is divided into three categories: internal doctors, specialist physicians and general practitioners. The profiles of internal doctors and medical specialists are the most representative with 46 and 36% respectively, followed by general practitioners with 18%. These professionals have more or less long experience in practice. An experience of less than 5 years is the most widespread (45%). Yet a good number of professionals have accumulated an experience between 5 and 10 years (17%) and the rest (32%) have experience between 10 and 15 years.

The results of our study showed that, Just 2% of prescribers have correctly esti-mated the dose delivered by standard lung x-ray examination for a 1 year old child, and no more than 7% of prescribers have correctly estimate the dose delivered during a chest CT scan for a 5 years old child.

These finding have confirmed in several previous studies for example, Quinn et al. [7], have reported that most of the participants to their study have greatly underesti-mated the actual dose of ionizing radiation received by patients [7]. 93% of doctors and intern doctors have underestimated the dose received during radiological examination as have reported by Arslanoğlu et al. [8]. The study of Hiltrud Merzenich et al. have reported that 32% of the physicians questioned have correctly estimated the average effective dose received during a standard chest X-ray radiograph for adults. Also, the doses received during a CT scan have underestimated per 47% for adults and per 66% for children [9]. For another study, Semghouli et al. [10] have reported that 46% of physicians interviewed have underestimated the dose during CT scan of abdomen pelvic for adult.

On the other hand our study demonstrated that 20% of prescribers assign a dose of irradiation higher than 0.01 mSv during a thoracic Magnetic Resonance Imaging (MRI) exam for child. This result is relatively lower than 28% reported by Jacob et al. [11] and is higher than 8% published by Shiralkar et al. [4].

5 Conclusions

The radiation doses in CT scan are poorly mastered, and indications of CT exams are not correctly applied by our clinicians. The risk of developing a malignant neoplasm after a radiological exposure in pediatric CT scan has been confirmed by all our prescribers. There is a need of arising awareness about this topic and both initial and continuing training in radiation protection for patients could significantly improve our practitioners' knowledge of radiation protection for patients.

References

1. Godet, J.-L.: Dossier radioprotection du patient, un nouveau cadre réglementaire pour les expositions des patients aux rayonnements ionisants (transposition des directives 96/29 et 97/43 Euratom/p48 (2002)
2. Etard, C., Sandra, S.-T., Bernard, A.: Institut de veille sanitaire, Exposition de la population française aux rayonnements ionisants liée aux actes de diagnostic medical/p3 (2007)
3. Shiralkar, S., Rennie, A., Snow, M., Galland, R.B., Lewis, M.H., Gower-Thomas, K.: Doctors' knowledge of radiation exposure: questionnaire study. BMJ **327**, 371–372 (2003)
4. Gerben, B., Charles, J.B.: Doctors knowledge of patient radiation exposure from diagnostic imaging requested in the emergency department (2010)
5. Semghouli, S., Amaoui, B., Maamri, A.: Estimated radiation exposure from medical imaging for patients of radiology service of Al Faraby Hospital, Oujda Morocco. Int. J. Cancer Therapy and Oncology. 3(3) (2015). http://dx.doi.org/10.14319/ijcto.33.25
6. Semghouli, S., Amaoui, B., Choukri, A., Hakam, O.K.: Physicians' knowledge about patient radiation exposure from ct examinations: case of Hassan II Hospital Agadir-Morocco. In: Proceedings of the 14th International Congress of the International Radiation Protection Association, vol. 2, pp 663-668 (2017). ISBN 978-0-9989666-2-5
7. Quinn, A.D., Taylor, C.G., Sabharwal, T., Sikdar, T.: Radiation protection awareness in non-radiologists. Br J Radiol (1997)
8. Arslanoğlu, A., Bilgin, S., Kubalı, Z., Ceyhan, M.N., İlhan, M.N., Maral, I.: Doctors' and intern doctors' knowledge about patients' ionizing radiation exposure doses during common radiological examinations. Diagn. Interv. Radiol. **13**, 53–55 (2007)
9. Hiltrud, M., Lucian, K., Gael, H., Melanie, K., Shunichi, Y., Hajo, Z., Paediatric CT scan usage and referrals of children to computed tomography in Germany-a cross sectional survey of medical practice and awareness of radiation related health risks among physicians (2012)
10. Semghouli, S., Amaoui, B., El Kharras, A., Bouyakhlef, K., Hakam, O.K., Choukri, A.: Establishment of a diagnostic reference level for brain CT procedures in Moroccan Hospitals. Int. J. Adv. Res. 5(5), 319–324 (2017)
11. Jacob, K., Vivian, G., Steel, J.R.: X-ray Dose Training: Are We Exposed to Enough (2004)

Automating IT Project Governance Lifecycle Through Semantic Technologies

Abir El Yamami[✉], Khalifa Mansouri, Mohammed Qbadou,
and Elhossein Illoussamen

Laboratory: Signals, Distributed Systems and Artificial Intelligence (SSDIA),
ENSET Mohammedia, Hassan II University of Casablanca, Casablanca,
Morocco
abir.elyamami@gmail.com

Abstract. This paper aims to propose an IT project governance system based on semantic technologies. The main feature of our system is the use of a combination of COBIT V5 and PMOBOK V3 good practices, to verify that the projects are continuously controlled and maintained. The purpose is to provide a machine readable document for IT project Governance domain model based on semantic ontology. The proposed ontology has been evaluated for its inconsistency, incompleteness, and redundancy. IT project governance ontology has been implemented in protégé software, and the integrity has been validated through the inference engines: Fact ++ 1.6.5 and Pellet. The proposed domain knowledge has been validated through SPARQL queries using JENA FUSEKI Server. It is found that the integrated use of COBIT and PMBOK practices can help organizations to better control their IT project, achieving in that way a better alignment between the business and the IT.

Keywords: IT project management · IT governance · PMBOK
Ontology · Protégé · SPARQL

1 Introduction

Project Governance has emerged as a strategic tool for coordinating and maintaining any type of projects, and can be effectively applied to IT projects [1]. IT project governance aims to align the information system with the company's strategic priorities, it provides a global vision of all the projects and allows not only to standardize management processes and rules, but also to be able to revise priorities if necessary [2]. It ensures that not only, high value-added projects are funded and launched in a secure manner, but also executed according to the needs and priorities.

The issues of IT governance mostly addressed in management reviews have led to the creation of good practices frameworks such as COBIT and the publication of standards such as ISO 27001 that deals with IT security management. Although these frameworks seem to be useful for organizations, the problematic of their multiplicity seems to be crucial, managers tend to implement simultaneously many frameworks without taking into consideration the possible effects of their multiplicity. In this context,

© Springer Nature Switzerland AG 2019
M. Ben Ahmed et al. (Eds.): SCA 2018, LNITI, pp. 542–556, 2019.
https://doi.org/10.1007/978-3-030-11196-0_46

CIGREF points out in a professional publication that the multiplicity of IT governance standards and frameworks is often a source of contradiction and confusion [3].

COBIT V5 (Control Objectives for Information and Related Technology) is the most global framework for IT governance, the first version of COBIT has been created in 1996 by ISACA and ITGI, as a reference system for governance and audit of information systems [4]. It aims to link business risks, control needs and technical issues based on best practices in IT auditing.

The major limitation of COBIT framework is that it does not provide indications about the implementation of the proposed practices. Thereby, this contribution is an attempt to implement the sub-process "APO05-04" «Monitor, optimize and report on investment portfolio performance».

To overcome this problem, this paper provides an integrated framework of IT project governance (ITPG) based on the most popular good practices frameworks COBIT and PMBOK practices. We have selected this framework for several reasons:

- PMI (Project Management Institute) guides have achieved extensive exposure and global acceptance [5].
- PMI guides provide more details, tools and techniques on project management.
- The PMI guide for project risk management is generic and can be applied to any organization.
- Finally, the PMI is involved in the ISO/PC236 Project Committee and the ISO/TC258 Committee, so that the risk management practices of ISO and PMI will be more closely aligned [5].

Thus, this contribution presents a new approach for automating IT project governance lifecycle through semantic technologies; it provides a machine readable document for ITPG domain model. Ontology, which is a formal and explicit specification of a shared conceptualization [6], can be used to establish a formal specification of domain semantics.

The rest of this paper is organized as follow: Sect. 2 presents a literature review about the concepts of IT governance and project portfolio governance, Sect. 3 depicts our proposed methodology, Sect. 4 presents the evaluation of the proposed system, and finally the results of our contribution are presented in Sect. 5 to sum up with a conclusion in Sect. 6.

2 Literature Review

2.1 IT Governance

Although still relatively new, IT governance field has its own bodies of knowledge and its various techniques and methodologies [7]. The number of IT governance good practices frameworks has been increased, each framework is coming from a professional community that has its own issues and its own culture. According to [8], «*the field of information systems is a real challenge. Companies want to develop their value creation capacity, but they need to transform. This transformation will have to go through an integrated IT governance and by in-depth modifications of the information system.*»

IT governance has its origins, first of all in the production of structured and meaningful information, then in the roles of information systems in organizations and finally in the communication between stakeholders. It appeared necessary to introduce radical reforms in corporate governance, particularly through mechanisms that ensure the reliability and completeness of the information provided to shareholders.

The governance, in general, refers to the ability of an organization to control and regulate its operations in order to avoid conflicts of interest between its various stakeholders. More specifically, IT governance concerns the management and regulation of the information system implemented in an organization. Like corporate governance, the IT governance has attracted a lot of attention and controversy over the past decade.

- According to [9]: "governance is the association of steering and control, steering to ensure that today's decisions prepare properly tomorrow, and control to measure the gap with respect to what was planned."
- [10] define IT governance as a term used to describe how a responsible of the governance of an entity will take into account the supervision, monitoring, control and direction of this entity.
- [11] guide the definition of IT governance by focusing on the concept of decision: They define IT governance as a steering process that aims to control decisions and the underlying risks in order to minimize the risks to the company.
- [12]: "IT governance is seen as a management process, based on best practices, that allows the company to optimize its investments in information system in order to achieve a set of objectives (contribute to its value creation objectives, increase the performance of IT processes and their customer orientation, manage the investment in information system, develop solutions and skills aspects in information system that organization will need in the future, to ensure that the risks related to the information system are under control) while developing transparency."
- [13]: "IT governance is the transposition of the principles of corporate governance to the information system level". It allows transparent communication between the shareholders (the patrimonial power), which fix the objectives and ensure the monitoring; the administrators (the managerial power) for the achievement of the objectives designed in a transparent way, and the information system (IT department), which manages resources and IT processes.

2.2 Project Portfolio Governance

Project portfolio is defined as a group of single projects and programs that are under a single sponsorship for facilitation of effective management to meet strategic business objectives. The differences between project portfolio management and project management approaches has been highlighted in [14].

Effective portfolio management is achieved through the application of a carefully designed set of processes aimed at achieving the overall growth and success of the organization [15]. Indeed, project management frameworks have become an obligation to meet the demands of the environment and to achieve the strategic objectives designed by the organization. In this sense, [16] states that project management is the primary ability of an organization to respond to change and gain a competitive advantage. Thus, [17] defines project management as a critical skill for organizational strategy.

Despite the proliferation of project management frameworks/methodologies/software, many projects fail. According to a study of the international Standish group, the failure rate of projects is high in IT sector, most projects exceed the budget/deadline with a lower quality than expected, and the success rate of projects has decreased from 34% in 2004 to 32% in 2009. In addition, a second survey was carried out by [18] to study the links between the project maturity factors, the project success factors and the performance factors. The result of this study suggests that organizations encouraging the project manager to obtain Project Management Professional (PMP) certification are likely to have procedures and policies in place [2].

3   Methodology

In order to develop an ontology for the project portfolios management domain, a collection, organization, analysis and presentation of data has been done. The official documentation of COBIT 5 [4] and PMBOK V5 guides were used as sources of information. The aim is to solve the problem of the effective decision-making within a project portfolio. In other words, it attempts to identify and formalize a decision-maker's value system, on which both decision support and decision-making will be based. This representation must be formal, complete and reusable for use by intelligent applications, other areas of knowledge or IT experts. Hence the choice of representation by ontological meta-model.

An ontology is a computational model of some portions of the world. It is a collection of key concepts and their inter-relationships, collectively giving an abstract view of an application domain [19]. It aims at interweaving human understanding of symbols with their machine process ability.

Figure 1 presents the proposed model for the governance of IT projects, it is based on a combination of COBIT processes: APO05-04 «Monitor, optimize and report on investment portfolio performance» and PMBOK knowledge domains «content management, time management, cost management, quality management and risk management».

Figure 2 depicts the proposed global ontology for of IT projects portfolios management and we present below the ontological description of each process:

3.1   Scope Management

Domain Objective: to ensure that all the work required by the project is carried out: it consists of creating a scope management plan that documents the definition and mastery of the content.

Domain description: The scope management plan describes how scope will be defined, developed, monitored, controlled and verified. It aims to:

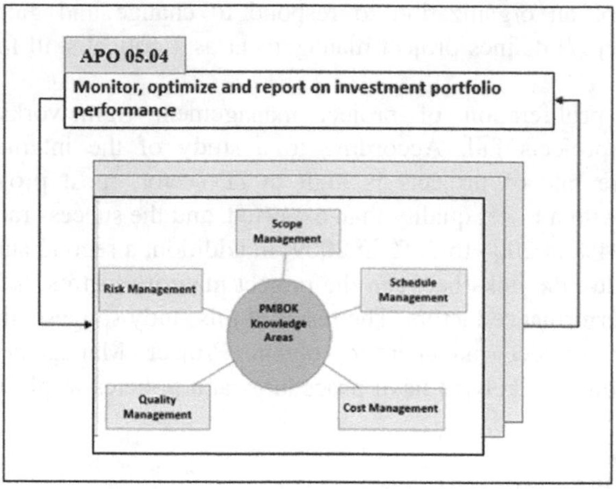

Fig. 1. IT project governance artefact

- Develop a detailed statement of project content by specifying
- A set of measurable project objectives,
- Project deliverables: there is two types of deliverables: direct output from the project such as Product, service or result and additional outcomes such as all plans, reports and other documentations.
- Project exclusions: defines stakeholder expectations about what is not part of the deliverables.
- Project Constraints: refers to any constraints that may limit project manager's options to get the work done; it can be cost constraints, milestone or schedule expectations or contractual penalties.
- Project assumptions defines all the assumptions made with respect to the scope and what may happen if assumptions are proven to be false.
- to specify how formal acceptance of completed project deliverables will be achieved by defining the acceptance criteria for the finished product, service or project result, thereby answering the question "how do we know when is it done?"
- To understand how requests for changes to the detailed statement of project content will be processed (scope Change).

3.2 Schedule Management

Domain Objective: to manage the completion of the project in due time.

Domain Description: establishes the procedures and documentation for the planning, development, management, execution and control of the project schedule. The schedule management plan establishes the following elements:

- Schedule Model: precises the scheduling methodology and scheduling tool required for the development of schedule model.
- Level of Accuracy: precises the acceptable range to be used for activity duration estimates.

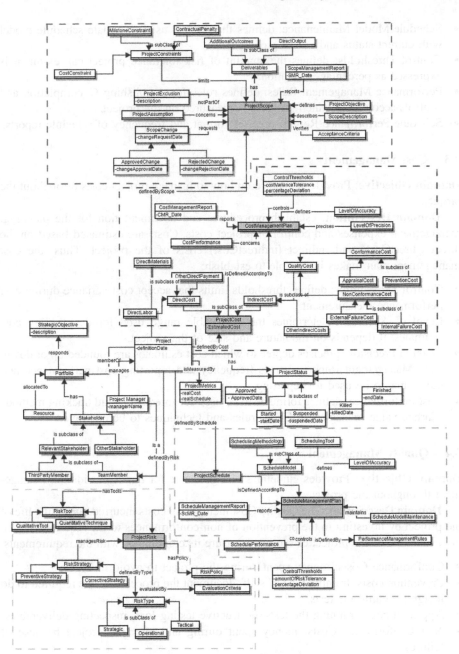

Fig. 2. Ontological Metamodel of IT project governance

- Schedule Model Maintenance: defines the process used to update schedule model with current status and record process.
- Control Thresholds: defines the amount of risk tolerance project can exibit, it is expressed as percentage deviation.
- Performance Management Rules: defines rules for establishing % completion and controls accounts at which progress is measured on the project.
- Schedule Performance Report: precises format and frequency of schedule reports.

3.3 Cost Management

Domain Objective: Provides guidance on how to manage project costs throughout the project.

Domain Description: establish procedures and documentation for the planning, management, expenses and control of project costs. Costs are estimated based on the direct (directCost) and indirect (indirectCost) costs of the project. Thus, the cost management plan makes it possible to establish:

- Control Threasholds: defines thresholds limits that accept cost variance during cost performance measurement.
- Level of Accuracy: determines the acceptable range used for determining cost estimates. It depends on the nature and size of the project.
- Level of Precision: precises degree to which cost estimates are rounded up or down.
- Cost Management Report: defines frequency and reporting cost performance and the format to be used for reporting.
- Cost Performance: defines formulae to be used for forecasting and tracking methods (Earned value management EVM) rules and techniques to be employed.

3.4 Quality Management

Domain Objective: Provides guidelines and guidance for quality validation management throughout the project.

Domain Description: The Quality Cost includes all costs incurred during the life of the project by investing in the prevention of non-conformances to the requirements, in the evaluation of the project or service to ensure its compliance with the requirements.

- Conformance Costs: money spent during the project to avoid failures.
- Prevention costs: training the QC staff, document the processes; time required to do it right; running the tests.
- Appraisal costs: running the tests, destructive testing loss, inspecting deliverables.
- Non Conformance Costs: money spent during and after the project because of failures.
- Internal failure costs: rework, scrap
- Exernal failure costs: liabilities, law suits, product recalls, warranty work, lost business and lost credibility.

3.5   Risk Management

Domain Objective: Provides guidelines and guidance for managing project risks. The goal is to ensure that the level, type and visibility of the risk management will be properly adapted to the importance of the project for the organization.

   Domain description: The PMI defines risk as a threat probability, that its occurrence may affect resources and/ or activities. For each risk, the PMI specifies a set of risk policies (RiskPolicy), evaluation criteria (EvaluationCriteria) and one or more tools that can be qualitative techniques or quantitative tools (QualitativeTools, QuantitativeTechnique). A risk can be operational, tactical or strategic. The risk strategy can be corrective or preventive (PreventiveStrategy, CorrectiveStrategy).

4   Artefact Evaluation

4.1   Ontology Taxonomy Evaluation

In order to evaluate our proposed artefact for its inconsistency, incompleteness, and redundancy, PMBOK ontology has been implemented in protégé software [20], and the integrity has been validated trough the inference engines: Fact ++ 1.6.5 [21] and Pellet

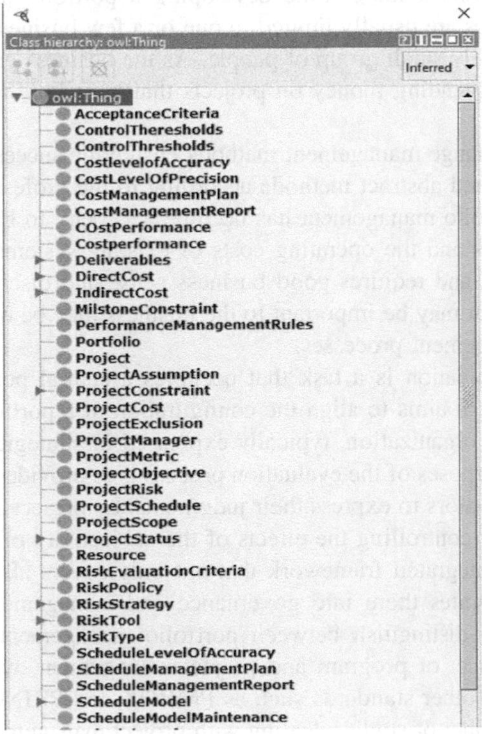

Fig. 3. protégé generated classes

[22] integrated as plug-in in protégé software. Figure 3 represents the generated inferred classes.

4.2 Domain Knowledge Validation

The objective of the proposed ontology is to develop a governance system of IT projects by integrating PMBOK knowledge domains (management of costs, deadlines, content, quality and risks). In order to validate our knowledge of the field, we will calculate the project portfolio management KPIs through the proposed project governance system.

For that aim, Jena Fuseki Server 3.4.0 has been used to demonstrate our ontology-based approach. Jena Fuseki is a SPARQL server that provides REST-style SPARQL HTTP update, SPARQL query, and SPARQL update using the SPARQL protocol over HTTP [23]. Protégé software has been used to populate the ontology data. This later is then exported into XML/OWL format and uploaded to be sorted as a dataset on Jena Fuseki Server. Table 1 expounds the SPARQL queries used for the calculation of IT Portfolio Governance KPIs.

5 Results and Discussion

Companies rarely spend a lot of time developing a portfolio management strategy because their activities are usually limited to one or a few business sectors and can be managed by a relatively small group of people. As the business grows, it runs the risk of losing focus and spending money on projects that have little to do with long-term goals.

Organizational change management methods ensure the success of the project, but they are still considered abstract methods according to the professional staff [24].

As a result, portfolio management has become necessary to balance and prioritize between new projects and the operating costs of existing systems. Portfolio management can be difficult and requires good business sense and disciplined management, otherwise projects that may be important to the business may be overlooked or missed in the detailed management processes.

Thus, project evaluation is a task that occupies a central position in the project management activity. It aims to align the configuration of a portfolio of projects with the objectives of the organization, typically expressed in strategic or technical terms. Indeed, one of the purposes of the evaluation process is to provide a common language that would help evaluators to express their judgments, by preserving certain principles of rationality, and by controlling the effects of the subjectivity of the judgments.

COBIT 5 is an integrated framework that not only covers all the company's processes, but also separates them into governance and management processes, which makes it possible to distinguish between portfolio management, which is rather a function of governance, of program and project management, which is more operational, by integrating other standards such as PMBOK and PRINCE2.

Although the number of studies dealing with project management is increasing, the literature shows the absence of tools considering the governance activities of IT

Table 1. SPARQL queries for calculating project portfolio management KPIs

KPI 1	Time between defining and starting projects
SPARQL Query	```SELECT ?project ?timeToStart
WHERE {
?project o:memberOf o:portfolio1;
o:projectDefinitionDate ?dfDate;
o:has ?status.
?started o:started ?status;
o:startedDate ?startedDate.
BIND (?startedDate - ?dfDate as ?timeToStart).
}```
Result	

KPI 2	Time to approve projects
SPARQL Query	```SELECT ?project ?timeToApprove
WHERE {
?project o:memberOf o:portfolio1;
o:projectDefinitionDate ?dfDate;
o:has ?status.
?started o:approved ?status;
o:approvedDate ?approvedDate.
BIND (?approvedDate - ?dfDate as ?timeToApprove).
}```
Result	

KPI 3	% of killed projects
SPARQL Query	```SELECT ?portfolio (COUNT(DISTINCT ?project) *100 / (COUNT(
DISTINCT?p)) as ?percentOfKilledProjects)
WHERE {
?p o:memberOf ?portfolio.
?project o:memberOf ?portfolio;
o:has ?status.
?killed o:killed ?status;
} group by ?portfolio```

Result	
KPI 4	**Average number of projects per project manager**
SPARQL Query	```Select (avg (?nb) as ?AVGProjectPerManager)
WHERE {
Select ?manager (count (?project) as ?nb)
WHERE {
?manager o:manages ?project.
?project o:has ?status.
optional{
?finished o:finished ?status.
}
optional{
?killed o:killed ?status.
}
filter (!bound (?finished)).
filter (!bound (?killed)).
}
group by ?manager
}```
Results	
KPI 5	**% of projects meeting stakeholder expectations**
SPARQL Query	```SELECT ?portfolio (COUNT(DISTINCT ?project) *100 / (COUNT(DISTINCT ?p)) as ?percentonStakeholderNeeds)
WHERE {
?p o:memberOf ?portfolio.
?project o:memberOf ?portfolio.
?project o:has ?status.
?finished o:finished ?status.
FILTER (?finished = o:accepted). }
group by ?portfolio```
Results	
KPI6	**% of projects meeting deadlines and budget**
	``SELECT ?portfolio (COUNT(DISTINCT ?project) *100 / (COUNT(DISTINCT ?p)) as ?PercentOfProjectOnTimeAndBudget)
WHERE {
?p o:memberOf ?portfolio.
?project o:memberOf ?portfolio;```

| | ```
                              o:definedByCost ?cost;
                              o:definedBySchedule ?schedule.
         ?cost o:estimatedCost ?estimatedCost.
         ?schedule o:estimatedSchedule ?estimatedSchedule.
         ?project o:isMeasuredBy ?metric.
         ?metric o:realCost ?realCost;
                              o:realSchedule ?realSchedule.
            FILTER (?realSchedule < ?estimatedSchedule ).
            FILTER (?estimatedCost <?realCost).

}
group by ?portfolio
``` |
| **Results** |

QUERY RESULTS
:: Table Raw Response ⬆
Showing 1 to 1 of 1 entries   Search: [   ]   Show 50 ▾ entries

| portfolio | ◊ PercentOfProjectOnTimeAndBudget | ◊ |
| 1   o:portfolio1 | "20.0"^^xsd:decimal | |

Showing 1 to 1 of 1 entries |
| **KPI7** | **% of projects with scope changes** |
| **SPARQL Query** | ```
SELECT ( COUNT( DISTINCT ?project) *100  / ( COUNT( DISTINCT?p)  ) as
?percentOfSC) ?portfolio
    WHERE {
        ?p o:memberOf ?portfolio.
            ?project o:memberOf ?portfolio.
            ?project o:definedByScope ?scope.
            ?changereq o:requests ?scope.
            ?SCAapproval o:approves ?changereq.

}
Group By ?portfolio
``` |
| **Results** |

QUERY RESULTS
:: Table Raw Response ⬆
Showing 1 to 1 of 1 entries   Search: [   ]   Show 50 ▾ entries

| percentOfSC | ◊ portfolio | ◊ |
| 1   "40.0"^^xsd:decimal | o:portfolio1 | |

Showing 1 to 1 of 1 entries |
| **KPI8** | **Deviation of the estimated budget** |
| **SPARQL Query** | ```
SELECT ?portfolio ( (?real- ?estimated) as ?DeviationOfPlannedBudget)
WHERE {
   select (SUM ( DISTINCT ?realCost) as ?real) (SUM ( DISTINCT ?estimat-
edCost) as ?estimated)
     where {
                 ?project o:memberOf ?portfolio;
                              o:definedByCost ?cost.
            ?cost o:estimatedCost ?estimatedCost.
            ?project o:isMeasuredBy ?metric.
            ?metric o:realCost ?realCost.
        }    group by ?portfolio
     }
``` |
```

Results	

projects as a whole; like many activities of organizations, project portfolio management must find a tooled response through information system applications. To this end, this work aims to provide a global approach for the control of IT projects within a project portfolio.

Since a complete ontology for the management of project portfolios does not exist in the literature, it seems necessary to construct a shared representation of the manipulated concepts. The objective is to contribute to the professional literature by providing a machine-readable document for the governance of IT projects domain knowledge, and on the other hand to the scientific literature interested in the improvement of IT governance standards by increasing the understanding of the architectures of these repositories. For this purpose, COBIT and PMBOK practices have been integrated.

The proposed artefacts are built according to the practices of the COBIT process. (APO05.04: Monitor, optimize and report on investment portfolio performance.) And PMBOK knowledge domains (content management, time management, cost management, quality management, risk management). It has been implemented in the protected software to check for inconsistency, completeness and redundancy. Integrity has been validated by inference engines: Fact ++ and Pellet.

COBIT outlines a set of key performance indicators for the APO5 process, but does not prescribe how to calculate these indicators; In order to validate the knowledge domain of the proposed ontology, we used SPARQL queries using the Jena Fuseki Server. With the proposed alternative, project portfolio management indicators can be effectively calculated, demonstrating that a combination of PMBOK and COBIT practices can lead to a significant improvement in Business-IT alignment.

6 Conclusion

Although the number of research dealing with project management is increasing, the literature shows the absence of tools considering the governance activities of IT projects as a whole; like many activities of organizations, project portfolio management must find a tooled response through information system applications. To this end, this work is an attempt to provide a global approach for the control of IT projects within a project portfolio based on a combination of COBIT and PMBOK frameworks.

Yet the adoption of PMBOK knowledge domains is not an easy task as their definition and role is unclear. This paper presented an alternative for the system of governance of IT projects, it provides an ontology for the formalization of this domain knowledge. The aim is to eliminate misunderstanding of the objective behind the implementation of IT project portfolio management models. Protégé software was used in the design and evaluation of the proposed ontology. The latter has been validated via SPARQL queries using JENA FUSEKI Server.

References

1. El Yamami, A., Ahriz, S., Mansouri, K., Qbadou M., Illoussamen, E.H.: Rethinking IT project financial risk prediction using reference class forecasting technique. In: 2018 4th International Conference on Optimization and Applications (ICOA), Mohammedia, Morocco, 2018
2. El Yamami, A., Mansouri, K., Qbadou M., Illousamen, E.H.: Multi-objective IT Project Selection Model for Improving SME Strategy Deployment. Int. J. Electr. Comput. Eng. **81** (2), 1103 (2018)
3. CIGREF: Gouvernance du Système d'Information (2011)
4. ISACA: COBIT 5 A Business Framework for the Governance and Management of Enterprise IT, USA (2012)
5. Dunne, E.S.: Project Risk Management: Developing a Risk Framework for Translation Projects (2013)
6. Gruner, T.: A translation approach to portable ontology. Knowl. Acquis. **5**(2) (1993)
7. El Yamami, A., et. al.: representing IT projects risk management best. Eng. Technol. Appl. Sci. Res. **7**(5), 2062–2067 (2017)
8. CIGREF: CIGREF (2009) [Online]. Available: http://www.cigref.fr/cigref_publications/ RapportsContainer/Parus2009/Referentiels_de_la_DSI_CIGREF_2009.pdf. [Accessed 10 4 2016]
9. CIGREF: IT frameworks, CIGREF (2009) [Online]. Available: http://www.cigref.fr/cigref_ publications/RapportsContainer/Parus2009/Referentiels_de_la_DSI_CIGREF_2009.pdf. [Accessed 10 4 2016]
10. Roussey, R.S.: Broad Briefing on IT Governance (2004)
11. Weill, J.R.P.: IT Governance: How Top Performers Manage IT Decision Rights for Superior Results. Harvard Business School Press, Boston (2004)
12. Leignel, J.-L.: Gouvernance du système d'information, CIO Stratégie, Nice, France (2006)
13. Anica-Popa, L., Anica-Popa, I., Florescu, V.: Governance of information system and audit. In: 1'st International Conference on Accounting and Auditing BCAA, Edirne, Turcia, 2007
14. El Yamami, A., Mansouri, K., Qbadou M., Illoussamen, E.: Toward a new model for the governance of organizational transformation projects based on multi-agents systems. In: 2016 11th International Conference on Intelligent Systems: Theories and Applications (SITA), Mohammedia, 2016
15. Levine, R.: Finance and Growth: Theory and Evidence. Handbook of economic growth (2005)
16. Amaral, A.A.M.: Project Portfolio Management Phases: A Technique for Strategy Alignment. World Academy of Science, Engeneering and Technology (2009)
17. Hurt: Building value through sustainable project management offices. Project Manag. J. (2009)
18. Vittal Anantatmula, P.R.: Linkages among project management maturity, PMO, and project sucess (2013)
19. Lee, C.S., Jian, Z.W., Huang, L.K.: A fuzzy ontology and its ap- plication to news summarization. IEEE Trans. Syst. Man Cybern. Part B: Cybern. **35**(5), 859–880 (2005)

20. Musen, M.: The Protégé project: a look back and a look forward. AI matters., Assoc. Comput. Mach. Specif. Interest Group Artif. Intell., Intell. **1**(4), (June 2015). https://doi.org/10.1145/2557001.25757003
21. Building, K.: "Fact ++," School of Computer Science University of Manchester (2017) [Online]. Available: http://owl.man.ac.uk/factplusplus/
22. Pellet: http://pellet.owldl.com/, Clark & Parsia, LLC, 2011. [Online]
23. Jena, A.: JENA FUSEKI SERVER, Apache Software Foundation (2017) [Online]. Available: https://jena.apache.org/documentation/fuseki2/index.html
24. Jern, A.: Introducing information systems in organizations. In: Faculty of Electronics, Communication, and Automation, 2009

Comparative Study of Batch and Stream Learning for Online Smartphone-based Human Activity Recognition

Ilham Amezzane[1]([⊠]), Youssef Fakhri[1], Mohamed El Aroussi[1], and Mohamed Bakhouya[2]

[1] LaRIT Lab, Faculty of Sciences, Ibn Tofail University, Kenitra, Morocco
{ilham.amezzane,fakhri}@uit.ac.ma, mohamed.
elaroussi@ieee.org
[2] Faculty of Computer and Logistics, TIC Lab Sala Aljadida, International
University of Rabat, Rabat, Morocco
Mohamed.bakhouya@uir.ac.ma

Abstract. The availability of diverse embedded sensors in modern smartphones has created exciting opportunities for developing context-aware services and applications, such as Human activity recognition (HAR) in healthcare and smart buildings. However, recognizing human activities using smartphones remains a challenging task and requires efficient data mining approaches due to the limited resources of the device. For example, the training process is usually performed offline but rarely online on the mobile device itself, because traditional batch learning usually needs a large dataset of many users. Therefore, building models using complex multiclass algorithms is generally very time-consuming. In this paper, we present a comparison study using two approaches in order to reduce training time and memory usage while maintaining significant performance. In the first approach, we conducted experiments using batch learning on a GPU platform. Results showed that High Performance Extreme Learning Machine (HPELM) offers the best compromise: accuracy, memory usage and training time. Moreover, it achieved better performance on two dynamic activities, outperforming the best SVM model obtained in our previous study. In the second approach, we conducted experiments using online stream learning on the MOA platform. Unlike the first approach, experiments were performed using accelerometer data only. We also studied the effects of user/device dependency and feature engineering on the classification performance and memory usage by comparing five constructed real data streams. Simulation results showed that Hoeffding Adaptive Tree has comparable performance to batch learning, especially for user and device dependent data streams.

Keywords: Human activity recognition · GPU · Online stream learning

© Springer Nature Switzerland AG 2019
M. Ben Ahmed et al. (Eds.): SCA 2018, LNITI, pp. 557–571, 2019.
https://doi.org/10.1007/978-3-030-11196-0_47

1 Introduction

Smartphone-based Human Activity Recognition (SHAR) involves the use of different mobile embedded sensors available on modern mobile phones. It also involves Machine Learning (ML) techniques to automatically collect and infer user activities for different domains such as healthcare monitoring, assisted living for elderly, sports and wellbeing applications [1–3]. Typical batch learning is generally performed on a large dataset collected from many users, and thus having long computation times. In the supervised learning approach, the classifier usually needs to be trained offline first (in desktop PCs, servers, or cloud systems) using labeled data, before online implementation where the trained model classifies each block (batch) of incoming data as one of the targeted activities within a small-time window. This is made possible because data collected from sensors are provided in a time-series manner (i.e., data streams). Traditional batch classification approaches are based on a static data scenario and assume that data arrive in form of blocks and must be processed after they become available. Batch mode offers better robustness to local fluctuations of the stream and a broader outlook on the incoming data [4]. Nonetheless, in the case when training is implemented online, authors usually consider using limited training data that could be collected only in a few minutes due to the limited memory available on the devices. There are limitations that can result from using these two kinds of implementation: for the offline mode, the obtained model is static; once a model is generated, it does not adapt to the user's activity profile changes and it is not subject specific. However, for the online mode, training of single user small data batches usually requires cross-validation (CV) techniques in order to avoid overfitting, and thus the entire dataset is allocated into the main memory. In addition, spending few minutes or even seconds for model training may discourage users from adopting SHAR applications in their everyday life, especially if they cannot tolerate long waiting each time they need to update the application to profile's changes or new activities. Subsequently, there is a need towards implementing fast training models on the mobile devices in an online manner, since this can help developing fast user's dependent applications. Moreover, online learning avoids transmission of sensitive user information, such as location, activities and health to a server or a cloud. Robustness and responsiveness are also strengthened in this case, since training on-device is independent from unreliable wireless communication links or Internet connection.

The rest of this paper is organized as follows. In Sect. 2 we present an overview of related works from literature. Materials and methods used in the present study are introduced in Sect. 3. Experimental results are discussed in Sect. 4. Section 5 presents conclusions.

2 Related Work

According to the survey conducted in [5], only 6 out of 30 online recognition systems have used online learning. For instance, in [6], real-time SHAR was performed using the clustered kNN (k-Nearest Neighbors) classification method. Authors considered only limited training data, and consequently, the amount of computation was reduced.

However, no information is provided regarding the training time. In [7], authors demonstrate the Mobile Activity Recognition System (MARS) where for the first time the model is built and continuously updated on-board using data stream mining. Features were extracted from the accelerometer sensor and incremental Naive Bayes (NB) was used to update an anytime model in order to accommodate changes in the data stream. Weighting solution consisting in multiplying the stored statistics of the NB algorithm by a fading factor *alpha* (0<*alpha*<1) was also considered in order to avoid any additional memory cost and to update the model with the new information. However, no information is provided regarding the classification accuracy. In [8], authors proposed a framework for online SHAR called Mobile Online Activity Recognition System (MOARS), which integrates data collection, training and inference. The framework dynamically considers real-time user's feedbacks in order to increase prediction accuracy, at the cost of additional time. However, no information is provided regarding the training time. In [9], authors compared Decision Tree (DT), kNN, and Neural Networks (NN) training times. The faster was kNN, with only 0.1 s. The DT was also fast, with 2.88 s. The training process of the NN was much slower, as it needed 967.16 s. However, both DT and NN produced a model of ≈ 1 MB size while kNN produced a model of ≈ 10 MB. Finally, in [10], authors present an Online-Independent Support Vector Machine (OISVM), which utilizes a small portion of data from the unseen data to update on-the-fly the parameters of the SVM algorithm, in order to eliminate the effects of the orientation, the placement and the subject variations. However, the standard SVM algorithm operates offline in a batch manner. Authors state that the method dramatically reduces time and space requirements, at the price of a negligible loss in accuracy. In [11], an adaptive stream learning framework called STAR, proposes an active learning technique well suited for choosing only a small amount of data to be labeled. Then the system is further refined with the selected true labeled data. Additionally, the framework uses incremental learning approach to handle concept drift over evolving streaming data in order to fit a particular user or context. The experiments showed low computational cost and real time recognition. In [4], a novel online scheme, combining principles of active and adaptive ensemble learning has been proposed. The system architecture uses a weighted NB that can adapt to the incoming data of the stream without a need for an explicit concept drift detector. The proposed ensemble is enhanced with an active learning module to reduce the labeling cost over real-time senor data streams.

On the other hand, many attempts exist in the literature in order to speed-up the batch training phase by means of hardware (HW) accelerators such as multi-core Central Processing Unit (CPU) and Graphics Processing Unit (GPU). There are currently many accelerated libraries designed for multi-core CPUs. However, one of the increasingly popular trends is the use of GPUs because they allow for a distribution of small single tasks among a large number of GPU cores, which should result in higher performance compared to CPU computation. Generally, the codes of the learning algorithms need to be modified in order to exploit their inherent parallelism using appropriate computing frameworks like CUDA or OpenCL. Consequently, many new libraries have been introduced such as Google's TensorFlow (TF) [12], which implements many ML algorithms using CUDA framework. In the context of SHAR, few works are using GPUs in order to facilitate faster computations for the online training phase. For example, in [13], a Deep Recurrent NN (DRNN) was constructed using raw time series data of acceleration sensors. Although parallel processing using the GPU reduced the training time, it was still very large (116.39 s per epoch on average). In [14], authors designed an automatic user-adapted physical activity classifier. The data used were obtained from acceleration sensors and gyroscopes. To prevent the training dataset from becoming too large during each update and occupying a large amount of storage space, they included the confirmed new samples in the training set and removed the samples, which are farthest from new samples. In addition, a parallel algorithm based on the OpenCL framework was designed, and this enhanced the computing efficiency approximately nine-fold. Although the average power for the parallel design based on OpenCL is slightly higher than the serial design, the run-time for the parallel design is much shorter, resulting into substantially lowered total energy consumption. In our previous work [15], we made a comparison study exploiting Feature Selection (FS) approaches to reduce the computation and training time needed for the discrimination of targeted activities while maintaining good accuracy. The combination of the Recursive Feature Elimination (RFE) method with the RBF-SVM algorithm provided us with a feature subset (50 features), which offered the best compromise accuracy/training time. With FS approaches and parallel execution over two CPU cores, we managed to reduce the number of data samples needed to reach the same significant accuracy in less time than sequential execution. In the present work, we conduct a comparative study of the SHAR training phase, considering two approaches, in order to reduce training time and memory usage while maintaining significant classification accuracy. In the first approach, we conduct experiments using a GPU platform for traditional batch learning. In the second approach, we conduct experiments using a software platform for online data stream learning.

3 Materials and Methods

3.1 GPU-Based Approach

Basically, TF is a framework that was developed to perform computations based on the concept of data flow graphs. The graphs can be executed in parts or fully on available low-level devices such as CPU and GPU. This may help alleviate the CPU computation burden for calculations that can be run in parallel over GPU. Figure 1 shows the TF architecture [16]: the core is written in C^{++} and the graph is defined in a high-level

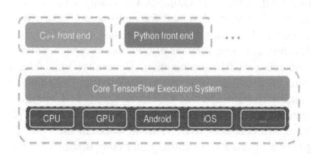

Fig. 1. Architecture of tensorflow [16]

language such as Python. TF offers other particularities: it is open source, can be implemented on Android smartphones and computes gradients automatically.

For the present section, we first trained and tuned the selected TF classifiers over CPU only, then over GPU only, in order to evaluate any improvements in training time. When running TF-based graph over multiple CPU cores, it is possible to execute operations that can be parallelized internally by scheduling tasks via a thread pool configuration, and therefore to control the maximum parallel speedup for a single operation. In our CPU-based experiments, we configured the thread pool to run over 2 CPU cores. We have conducted the CPU-based experiments using a laptop with intel core i5-3320M CPU, Windows 10, 4G of RAM and a processor x64/ 2.6 GHz. We have conducted the GPU-based experiments using Tesla K80 GPU, provided by Google Colaboratory (a jupyter notebook environment) [17]. The GPU has a compute capability equal to 3.7 and all codes are written in Python 3.5. It is worth noting that in all the experiments of this section, we used the same training and validation sets used in our previous study with the same 50 selected features [15]. 10-fold CV techniques have also been performed.

The TF supported classifiers selected in the present study are Deep Neural Network (DNN) and Softmax Regression (SMR). The tuned DNN consists of 3 layers containing 30, 18 and 11 units respectively from the input layer to the output layer. The CV technique was too time-consuming; therefore, we decided to train the DNN without any re-sampling techniques and measure its classification accuracy over validation datasets. SMR is a generalization of logistic regression that we can use for multiclass classification (under the assumption that the classes are mutually exclusive). Regarding the multi-class RBF-SVM classifier, although it is supported by TF, the CV techniques of this algorithm are still not supported at the time of writing this paper. Therefore, we used the sickit-Learn package in Python, and we simply executed the algorithm over 2 CPU cores then over the GPU.

Extreme Learning Machine (ELM) is an algorithm that we also tested, because it offers certain advantages in generalization and computational efficiency [18]. ELM works for the generalized Single-hidden Layer Feedforward Networks (SLFNs). SLFN has three layers of neurons, but the only layer of non-linear neurons in the model is the hidden layer. Input layer provides data features and performs no computations, while an output layer is linear without a transformation function and without bias. The structure of an SLFN, presented in Fig. 2 [19], contains L hidden nodes, being x the input vector and G (a_i, b_i, x) denotes the output function of the i-th hidden node, being a_i, b_i the hidden node parameters and β_i the weight vector connecting the i-th hidden node to the output nodes.

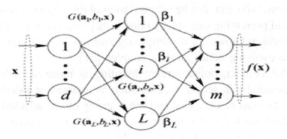

Fig. 2. Structure of an SLFN [19]

In the ELM method, input layer weights and biases are set randomly and never adjusted. Because the input weights are fixed, the output weights are independent of them and have a direct solution without iteration. This is the reason why such model is very fast to compute [20]. Setting randomly the layer-one weights can possibly be effective because our labeled dataset is small. We first trained and tuned an ELM model on CPU. The final model consists of 120 *sine* hidden neurons. First results showed similar performance as our previous RBF-SVM model, with 92% for accuracy and 93% for averaged F1-score and total CPU time of 11.15 s. This is the reason why we decided to train a version of ELM, which can be run on a GPU, called High Performance ELM (HPELM) [20]. Our final model consists of 150 *sigmoid* and 50 *tanh* hidden layer neurons.

3.2 Online Stream Learning Approach

Until recently, online learning inside mobile environments has been a challenging task for ML techniques considering the limited resources as well as time constraints. These challenges pose the need for different algorithms than those used for classical batch learning where data are stored in finite and persistent data repositories, or those used for incremental learning, as they do not focus on computational restrictions and do not consider dynamic changes [21]. To tackle these challenges, several algorithms have been introduced in the literature depending on the two basic models of data streams: (i) stationary: where examples are drawn from a fixed although unknown probability distribution, and (ii) non-stationary: where data can evolve over time. In [22], the Hoeffding Tree (HT) algorithm for stationary data streams was proposed. Its novelty consisted of waiting for new instances to arrive instead of reusing instances as it is the case for batch learning. The strength of HT is that it has theoretical proofs guaranteeing that it can build trees of the same quality as batch-learned trees if the number of instances needed at a node when selecting a splitting attribute is sufficient. The name of this algorithm is derived from the Hoeffding Bound [23]. Moreover, HT contributes in solving the uncertainty in learning time since it consumes constant time per instance. On the other hand, the Hoeffding Adaptive Tree (HAT) is, an adaptive extension to the HT [27] that also has theoretical guarantees and uses the ADWIN algorithm [21] as a drift detector and error estimator. Finally, it is worth noting that several traditional incremental classifiers were also adapted to concept drift requirements; where sliding windows are usually employed as a forgetting mechanism. For example, the kNN is naturally transformed to incremental versions for use in the streaming case with different techniques for selecting the limited subset of the most useful examples for accurate predictions. NNs can also be adapted to evolving data streams by dropping the epoch protocol and presenting examples in one pass [24]. NB is a naturally incremental classification algorithm known for its simplicity and low computational cost, however, it does not tackle evolving nature of data sources [21].

For the SHAR context, online stream learning is more suitable than traditional batch learning, especially in sensitive cases such as assisted or elderly supervision [4] since incoming data must be classified and the learning model adapted in real-time. The other benefit is that the training samples do not need to be stored on the phone nor scanned more than one time as they arrive [26].

In our experiments, we constructed five real data streams in order to study the effects of feature engineering, user dependency and device dependency on the classification accuracy, processing speed and memory usage. The steams are as follows: (1) Stream 1: constructed from the same dataset and feature subset (50 features) used in the previous section. The dataset is the combination of the training and test datasets of 9 users, collected from two sensors (accelerometer and gyroscope) of the same device. (2) Streams 2–5: extracted from a large public HAR dataset [25]. The original dataset contains data collected from embedded accelerometers and gyroscopes of different smart phones and watches that were held by 9 users. Different classes of physical activity were targeted such as Biking, Sitting, Standing, Walking upstairs, Walking downstairs and Walking. It's worth noting that each instance in the streams 2–5 has the attributes of x, y, z accelerometer data only activity. Therefore, no feature engineering has been performed. The dimensions of the five streams are as follows:

- *Stream* 1: contains 10,299 instances.
- *Stream* 2: contains the data of user ("a") collected from different devices, with 1,362,520 instances.
- *Stream* 3: contains the data of user ("a") collected from the same device (Nexus 4), with 646,825 instances.
- *Stream* 4: contains the data of users ("a" and "b") collected from different devices, with 2,911,288 instances.
- *Stream* 5: contains the data of users ("a" and "b") collected from the same device (Nexus 4), with 1,376,743 instances.

The classification performance of data streams was evaluated on stream by the Test-then-Train method, where the accuracy can be incrementally updated because the model is always being tested on recent examples [28]. Two different evaluators exist for this method: (i) Basic Classification: which gives the accuracy computed from the beginning of the stream, and (ii) Window Classification: which gives the up-to-date accuracy computed using the last window of instances. We have conducted all the experiments using the MOA framework [29] which is written in Java, allowing portability in many platforms such as Android Smartphones. Finally, it is worth noting that according to [30], processing large data streams may easily exhaust memory if there is no intentional limit set on its use. In MOA, various memory sizes are available depending on the given application scenario. In the present context, we set memory limit to 32 Mbytes which simulates handheld devices such as smartphones [30].

4   Results and Discussion

4.1   GPU Acceleration Approach

Best results obtained in our previous work are presented in Table 1 for comparison. They may vary depending on hardware platforms, the written code, the libraries and the programming languages used. Experimental results presented in Table 1 show clearly that, in all cases, GPU offers better training times than CPU. It can be seen that the longest training time was given by DNN for both CPU and GPU platforms, although we did not use CV techniques in that case. Nonetheless, the validation accuracy is above 90%. On the opposite, the shortest time was given by HPELM (less than one second), which is about 1 order of magnitude less than the SVM's CPU time reached in our previous work [15]. Moreover, this gain is obtained without any decrease in validation accuracy over the same feature subset. Therefore, the HPELM classifier gives the best compromise accuracy/time for both CPU and GPU platforms. After getting better results with the use of GPU, we retrained all the classifiers using the original feature set in order to find out the best performance that can be obtained without using FS methods. We tuned the number of hidden layer neurons of the new HPELM in order to adapt to the large number of features. Our best model consists of 500 *sigmoid* and 50 *tanh* hidden layer neurons. Experimental results are also presented in Table 1. It can be seen that the least execution time is given again by HPELM. The result is about 15 times less than the SVM's CPU time reached in our previous work for the same feature set. It was also obtained without any decrease in validation accuracy. Therefore, the HPELM classifier gives the best compromise accuracy/time for both CPU and GPU platforms even with no FS methods.

Table 1. Training times over CPU and GPU with validation accuracies

Feature set	Best performance of [15]		Present work				
	CPU time	Acc.	Classifier	CPU time	Acc.	GPU time	Acc.
50	7.92	0.92	DNN	476.45	0.91	62.54	0.93
			SMR	5.15	0.88	1.95	0.88
			HPELM	4.56	0.93	0.73	0.92
			SVM	6.34	0.92	2.61	0.92
561	50.71	0.95	DNN	1305.84	0.95	133.86	0.94
			SMR	21.23	0.85	13.06	0.85
			HPELM	32.56	0.95	3.26	0.95
			SVM	36.31	0.94	14.61	0.94

Confusion matrices shown in Tables 2 and 3 contain information about True activities versus Predicted ones of the HPELM model run on GPU, using the validation data over both feature sets. Overall, it can be seen that there is almost a perfect classification between "dynamic" and "static" activities. Moreover, compared to SVM in the previous work, HPELM is better in discriminating "dynamic" activities, especially "Walking" and "Walking upstairs".

In Tables 4 and 5, we present Precision, Recall and F1-score values. It can be seen that the least significant F1-scores are those corresponding to the "Standing" and "Sitting" activities. It was the same case with SVM in the first study. On the contrary, "Laying" is less recognized with HPELM. Nonetheless, regarding "dynamic" activities, "Walking Upstairs" and "Walking" are better recognized.

Table 2. Confusion matrix of HPELM over GPU (validation dataset/50 features)

True activities	Predicted activities					
	Laying	Sitting	Standing	Walking	W-Dn	W-Up
Laying	477	13	24	0	0	0
Sitting	0	422	53	0	0	0
Standing	0	69	429	1	0	0
Walking	0	0	2	418	0	1
Walking_Dn	0	0	0	37	312	13
Walking_Up	0	0	0	1	1	385

Table 3. Confusion matrix of HPELM over GPU (validation dataset/561 features)

True Activities	Predicted activities					
	Laying	Sitting	Standing	Walking	W-Dn	W-Up
Laying	511	1	1	0	0	1
Sitting	0	446	29	0	0	0
Standing	0	44	455	0	0	0
Walking	0	0	2	417	2	0
Walking_Dn	0	0	0	21	317	24
Walking_Up	0	0	0	1	0	386

Table 4. Performance metrics for each activity over GPU (validation dataset/50 features)

Activities	Precision	Recall	F1-score	Support
Laying	1	0.93	0.96	514
Sitting	0.84	0.89	0.86	475
Standing	0.84	0.86	0.85	499
Walking	0.91	0.99	0.95	421
Walking_Dn	1	0.86	0.92	362
Walking_Up	0.96	0.99	0.98	387
Avg/Total	0.92	0.92	0.92	2658

Table 5. Performance metrics for each activity over GPU (validation dataset/561 features)

Activities	Precision	Recall	F1-score	Support
Laying	1	0.99	1	514
Sitting	0.91	0.94	0.92	475
Standing	0.93	0.91	0.92	499
Walking	0.95	0.99	0.97	421
Walking_Dn	0.99	0.88	0.93	362
Walking_Up	0.94	1	0.97	387
Avg/Total	0.95	0.95	0.95	2658

It can also be interesting to compare our results with the results of a recent work presented in [31], in which authors used a dynamic feature extraction from time series HAR data, based on the same public dataset. The objective of their study was to improve the accuracy of "dynamic" activities. They compared results of three different classifiers over the 561 static features with simulation results of Convolutional Neural Network (CNN) over dynamic features. They found that SVM is the most effective algorithm for static features, while the performance of CNN with dynamic features is better, although high computational. In Table 6, we compare the classification accuracy for each activity (derived from Table 3) with the corresponding values presented in [31]. It can be seen HPELM with static features offers the best average for all activities although the average of dynamic activities is slightly in favor of CNN with dynamic features. It's worth noting that the accuracy of CNNs is, however, generally time demanding under time constraint cost [32].

Table 6. Comparison of classification accuracy for each activity using static and dynamic features

Activities	Class accuracy	
	HPELM with 561 static features	CNN with dynamic features [31]
Laying	0.994	0.8100
Sitting	0.938	0.7739
Standing	0.911	0.8909
Walking	0.99	0.998
Walking_Dn	0.875	0.9905
Walking_Up	0.997	0.9703
Avg of last 3	0.954	0.9862
Avg of all	0.9508	0.9056

Since memory is an important parameter that should be taken into consideration when implementing any training algorithm on a mobile device, we have measured the memory usage of each training operation in the CPU-based mode. Our goal is to find the classifier that offers the best compromise accuracy/time/memory. Estimated results are presented in Table 7 where it can be clearly seen that here again, HPELM shows the best memory usage consuming only 32 Bytes in the training phase.

Table 7. Comparison of memory usage for each training algorithm

Classifier	Features	Memory usage (MBytes)	Memory usage (bytes)
DNN	50	144.78	151.81×10^6
	561	157.65	165.3×10^6
SMR	50	490×10^{-6}	512
	561	490×10^{-6}	512
HPELM	50	30×10^{-6}	32
	561	30×10^{-6}	32
SVM	50	0.23	0.24×10^6
	561	4.46	4.67×10^6

4.2 Online Stream Learning Approach

The experimental results related to online stream learning are presented in Table 8. Stream 1 results show that the classification accuracy is similar for all algorithms. However, it is 5% less than the value reached by traditional batch learning in the previous section.

Table 8. Performance results of the stream learning approach

Data stream	Classifier	Sample Freq.	Interleaved test and train	
			Basic	Window (=1000)
Stream 1	N.Bayes	10	86.74	87.44
	H.T		86.71	87.44
	Adaptive H.T		86.33	87.46
Stream 2	N.Bayes	1000	66.95	54.53
	H.T		79.59	71.66
	Adaptive H.T		92.67	89.61
Stream 3	N.Bayes	1000	74.99	61.64
	H.T		85.53	76.03
	Adaptive H.T		94.86	90.83
Stream 4	N.Bayes	1000	56.20	42.93
	H.T		74.28	67.01
	Adaptive H.T		91.37	89.88
Stream 5	N.Bayes	1000	63.39	46.42
	H.T		79.70	70.46
	Adaptive H.T		93.10	91.15

It is worth noting that although the algorithms showed the same accuracy, they did not consume the same amount of time. In Table 9, it can be seen that the simplest algorithm which is NB, consumed the shortest time (3.62 s), while the most complex one which is HAT, consumed the longest time (5.12 s). The differences are not significant however for this stream. We have also measured the number of bytes per hour (RAM/Hour) consumed by each of the training algorithms. Results are also presented in Table 9 where it can be seen that although NB offered the shortest execution time, it consumed more Bytes/Hour than HT.

Table 9. Comparison of model costs for streams 1 for each algorithm

Stream	Classifier	Memory usage (bytes/h)	Training time (s)
Stream 1	NB	66.85	3.62
	HT	56.13	4.06
	Adaptive HT	86.38	5.12

Streams 2–5 results in Table 8, show that NB generally performs poorly, and HAT offers the best accuracies in all experiments, especially for Stream 3, possibly because the data are user-dependent and device-dependent at the same time. The second best performance is obtained with Stream 5, where data are user-independent and device-dependent. It can be concluded that device dependency has slightly greater impact on the HAT performance than user dependency. Finally, it can be also seen that online stream learning using only 3-axis accelerometer signals showed comparable performance to batch learning with engineered features extracted from the signals of two sensors. This is a significant advantage in favor of the online stream learning approach regarding energy efficiency in the SHAR context, since feature engineering and the use of multiple sensors are generally more energy consuming than raw sensor signals of one sensor as discussed in [15]. Finally, since HAT offered the best accuracy for Stream 2–5, we present the results of its memory consumption for each stream in Table 10, where it can be seen that it showed the best Bytes per Hour for Stream 3. It may be explained by the fact this stream contains less data than the others because restricted to one user, holding the same device.

Table 10. Comparison of model costs for streams 2–5 for the best classifier

Classifier	Data stream	Memory usage	
		(MB/h)	(bytes/h)
Adaptive HT	Stream 2	0.58×10^{-4}	60.81
	Stream 3	0.33×10^{-4}	34.60
	Stream 4	1.25×10^{-4}	131.07
	Stream 5	0.79×10^{-4}	82.83

Since classes (or activities) are quite balanced in our constructed streams, we measured the F1-score to get an idea on its evolution over processing time for each activity. For the sake of concision, we present the results of Stream 3 only, since it is the one showing the best accuracy. In Fig. 3: (a) to (f), it can be seen that "Sitting", "Biking" and "Walking" activities have the highest F1-score, with smooth curves over time. "Standing" also shows high F1-score but with less smooth curve. On the opposite, "Walking upstairs" and "Walking downstairs" present the least F1-scores, with fluctuating curves. Curves are colored with the number of learning instances.

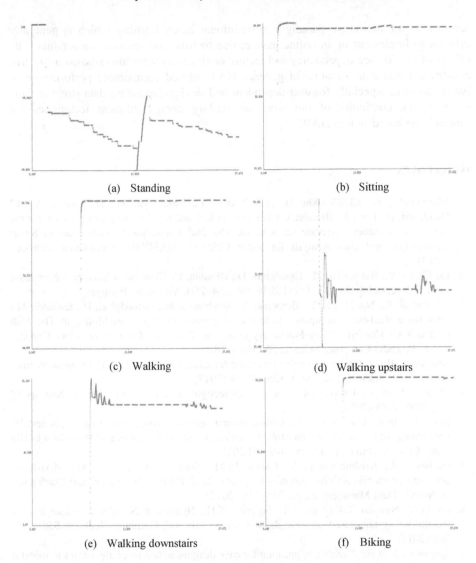

(a) Standing

(b) Sitting

(c) Walking

(d) Walking upstairs

(e) Walking downstairs

(f) Biking

Fig. 3. Evolution curves of F1-score for each activity of stream 3

5 Conclusion

In this paper, we have evaluated two approaches using smartphone sensors data in order to improve the HAR training time obtained in our previous study. In the first approach, we conducted experiments using CPU and GPU platforms in order to compare different classifiers. Results show that HPELM offers the best compromise: accuracy, training time and memory usage for both platforms. HPELM also achieved better performance on two dynamic activities compared to the best SVM model obtained in our previous study. In the second approach, we adopted online stream learning as an alternative paradigm to traditional batch learning which is generally difficult to implement in an online manner due to time and resource constraints. The effects of user/device dependency and feature engineering were also studied using five constructed real data streams. In general, HAT offered comparable performance to batch learning, especially for user-dependent and device-dependent data streams. This study is the continuity of our work on finding energy efficient techniques for Smartphone-based online HAR.

References

1. Kharboucha, A., El Khoukhi, H., Nait Malek, Y., Bakhouya, M., De Florioa, V., El Ouadghiri, D., Latre S., Blondia, C.: Towards an IoT and big data analytics platform for the definition of diabetes telecare services. In: The 2nd International Conference on Smart Applications and Data Analysis for Smart Cities (SADASC'18), Casablanca, Morocco (2018)
2. De Florio, V., Bakhouya, M., Eloudghiri, D., Blondia, C.: Towards a Smarter organization for a Self-servicing Society. DSAI 2016, pp. 254–260, Vila Real, Portugal
3. Elkhoukhi, A., NaitMalek, Y., Berouine, A., Bakhouya, M., Elouadghiri, D., Essaaidi, M.: Towards a Real-time Occupancy Detection Approach for Smart Buildings, in The 15th International Conference on Mobile Systems and Pervasive Computing, Gran Canaria, Spain, Procedia Computer Science, Elsevier (2018)
4. Krawczyk, B.: Active and adaptive ensemble learning for online activity recognition from data streams. Knowl.-Based Syst. **138**, 69–78 (2017)
5. Shoaib, M., et al.: A survey of online activity recognition using mobile phones. Sensors **15** (1), 2059–2085 (2015)
6. Kose, M., Incel, O.D., Ersoy, C.: Online human activity recognition on smart phones. In: Proceedings of the Workshop on Mobile Sensing: From Smartphones and Wearables to Big Data, Beijing, China, pp. 11–15, 16 April 2012
7. Gomes, J.B., Krishnaswamy, S., Gaber, M.M., Sousa, P.A., Ruiz, E.M.: MARS: A personalised mobile activity recognition system. 2012 IEEE 13th International Conference on Mobile Data Management, pp. 316–319 (2012)
8. Lu, D.N., Nguyen, T.T., Ngo, T.T., Nguyen, T.H., Nguyen, H.N.: Mobile online activity recognition system based on smartphone sensors. Adv. Inf. Commun. Technol. **538**, 357–366 (2016)
9. Spinsante, S. et al.: A mobile application for easy design and testing of algorithms to monitor physical activity in the workplace. Mobile Information Systems, 5126816:1–5126816:17 (2016)
10. Chen, Z., Zhu, Q., Soh, Y.C., Zhang, L.: Robust Human Activity Recognition Using Smartphone Sensors via CT-PCA and Online SVM. IEEE Trans. Industr. Inf. **13**, 3070–3080 (2017)
11. Abdallah, Z.S., Gaber, M.M., Srinivasan, B., Krishnaswamy, S.: Adaptive mobile activity recognition system with evolving data streams. Neurocomputing. **150**, 304–317 (2015)
12. Abadi, M., Barham, P., Chen, J., Chen, Z., Davis, A., Dean, J., Devin, M., Ghemawat, S., Irving, G., Isard, M., Kudlur, M.: Tensorflow: large-scale machine learning on heterogeneous distributed systems. arXiv preprint arXiv:1603.04467 (2016)
13. Inoue, M., Inoue, S., Nishida, T.: Deep recurrent neural network for mobile human activity recognition with high throughput. Artif. Life. Rob. **23**, 173–185 (2017)

14. Li, P., Wang, Y., Tian, Y., Zhou, T.S., Li, J.S.: An automatic user-adapted physical activity classification method using smartphones. IEEE Trans. Biomed. Eng. **64**(3), 706–714 (2017)
15. Amezzane, I., Fakhri, Y.El, Aroussi, M., Bakhouya, M.: Towards an efficient implementation of human activity recognition for mobile devices. EAI Endorsed Trans. Context-aware Syst. Appl. **18**(13), e3 (2018)
16. Introduction to Machine Learning with Tensorflow Homepage.: https://www.slideshare.net/PTomeo1/introduction-to-machine-learning-with-tensorflow. Last accessed 17 July 2018
17. Colab   Homepage.   https://colab.research.google.com/notebooks/welcome.ipynb.   Last Accessed 15 May 2018
18. Albadra, M.A.A., Tiuna, S.: Extreme learning machine: a review. Int. J. Appl. Eng. Res. **12** (14), 4610–4623 (2017)
19. López-Fandiño, J., Quesada-Barriuso, P., Heras, D.B., Argüello, F.: Efficient ELM-based techniques for the classification of hyperspectral remote sensing images on commodity GPUs. IEEE J. Sel. Top. Appl. Earth Obs. Remote Sens. **8**(6), 2884–2893 (2015)
20. Akusok, A., Björk, K.M., Miche, Y., Lendasse, A.: High-performance extreme learning machines: a complete toolbox for big data applications. IEEE Access **3**, 1011–1025 (2015)
21. Krawczyk, B., Minku, L.L., Gama, J., Stefanowski, J., Woźniak, M.: Ensemble learning for data stream analysis: a survey. Inf. Fusion **37**, 132–156 (2017)
22. Domingos, P., Hulten, G.: Mining high-speed data streams. In: Knowledge Discovery and Data Mining, pp. 71–80 (2000)
23. Bifet, A., Holmes, G., Kirkby, R.: Data Stream Mining: A Practical Approach, May 2011. http://jwijffels.github.io/RMOA/MOA_2014_04/doc/pdf/StreamMining.pdf
24. Stefanowski, J., Brzezinski, D.: Stream Classification. In: Sammut, C., Webb, G.I. (eds.) Encyclopedia of Machine Learning and Data Mining. Springer, Boston, MA (2017)
25. Stisen, A., Blunck, H., Bhattacharya, S., Prentow, T.S., Kjærgaard, M.B., Dey, A., Sonne, T., Jensen, M.M.: Smart devices are different: Assessing and mitigating mobile sensing heterogeneities for activity recognition. In: Proceedings of the 13th ACM Conference on Embedded Networked Sensor Systems. ACM (2015)
26. Bifet, A., Zhang, J., Fan, W., He, C., Zhang, J., Qian, J., Holmes, G., Pfahringer, B.: Extremely fast decision tree mining for evolving data streams. In Proceedings of the 23rdACM SIGKDD International Conference on Knowledge Discovery and Data Mining, pp. 1733–1742. ACM (2017)
27. Bifet, A., Gavaldà, R.: Adaptive learning from evolving data streams. In 8th InternationalSymposium on Intelligent Data Analysis, pp. 249–260 (2009)
28. MOA Tutorial 1. https://moa.cms.waikato.ac.nz/tutorial-1-introduction-to-moa/. Last Accessed 15 May 2018
29. Bifet, A., Zhang, J., Fan, W., He, C., Zhang, J., Qian, J., Holmes, G., Pfahringer, B.: Extremely fast decision tree mining for evolving data streams. In Proceedings of the 23rd ACM SIGKDD International Conference on Knowledge Discovery and Data Mining, pp. 1733–1742. ACM (2017)
30. Bifet, A., Gavaldà, R.: Adaptive learning from evolving data streams. In 8th International Symposium on Intelligent Data Analysis, pp. 249–260 (2009)
31. Nakano K., Chakraborty B.: Effect of dynamic feature for human activity recognition using smartphone sensors. In: IEEE 8th International Conference on Awareness Science and Technology (iCAST), pp. 539–543, 8 November 2017. IEEE
32. He, K., Sun, J.: Convolutional neural networks at constrained time cost. In Proceedings of the IEEE Conference on Computer Vision and Pattern Recognition, pp. 5353–5360 (2015)

Designing and Developing Multi-agent Systems for Management of Common Renewable Resources

Mohamed Kouissi[1], El Mokhtar En-Naimi[1(✉)],
Abdelhamid Zouhair[2], and Mohammed Al Achhab[3]

[1] LIST Laboratory, FST of Tangier, Tangier, Morocco
{mohamed.kouissi, ennaimi}@gmail.com
[2] ENSA of Al-Houceima, Al Hoceima,, Morocco
zouhair07@gmail.com
[3] ENSA of Tetuan, Tetouan, Morocco
alachhab@gmail.com

Abstract. In this paper, we present a new approach of design and developing multi agent systems. Our approach is based on Model Driven Architecture (MDA), which aims to establish the link between existing agent architectures and the models or meta-models of multi-agent systems that we build from AUML. We have designed a generic and scalable class diagram to develop complex multi-agent systems. The source code of the models is generated by an open source tool called AndroMDA [Elallaoui et al. in Automated model briven testing using AndroMDA and UML2 testing profile in scrum process. Procedia Comput. Sci. 83, 221–228, 2016, 13]. The model and source code will facilitate the design and development of applications to implement and simulate multi-agent models for Management of Common Renewable Resources. This approach allows reuse of the model and generated source code to develop new applications.

Keywords: Multi agents systems · Simulation · Models · Common renewable resources · Model driven architecture (MDA) · Decision making Jade

1 Introduction

Currently, decision-makers still face problems for decision-making. Problem solving using decision aid is based on the use of computer models that enable decision makers to evaluate, analyze and simulate data to streamline their choice. The decision aid system allows to reason about a static or dynamic situation.

The modeling and simulation of interactions between actors concerning renewable natural resources is quite complex in order to find the appropriate models to implement the system. Hence the need for a proposal for a new approach for analysis, design and development of multi agent systems!

In Sect. 2 we present some related works. Section 3 deals with the description of the problem. Section 4 is devoted to the state of the art of which we present some concepts related to our research. Section 5 describes the MDA approach and our

© Springer Nature Switzerland AG 2019
M. Ben Ahmed et al. (Eds.): SCA 2018, LNITI, pp. 572–587, 2019.
https://doi.org/10.1007/978-3-030-11196-0_48

proposed approach. In Sect. 6, we present the architecture of our approach as well as the design and modeling of the model. The last part is devoted to the conclusion and presentation of the perspectives of our future research.

2 Related Work

There is some work that proposes solutions for the management of common renewable resources. Urbani [1] proposes a new approach for the definition of a decision aid system for water management according to the multi-agent paradigm and a geographic information system. The author has created a platform for modeling and simulating interactions between actors regarding renewable natural resources in order to find adequate models of reality. This approach is not generic and can't be applied to other types of renewable resources.

Maalal [2] proposes an approach for designing multi agent systems. It is an application that from a given AUML model [3] allows the generation of source code using the AndroMDA tool to create applications for multi-agent systems. The application is generic and scalable. This approach does not contain a module for modeling and simulation of common renewable resources.

Becu [4] presents the new functionalities of Cormas, a multi agents modeling platform to the management of renewable resources. Cormas facilitate the design of MAS and analysis of scenarios. The author has developed new functionalities such the automatic generation of code to facilitate the implementation of the models, also has developed an extension to control the movement of agents on the spatial grid through real objects.

The combination of the three approaches allows us to propose an adaptive and generic solution for modeling multi agent systems designated for the management of common renewable resources.

3 Problematic

Nowadays, more countries are turning green and they turned to the renewable resources to produce power. So, the demand on the use of renewable resources increases. The management of these resources has become mandatory. For this, we need to design a model that respect the analysis and design phases which is essential to build a multi agent system to manage renewable resources, this consists of using existing analysis methodologies for the design of our model. On the other hand, the current methodologies are limited by a very specific model that is not scalable, and they do not allow the generation of the source code in order to reuse it to implement other multi agent systems, hence the proposal of a new adaptive approach for the design and development of multi agent systems for Management of Common Renewable Resources?

4   State of the Art

4.1   Decision AID Systems

The decision is an act of choice between several alternatives and options to solve a given problem. A Decision AID System is an interactive system that helps decision-makers to extract useful information from raw data, documents, personal knowledge and business models to identify and resolve incidents and take good action and decisions.

The construction of a decision aid system is based on modeling and simulation of computer models allowing to solve decision-making problems.

4.2   Agents

Agent. There is still no consensus on the definition of an agent. But we propose a first definition: "An agent is a software or physical entity to which is assigned a certain mission that it is able to accomplish independently and in cooperation with other agents".

According to Ferber "An agent is an autonomous entity, real or abstract, that is able to act on itself and on its environment, which, in a multi-agent universe, can communicate with other agents and whose behavior is the consequence of observations, knowledge and interactions with other agents".

Agents classification. Agents can be classified into three categories: Reactive agents, Cognitive Agents, and Hybrid Agents. So, the Reactive agents are very simple agents, they don't have a complete representation of their environment. They simply react "reflexively" according to their immediate needs. Concerning the cognitive agents, they are very intelligent with an important reasoning ability. They have a representation of their environment and can communicate with each other by sending messages.

Some problems can't be resolved using only cognitive or reactive agents, hence the proposition of a new category that is a compromise between the two kinds of agents. We are talking here about hybrid architecture.

4.3   Multi Agents Systems (MAS)

A multi-agent system (MAS) is generally defined as a set of agents, operating in an environment, interacting each other and with the environment. Agents interact with each other and may behave towards each other as collaborators, competitors, enemies or strangers (Chaib-Draa et al. [5]).

Agents can act autonomously and asynchronously in a multi agent system, they can communicate, move and modify their behavior.

Multi agent systems can be used in several application fields such as e-commerce, health, industry, distributed information systems, etc.

4.4 Analysis and Design Methodologies for MAS

Quality software depends on several criteria. First, it must respect the specifications and respond the functional needs of the customer, in addition to that, it must be reliable, robust, maintainable, powerful, and easy to use. Software engineering ensures that these criteria are respected according to methodologies that facilitate the construction process of software for its proper functioning. For this, several models for software engineering have been proposed. In addition, when dealing with distributed complex systems, these models are often abandoned for a few reasons: The mechanisms available to represent the organizational structure of the system are insufficient, and the interactions between the different entities are defined in a too rigid way [6]. The use of agents models and multi-agent systems can be a good answer to these problems, because agent-oriented approaches increase our ability to analyze, model, design, and build complex distributed systems.

Currently, there are many methodologies for the analysis and design of multi-agent systems. Here are some examples of existing methodologies (non-exhaustive list):

AUML (Agent Unified Modeling Language): It is an extension of UML which aims to propose a set of notations better adapted to the multi-agent paradigm. It supports all diagrams of the analysis and design phases with a modification of these diagrams to represent the interaction between agents and the internal state of an agent.

In AUML, objects are replaced by agents and roles, notations have been added to represent complex interactions, such as new types of branching in the sequence diagram to take into account the indeterminism of behavior of an agent. AUML is consistent with existing FIPA specifications.

The Agent Modeling Language (AML) [7]: The Agent Modeling Language (AML) is a semi-formal visual modeling language for specifying, modeling, and documenting systems that incorporate features derived from the theory of multi-agent systems. It is specified as an extension to UML 2.0 according to the main OMG modeling frameworks (MDA, MOF, UML and OCL). The ultimate goal of AML is to provide software engineers with a complete, highly expressive and adaptive modeling language for developing commercial software solutions based on multi-agent technologies.

ASPECS (Agent-oriented Software Process for Engineering Complex Systems): It is a step-by-step requirements to code software engineering process based on a holonic organisational metamodel. It integrates design models from both object- and agent-oriented software engineering (OOSE and AOSE) [8]. The main goal of ASPECS is to develop societies of holonic (or not) multi agent systems. ASPECS uses UML as a modelling language.

Voyelles: It was proposed by Yves Demazeau, and it is based on the decomposition of the view of a multi-agent system according to five dimensions or models: Agent, Environment, Interaction, Organization and user. A multi agent system according to the Voyelles methodology is a set of agents (A) that evolves in an environment (E), subject to an organization (O), interacting with each other (I), with their roles, and which is centered on the user (U).

Gaia: Is a methodology where the multi agent system is seen as an organization composed of interacting roles. It is designed for the analysis and design of all types of multi-agent systems and highlights two levels, a macro-level that focuses on modeling a group of agents and a micro-level that focuses on the agent. GAIA considers a multi-agent system as a structure composed of one or more organizations.

INGENIAS: The INGENIAS methodology, is a general and flexible approach for the development of multi-agent systems. It is based on the division of the problem into several aspects that form the different views of the system. INGENIAS defines meta-models that describe the different aspects of a multi-agent system and their relationships. The INGENIAS methodology provides researchers with a development environment called IDK (INGENIAS Development Kit).

MaSE [9]: (Multi-agent System Engineering) is modeling MAS in terms of objectives, roles, agents, tasks and interactions. it considers an agent as a particular object, intelligent or not, with a coordinating ability through the conversation with other agents. The MaSE methodology consists of an analysis phase and a design phase. The analysis focuses on producing the roles that the system must fulfill in order to achieve its goals.

ADELEFE: (Toolkit to develop software with emergent functionality) The aim of the ADELFE toolkit is to guide developers during the development of adaptive multi-agent systems. ADELEFE offers a process, notation and tools based on the UML notation. These systems are based on cooperative agents that exhibit or possess social behaviors: composed by one or more cooperative agents who observe cooperation rules and who have a representation of the world. Dynamic aspects can be modeled using interaction diagrams.

All these methodologies presented above are recent. They consider only certain aspects of the life cycle, usually analysis and design. For the implementation, most of the methodologies are conditioned by the use of a particular agent architecture. However, there isn't yet a sufficiently mature standard methodology that addresses all necessary aspects to define an MAS, nor the entire development cycle for this kind of software.

The vast majority of agent-based applications are built without using reusable agent components and can't be generalized [10, 11].

5   The MDA Approach

MDA (Model Driven Architecture) is a model-driven engineering process. It is a set of techniques for modeling and transformation of computation independent business models (computation independent model, CIM) into a platform independent model (PIM) and finally the transformation into a specific platform model (PSM) for the concrete implementation of the system. The MDA allows to take into account, separately, both the business and the technical aspect of the system to be developed, through modeling. The source code of the system will be generated automatically from the models.

Model Driven Architecture provides long-term Flexibility for the Implementation, Integration, Maintenance, Testing, and Simulation, Fig. 2 Represents the MDA Approach [12] (Fig. 1).

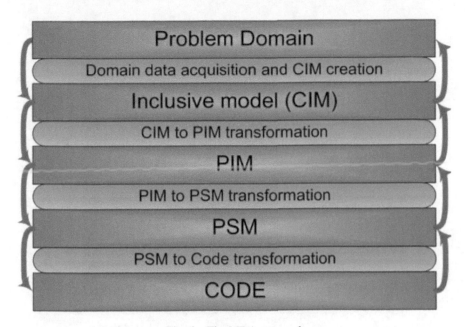

Fig. 1. The MDA approach

The CIM is independent of any computer system, it describes the business concepts, the management rules, the processes to follow and the situation in which the system is used. PIM is a design model that represents the business logic of the system and describes it independently of any platform and any used technology to deploy an application. The PSM is platform dependent and is used to generate the source code for the target technical platforms.

Among the MDA tools we find AndroMDA which is a source code generation framework, it generates the source code for any desired target platform from a business model written in UML language [13]. AndroMDA can automatically translate high-level business specifications into a robust and quality code and save time implementing applications.

The diagram below illustrates the different application layers, for example, Java technologies supported by AndroMDA [2].

6 Our Proposed Approach

Our approach is based on Model Driven Architecture (MDA) which aims to establish the link between existing agent architectures and the models or meta-models of multi-agent systems that we build from AUML. Our idea is to create an application to generate code source as an API from a model to develop multi agent systems applications for the management of renewable resources. The model and source code will facilitate the design and development of applications to implement and simulate multi-agent models, they are also generic and scalable.

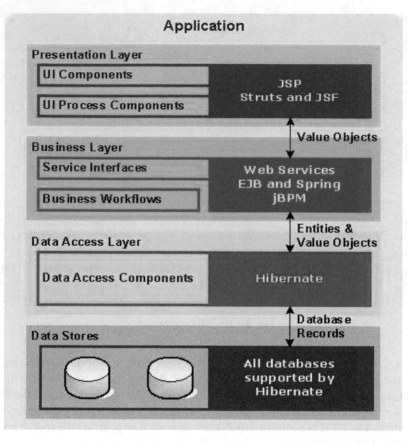

Fig. 2. Application layers supported by AndroMDA

Our approach offers positive points as follows:

✓ Reusing the model and source code for new applications.
✓ Reduce development costs.
✓ Reduce analysis and modeling time.
✓ A generic code to support any type of renewable resources.

The figure below presents our proposed architecture to implement the model (Fig. 3).

This architecture allows us to build a decision aid system for the management of Common Renewable Resources. The system contains different modules which allow to:

• Build a multi-agent model of the real system.
• Start the simulation of a given model following specific scenarios.
• Create and manage scenarios.
• Manage all developed multi agents models thanks to "Manage Models" Module.

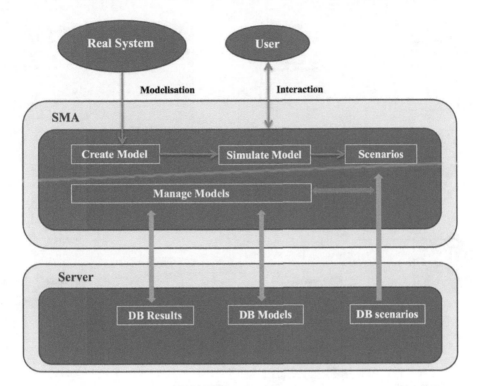

Fig. 3. General architecture

- Store the obtained results and scenarios during the simulation into a remote data base.

7 Design and Modeling

7.1 Description of the AUML Class Diagram

Entity: It is a class that represents the spatial entity that contains the agent or a set of agents, it takes two sections, Environment Perception and Attributes. The attributes section represents all information about the environment (Fig. 4).

The entity class contains several methods to start, perceive information from agents linked to it, and also modify their status.

Agent Class: This is a main class of our model, which allows to instantiate an Agent with its properties. This class takes three sections, Roles, Attributes, and Perception. Roles represent Agent functionalities, attributes represent all information that an agent should possess, and finally the perception of an agent in relation to its environment and also to other agents.

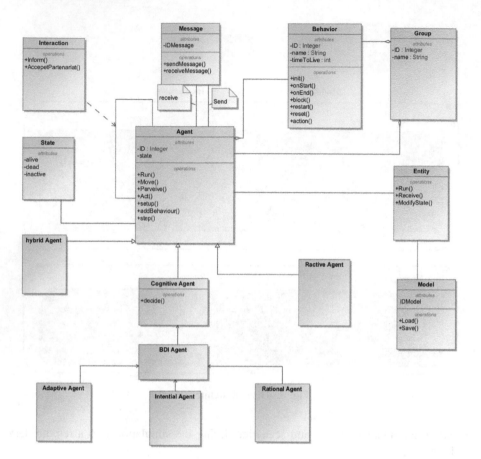

Fig. 4. Generic class diagram for MAS

An agent interacting with its environment and also with other agents contains several methods to start, collect information and perform actions. Run(), Perceive() and Act().

Reactive agent: This class extends the Agent class.

Cognitive agent: This class extends the Agent class, it possess a representation of its environment. It contains a method called Decide() which allow to perform actions to make decisions based on past experiences.

Hybrid Agent: This class extends the Agent class, it represents an hybrid agent behavior that is a compromise between both reactive and cognitive agents.

BDI Agent: This class extends the Cognitive Agent class,

Message: It saves exchanged messages between agents.

State: This class contains agent status, an agent can be active, inactive, alive, etc....

Behavior: It is used to model several tasks which an agent can achieve, it allows to perform many actions « start() » « block() » « restart() » « reset() » .

Interaction: This class save actions that occur between agents, such as the collaboration request, also information about the environment.

Group: A group is a set of agent. An agent may have several behaviors, and each behavior is local to a group.

Model: This class is used to save models and data obtained after simulation.

7.2 Use Case Diagram Example

This use case defines the interaction between the user and our system to achieve a specific goal. The user is a decision maker. He can create spatial entities or environment, create and affect agents into entities. The user can also create models and start the simulation of a given model following scenarios and store the results (Fig. 5).

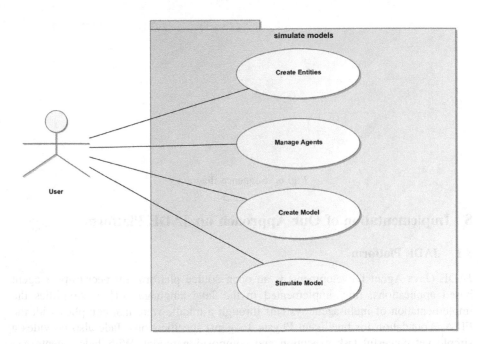

Fig. 5. Use case diagram

7.3 Sequence Diagram

This Sequence Diagram illustrates the interaction and the message exchanged between the user and different objects in the system. The user is a decision maker. First of all, he creates an Entity, then he can create and manage a set of agents in this Entity. After that, the user creates a model and start the simulation. The data obtained during the simulation is compared with a real data source which is stored in a remote data base. The user can validate or abort the model according to the obtained results (Fig. 6).

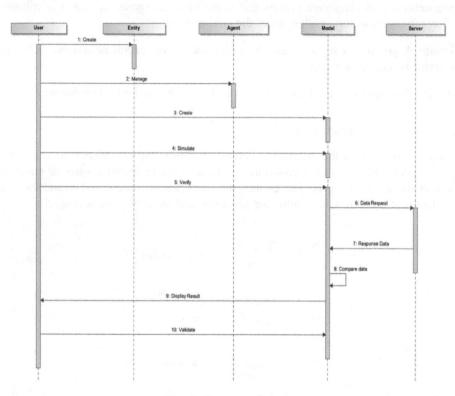

Fig. 6. Sequence diagram

8   Implementation of Our Approach on JADE Platform

8.1   JADE Platform

JADE (Java Agent DEvelopment) is an open source platform for peer-to-peer agent based applications, fully implemented in the Java language [14]. It simplifies the implementation of multi-agent systems through a middle-ware that complies with the FIPA (Foundation for Intelligent Physical Agent) specifications. Jade also provides a simple yet powerful task execution and composition model. With Jade, agents can communicate with each other using asynchronous messages. JADE agents can be deployed transparently on different Java environments such as Android devices.

To develop our model, we opted for Jade platform, it offers a lot of benefits. We cite below some of them:

- Jade is open source.
- It is a free software.
- It is developed in Java and it works in all operating systems.
- It contains a graphical toolkit to manage the platform of agents.
- It provides several methods to perform the basic tasks of the agents.

8.2   Implementation of Our Approach

To implement our approach, we had first set up our development environment for AndroMDA. we have installed various programs like Java, Eclipse as we will focus on generating a Java application, Maven to build and deploy applications generated by AndroMDA. After that, we have downloaded and installed the AndroMDA application plugin.

The screen bellow presents the installation and the generation of AndroMDA application (Fig. 7).

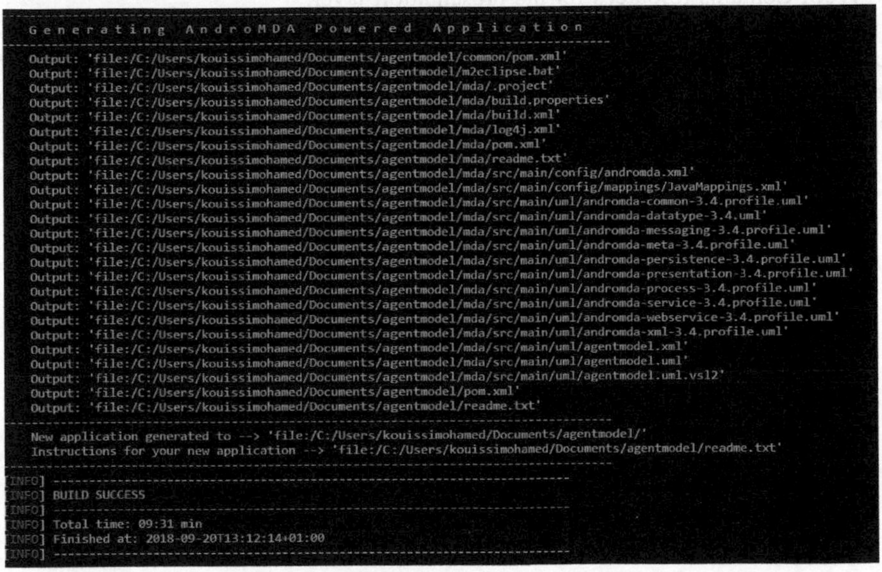

```
Generating  AndroMDA  Powered  Application
-------------------------------------------------------------------
Output: 'file:/C:/Users/kouissimohamed/Documents/agentmodel/common/pom.xml'
Output: 'file:/C:/Users/kouissimohamed/Documents/agentmodel/m2eclipse.bat'
Output: 'file:/C:/Users/kouissimohamed/Documents/agentmodel/mda/.project'
Output: 'file:/C:/Users/kouissimohamed/Documents/agentmodel/mda/build.properties'
Output: 'file:/C:/Users/kouissimohamed/Documents/agentmodel/mda/build.xml'
Output: 'file:/C:/Users/kouissimohamed/Documents/agentmodel/mda/log4j.xml'
Output: 'file:/C:/Users/kouissimohamed/Documents/agentmodel/mda/pom.xml'
Output: 'file:/C:/Users/kouissimohamed/Documents/agentmodel/mda/readme.txt'
Output: 'file:/C:/Users/kouissimohamed/Documents/agentmodel/mda/src/main/config/andromda.xml'
Output: 'file:/C:/Users/kouissimohamed/Documents/agentmodel/mda/src/main/config/mappings/JavaMappings.xml'
Output: 'file:/C:/Users/kouissimohamed/Documents/agentmodel/mda/src/main/uml/andromda-common-3.4.profile.uml'
Output: 'file:/C:/Users/kouissimohamed/Documents/agentmodel/mda/src/main/uml/andromda-datatype-3.4.uml'
Output: 'file:/C:/Users/kouissimohamed/Documents/agentmodel/mda/src/main/uml/andromda-messaging-3.4.profile.uml'
Output: 'file:/C:/Users/kouissimohamed/Documents/agentmodel/mda/src/main/uml/andromda-meta-3.4.profile.uml'
Output: 'file:/C:/Users/kouissimohamed/Documents/agentmodel/mda/src/main/uml/andromda-persistence-3.4.profile.uml'
Output: 'file:/C:/Users/kouissimohamed/Documents/agentmodel/mda/src/main/uml/andromda-presentation-3.4.profile.uml'
Output: 'file:/C:/Users/kouissimohamed/Documents/agentmodel/mda/src/main/uml/andromda-process-3.4.profile.uml'
Output: 'file:/C:/Users/kouissimohamed/Documents/agentmodel/mda/src/main/uml/andromda-service-3.4.profile.uml'
Output: 'file:/C:/Users/kouissimohamed/Documents/agentmodel/mda/src/main/uml/andromda-webservice-3.4.profile.uml'
Output: 'file:/C:/Users/kouissimohamed/Documents/agentmodel/mda/src/main/uml/andromda-xml-3.4.profile.uml'
Output: 'file:/C:/Users/kouissimohamed/Documents/agentmodel/mda/src/main/uml/agentmodel.xml'
Output: 'file:/C:/Users/kouissimohamed/Documents/agentmodel/mda/src/main/uml/agentmodel.uml'
Output: 'file:/C:/Users/kouissimohamed/Documents/agentmodel/mda/src/main/uml/agentmodel.uml.vs12'
Output: 'file:/C:/Users/kouissimohamed/Documents/agentmodel/pom.xml'
Output: 'file:/C:/Users/kouissimohamed/Documents/agentmodel/readme.txt'
-------------------------------------------------------------------
New application generated to --> 'file:/C:/Users/kouissimohamed/Documents/agentmodel/'
Instructions for your new application --> 'file:/C:/Users/kouissimohamed/Documents/agentmodel/readme.txt'

[INFO] -----------------------------------------------------------
[INFO] BUILD SUCCESS
[INFO] -----------------------------------------------------------
[INFO] Total time: 09:31 min
[INFO] Finished at: 2018-09-20T13:12:14+01:00
[INFO] -----------------------------------------------------------
```

Fig. 7. Generating AndroMDA application

AndroMDA plugin generate a various folders and files as shown in Fig. 8.

MDA: The MDA folder contains our UML model. It is configured to generate the needed files to build the application.

Common: The Common folder contains resources and classes shared between other sub-projects.

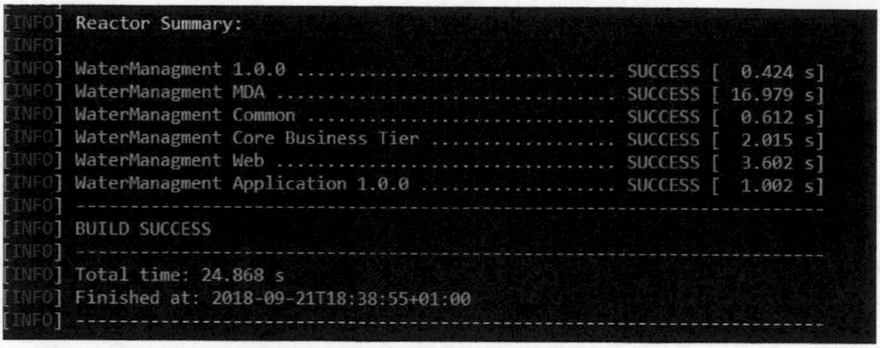
[INFO] Reactor Summary:
[INFO]
[INFO] WaterManagment 1.0.0 .... SUCCESS [0.424 s]
[INFO] WaterManagment MDA .... SUCCESS [16.979 s]
[INFO] WaterManagment Common .... SUCCESS [0.612 s]
[INFO] WaterManagment Core Business Tier .... SUCCESS [2.015 s]
[INFO] WaterManagment Web .... SUCCESS [3.602 s]
[INFO] WaterManagment Application 1.0.0 .... SUCCESS [1.002 s]
[INFO] --
[INFO] BUILD SUCCESS
[INFO] --
[INFO] Total time: 24.868 s
[INFO] Finished at: 2018-09-21T18:38:55+01:00
[INFO] --

Fig. 8. Folders generated by AndroMDA

Core: Contains resources and classes that use frameworks like Spring and Hibernate. It contains also the entity classes and data access object.

Web: Contains resources for the presentation layer.

To model our approach in UML, we have installed MagicDraw19.0 [15], this tool can export our model into a format that AndroMDA can understand. so, we started the process to design and validate our class diagram to generate then the source code of our application. The result is shown in the Fig. 9.

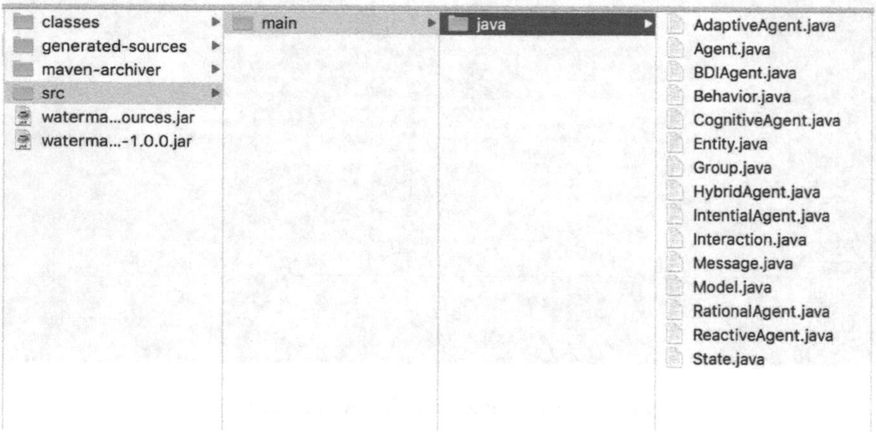
classes
generated-sources
maven-archiver
src
waterma...ources.jar
waterma...-1.0.0.jar
main
java
AdaptiveAgent.java
Agent.java
BDIAgent.java
Behavior.java
CognitiveAgent.java
Entity.java
Group.java
HybridAgent.java
IntentialAgent.java
Interaction.java
Message.java
Model.java
RationalAgent.java
ReactiveAgent.java
State.java

Fig. 9. Generated code source by AndroMDA

The source code can be easily modified by the developer to implement the desired classes and methods by using Eclipse (Fig. 10).

After import of our project into Eclipse, we added the Jade library to the CLASSPATH environment variable, so some of our classes can extend the Jade classes as presented in Fig. 11.

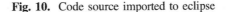

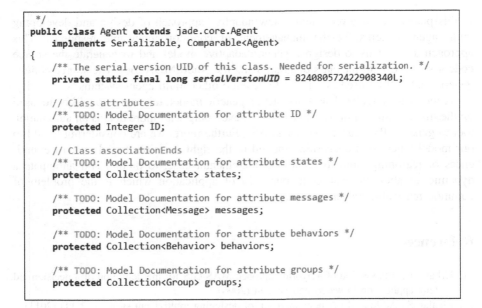

Fig. 10. Code source imported to eclipse

```java
*/
public class Agent extends jade.core.Agent
    implements Serializable, Comparable<Agent>
{
    /** The serial version UID of this class. Needed for serialization. */
    private static final long serialVersionUID = 824080572422908340L;

    // Class attributes
    /** TODO: Model Documentation for attribute ID */
    protected Integer ID;

    // Class associationEnds
    /** TODO: Model Documentation for attribute states */
    protected Collection<State> states;

    /** TODO: Model Documentation for attribute messages */
    protected Collection<Message> messages;

    /** TODO: Model Documentation for attribute behaviors */
    protected Collection<Behavior> behaviors;

    /** TODO: Model Documentation for attribute groups */
    protected Collection<Group> groups;
```

Fig. 11. Extend the agent jade class

Below the result of the creation of three agents (AgentOne, AgentTwo, Agent-Three) added to the main container of Jade Platform (Fig. 12).

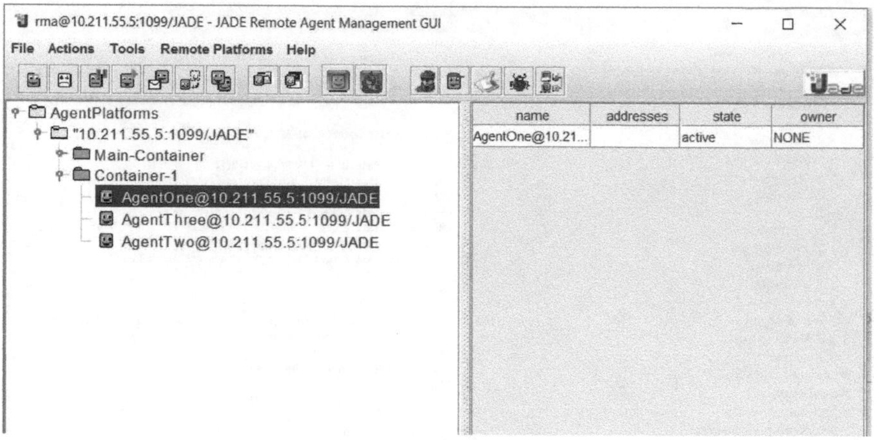

Fig. 12. Jade container

9 Conclusion and Perspectives

In this paper, we have presented a new adaptive approach of design and developing multi agent systems for the management of common renewable resources. This approach allowed us to design a generic/adaptive model and to generate the source code associated with the class diagram using the AndroMDA tool. The source code is generic and will be used as an API to develop other multi agent systems.

After we have realized, as a first step, a generic model, the next step will be focused on the implementation of our approach by developing with more details an automation tool to generate the source code as an API. Furthermore, we are considering to add to our model a new part concerning the aid to the right decision, based on past experiences or reasoning from dynamic and incremental case, to predict and anticipate a dynamic situation in relation to our field of application which is the problem of common renewable resources management.

References

1. Urbani, D.: Elaboration of a hybrid approach MAS-GIS for the definition of a decision aid system, application to water management (2006)
2. Maalal, S., Addou, M.: A new approach of designing multi-agent systems. 2(11) (2011)
3. Bouquet, F., Sheeren, D., Becu, N., Gaudou, B., Lang, C., Marilleau, N., Monteil, C: Formalism of Description for Agents Models, pp. 37–72 (2015)
4. Becu, N., Bommel, P., Le Page, C., Bousquet, F.: Cormas, A Multi-agent Platform for Collectively Designing Models and Interacting with Simulations (2016)
5. Chaib-Draa, B., Jarras, I.: Overview on Multi-agent Systems (2002)
6. Jennings, N.R.: On agent-based software engineering. Artif. Intell. **117**, 277–296 (2000)
7. Trencansky, I., Cervenka, R.: Agent Modeling Language (AML): A comprehensive approach to modeling MAS. Informatica **29**(4), 391–400 (2005)

8. Cossentino, M., Gaud, N., Hilaire, V., Galland, S., Koukam, A.: ASPECS: An Agent-Oriented Software Process for Engineering Complex Systems, vol. 20, pp. 260–304 (2010)

9. Fethi, M., Kamel, Z., Khaled, G., Pierre, B.: Modeling of a multi-agent system for the resolution of a problem of vehicle tours in an emergency situation. In: 9th International Conference on Modeling, Optimization and Simulation—MOSIM'12

10. En-Naimi, E.M.: Module: Multi-Expert Systems & Multi-Agents Systems (MAS). Masters: SIAD, MBD and SIM in S3, from 2015/2016 to 2018. FST of Tangier, UAE, Morocco

11. Argente, E., Julian, V., Botti, V.: Multi-agent system development based on organizations. Electron. Notes Theor. Comput. Sci. **150**, 55–71 (2006)

12. Lopata, A., Ambraziunas, M.: Knowledge Subsystem's Integration Into MDA Based Forward and Reverse is Engineering (2010)

13. Elallaoui, M., Nafil, K., Touahni, R. Messoussi, R.: Automated model driven testing using AndroMDA and UML2 testing profile in scrum process. Procedia Comput. Sci. **83**, 221–228 (2016)

14. Jade Framework, http://jade.tilab.com

15. MagicDraw, https://www.nomagic.com/services/training

16. Nfaoui, E.H.: Distributed Decision Support Architecture and Proactive Simulation in Supply Chains: A Multi Agent Approach (2008)

17. Model Driven architecture, http://www.omg.org/mda/

18. Unified Modeling Language, http://www.omg.org/spec/UML/2.0/

19. AndroMDA, http://www.andromda.org/

20. Kaliappan, P.S., Koenig, H.: Designing and verifying communication protocols using model driven architecture and spin model checker. J. Softw. Eng. Appl. **1**, 13–19 (2008)

21. Ferber, J.: Multi-agents systems: general view. Tech. Comput. Sci. **16**(8), 979–1012 (1997)

22. Bousquet, F.: Accompaniment Modeling, Multi-Agents Simulation and Managment of Natural and Renewable Ressources

Novel Network IDS in Cloud Computing Based on Optimized Back Propagation Neural Network Using a Self-adaptive Genetic Algorithm

Zouhair Chiba[✉], Noreddine Abghour, Khalid Moussaid,
Amina El omri, and Mohamed Rida

LIMSAD Labs, Faculty of Sciences, Hassan II University of Casablanca, 20100
Casablanca, Morocco
chiba.zouhair@gmail.com, {NOREDDINE.ABGHOUR, KHALID.
MOUSSAID, AMINA.ELOMRI, MOHAMED.RIDA}@univh2c.ma

Abstract. Nowadays, Cloud Computing (CC) had become an integral part of IT industry. It represents the maturing of technology and is a pliable, cost-effective platform which provides business/IT services over the Internet. Although there are several benefits of adopting this paradigm, there are some significant hurdles to it and one of them is security. In fact, due to the distributed and open nature of the cloud, resources, applications and data are vulnerable and prone to intrusions that affect confidentiality, availability and integrity of Cloud resources and offered services. Network Intrusion Detection System (NIDS) has become the most commonly used component of computer system security and compliance practices that defends network accessible Cloud resources and services from various kinds of threats and attacks, while maintaining performance and service quality. In this work, in order to detect intrusions in CC environment, we propose a novel anomaly NIDS based on Back Propagation Neural Network (BPNN) classifier optimized using a Self-Adaptive Genetic Algorithm (SAGA). SAGA consists of a standard Genetic Algorithm improved by means of an Adaptive Genetic Algorithm, namely Adaptive Mutation Algorithm. Since, Learning rate and Momentum term are among the most relevant parameters that affect the performance of BPNN classifier, we have employed SAGA to find the optimal values of these two critical parameters, which ensure high detection rate, high accuracy and low false alarm rate. Our novel NIDS is called "ANIDS BPNN-SAGA" (Anomaly NIDS optimized by using Self-Adaptive Genetic Algorithm). The CloudSim simulator and KDD CUP' 99 dataset are used to verify the proposed system. The obtained experimental results have demonstrated the superiority of the proposed approach in comparison with state-of-the-art methods.

Keywords: Cloud computing · Anomaly detection · Network intrusion detection system · Back propagation neural network · Optimization
Genetic algorithm · Adaptive genetic algorithm · Adaptive mutation algorithm
Learning rate · Momentum term

© Springer Nature Switzerland AG 2019
M. Ben Ahmed et al. (Eds.): SCA 2018, LNITI, pp. 588–602, 2019.
https://doi.org/10.1007/978-3-030-11196-0_49

1 Introduction

Nowadays, one of the fastest growing and most used technologies in the IT field is Cloud computing. Cloud computing (CC) refers to an information technology (IT) paradigm that delivers convenient, on-demand ubiquitous network access to shared pools of configurable computing resources (e.g., networks, servers, storages, and applications) as "service" via the internet for satisfying computing demands of users.

The architecture of CC characterized by distributed technology and big data applies such technology as virtualization and multi-tenancy, which bring vulnerabilities, security and sharing risks specific to cloud computing. Also, uploading sensitive data to public cloud storage services poses security risks such as accessibility, confidentiality and integrity to organizations that use CC. Moreover, non-stop cloud services have caused high levels of abuse and intrusion. Therefore, to overcome these issues, intrusion detection system (IDS) besides firewall are used to protect resources and data in the cloud environment from attacks and malicious activities, while maintaining performance and service quality. Firewall cannot be used to detect insider attack, and some of attacks, like denial of service (DoS) and distributed denial of service (DDoS) are too complex for firewalls. Therefore, it is the inevitable to a developed intrusion detection system in a cloud environment [1]. Since the network is the backbone of Cloud, and hence vulnerabilities in network directly affect its security, the presence of network IDS (NIDS) in a Cloud network plays a vital part in attack mitigation, and it has become an integral part of a secure Cloud. NIDS represents a protection layer that detects real-time aggressive behavior, malicious activity or suspicious pattern by monitored the network traffic and takes corrective measure to avoid or minimize the occurrence of attacks. As result for that, data integrity will be preserved from attacks, security and safety of the network will be maintained, the administrative capacity of the system administrator's security will be reinforced and operational performance of the system will be optimized [2].

In recent years data mining techniques are used for intrusion detection in wide range because the automation of intrusion detection. One of the data mining techniques that has successfully used in solving complex practical problems is neural network. Artificial neural networks have the ability to solve several problems confronted by the other present techniques used in intrusion detection [2].

There are three advantages of intrusion detection based on neural network [3]:

- Neural network provides elasticity in intrusion detection process, where the neural network has the ability to analyze and ensure that data right or partially right. Likewise, neural network is capable of performing analysis on data in nonlinear fashion;
- Neural network has the ability to process data from a number of sources in a nonlinear fashion. This is very important especially when coordinated attack by multiple attackers is conducted against the network;
- Neural network is characterized by high speed in processing data.

According to survey in [4] Back Propagation Neural Network (BPNN) has good detection rate as compared to other neural network techniques.

In this work, BPNN classifier was used to classify attacks and a Self-Adaptive Genetic Algorithm was employed to optimize and increase the accuracy of this classifier. Self-Adaptive Genetic Algorithm constitutes of a standard Genetic Algorithm improved by incorporating to it an Adaptive Genetic Algorithm, namely Adaptive Mutation Algorithm.

Our major purpose is to build a powerful and an efficient network intrusion detection system called "ANIDS BPNN-SAGA", based on anomaly approach using a very popular soft computing tool widely used for intrusion detection namely Back propagation Neural Network (BPNN), optimized by means of a Self-Adaptive Genetic Algorithm (SAGA). The main goals of our developed system is to reduce impact of network attacks (known attacks, and unseen attacks), while ensuring higher detection rate, lower false positive rate, higher accuracy and higher precision with an affordable computational cost. Further, we have chosen to position the proposed IDS on Front-End and Back-End of the Cloud, to detect and stop attacks in real time impairing the security of the Cloud Datacenter.

The rest of this paper is organized as follows: Sect. 2 introduces the literature surrounding network IDS for the Cloud Computing. Next, Sect. 3 gives positions of the proposed system in a Cloud network. Section 4 explains the background of this chapter such BPNN, GA and AGA. Section 5 presents the proposed system in detail, describes its work, and provides the framework of our model. Experimental results and analysis are given in Sect. 6. Finally, Sect. 7 ends with the conclusion and the scope of future work.

2 Literature Review

Lo et al. [5] have developed the co-operative intrusion detection model for the grid and cloud computing in which the IDS are distributed among the nodes of the grid and alert other nodes when an attack occurs. There are two principal advantages of the proposed system; in one hand, this approach helps other nodes in avoiding the same attacks from occurring, and on the other hand, it prevents a single point of failure since the IDSs are distributed across the cloud.

Modi et al. [6] have integrated a signature Apriori based NIDS to Cloud. Signature Apriori takes network packets and known attack signatures as input and generates new derived rules that are updated in the Snort. Therefore, Snort is able to detect known attacks and derivative of known attacks in the Cloud. This approach improves the efficiency of Snort. However, it cannot detect unknown attacks.

Navimipour and Hajimirzaei [7] have developed new intrusion detection system (IDS) based on a combination of a multilayer perceptron (MLP) network, artificial bee colony (ABC) and Fuzzy C-means (FCM) clustering algorithm. An ANN can operate alone in an IDS, but the combination of ANN, ABC and fuzzy clustering makes an IDS more powerful and efficient. The proposed method involves three phases, which are training, validation and testing. The homogeneous subsets of training data are prepared with fuzzy clustering. Consequently, the training speed rate is enhanced by separating the dataset into uniform subsets. During training phase, after performing the clustering, MLP network with backpropagation (BP) algorithm is used to build and train the IDS model. The steepest descent method is adapted to the BP learning rule. The weight and threshold value of the network are adjusted by BP to reach a low-error sum of squares. With BP, the gradient descent method is used to balance the weight values of all layers. Generally, the initial weights of the network are generated in random way within a certain interval; the training starts with this starting point and proceeds step by step to a minimum error. The ABC helps the MLP to determine ideal/optimal values for linkage weights and biases more rapidly. The performance of the system is precisely assessed in the validation phase. Finally, in the testing phase, intrusion detection is processed by passing the test data through the previously trained model. The CloudSim simulator and NSL-KDD dataset were used to verify the proposed model. Various evaluation criteria, such Mean absolute error (MAE), root mean square error (RMSE) and the kappa statistic were employed to compare similar IDSs with the proposed system. The obtained results have demonstrated the superiority of the proposed approach in comparison with other state-of-the-art methods.

Sharma et al. [8] have presented an intrusion detection system called Hypervisor Detector implemented at the hypervisor layer for detecting the intrusion behavior of the cloud network. The Hypervisor Detector is designed with a hybrid algorithm which is a fusion of WLI-FCM clustering algorithm and Back Propagation Artificial Neural Network to improve the detection accuracy of this cloud intrusion detection system. This model works in three phases. In first phase, the fuzzy clustering (WLI-FCM) module is used to split the large dataset into small distinctive clusters in order to improve the learning capability of ANN (Artificial Neural Network). Then, in second phase, the resultant clustered result is given as input to the training algorithm for learning process. A back propagation neural network is used for the training goal. Thus, different ANN modules are trained according to their cluster values. In the last phase, the results of various ANN from the second phase are combined to obtain the result. The proposed system is implemented in Cloudsim 3.0 and compared with K-means and classic FCM by means of different evaluation criterions such as number of clusters, number of features used. The DARPA's KDD cup dataset 1999 was used for simulation. The obtained performance results demonstrate that the proposed WLI-ANN outperforms the K-means and the classic FCM algorithms in terms of TPR, Accuracy and FPR. Hence, the proposed Hypervisor Detector is more convenient for detecting the anomalies and various attacks with high detection accuracy and low false alarm rate.

Ghosh et al. [9] have designed a network intrusion detection system to detect attacks and malicious activities in Cloud environment. The proposed IDS includes two stages. The first stage consists of creation of a feature subset by using a novel algorithm called BCS-GA which combines the advantages of Binary Cuckoo search algorithm (BCS) and Genetic Algorithm (GA). The proposed BCS-GA algorithm was applied on NSL-KDD training dataset to remove several irrelevant features, in order to reduce the training time and memory storage space required for such high dimensional dataset. Thus, initial NSL-KDD dataset contains 41 features, but after applying BCS-GA algorithm, it successfully reduced to 16 features. In the second stage, Neural Network classifier was trained by the reduced training dataset using 16 features. Thereafter, classification accuracy of that classifier was tested by means of a separate reduced testing dataset. Experimental results indicate that the proposed IDS produces 78.229% of accuracy.

Mehibs and Hakims [1] have proposed a network intrusion detection system in cloud computing environment based on Fuzzy c mean (FCM) algorithm to detect intrusion events from normal behavior, and thus, defend network accessible Cloud resources and services from various kinds of intrusions. The aim of fuzzy clustering algorithm is partition the dataset into two clusters, one for attack and another for normal behavior. The proposed module consists of two-phases; the first phase is training phase where the optimum cluster centers obtain. The second phase is testing phase which used the cluster centers result from training stage to determine the cluster of new unseen samples. The KDD cup 99 dataset was used to training and testing FCM algorithm. The experiences conducted by authors' show that the proposed system is characterized by a high detection rate with low false positive alarm.

3   Positioning of the Proposed System in a Cloud Network

The aim of our proposed IDS is to detect intruders and suspicious activities in and around the Cloud Computing environment by monitoring network traffic, while maintaining confidentiality, availability, integrity and performance of cloud resources and offered services. It allows detecting and stopping attacks in real time impairing the security of the Cloud Datacenter.

As shown in Fig. 1, we propose to place our NIDS on two strategic positions:

- **Front-End of Cloud**: Placing NIDS on front end of Cloud helps to detect network intrusions or attacks coming from external network of Cloud, launched from zombie hosts or by hackers connected to the Internet who attempt to bypass the firewall in order to access the internal cloud, which can be a private one. Therefore, NIDS plays the role of the second line of defense behind the firewall to overcome its limitations and acts as an additional preventive layer of security.
- **Back-End of Cloud**: Positioning NIDS sensors on processing servers located at back end of the Cloud helps to detect intrusions occurring on its internal network. In a virtual environment, we have many virtual machines on the same physical server, and they can inter-communicate through the virtual switch without leaving the physical server. Thus, network security devices on the LAN can't monitor this network traffic; if the traffic does not need to pass through security appliances primarily a firewall, therefore, a loophole for all kinds of security attacks will be opened. Hence, the starting point of an attacker/hacker is compromising only one VM, and using it as a springboard to take control of the other VMs within the same hypervisor. This is generally done without being monitored or detected, giving the attacker a huge hack domain. Moreover, the virtual environment is exposed to various threats and risks, centered mostly on the hypervisor; Hyper jacking, VM escape, VM migration, VM theft and Inter-VM traffic.

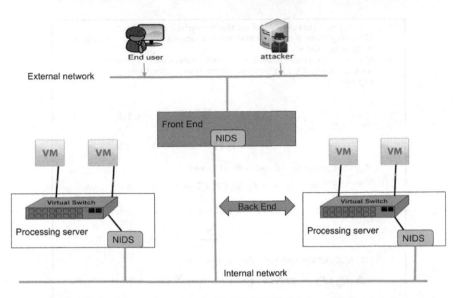

Fig. 1. Positions of proposed ANIDS BPNN-GA in a cloud network

4 Related Background

This section provides the necessary background to understand the problem in hand. First subsection shed the light on BPNN. The second subsection introduces and describes the operation of a standard GA and the last subsection briefly presents Adaptive Genetic Algorithms, especially Adaptive Mutation Algorithm used in our work.

4.1 Back Propagation Neural Network (BPNN)

Back Propagation Neural Network (BPNN) is a special type of neural network. It is also named error back propagation neural network, and it is a multilayer feed forward neural network, which uses Multi-layer Perception as network architecture and Back Propagation Learning Algorithm as training or learning algorithm. Back Propagation network learns by example. You provide the algorithm examples of what you want the network to do and it changes the network's weights so that, when training is achieved, it will give you the required output for a particular input [10]. BPNN with a strong self-learning ability, be able to do adaptive calculations, it is a large scale nonlinear adaptive systems. For every kinds of neural networks, BPNN is relatively mature [11].

Role of Learning rate and Momentum term in Learning phase of BPNN

As it is shown in Fig. 2, to adjust the weights of connections between neurons in neural network, Back Propagation Learning Algorithm uses Eqs. (5) and (6).

In those equations, this algorithm uses two prominent and crucial parameters namely Learning rate and Momentum term.

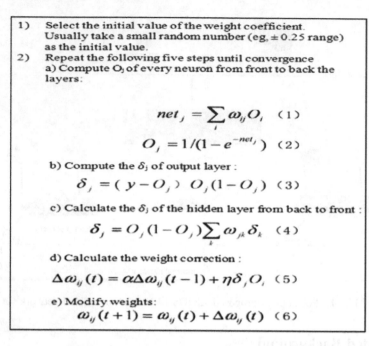

1) Select the initial value of the weight coefficient. Usually take a small random number (eg. ± 0.25 range) as the initial value.

2) Repeat the following five steps until convergence
   a) Compute O_j of every neuron from front to back the layers:

$$net_j = \sum_i \omega_{ij} O_i \quad (1)$$

$$O_j = 1/(1 - e^{-net_j}) \quad (2)$$

b) Compute the δ_j of output layer :

$$\delta_j = (y - O_j) \, O_j (1 - O_j) \quad (3)$$

c) Calculate the δ_j of the hidden layer from back to front :

$$\delta_j = O_j (1 - O_j) \sum_k \omega_{jk} \delta_k \quad (4)$$

d) Calculate the weight correction :

$$\Delta \omega_{ij}(t) = \alpha \Delta \omega_{ij}(t - 1) + \eta \delta_j O_i \quad (5)$$

e) Modify weights:

$$\omega_{ij}(t + 1) = \omega_{ij}(t) + \Delta \omega_{ij}(t) \quad (6)$$

Fig. 2. Operation of BPNN learning process [10]

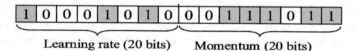

Learning rate (20 bits)   Momentum (20 bits)

Fig. 3. Structure of GA chromosome

- **Learning rate (η):** The learning rate is a relatively small constant that indicates the relative modification in weights. If the learning rate is too low, the network will learn very slowly. However, if the learning rate is too high, the network may oscillate around minimum point, overshooting the lowest point with each weight adjustment, but never actually reaching it. Habitually, the learning rate is very small, located in the interval [0; 1].

- **Momentum term (α):** The introduction of the momentum term is used to accelerate the learning process by "encouraging" the weight changes to continue in the same direction with larger steps. Furthermore, the momentum term prevents the learning process from settling in a local minimum by "over stepping" the small "hill". Typically, the momentum term has a value between 0 and 1 [12].

4.2 Standard Genetic Algorithm

John Holland in 1970s [13] has introduced familiar problem solving algorithms called Genetic Algorithms (GAs) which are based on the principles of biological development, natural selection and genetic recombination. GAs are computational intelligence techniques and search procedures often used for optimization problems. A potential solution to the problem is encoded in chromosome (each chromosome represents an individual) like data structure. Each parameter in a chromosome is called as gene. Genes are selected according to our problem definition. These are encoded on bits, character or numbers. The set of generated chromosome is called a population. An evaluation function is used to calculate the goodness of each chromosome according to the desired solution; this function is known as "Fitness Function". From one generation to the next, GA evolves the group of chromosomes to a new population of quality individuals through selection, crossover (recombination to produce new chromosomes), and mutation operators until a global optimum solution is found at the end of GA process (convergence).

4.3 Adaptive Genetic Algorithms: Adaptive Mutation Algorithm

Adaptive genetic algorithms (AGA) [14] are a popular subset of genetic algorithms, which can provide significant performance improvements over standard implementations when utilized in the suitable circumstances. A key factor that determines how well a genetic algorithm (GA) will perform is the manner in which its parameters are configured. Thus, finding the right values for the mutation rate and crossover rate plays in substantial role when building an efficient and effective GA. Typically, configuring the parameters will require some trial and error, together with some intuition, before eventually attaining a satisfactory configuration. AGA are useful because they can help in the tuning of these parameters automatically by adjusting them based on the state of the algorithm. These parameter adjustments take place while GA is running, hopefully resulting in the best parameters being used at any specific time during execution. It is this continuous adaptive adjustment of GA parameters that will often result in its performance improvement. AGA used in this work uses information such as the average population fitness and the population's current best fitness to calculate and update its parameters in a way that best suits its present state. For example, by comparing any specific individual to the current fittest individual in the population, it's possible to gauge how well that individual is performing in relation to the current best. Typically, we want to augment the chance of preserving individuals that are performing well and reduce the chance of preserving individuals that don't perform well. One way we can do this is by allowing the algorithm to adaptively update the *mutation rate*. We can determine if the algorithm has started to converge by calculating the difference between the current best fitness and the average population fitness. When the average population fitness is close to the current best fitness, we know the population has started to converge around a small area of the search space. When calculating what the mutation rate should be for any given individual, two of the most important factors/characteristics to consider are how well the current individual is performing and how well the entire population is performing as a whole. The algorithm we had used in this work to assess these two characteristics and update the mutation rate is called **Adaptive Mutation Algorithm**, and it is defined as follows:

$$p_m = (f_{max} - f_i) / (f_{max} - f_{avg}) \times m, f_i > f_{avg} \qquad (1)$$

$$p_m = m, \quad f_i \le f_{avg} \qquad (2)$$

When the individual's fitness is higher than the population's average fitness, we take the best fitness from the population (f_{max}) and find the difference between the current individual fitness (f_i). We then find the difference between the max population fitness (f_{max}) and the average population fitness (f_{avg}) and divide the two values as shown by Eq. 1. We can use this value to scale our mutation rate that was set during initialization. As indicated by Eq. 2, if the individual's fitness is the same or less than the population's average fitness, we simply use the mutation rate as set during initialization. Adaptive genetic algorithm can be employed to adjust more than just the mutation rate however. Similar technique can be applied to adjust other parameters of the genetic algorithm like the crossover rate to get further improvements as needed.

5 The Proposed System

This section presents the proposed system in detail, describes its work and provides the framework of our model.

5.1 Approach of Our Novel Proposed System

In order to enhance the performance of our previous and best obtained ANIDS [10] based on BPNN classifier presented in Table 1, we have developed and integrated to that ANIDS a module of optimization based on Self-Adaptive Genetic Algorithm (SAGA). As mentioned previously, SAGA constitutes of a standard GA improved by an Adaptive Mutation Algorithm as explained in detail in Sect. 4.3. The role of that optimization module is searching the optimal values of learning rate (LR) and Momentum term (MT) parameters of BPNN. LR and MT are two prominent parameters in the learning phase of BPNN (Sect. 4.1), which influence the convergence and performance of that classifier. Since their values are between 0 and 1, we thought to use SAGA in order to find the optimal or near-optimal values of those parameters in the interval [0; 1], which is a large space of real numbers.

As shown in Fig. 4, illustrated our approach, SAGA process begins with a randomly generated population of individuals (potential solutions) represented by their chromosomes; each chromosome takes the form of a pair of values (Learning rate, Momentum). Then, this population evolves through several generations by means of genetic operations such elitism, selection, recombination (crossover), and mutation until stopping or optimization criteria of SAGA is met. SAGA uses a constant crossover rate, but mutation rate is adapted automatiqually by Adaptive Mutation Rate (Sect. 4.3). At each generation, each chromosome (chromosome = pair (Learning rate, Momentum)) of current generation is evaluated by passing it as parameter to ANIDS obtained in our precedent work [10], after using Min-Max normalization to convert the two substrings representing Learning rate and Momentum term into values between 0 and 1. This IDS firstly goes through the learning phase, then passes to the

Table 1. Parameters values and performance of our best previous ANIDS [10]

		Value
Parameters	Number of attributes	12
	Normalization	Min-Max
	Architecture (I–H–O)	12–10–1
	Method of calculating number of nodes in hidden layer	H = 0.75*Input + Output
	Activation Function	Sigmoid
Performance	Accuracy	98.66%
	Precision	99.62%
	FPR	1.13%
	FNR	1.41%
	TPR (DR)	98.59%
	TNR	98.87%
	F-score	0.99
	AUC	98.73%

Table 2. Parameters of our module of optimization based on Self-Adaptive Genetic Algorithm (SAGA) for our previous ANIDS

	Parameters	Value
Self-Adaptive GA	Length of chromosomes	40 bits
	Elitism number: the number of best chromosomes which will be copied without changes to a new population (next generation)	5
	Population size	50
	Maximum number of generations	100
	Crossover rate (constant value)	0.95
	Initial Mutation rate (Adaptive value)	0.1

test/evaluation phase and returns the values of performance metrics calculated at the end of the last phase (test). Among those performance metrics, we have selected the relevant of them to serve as "Fitness Function" for evaluation of goodness of chromosomes. From one generation to the next, SAGA converges towards the global optimum through genetic operations mentioned earlier. Finally, the best individual (chromosome) is picked out as the final result once the optimization criterion is met. In our work, termination condition adopted for SAGA is the maximum number of generations (100 generations). Hence, the best chromosome resulted correspond to the optimal or near-optimal values of the pair (Learning rate, Momentum), which ensuring high detection rate and low false alarm rate. For successful use of SAGA, two key elements must be well defined; *the representation/encoding of chromosomes* and the *Fitness Function*:

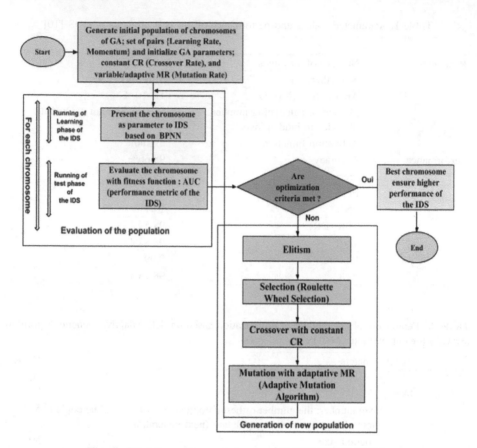

Fig. 4. Workflow of proposed system ANIDS BPNN-SAGA

- **Chromosome encoding/representation**: In our study as indicated by Fig. 3, we have chosen the binary representation for chromosomes. At the beginning of SAGA process; when the initial population of individuals was created, 50 binary strings of length 40 bits representing the chromosomes were generated randomly. Afterwards, each string is divided into two substrings of 20 bits. The first binary substring is then converted in decimal value, and this value is normalized using the Min Max normalization to get a value between 0 and 1, which will serve as the Learning rate of our IDS. Similarly, the foregoing approach is applied to the second binary substring to obtain Momentum term of our IDS, with a value within the interval [0; 1].
- **Fitness Function or Evaluation Function**: Among the values returned by the IDS at the end of the test/evaluation phase, we have chosen the AUC metric as a score (fitness function) of individuals to assess their adaptability to the optimization problem. The AUC [10] metric is by definition the ability to avoid misclassifications of network packets, and from our point of view, it represents a good compromise between the DR (Detection Rate) metric and the FPR (False Positive Rate) metric.

In effect, this is due to the fact that AUC is the arithmetic mean of the DR and TNR (1-FPR) as shown by Eq. 3 of the AUC:

$$AUC = (DR + TNR)/2 = (DR + (1 - FPR))/2 \qquad (3)$$

As it is known, a good IDS is one that achieves a high detection rate (DR) and a low false positive rate (FPR). In fact, as the value of the DR measure increases and that of FPR measure decreases, consequently, the value of AUC increases. Therefore, from our point of view, AUC is the best metric for evaluating an IDS [10]. That is the reason of choice of AUC as fitness function (Fig. 4).

In our optimization module based on Self-Adaptive GA, we have employed the following algorithms/methods: Elitism, Roulette Wheel Selection, Single Point Crossover, Bit Flip Mutation and Adaptive Mutation Algorithm.

5.2 Framework of Proposed System

Our optimized ANIDS Based BPNN-SAGA passes firstly through an optimization stage in order to be optimized. Consequently, it becomes ready to operate in operation/normal mode. The framework of our system in optimization mode consists of four modules as illustrated in Fig. 5.

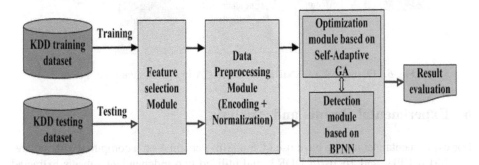

Fig. 5. Framework of ANIDS BPNN-SAGA in optimization stage

- **Feature selection module**: Feature selection is the most critical stage in building intrusion detection models. Our intrusion detection model incorporates a feature selection module mainly to select useful features for intrusion detection. This module is based on *A modified Kolmogorov–Smirnov Correlation Based Filter Algorithm*, which allows selection of a set of 12 relevant features among 41 features of KDD dataset.
- **Data Preprocessing module**: Data Preprocessing includes two operations; data conversion (Categorical encoding) and Normalization. "Categorical encoding" refers to the process of assigning numeric values to nonnumeric features/attributes, so as to make the processing task much simpler, as numeric data can be easily handled upon. Whereas, "Normalization or Scaling" refers to the process of scaling

the feature values to a small range that can help to obtain better detection results and avoid numerical difficulties during the calculation. Our data preprocessing module uses Min Max normalization method.

- **Detection module based on BPNN and optimization module based on SAGA**: In order to enhance the performance of the detection module based on BPNN, this module interacts with an optimization module based on SAGA as explained in detail in Sect. 5.1 with the goal to search the optimal values of Learning rate and Momentum term. The period of optimization is called "optimization stage". This period is finished at the end of SAGA process. Thus, the optimal values of Learning rate and Momentum term are found.

After passing through the optimization phase and finding the optimal values of Learning rate and Momentum term, the proposed IDS "**ANIDS BPNN-SAGA**" operates in operation/normal mode as shown in Fig. 6 to classify connection instances extracted from KDD' 99 cup test dataset.

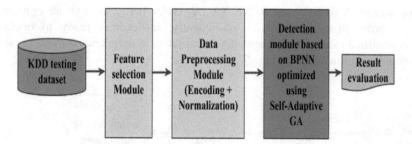

Fig. 6. Framework of ANIDS BPNN-SAGA in operation/normal mode

6   Experimental Results and Analysis

For experimental set up, we have used CloudSim version 4 on a computer with a Core-i7 2700 K CPU, and 16GB of DDR3, and utilized two independent subsets extracted from two KDD cup 99 datasets, namely *kddcup.data_10_percent.gz 10% dataset* (training dataset) and *corrected.gz KDD dataset* (testing dataset). Table 3 shows the details of our datasets and Table 2 presents the parameters of our module of optimization based on SAGA for our previous ANIDS.

Table 3. Distribution and size of training and testing datasets

Category dataset	Normal	DoS	Probe	R2L	U2R	Total
Training dataset	15,000	60,000	2490	165	30	77,685
Testing dataset	20,000	59,245	585	170	260	80,260

The experiments conducted on our proposed system show that at the end of SAGA process that is to say after 100 generations, the best individual (chromosome) generated by SAGA, contains the value **9.536752259018191E-7** for Learning rate and the value **9.536752259018191E-7** for Momentum term. As demonstrated by Table 4, this chromosome or this pair of values (Learning rate, Momentum term) allows our optimized ANIDS "**ANIDS BPNN-SAGA**" to outperform the original ANIDS BPNN, the optimized ANIDS using only GA (ANIDS BPNN-GA) and other state-of-art methods.

Table 4. Comparison of performance between original ANIDS BPNN, ANIDS BPNN-GA and ANIDS BPNN-SAGA

Work	Our original ANIDS BPNN [11]	Our ANIDS BPNN-GA	[8]	[15]	[16]	[17]	Our proposed IDS ANIDS BPNN-SAGA
Accuracy	98.66%	98.82%	93.88%		96.53%	95.29%	99.03%
Precision	99.62%	99.96%		97.2%	95.23%	99.45%	99.87%
FPR	1.13%	0.11%	18.8%	1.7%	0.56%		0.38%
FNR	1.41%	1.54%					1.17%
TPR (DR)	98.59%	98.46%	96.29%	97.3%	76.50%	95.76%	98.83%
TNR	98.87%	99.89%					99.63%
F-score	0.99	0.99		0.972	0.8484		0.99
AUC	98.73%	99.18%		97.2%	87.97%		99.23%
Average time of classification	8.90854 E-5 s	0.897584 E-5 s					8.644391E-5 s

7 Conclusion and Future Work

In order to enhance the performance of our previous ANIDS based BPNN [10], we have developed for it an optimization module based on Self-Adaptive Genetic Algorithm (SAGA) with the goal of searching the optimal values of critical parameters for BPNN classifier, namely Learning rate (LR) and Momentum term (MT). We have used binary encoding for chromosomes; where the first substring (half part) represents LR, whereas the second half part represents MT. Concerning the "Fitness Function", we have adopted AUC metric. Each chromosome generated by SAGA is introduced to the ANIDS, which thereafter goes through learning phase and a test phase. At the end of the last phase, AUC measure is computed. Finally, after production of 100 generations of SAGA, the best value of LR and MT are found. SAGA constitutes of a standard Genetic Algorithm improved by incorporating to it an Adaptive Genetic Algorithm (AGA) that is Adaptive Mutation Algorithm. Experimental results conducted using CloudSim 4 and KDD CUP' 99 datasets demonstrate that our novel IDS "ANIDS BPNN-SAGA" outperforms the original ANIDS BPNN, ANIDS BPNN optimized by only GA and several recent works.

Our future work is to apply AGA to adjust automatically crossover rate parameter of GA to provide further enhancement of our IDS.

References

1. Mehibs, S.M., Hashim, S.H.: Proposed network intrusion detection system based on fuzzy c mean algorithm in cloud computing environment. J. Univ. Babylon **26**(2), 27–35 (2018)
2. Mehibs, S.M., Hashim, S.H.: Proposed network intrusion detection system in cloud environment based on back propagation neural network. J. Univ. Babylon Pure Appl. Sci. **26** (1), 29–40 (2018)
3. Wu, S.X., Banzhaf, W.: The use of computational intelligence in intrusion detection systems: a review. Appl. Soft Comput. **10**(1), 1–35 (2010)
4. Shah, B., Trivedi, B.H.: Artificial neural network based intrusion detection system: A survey. Int. J. Comput. Appl. **39**(6), 13–18 (2012)
5. Lo, C.C., Huang, C.C., Ku, J.: A cooperative intrusion detection system framework for cloud computing networks. In: 39th IEEE International Conference on Parallel Processing Workshops (ICPPW), pp. 280–284. IEEE, San Diego (2010). https://doi.org/10.1109/icppw. 2010.46
6. Modi, C.N., Patel, D.R., Patel, A., Rajarajan, M.: Integrating signature apriori based network intrusion detection system (NIDS) in cloud computing. Procedia Technology, **6**, 905–912
7. Hajimirzaei, B., Navimipour, N. J.: Intrusion detection for cloud computing using neural networks and artificial bee colony optimization algorithm. ICT Express. 2018. In press. https://doi.org/10.1016/j.icte.2018.01.014
8. Sharma, P., Sengupta, J., Suri, P.K.: WLI-FCM and artificial neural network based cloud intrusion detection system. Int. J. Adv. Netw. Appl. **10**(1), 3698–3703 (2018)
9. Ghosh, P., Jha, S., Dutta, R., Phadikar, S.: Intrusion detection system based on BCS-GA in cloud environment. In: Shetty, N., Patnaik, L., Prasad, N., Nalini, N. (eds.) Emerging Research in Computing, Information, Communication and Applications (ERCICA 2016), pp. 393–403. Springer, Singapore (2018). https://doi.org/10.1007/978-981-10-4741-1_35
10. Chiba, Z., Abghour, N., Moussaid, K., El Omri, A., Rida, M.: A novel architecture combined with optimal parameters for back propagation neural networks applied to anomaly network intrusion detection. Comput. Secur. **75**, 36–58 (2018)
11. Qian, Q., Cai, J., Zhang, R.: Intrusion detection based on neural networks and Artificial Bee Colony algorithm. In: 2014 IEEE/ACIS 13th International Conference on Computer and Information Science (ICIS), pp. 257–262. IEEE, Taiyuan (2014)
12. Multi-Layer Perceptron. http://www.cse.unsw.edu.au/~cs9417ml/MLP2
13. Uppalaiah, B., Anand, K., Narsimha, B., Swaraj, S., Bharat, T.: Genetic algorithm approach to intrusion detection system. IJCST **3**(1), 156–160 (2012)
14. Jacobson, L., Kanbe, B.: Genetic Algorithms in Java basics, pp. 143–144. Apress, New York (2015). https://doi.org/10.1007/978-1-4842-0328-6
15. Aslahi-Shahri, B.M., Rahmani, R., Chizari, M., Maralani, A., Eslami, M., Golkar, M.J., Ebrahimi, A.: A hybrid method consisting of GA and SVM for intrusion detection system. Neural Comput. Appl. **27**(6), 1669–1676 (2016)
16. Hamamoto, A.H., Carvalho, L.F., Sampaio, L.D.H., Abrão, T., Proença Jr., M.L.: Network anomaly detection system using genetic algorithm and fuzzy logic. Expert Syst. Appl. **92**, 390–402 (2018). https://doi.org/10.1016/j.eswa.2017.09.013
17. Sharma, R., Chaurasia, S.: An enhanced approach to fuzzy c-means clustering for anomaly detection. In: Somani, A., Srivastava, S., Mundra, A., Rawat, S. (eds.) Proceedings of First International Conference on Smart System, Innovations and Computing. Smart Innovation, Systems and Technologies, vol. 79, pp. 623–636. Springer, Singapore (2018)

State of the Art in the Contribution of an Ontology-Oriented Knowledge Base to the Development of a Collaborative Information System

Meryam El Mrini[✉], El Hassan Megder, and Mostafa El yassa

IRF-SIC Laboratory, Ibn Zohr University, Agadir, Morocco
{elmrinimeryam,melyass}@gmail.com, Megderel@yahoo.fr

Abstract. This article reviews the state of the art of implementing collaborative information systems that guarantee interoperability using ontologies as a semantic dimension. It presents the implementation of different systems of collaborative information and collaborative platform. It also presents the various limitations that this type of information systems still suffers from and works that have tried to overcome its limitations and finally it gives our perspective to this topic.

Keywords: Base of knowledge · Ontologies · Collaborative information system · Alignment · Merging · Interoperability

1 Introduction

Market conditions nowadays force companies to open up to several industrial collaboration networks in order to guarantee their commercial efficiency. The survival of these collaborative networks is conditioned by their ability to be flexible and adapt quickly to market change.

Collaboration networks imply requirements for collaborative platform development that will support collaboration between multiple companies. Each company is represented in collaborative networks by its information system, but the problem of collaboration between several information systems is the ability of each of them to communicate with others at low transaction costs and as quickly as possible.

According to the standard [1] there are several levels of collaborative maturity that can be used to characterize a company: communication (able to exchange and share information), open (capable of sharing business services and functions with others), federated (able to work with others by following collaborative processes that have a common goal and to ensure their own) and interoperable (able to work with other people without a special effort), the most important level is the interoperability that ensures connectivity between different collaboration partners.

By definition, interoperability is the ability of two or more systems or components to exchange information and use the information exchanged [2]. Interoperability is considered as the ability of companies to structure, formalize and present their

M. Ben Ahmed et al. (Eds.): SCA 2018, LNITI, pp. 603–617, 2019.
https://doi.org/10.1007/978-3-030-11196-0_50

knowledge and know-how to exchange or share them, for this reason it is considered an essential requirement for companies that must integrate dynamically. Interoperability is therefore essential for collaboration between several information systems.

Several solutions have been proposed to solve the problem of interoperability: European Interoperability Framework(EIF) [3], ATHENA Interoperability Framework (AIF) [4], Interoperability Development for Enterprise Applications and Software (IDEAS) [5], e-Government Interoperability Framework (e-GIF) [6], a mediation information system [7] and the project PIM4SOA [8].

The previous solutions did not take into account the semantic level of the collaboration which created a gap between the business and technical level, hence the use of knowledge management concretized by ontologies [9].

Several techniques have been developed and a set of models have been proposed in the literature. This article reviews the state of the art of setting up collaborative information systems based on ontologies.

This article is organized as follows, Sect. 2 presents the collaborative information system, the different solutions for setting up mediation information system, Sect. 3 discusses ontology-oriented knowledge bases, Sect. 4 presents the ontologies and the development of collaborative information systems, Sect. 5 deals with the alignment and merging of ontologies, Sect. 6 represents a comparative analysis of the various works already mentioned and finally our perspective on the extension of proposed solutions.

2 The Collaborative Information System

2.1 Information System

We will present different definitions of an information system: "An information system is an organized set of resources: material, software, employees, data, procedures, in order to acquire, to process, to store, to disseminate information (data, documents, image, sound, etc.) in organization" [10]; "An information system is the set of all elements that contribute to the process and the circulation of information in an organization (data base, software, procedures, documents) including Information Technology" [11]; "Technically, we can define an information system as a set of elements interconnected which collect (or recover), process, store and disseminate information in order to support decision and process control in organization" [12].

All these definitions underline several invariants regarding the information system: the information system supports the activity of the company by managing its resources, contributing to its processes and the flow of information as it interconnects the different elements of the enterprise.

2.2 The Information System in a Collaborative Context

In a context of information systems collaboration, companies must determine the degree of openness of their information system towards other partners. For strategic reasons, most companies in a collaborative network are careful not to disclose their know-how and hide knowledge related to their internal business processes [13]. It is a public part that represents an interface accessible by the other partners, and a private part which remains accessible just for the authorized partners. The private/public relation is primordial in the conception of an information system in an interoperability context [14, 15].

A corporate interoperability framework proposed in [16], which defines three interoperability barriers: conceptual, technological and organizational. Considering that enterprise information systems are the practical and operational parts of a business, an important requirement is to remove the technological barriers between them. The possibility of breaking down conceptual barriers by removing technological barriers is also taken into consideration, but several problems arise: an agility problem, an interoperability problem and a business/technical correspondence problem [17].

A Problem of Agility
In the face of unexpected developments and changes in the internal and/or external environment, it becomes essential to design the information system in an agile manner. The notion of agility became a well-known property at the end of the 90s. It is mainly correlated with a dynamic of interorganizational collaboration. This evolution is described in [18] as "the transformation of a fixed structure into a fluid environment". It is also described by "the transformation of a static construction of lego to a living organism" [19].

Faced with the needs of permanent adaptability, the need for agility appears obvious and indispensable in order to allow, among others, the quality, the integration, the interoperability and the coherence between the business world and the technical world. Therefore, the agility of the information system becomes a major goal and must be a quality that any company must have to meet the demands of its customers, competition and rapidly evolving technologies [20–22].

An Interoperability Problem
Companies also need to have several components of their information system in interaction to achieve their goals. It is necessary that they can communicate in a transparent way and with less effort. This mechanism is generally called interoperability.

The authors in [23] define interoperability as "the ability of systems to work together without particular effort for the users of these systems". In [24] the authors clarify this definition by adding that "interoperability refers to the ability of systems, natively foreign to each other, to interact in order to establish harmonious and finalized collective behaviors, without having to modify in depth their structure or their individual behavior".

Business/Technical Correspondence Problem
Agility and interoperability are needed for managing collaboration within the information system but it is also essential to ensure effective correspondence (also known as reconciliation), between the business and technical worlds, which is a major objective for any business. Indeed, a misunderstanding between these two worlds can stop its evolution. Reconciliation is about providing agile solutions by finding the right services that meet the needs of the business activities of the collaborative process. We distinguish three correspondence problems that, once solved, can allow a relevant connection

between business processes and partner systems: an information problem, a functional problem and a behavioral problem [25]. These problems can be expressed differently: how to ensure the communication between the different components? How to ensure reconciliation between business activities and technical services? How to obtain executable processes from business processes? [26].

2.3 Mediation Information System

An agile, interoperable information system that supports business/technical correspondence is called the Mediation Information System (MIS). This is the starting point of the MISE project (Mediation Information System Engineering) [27] which proposes a solution for the design and realization of a MIS. This project aims to develop an approach and methods for the design of collaborative information systems from interoperable information systems following a principle of mediation between these systems.

The concept of mediation as presented in [28] is the origin of the idea of a system with federated behavior, for more than 26 years ago, by various works that were interested in searching distributed information in independent databases. The MISE project has adopted and extended these ideas to the problem of companies wishing to collaborate. From the perspective of the MISE project, according to [29], we ask the mediating component of the information system to deal with these resistances. He did the role of the intermediary, he manages the characteristics of the collaboration, convert the data and connect the information systems of the various partners involved.

A first definition of a mediation-based collaborative information system (MIS) and the mediator, as a software component for implementing the mediation-based system, is a result of the MISE project (see Fig. 1). An engineering approach directed by the mediator models and coupled by a business process management approach based on the SOA (Service oriented architecture) were the focus of the work of [30] who sets the design of this collaborative information system based on the Model Driven Architecture (MDA). He was interested in the passage of a Computer Independent Model

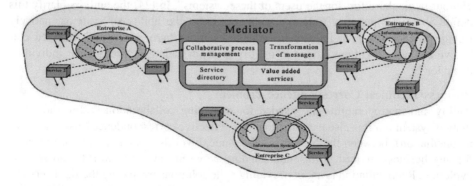

Fig. 1. The structure of the mediation information system proposed by the MISE project (2007)

(CIM) model [31, 32] where the partners provide their collaboration process to a Service Oriented Architecture (SOA) based Platform Independent Model (PIM) a model that describes a response to the specifications defined in the CIM model.

The work of [30] used two metamodels: a BPMN metamodel and an SOA metamodel to make the transition from a CIM level model expressed as a Business Process Model (BPMN) to another SOA-based PIM model.

3 Ontology-Oriented Knowledge Bases

3.1 Knowledge Bases

First of all, we are going to define knowledge according to [33] knowledge includes implicit and explicit restrictions between objects as well as operations and relationships, which allow defining general and specific heuristics such as inference processes related to the situation to be modeled.

According to [34] It is first of all necessary to characterize the knowledge in relation to several terms to which it is abusively assimilated. Even if there are no clearly defined boundaries between the concepts of data, information, process, and knowledge, each of these notions plays a proper role according to its level of entry into a system's action process.

Before the emergence of the field of knowledge engineering, the knowledge used by expert systems was that of an expert who coded it directly in a representation language. An expert system is a set of software modeling, in a specific area, the skills and reasoning of one or more experts.

The expert system is designed to help a user in a particular area find the right solution for his problem; This expert system consists of a knowledge base concerning the studied domain and an inference engine to deduce new knowledge.

However, the expert systems had drawbacks concerning the acquisition of knowledge, hence the advent of knowledge engineering to better model the reasoning and better model the domain.

3.2 Ontologies

Ontologies are a means of capturing and representing knowledge, [35] introduces the notion of ontology as "an explicit specification of a conceptualization". This definition has been a little bit modified by [36]. A combination of the two definitions can be summarized as "an explicit and formal specification of a shared conceptualization". This definition is explained in [37]: explicit means that the "type of concepts and constraints on their uses are explicitly defined", formal refers to the fact that the specification must be machine readable, shared refers to the notion that an ontology "captures consensual knowledge, which is not specific to an individual but validated by a group", conceptualization refers to "an abstract model of a certain phenomenon of the world based on the identification of relevant concepts of this phenomenon".

4   Ontologies and the Development of Collaborative Information Systems

Ontologies are a more abstract level of a collaborative information system, they represent the business level, the works cited in Sect. 2 did not take into account the business and semantic level which created a big gap between the business and technical level at the level of the collaborative information system which does not give a satisfactory collaboration.

The work of [38] was interested in a more abstract level of the MISE project: the business level. She defined a knowledge-based system (Kbs) to generate automatically the CIM model and this by offering the different partners the opportunity to describe the desired collaboration.

The main goal of this work is to be able to capture, adapt and transform all knowledge concerning the collaboration in question, with the intention of producing a collaborative business process compatible with the CIM model.

For this reason, a knowledge-based system (kbS) has been developed in order to support the modeling of the business collaborative process. The system consists of four functionalities (see Fig. 2).

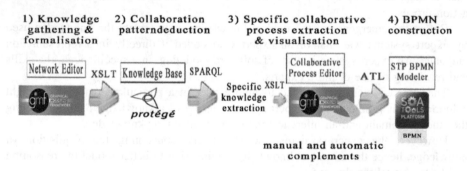

Fig. 2. The technical architecture of the solution (Rajisri 2009)

This work of [38] was able to develop an approach to develop a knowledge-based system dedicated to the specification of a valid collaborative process model to be executed under a collaborative platform.

The thesis supported by [39] continued the work of [38] to be able, not only to design a collaborative information system based on mediation, but also to make an evolutionary maintenance in a double movement of reverse engineering and engineering adapted to the recurring evolution of the need and which provides agility in operation.

The thesis research of [40] enriched the works of [38] by automating the generation of the characterization and the transformation of a model of collaborative situation in a model of mapping (cartography) of collaborative business processes.

As a supplement to these works, [41] developed the transformation of business processes in feasible technical processes. This passage consists in selecting among the available services those who cover the features of the activities jobs modeled in the various processes and annotated semantically.

By treating a particular type of collaboration that of the case of crisis by [42], within the framework of the project ISyCri conceived an information system for several partners who have to solve, or at least reduce, a crisis in which they are involved, this proposed solution is an information system of mediation.

The thesis of [43] based itself on the works [38], while enriching the ontology quoted in the work of Rajisri, adding dynamic concepts in the ontologies and by adding rules of transformation of the collaborative process there a set of services.

The work of [44] combined to the work of [38] and that of [43], to propose a more generic approach allowing a metamodelisation of the interorganizational collaborative process. This approach starts of the principle that within an interorganizational collaboration, the various actors do not still have the same point of view on the notion of business process, each evolving in the environment, its universe of skill and each having sound his own model of business process. The points of view of every partner were apparently completely separated but, in reality, they establish a representation of the various aspects of a same and unique system. Within the framework of an interorganizational collaboration; the need to reach a common purpose and the need for information exchange between the various actors lead to opt on second thought in meta-modeling of the collaboration.

This work been a part of global architecture presented in the work "Towards a Platform in Cloud for the Integration of the Interorganizational Workflows" makes by [45].

The work of [38], as we have already said, completes the approach proposed by [30] by providing a model of collaborative process, the latter, which can be constrained by shared resources or by processing times, must be checked before being handed over to the collaboration execution platform. In the literature several techniques have been adopted, for verification of the process model, [46, 47] use the technique of model checking for the verification of business processes. Reference [46] explored the structural theory of Petri nets to approach the modeling and verification of business processes by improving a chosen algorithm in the literature. The work of [48] adopts an approach based on automatons as well as formal composition and [49] uses an approach of check based on the transformation (processing) of graphs.

After having analyzed all the works we have already mentioned in the previous sections it has been found that there is always a need to enrich the collaborative ontology, there was a lack of concepts, that's why we thought about using ontology alignment by aligning ontologies representing the different partners in the collaboration, which will give us as a result a richer collaborative ontology.

5 The Alignment and Merging of Ontologies

5.1 The Definition of the Alignment of Ontologies

The alignment in its general sense refers to the adjustment of objects to others in a desired orientation. In computer science, we talk about the alignment of ontologies, which represents a crucial task in several Sub Application domains: communication in multiagent systems, data warehouse, etc. In the same field or related fields, there is a risk of having several different ontologies (heterogeneity). This is why the notion of ontology alignment has been introduced to pare two ontologies and to manage as automatically as possible, pairings ontologies, which consists of finding correspondences each linking two entities (by examples, concepts, instances, terms, etc.) by a relation (equivalence, sub Sumption, incompatibilities.), possibly provided with a degree of confidence. All of these correspondences are called alignment and are exploited after-wards in the same system.

5.2   The Alignment Process

The alignment process combines three dimensions: input, alignment process and output. The input or the input that is constituted structures to be aligned, they can be XML schemas, relational schemas, ontologies. In our case, they are the last ones. The alignment process: is a task during which, from ontology O and another O', it determines an alignment A' between these two ontologies, this task is carried out using a strategy or a combination of basic alignment techniques (see Fig. 3).

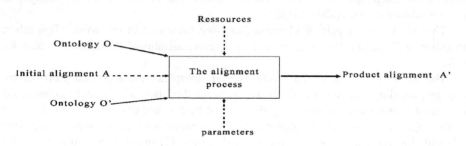

Fig. 3. Diagram summarizing the alignment process

5.3   The Goals of Ontology Alignment

The objective of the alignment of ontologies is to realize the semantic interoperability. The semantic interoperability is the capacity of two or several information systems to find a common understanding, from the exchanged data, to produce useful results. It by putting two heterogeneous ontologies in a mutual agreement by detecting a set of correspondences between the entities of these semantically bound ontologies, to allow the data to be exchanged, handled, integrated … thanks to the use of various methods and approaches.

There is an initiative every year to define the best matching systems between ontologies, OAEI (ontology alignment assessment initiative) [50] is a coordinated international initiative that organizes the evaluation of a number growing ontology matching systems. Its main purpose is to openly compare systems and algorithms on the same basis, allowing anyone to draw conclusions about the best ontology matching systems. In addition, allow developers, from such assessments, to improve their systems (see Fig. 4).

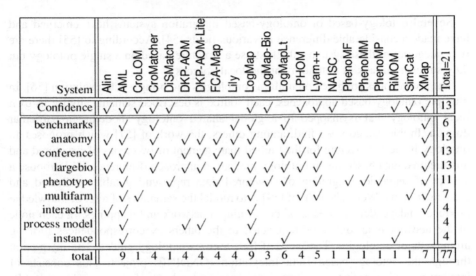

Fig. 4. The list of OAEI participants of year 2016

The systems that have already participated (AML, CroMatcher, Lily, LogMap, LogMapLite, XMap) always get the best results with Lily and CroMatcher still getting an impressive result.

This initiative allows choosing the best system of correspondence of ontologies, allowing obtaining from better results, what is going to help us apply our prospect which we are going to detail in the following section.

5.4 The Merging of Ontologies

The merging of the ontologies is a sector of the domain of the knowledge management which takes a part more and more stressed in the management of the ontologies while considering the fast evolution of the technology of the knowledge [51]. It is recognized as being an essential operation occurring at several levels of the engineering ontologies including the integration of several modules so assuring the interoperability between the systems, the reusability and the division of the knowledge [52].

One of the motivations of the merging of the ontologies holds the construction of an ontology in leave different sources. This merger is activated further to a need which could require the integration of several ontologies modeling various parts of a domain.

The Different Architectures of Merging Ontologies

Data integration is a process of repatriating data from different heterogeneous sources to either treat them locally mediating approach [53] or store them in a common database (approach data ware-house) [54]. In the mediator approach, the integration system generates, from a request from the user, as many subrequests as there are data sources to interrogate; then he builds the final answer from the result of each subquery and passes it to the user.

Several ontology-based or ontology-based integration systems have emerged and have made a considerable difference in various fields [54]. According to [55] there are two categories of these systems: the first use a structure based on a single ontology but they suffer from lack of autonomy at the local sources.

The second ones are based on multiple ontologies and bring a better solution [56]. In multiple ontology-based approaches, each source is described semantically by its own local ontology that is mapped to a global shared ontology modeling a particular domain. In this category we find, among others, the work of [57] which proposes an ontology-based architecture for the unified management of two types of structured and unstructured data based on an ontology mediation approach. Reference [58] propose a system of semantic integration of structured data representing tables collected and extracts from the Web. The goal of [59] is to model the semantics of a set of knowledge produced independently of each other, forming a network and mappings. In each node of the network is an ontology, connected to the others by correspondences forming alignments of ontologies. The work of [60] focuses ontology-guided data integration through the ontology-based Data Integration (ONDINE) project. This consists of integrating and querying a relational database and a base of conceptual graphs. Its integration system is based on a domain ontology that is built from local databases. The ontologies merge Web Language (OWL) using the hierarchical classification technique is proposed by [61]. For each ontology, the set of concept pairs, of which the first component of the pair subsumes the second, is calculated. The advantage of this approach is its scalability, the classification algorithm that is used supports input more than two ontologies but requires reduplication of redundant concepts.

The work of [62] proposed a cellular algorithm for the automatic fusion of several input ontologies, they proposed to provide an original solution to the problems of semantic integration of data. Their solution lies in the optimization of storage space and processing time. For this, the experimental idea that they proposed is to use the CASI cellular machine [63] to simulate the ontology fusion process. The principle of this method is to consider that the states of the cellular automaton will be able to contain the description of ontologies in the form of rules of association. Then, they adapted local transition functions so that ontologies form in groups when they represent similar concepts.

6 Comparative Analysis

This section reports a comparative analysis of examined works. It also highlights the benefits and drawbacks of those examined works. In the previous sections, we analyzed in depth the development of the implementation of collaborative information system.

Firstly, we examined the corporate interoperability framework proposed by [16] who tried to break the barriers between the conceptual and technological level, but several problems arise: an agility problem, an interoperability problem and a problem of the correspondence between the business and technical level, we also examined the solution proposed for the previous problems, which is a mediation information system, this concept of mediation was firstly addressed by [28] who were interested in searching distributed information in independent databases, and it is the start point of the MISE project [27] which proposes a solution for the design and realization of a MIS, as part of this project the work of [30] focused on the design of this collaborative information system based on MDA, he was interested in the passage from a CIM model to an SOA architecture, but this solution assumes that the partners of the collaboration will provide the CIM level, but this hypothesis seemed too ambitious the practice, specifically because of the requirements of the CIM model in order to be exploitable for the logical level transformation, in addition it doesn't take into account the semantic level.

Secondly, we survey the development of collaborative information system using ontologies, the work of [38] was interested in a more abstract level of the MISE project: the semantic level using ontologies. Several works rely on the work of [38], by automating the generation of the characterization and the transformation of a model concerning a collaborative situation in a model of mapping, developing the transformation of business processes in feasible technical processes, treating a particular type of collaboration which is the case of crisis, enriching the ontology quoted in this work by adding dynamic concepts. After having analyzed all the previous works, we found that there is always a need to enrich the collaborative ontology, there was a lack of concepts, because of this of problem we thought about using the alignment and merging of ontologies representing the different partners in the collaboration, as a result we will have a richer collaborative ontology.

Thirdly, we choose the alignment and the merger of the ontologies as the solutions to the gaps noticed in the previous works, we based ourselves on the results of the OAEI which organizes every year a competition which defined the best algorithms of the alignment. The alignment is the base of the merger of the ontologies, we quoted several architectures of merger and we opted for an architecture which bases itself on the artificial intelligence [61] which proposed to use the CASI cellular machine to simulate the ontology fusion process, this work proposed an original solution to the problems of semantic integration of data found in the other architecture of merger.

7 Conclusion

The implementation of collaborative information systems has made significant progress over the last years, from the first classical collaborative systems to those taking into account the semantic dimension. They have been heavily invested in various fields such as industry, e-commerce, medicine, etc.

However, problems still remain, among which we can mention that there are always perspectives of enrichment of the ontologies used, which show that there is a lack concerning the concepts of the ontologies and do not cover in a way the network of collaborative partners and the collaborative process. This implies the development of a new ontology construction methodology.

In this article, we put the point on the state of the art of works on collaborative information systems and the use of ontologies for the implementation of these systems. We have also exposed the different algorithms and ontology alignment matching systems in a vision of enrichment of the ontology used for the elaboration of a collaborative information system.

To be more precise, each company has specificities regarding its information system and its own processes, when we try to achieve a collaborative ontology that will encompass specificities of several companies, we always have a lack of concepts of ontologies. The solution for this problem is to represent each enterprise by an ontology and to make the alignment and the merging of these different ontologies to have a collaborative global ontology that will be transformed to an executable process using metamodels.

References

1. IEC, Kosnake. IECTC65/290/DCstand (2005)
2. IEEE. Standard Computer Dictionary—A Compilation of IEEE Standard Computer Glossaries (1990)
3. EIF. European Interoperability Frame-work. White Paper, Brussels. Available on http://www.comptia.org (2004)
4. Athena Consortium. Public document: ATHENA General description v10. Available on http://www.athena-ip.org/ (2004)
5. IDEAS. A Gap Analysis–Required activities in Research, Technology and Standardization to Close the RTS Gap-Roadmaps and Recommendations on RTS Activities, IDEAS Deliverables (2003)
6. e-Gov. e-Government Unit, e-Government Interoperability frame-work, Version 6.1 (2005)
7. Touzi, J.: Aide à la conception de Système d'Information Collaboratif support de l'interopérabilité des entreprises. Ph.D. thesis. INPT (2007)
8. Benguria, G., Larrucea, X., Elveseater, B., Neple, T., Beardsmore, A., Friess, M.: Platform independent model for service oriented architectures. Enterprise Interoperability: New Challenges and Approaches, pp. 23–32. Springer, Berlin. ISBN-10: 1846287138 (2006)
9. Rajsiri, V.: Knowledge-based system for collaborative process specification. Ph.D. thesis. Department, Industrial Engineering, Toulouse University (2009)
10. Reix, R.: Systèmes d'Information et Management des Organisations, 3rd edn., p. 75. Librairie Vuibert, Paris (2000)
11. Educne: Information system definition. Retrieved Oct 2006 from http://www2.educnet.education.fr/sections/superieur/glossaire/ (2006)
12. Laudon, K.C., Laudon, J.P: Management Information Systems: Managing the Digital Firm. Pearson Education, Upper Saddle River, New Jersey (2006)
13. Vanderhaeghen, D., Zang, S., Hofer, A., Adam, O.: XML-based transformation of business process models—enabler for collaborative business process management. In: Proceedings of the 2nd GI Workshop XML4BPM, XML Interchange Formats for Business Process Management (2004)
14. Bauer, B., Müller, J.P., Roser, S.: A Decentralized Broker Architecture for Collaborative Business Process Modelling and Enactment, Enterprise Interoperability: New Challenges and Approaches. Springer. ISBN-10: 1846287138 (2006)
15. Vanderhaeghen, D., Werth, D., Kahl, T., Loos, P.: Service and Process Matching—An Approach towards Interoperability Design and Implementation of Business Networks, Enterprise Interoperability: New Challenges and Approaches. Springer. ISBN: 978-1-84628-713-8, pp. 189–198 (2006)
16. Chen, D., Doumeingts, G., Vernadat, F.: Architectures for enterprise integration and interoperability: past, present and future. Comput. Ind **59**(7), 647–659 (2008)

17. Zribi, S.: La gouvernance SOA pour une approche de conception de Système d'information de Médiation: réconciliation non-fonctionnelle de services pour mettre en œuvre les processus métier. Ph.D. thesis. Industrial Engineering Center, School of Mines in Albi, France (2014)
18. Bénaben, F., Touzi, J., Rajsiri, V., Pingaud, H.: L'interopérabilité des systèmes d'information comme moyen vers l'intégration de l'écosystème industriel. In: 7 th International Congres of Industrial Engineering, Trois-Rivières, Québec (2007)
19. Luzeaux, D., Ruault, J.R.: Systèmes de Systèmes, Concepts et Illustrations Pratiques. Hermes Science. ISBN 978-2-7462-1875-8 (2008)
20. Goranson, H.T.: The Agile Virtual Enterprise—Cases, Metrics, Tools, Quo-rum Books, Westport CN (1999)
21. Desouza, K.: (ed.). Agile Information Systems. Routledge (2006)
22. Rouse, W.B.: Agile Information Systems for Agile Decision-Making. Agile Information Systems, pp. 16–30. Elsevier, New York (2007)
23. Konstantas, D., Bourrières, J.P., Léonard, M., Boudjlida, N.: Interoperability of Enterprise Software and Applications. Springer Science & Business Media (2006)
24. Pingaud, H., Rationalité du développement de l'interopérabilité dans les organisations, Management des technologies organisationnelles, pp. 19–30. Presses de l'école des mines de Paris, France (2009)
25. Bénaben, F., Mu, W., Truptil, S., Lorré, J.P., Pingaud, H.: Information Systems Design for Emerging Ecosystems: A Model Driving Approach Diving into Abstraction Layers, IEEE-DEST'10. Springer, Dubai, EAU (2010)
26. Bénaben, F., Boissel-Dallier, N., Pingaud, H., Lorre, J.: Semantic issues in model-driven management of information system interoperability. Int. J. Comput. Integr. Manuf. 26(11), 1042–1053 (2013)
27. MISE (Mediation Information System Engineering), le centre genie insudtriel ecole des mines d'albi carmaux
28. Wiederhold, G.: Mediators in the architecture of future information systems. Computer 25 (3), 38–49 (1992)
29. Bénaben, F., Touzi, J., Rajsiri, V., Truptil, S., Lorré, J.P., Pingaud, H.: Mediation information system design in a collaborative SOA context through a MDD approach. In: Proceedings of the First International Workshop on Model Driven Interoperability for Sustainable Information Systems (MDISIS'08) Held in Conjunction with the CAiSE, vol. 8 (2008)
30. Miller, J., Mukerji, J.: MDA Guide Version 1.0. 1. Object Management Group, vol. 234, p. 51 (2003)
31. Wack, M., Cottin, N., Mignot, B., El Moudni, A.: Dossiers numériques. In: Charlet, J. (Coordinateur) Dossiers Numériques, vol. 6. Hermès. Numéro Spécial de la Revue Document Numérique, Paris (2002)
32. Gruber, T.R.: A translation approach to portable ontologies. Knowl. Acquisit. 5(2), 199–220 (1993)
33. Sowa, J.F.: Conceptual Structures: Information Processing in Mind and Machine. Addison-Wesley Publishing Company, USA (1984)
34. Borst, W.N.: Construction of Engineering Ontologies. University of Tweenty. Enschede, Centre for Telematica and Information Technology (1997)
35. Studer, R.: Benjamins V.R., Fensel, D.: Knowledge Engineering: Principles and Methods. Data & Knowledge Engineering 25, 161–197 (1998)
36. Truptil, S.: Etude de l'approche de l'interopérabilité par médiation dans le cadre d'une dynamique de collaboration appliquée à la gestion de crise. Thèse en systèmes industriels, INPT—EMAC (2011)

37. Wenxin, M.U.: Caractérisation métier et logique d'une situation collaborative. Ph.D. thesis. School of Mines in Albi, France (2012)
38. Boissel-Dallier, N.: Aide à la conception d'un système d'information de médiation collaboratif: de la cartographie de processus métier au système exécutable. Ph.D. thesis. Institut National Polytechnique of Toulouse, France (2012)
39. Truptil, S., Bénaben, F., Salatgé, N., Hanachi, C., Chapurlat, V., Pignon, J.-P., Pingaud, H.: Mediation Information System Engineering for Interoperability Support in Crisis Management (2010)
40. Sara, S., Rachid, B., Kenza, B.: Modeling of mediation system for enterprise systems collaboration through MDA and SOA. Ph.D. thesis. Information Logistic and Production Systems Team, National School of Applied Sciences (ENSA), Cadi Ayyad University-Marrakech (2013)
41. Semar-Bitah, K., Boukhalfa, K.: Vers une architecture de modélisation des processus collaboratifs interorganisationnels. In: Conference Paper, Centre Algérien de Développement des Technologies Avancées (2016)
42. Abbassene, A., Alimazighi, Z.: Vers une Plateforme en Nuage pour l'intégration des Workflows Inter-Organisationnels. Journée de l'Etudiant ESI (JEESI'14), Alger (2014)
43. Sbai, Z., Missaoui, A., Barkaoui, K., Ayed, R.B.: On the verification of business processes by model checking techniques. In: International Conference on Software Technology and Engineering 2010, Puerto Rico, USA (2010)
44. Sbai, Z., Barkaoui, K.: Vérification Formelle des Processus Workflow Collaboratifs. Conférence Francophone sur les Systèmes Collaboratifs 2012, Sousse, Tunisie (2012)
45. Sbai, Z.: Contribution à la modélisation et à la vérification de processus workflow. Ph.D. thesis. CNAM (National Conservatory of Arts and Crafts), France (2010)
46. Luis, E. Mendoza, M.: Business process verification using a formal compositional approach and timed automata. In: XXXIX Latin American Computing Conference, Vargas, Venezuela (2013)
47. El Mansouri, R.: Modélisation et Vérification des processus métiers dans les entreprises virtuelles. Thèse en Informatique, Université Mentouri Constantine (2009)
48. Faria, D., Pesquita, C., Balasubramani, B.S., Martins, C., Cardoso, J., Curado, H., Couto, F. M., Cruz, I.F.: AML results for OAEI 2016. In: Ontology Matching Workshop. CEUR (2016)
49. Despres, S., Szulman, S.: Construction d'une ontologie du droit communautaire. In: IC—16èmes Journées francophones d'Ingénierie des Connaissances, pp. 85–96, May 2005. Nice, France, Presses Universitaires de Grenoble, Grenoble (2005)
50. Results of the alignment evaluation imitative (2016)
51. Kais, S.: La fusion des ontologies. Unpublished master's thesis. University of Quebec in Monteal (2014)
52. Lamarre, P., Cazalens, S., Lemp, S., Valduriez, P.: A flexible mediation process for large distributed information systems. In: On the Move to Meaningful Internet Systems 2004: CoopIS/DOA/ODBASE (1), Lecture Notes in Computer Science, vol. 3290, pp. 19–36. Springer (2004)
53. Kimball, R.: The operational data warehouse. DBMS 11(1), 14–16 (1998)
54. Mena, E., Illarramendi, A., Kashyap, V., Sheth, A.P.: Observer: an approach for query processing in global information systems based on interoperation across preexisting ontologies. Int. J. Distrib. Parallel Databases 8(2), 223–271 (2000)
55. Bellatreche, L., Xuan, D., Pierra, G., Dehainsala, H.: Contribution of ontology-based data modeling to automatic integration of EC within ED. Comput. Ind. J. 57(8–9), 711–724
56. Khouri, S.: Modélisation conceptuelle à base ontologique d'un entrepôt de données. Mémoire de Magistère. Université Oued-Smar Alger (2009)

57. Diallo, G.: Efficient building of local repository of distributed ontologies. Paper Presented at IEEE Signal-Image Technology and Internet-Based Systems (SITIS), 2011 Seventh International Conference, pp. 159–166. 28 Nov–1 Dec 2011, Dijon, France

58. Saïs, F., Pernelle, N., Rousset, M.-C.: L2R: a logical method for reference reconciliation. In: Proceedings of the Twenty-Second AAAI Conference on Artificial Intelligence, 22–26 July 2007, pp. 329–334. Vancouver, British Columbia, Canada (2007)

59. Zimmermann, A.: Logical formalisms for agreement technologies. In: Ossowski, S. (ed.) Agreement Technologies, pp. 69–82. Springer, The Netherlands. https://doi.org/10.1007/978-94-007-5583-3_5 (2005)

60. Dibie-Barthélemy, J., Haemmerlé, O., Salvat, E.: A semantic validation of conceptual graphs. Knowl.-Based Syst. **19**, 498–510. https://doi.org/10.1016/j.knosys.2005.04.002 (2006)

61. Boussaïd, O., Ben Messaoud, R., Choquet, R., Anthoard, S.: X-warehousing: an XmL based approach for warehousing complex data. In: Proceeding of the 10th East-European Conference on Advances Databases and Information System (ADBIS06). Thessaloniki, Greece. Lecture Notes in Computer Science, vol. 4152, pp. 39–54 (2006)

62. Abdelouhab, F.Z., Atmani, B.: Une approche cellulaire de fusion d'ontologies. J. Decis. Syst. **26** (2016)

63. Atmani, B., Beldjilali, B.: Knowledge discovery in database: induction graph and cellular automaton. Comput. Inf. J. **26**(2), 171–197 (2007)

The Impact of Quantum Computing
on Computer Science

H. Amellal[1(✉)], A. Meslouhi[1], and A. El Allati[2]

[1] University Mohammed V Faculty of Sciences, Av. Ibn Battouta, B.P 1014
Rabat, Morocco
amellal@yandex.ru
[2] Laboratory of Engineering Sciences, Faculty of Sciences and Techniques,
Al-Hoceima B.P.3, C.P. 32003 Ajdir, Morocco

Abstract. Contrary to classical computer which store data using the digital bits 0 or 1, quantum computer use quantum bits (qubits) for coding information as O, 1, or 01 at the same time. In fact this superposition of states with the other quantum phenomena enables quantum computers to operate vast combinations of states at once. In this study, we analyze the impact of quantum computing in the simplification of classical computing complexities using in computer science. In fact, we focus on the effectiveness of quantum algorithms in different filed classical computer.

Keywords: Security · Unstructured databases · Quantum algorithms
Cryptography

1 Introduction

Quantum computing is a very promising, multidisciplinary young field of science, based on mathematics, quantum mechanics and computer Science. It was born out of classical information theory. This new features technology can be exploited, not only to improve the performance of certain information-processing tasks, but also to accomplish tasks which are impossible or intractable in the classical realm. Since the emergence of Alan Turing machine calculator, computer science was adopted on classical computing which founded on mathematical basis. However, despite the significant evolution in computer efficiency since then, many arithmetical questions have become intractable either because of the inability of principles of classical information theory or length time processing via today's computer such as the factorization of large prime numbers. Recently, computing science is beginning to use quantum mechanics to solve the complexities in classical computing. In fact, quantum computing is fundamentally employing and exploiting some principles of quantum physics to process information such as: entanglement, superposition [1–3]. Accordingly, in classical computing the information is stored in bits, which take the discrete values 0 and 1. Therefore, if storing one number takes 64 bits, then storing N numbers takes N times 64 bits. Moreover data processing is made basically in the same approach as by Human. Consequently, the type of calculations that can be solved is the same as the problems can be solved by Human, with one difference is the speed of completion of

© Springer Nature Switzerland AG 2019
M. Ben Ahmed et al. (Eds.): SCA 2018, LNITI, pp. 618–628, 2019.
https://doi.org/10.1007/978-3-030-11196-0_51

calculations. Contrary to classical computing, in quantum process, the information is stored in quantum bits (qubits). In fact, a qubit can be in states $|0\rangle$ and $|1\rangle$, but it can also be in a superposition of these states, $|0\rangle$ and $|1\rangle$ in the same time, where for every extra qubit we get, we can store twice as many numbers (see Fig. 1). Moreover, calculations are performed by unitary transformations on the state of the qubits. Therefore, if we combined with the principle of superposition, this creates possibilities that are not available for Human calculations. This translates into more efficient algorithms for factoring, searching and simulation of quantum mechanical systems, which give more opportunities if we use quantum algorithms in computer science [4].

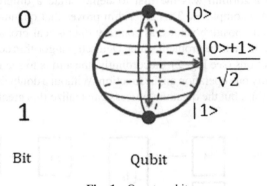

Fig. 1. Quantum bit

The performance of quantum computing against the classical one certainly infects the computer science. Accordingly, in this article, we will study the impact of quantum computing in general and more partially quantum algorithms on different filed of classical computer, where we focus on the effectiveness on unstructured databases and on classical cryptography.

The paper is organized as follows: In Sect. 2, Quantum computing. In Sect. 3, Effectiveness of quantum algorithms on classical computer sciences. Finally, conclusion is drawn in the last section.

2 Quantum Computing

2.1 Data Processing Based on Quantum Approach

Quantum algorithms is identical to the classic definition of algorithms, however the difference is in the principle of processing data. This will typically be of the form: Given an E, the algorithm finds an E_1 in time t. Accordingly, every algorithm will be considered as computing a sequence of vectors. Generally, the analysis of algorithms habitually will compose of giving an explicit description of what the vector is. The algorithm gives the operational description of the vector: it is the result of applying unitary transformations to the start vector. Therefore, when we identify what each vector is unambiguously, we will recognize what the last vector is. Then we will comprehend what the result of the last step of each algorithm does because in all situations the last step is a quantum measurement. Accordingly, several algorithms are gives the finally answer after the measurement is made, but other algorithms require the performance of a supplementary classical phase on the result of the measurement. Some are a bit more complicated, in that they need the quantum algorithm to be run a multiple number of times. Each run of the quantum algorithm gives a small amount of information about the answer that is most wanted [5].

2.2 Deutsch-Jozsa Algorithm Review

The Deutsch-Jozsa algorithm was the first to demonstrate a difference between the classical and quantum complexes. This algorithm proves that quantum amplitudes can take both negative and positive values, as contrary to classical probabilities which are definitively positive. In fact, the problem which Deutsch_s algorithm covers is declared in terms of binary functions (see Fig. 2). Accordingly, the aim is to see if the function is a constant by doing only one operation on the function. Without a doubt this is impossible in the classical computing, but the quantum process can realize this great performance [6].

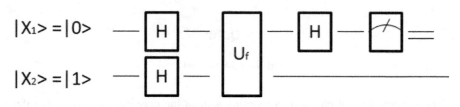

Fig. 2. Deutsch's algorithm circuit

The Deutsch-Jozsa algorithm will define computing a series of vectors v_0, v_1, v_2, v_3 each one of them is in the real Hilbert space $H1 \times H2$, where $H1$ and $H2$ are two dimensional spaces. We index vectors in this space by xy, where x and y are single bits. The algorithm always uses the same input vector as following:

- The initial vector is v_0 so that v_{01}
- To obtain v_1 we applying the Hadamard transform on each H_i of the space with $i = 1; 2$ individually
- The next vector v_2 is the result of applying U_f where $f'_{(xy)} = x(f(x \oplus y))$
- Finally we applying the Hadamard transform H_1 to obtain v_3

At the start of the circuit $|\alpha_0\rangle$, after the first Hadamard are applied $|\alpha_1\rangle$, after U_f is applied $|\alpha_2\rangle$, and after the last Hadamard is applied $|\alpha_3\rangle$. As a result:

$$|\alpha_0\rangle = |0\rangle|1\rangle \tag{1}$$

$$|\alpha_1\rangle = |+\rangle|-\rangle = \frac{1}{2}(|0\rangle|0\rangle - |0\rangle|1\rangle + |1\rangle|0\rangle - |1\rangle|1\rangle) \tag{2}$$

After the oracle U_f is applied, we have the next state:

$$|\alpha_2\rangle = \frac{1}{2}(|0\rangle|f(0)\rangle - |0\rangle|1 \otimes f(0)\rangle + |1\rangle|f(1)\rangle + |1\rangle|1 \otimes f(0)\rangle) \qquad (3)$$

We break the analysis into two cases: When f is constant and when f is balanced. We apply the final Hadamard gate. If f is constant then $f(0) = 1$, therefore, we can simplify $|\alpha_2\rangle$ as the following:

$$|\alpha_2\rangle = \frac{1}{\sqrt{2}}|+\rangle \otimes (|f(0)\rangle - |1 \otimes f(0)\rangle) \qquad (4)$$

Thus, the qubit 1 is now in state $|+\rangle$. We conclude that:

$$|\alpha_3\rangle = \frac{1}{\sqrt{2}}|0\rangle \otimes (|f(0)\rangle - |1 \otimes f(1)\rangle) \qquad (5)$$

Then, the qubit 1 is exactly in state $|\alpha_3\rangle$. Thus, measuring qubit 1 in the standard basis now yields outcome 0 with certainty. If f is balanced, then $f(0) \pm f(1)$. Since, f is a binary function, this means $f(0) \otimes 1 = f(1)$ and similarly $f(1) \otimes 1 = f(0)$. Therefore we can simplify $|\alpha_2\rangle$ by:

$$|\alpha_2\rangle = \frac{1}{\sqrt{2}}|-\rangle \otimes (|f(0)\rangle - |1 \otimes f(0)\rangle) \qquad (6)$$

The qubit 1 is exactly in state $|1\rangle$. Thus, measuring qubit 1 in the standard basis now yields outcome 1 with certainty. We conclude that, if f is constant, the algorithm outputs 0, and if f is balanced, the algorithm outputs 1. Thus, the algorithm decides whether f is constant or balanced, using just a single query, which impossible classically.

3 Effectiveness of Quantum Algorithms on Classical Computer Sciences

Deutsch-Jozsa algorithm played a significant role to prove the superiority of quantum computing against the classical one. Therefore, it began to show the importance to exploiting the advantages of quantum phenomena to solve the complexities of classical computing. In our study we focus on the impact of quantum algorithms on computer science in general and we focus on the impact on NoSQL databases and cyber security.

3.1 Impact on NoSQL Databases

NoSQL is a type of database management systems that do not follow all of the rules of relational databases and cannot use structured query langue (SQL). Unstructured databases are normally used in big data, which are principally prone to performance problems caused by the limitations of SQL and the relational model of databases. Many think of NoSQL as the modern database of choice that scales with Web requirements.

One of the biggest obstacles to this class of database is the length of data processing time. Accordingly, we discuss in the following to what extent can the quantum computing reduce data processing time in NoSQL databases using Grover's algorithm.

Grover's algorithm is a quantum algorithm was devised by Lov Grover in 1996. The algorithm performs a search over a NoSQL database of $N = n^2$ entries to find the exclusive element that satisfy the conditions of search. In classical information theory, the best algorithm for a search over unstructured database requires a linear search, which is N operation for N element. Grover's algorithm, reduce the number of operation necessary to solve a search from N operation to $\sqrt{N}$ Therefore, Grover's algorithm is the best possible quantum algorithm for searching in NoSQL database. Moreover, its effect appears more as the number of operations N increases. For example, if a data base contains 1400000000 elements, approximately the population of China, by using Grover's algorithm we can find the desired item in about 37417 steps on average. If it takes one second per iteration will be done in about 11 h. To search through the complete set of 1400; 000; 000 elements classically, at the same rate of one element per second, would take 45 years, assuming 24 h of work per day, 7 days per week. Accordingly, Grover's algorithm it very powerful algorithm in search and it can be base to form the data meaning for next generation of Big data especially NoSQL data bases [7, 8].

Mathematically Grover's algorithm solves the problem of unordered search. We consider that the data base include items represented by numbers from 0 through N − 1 forming a set $\beta = \{a_0, a_1 \ldots, a_{N-1}\}$, and given a boolean function $f = \beta \to \{0, 1\}$. The key will be represented by a function that is zero on all N numbers, with the exception of one, for which it is 1, on condition that:

$$\exists_{a\in\{0,1,\ldots N-1\}} f(a) = 1 \quad \text{and} \quad \forall_{a\in\{0,1,\ldots N-1\}_{a\neq x}} f(x) = 0 \tag{7}$$

Unordered search is frequently represented as a database search problem in which we are specified a database and we want to find an element that respects some conditions. For example, considering a database of N citizens, we might want to search where a specific citizen is located in our database. Classically, we communicate with our database by using an oracle O_f that implements a function f. Generally, O_f given an input of a, outputs $f(a)$ in invariable time. In fact, to evaluate any data meaning algorithm, we should focus on the time and the number of operation necessaries to find an item in unstructured data base. Classically, the most excellent result is $O(N)$ operations and $O(N)$ unities time, because the data is unordered, we should be look at every element in the list. Contrary, quantum search in unstructured data is based on Grover's algorithm, which begins with a quantum register of n qubits, where n is the quantity of qubits required to form the search space of size $2^n = N$.

3.2 The Impact of Quantum Computing on Cryptography

Cryptography is a necessary tool for protecting information in computer systems, including hardware, software and data, from cyberattacks. In the following we will discuss the impact of quantum algorithms on the cryptography.

Classical Cryptography

Symmetric encryption

Symmetric encryption uses a single key for encryption and decryption. Therefore, it is an obligation that the key remains confidential and should be known only by the interlocutors (Alice and Bob). In fact, users of this technique must agree previously on a key (K) to use, for this they are obliged to use a different network from the standard communication network that is susceptible to be attacked. Whenever, Alice wants to transmit a message (m) to Bob, she uses his secret key to encrypt, and she sends the result of this encryption through the same channel. Bob in turn uses the same secret key and the same public algorithm to decrypt the encrypted message (see Fig. 3). The data encryption standard (DES) is an example of the symmetric cryptography system it is widely used to protect sensitive data especially by the governments. Symmetric cryptography is characterized by a simple and fast implementation and short keys (128 or 256 bits). Moreover, this type of encryption shares a secret, so if the secret key is compromised by an attacker, it can decrypt all messages. Accordingly, there is a difficulty to secure the keys distribution, because the total number of keys increases very rapidly depending on the number total users.

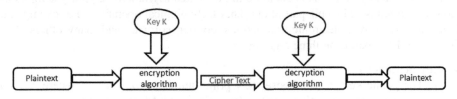

Fig. 3. Symmetric encryption principle

Asymmetric encryption

Asymmetric encryption was invented in 1975 by Whitfield Diffe and Martin Hellman of Stanford University [9]. The principle of asymmetric encryption is based on two different keys: the first is public used to encrypt and the second key is private, which never transmitted to anyone, reserved to decrypt the plaintext. Accordingly, in simple transmission the asymmetric cryptography ensures the confidentiality of a plaintext in a way that only the recipient can read it. Moreover, the recipient is certain that the message is sent by the right interlocutor (Sign a message). To solve this problem we use a public key that is usually published in a directory. In fact, Alice can send a plaintext to Bob without prior private communication and Bob is the only p son who can decrypt. The ciphertext based on the secret key (see Fig. 4). The RSA is an example of the asymmetric cryptography system. We can resume the public key cryptography in four steps:

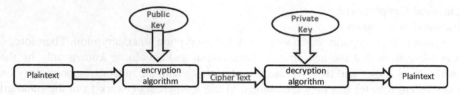

Fig. 4. Public key cryptography

1. Generate key: Alice generates a public key to encrypt and a private key to decrypt the message,
2. Distribute the key: Alice shares the public key with Bob,
3. Encrypt the message: Bob encrypts the plaintext using the public key and send the ciphertext to Alice,
4. Decrypt the message: Alice decrypts the ciphertext using the private key.

The security analysis of RSA

This algorithm is published by Ron Rivest, Adi Shamir and Leonard Adleman of the Massachusetts Institute of Technology (MIT), he is named by the initials of his three inventors. Actually, the RSA is the more implemented asymmetric cryptosystem in our days he surrounds us in a big part of our dairy technological operations. For example, it is in our bank cards, electronic transactions, courier, software and many others [10]. Generally, RSA based on three algorithms:

- Key generator
    - The public key: Alice chooses tow prime numbers "p" and "q" where $n = pq$ and consider a function: $f(n) = (p - 1)(q - 1)$ Moreover, Alice chooses an integer "e" where $\gcd(e, f(n)) = 1$.
    - The secret key: Alice Chooses a number "d" with no factor common with $f(n)$.
- Encryption algorithm
    - For a plaintext "M" and a ciphertext "C" we have: $C = M^e \bmod n$
- Decryption algorithm
    -

$$M = C^d \bmod n$$

In reel implementation of RSA, Alice should choice a very large numbers "p", "q" and "e". Therefore, the generation of public and secret key is generated by a computing program. For example, if we use a key size 4096 bit we generated the following keys:

```
-----BEGIN PUBLIC KEY-----
MIGMA0GCSqGSIb3DQEBAQUAA4GMADCBiAKBgGPilKfaHg7cafOdmbE6K8
I86agBckTk1L1NF3xa1rzhe2nSbqiwZJW2G+rjEFQ6l3pYC2AZJPkLNuL
bZuFe5ok85MCnB+BCWopn44ME15tjok4gEd0/tVxHU15G+ZLhwo1PsnPZ
xudlBsSvt/Ri5PG4PBEmwAFzW8JQHnZXBbAgMBAA
-----END PUBLIC KEY-----
-----BEGIN RSA PRIVATE KEY-----
MIICWwIBAAKBgGPilKfaHg7cafOGdmbE6K8I86agtBckTk1L1NF3xa1rz
he2nSbqiwZJW2G+rjEFQ6l3pYC2AZJPkLNuLbZuFe5ok85MCnB+BCWopn
44ME15tjok4gEd0/tVxHU15G+ZLhwo1PsnPZxudlBsSvt/Ri5PG4PBEmw
AFzW8JQHnZXBbAgMBAAECgYBXXbke8/Z0kzANNKBdlX/ckZzkMw4ya6wd
ieV07ik2XeQe87iTcH
T3o7RNU4ZzFCKxmRhm80t+mR99AjLzrWh6TreTnrNfTNRJS3q5ya2eU9K
63jCNlNSNZ/Z5h6YmKvEHOgsa2tydEjiweI7IhHmD3P+3RUic2SY+rZCd
+OVi8QJBAKGrlwOrM95sFI7YaBD4qUMQKmsGWDYXU5/VFnbqxdZCMdEfr
+uQaMdA3FgkeBBzs0bR74TOLgWwRjfyCTIOMaMCQQCeKj46/f3Knbcz7j
oA+Upc899fs96UVK7WA6gVtZcEfURZNQ4vIlj/eEmNmHcsPDNqIPJnYY9
u4b/Mg7MXXuHpAkEAixYSP6fEYfBd516qeR9Uj0uTEqxMq4x59yHdx8TH
G0OrlBxMvr5oRjO43WPSn2TKsXzfMUQoN+ClKHBkFNo9PQJAJVlsYJnh+
JrNzGGRKPG1ccXVFdBv2y80Ki2q35WGF/VM0pQI yXCe+5T4KYu7eKKus
X0eQPN034ZHuI/RFs+9IQJAaqsl50B+RsM33+gehZaCKDIKQGDYvIXPL4
YmI0QMEBveFiBh848sEE3l8R42o/8DGxsqgiWy9QSavmXSLuSVwQ==
-----END RSA PRIVATE KEY-----
```

From the previous example, it is clear how the key size is large. Therefore, in the implementation phase, there is a big difficulty to generate a new key for every new message, to respect the theoretical norms of security as defined by Claude Shannon (one time pad). For example, to encrypt a file its size 4 GB, the key size should be 4 GB [11]. In fact, to realize the principle of a one-time pad requires a very big financial investment and very long time to generate keys. Therefore, if we analyze the security level of RSA from information theory point of view, the absence of mathematical proof makes RSA a breakable system. On the contrary, if we analyze the security from engineering point of view, we find that all implemented algorithms and the present computers need years of calculus to break RSA encryption, which makes RSA one of the best cryptosystems in our days. In fact, with the appearance of quantum information the concept of calculation has changed and its complexity has become simpler. Actually, after the successful implementation of quantum key distribution protocols in the network security, we can confidently say that the entry into the quantum engineering era is only a question of a few years. Therefore, the implementation of quantum algorithms soon will be a reality. In classical computing, to factorize a large prime number takes a very long time, because the process to factorize an integer with n bits grows exponentially with n. Therefore, many classical cryptosystems exploit this complexity to encrypt data. Accordingly, in 1994, Peter Shor proposed a quantum algorithm to factor an n-digit number with a time complexity polynomial in n

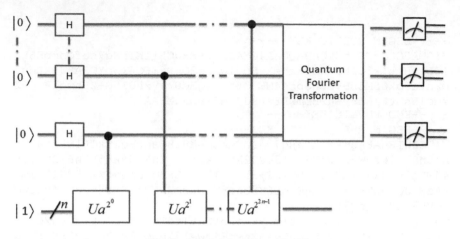

Fig. 5. Shor's algorithm principle

(see Fig. 5), based on the detection of a period r in a function f by using quantum Fourier transform [12–17].

We can resume the scenario processing of Shor_s algorithm by the following steps:

- Classical computer part
  - Sketch of various subroutines,
  - Reduction to period-finding problem,
  - Full classical algorithm.
- Period-finding on quantum computer
  - Quantum Fourier Transform
  - Period-finding algorithm.

We consider:

$$f : N \leftarrow 0, 1 \dots, M - 1 \tag{8}$$

For a period r:

$$f(x + r) = f(x) \tag{9}$$

For positive integers a and N, the order of a modulo N is defined as:

$$a^r \equiv 1 (\mathrm{mod} N) \Rightarrow f(x) = a^r (\mathrm{mod} N) \tag{10}$$

If the integer a has order r modulo N, then:

$$a^r \equiv 1(\mathrm{mod} N) \Rightarrow (a^{\frac{r}{2}} + 1)(a^{\frac{r}{2}} - 1) \equiv 0(\mathrm{mod} N) \tag{11}$$

If the integer a has order r modulo N, then:

$$a^r \equiv 1(\mathrm{mod}N) \Rightarrow (a^{\frac{r}{2}}+1)(a^{\frac{r}{2}}-1) \equiv 0(\mathrm{mod}N) \tag{12}$$

This implies that at least one of the following is a nontrivial factor of N:

$$\gcd\left(a^{\frac{r}{2}}+1, N\right) \text{ and } \gcd\left(a^{\frac{r}{2}}-1, N\right) \tag{13}$$

Unless r is odd gcd, or the calculation returns N.

Quantum factoring algorithm solves the factorization of an n bit number in polynomial time, approximately $O((\log N)2)$, instead of $O(p) \approx O(N)$ in classical computing. Accordingly, the mathematical prove of the cryptosystem RSA reposes on the complexity of factoring the large prime numbers. Therefore, the implementations of Shor's algorithm impact the basis of classical security which founded on cryptosystem RSA.

4 Conclusion

In this work we discussed the impact of quantum computing on classical computer sciences in general and more particularly in the field of unstructured databases and cybersecurity. In the first, we studied Deutsch-Jozsa algorithm, which demonstrated clearly the robust of quantum computing against the classical one. Moreover, we analyzed the impact Grover's algorithm in unstructured databases processing, when we showed that this approach reduce the searching time in NoSQL databases. Accordingly, we proved the effectiveness of quantum factoring algorithm to factorize the large prime numbers, which make it very influential in breaking the classical cyber security.

References

1. Dirac, P.A.M.: The Principles of Quantum Mechanics, 3rd edn. Clarendon Press, Oxford (1947)
2. Hirvensalo, M.: Quantum Computing. Springer (2010)
3. Kaye, P., Laflamme, R., Mosca, M.: An Introduction to Quantum Computing. Oxford University Press (2007)
4. Kitaev, A.Y., Shen, A., Vyalyi, M.: Classical and Quantum Computation, vol. 47. Graduate Studies in Mathematics. American Mathematical Society Press (2002)
5. Deutsch, D.: Quantum computational networks. Proc. R. Soc. Lond. Ser. A **425**(1868), 7390 (1989)
6. Deutsch, D., Jozsa, R.: Rapid solution of problems by quantum computation. Proc. R. Soc. Lond. Ser. A **438**(1907), 553–558
7. Grover, L.: Fast quantum mechanical algorithm for database search. In: Proceedings of the 28th Annual ACM Symposium on Theory of Computing (STOC 96), pp. 212–219 (1996)
8. Grover, L.: Quantum mechanics helps in searching for a needle in a haystack. Phys. Rev. Lett. **79**(2), 325328 (1997)

9. Diffie, W., Hellman, M.: New directions in cryptography. IEEE Trans. Inf. Theory **22**, 644654 (1976)

10. Rivest, R.L., Shamir, A., Adleman, L.: Method for obtaining digital signatures and public-key cryptosystems. Commun. ACM **21**, 120126 (1978)

11. Shannon, C.E.: Communication theory of secrecy systems. Bell Syst. Tech. J. https://doi.org/10.1002/j.1538-7305.194.tb00928

12. Poulin, D.: Classicality of quantum information processing. Phys. Rev. A **65**(4), 042319(10) (2002)

13. Shor, P.: Polynomial-time algorithms for prime factorization and discrete logarithms on a quantum computer. SIAM J. Comput. **26**(5), 14841509 (1997)

14. Shor, P.: Algorithms for quantum computation: discrete logarithms and factoring. In: Proceedings of the 35th Annual IEEE Symposium on the Foundations of Computer Science, pp. 124–134 (1994)

15. Lomonaco, Jr, S.J.: Shor_s quantum factoring algorithm (2000). arXiv: quantph/0010034

16. Shor, P.W.: Why haven_t more quantum algorithms been found? J. ACM **50**(1), 8790 (2003)

17. Abrams, Daniel S., Lloyd, Seth: Quantum algorithm providing exponential speed increase for finding eigenvalues and eigenvectors. Phys. Rev. Lett. **83**, 51625165 (1999)